Concrete Masonry
Designer's Handbook

Concrete Masonry Designer's Handbook

Second Edition

John Roberts, Alan Tovey
and Anton Fried

Routledge
Taylor & Francis Group

LONDON AND NEW YORK

First published 2001 by Routledge

2 Park Square, Milton Park, Abingdon, Oxfordshire OX14 4RN
52 Vanderbilt Avenue, New York, NY 10017

Routledge is an imprint of the Taylor & Francis Group, an informa business

First issued in paperback 2019

Typeset in Sabon by Amy Boyle

British Library Cataloguing in Publication Data
A catalogue record for this book is available from the British Library

Library of Congress Cataloging in Publication Data
A catalog record for this book has been requested

ISBN 978-0-419-19440-8 (hbk)
ISBN 978-0-367-86524-5 (pbk)

Publisher's Note
The publisher has gone to great lengths to ensure the quality of this reprint but points out that some imperfections in the original may be apparent.

Contents

The Authors

Professor John Roberts
BSc(Eng), PhD, CEng, FIStructE, FICE, FBMS,
FIMgt, FRSA, MICT

John Roberts is a Civil Engineering graduate of London University where he also studied for his PhD. Prior to his appointment as Professor of Civil Engineering at Kingston University in 1988 Professor Roberts worked for the Cement and Concrete Association and the British Cement Association. He is co-author of the *Concrete Masonry Designer's Handbook*, the *Handbook to BS 5628 Part 2*, *Efficient Masonry Housebuilding Design Approach*, *Efficient Masonry Housebuilding Detailing Approach* and the *Design Guide for Eurocode 6*. He has published over 70 papers related to various aspects of masonry design and construction.

Professor Roberts has extensive experience of work on British Standards and Codes of Practice and currently chairs or is a member of a number of the committees involved in the development of the European standards for masonry. He is a past President of the British Masonry Society and was awarded the Honorary Degree of Doctor of Science by MATI in Moscow.

Alan Tovey
CEng, FIStructE, ACIArb, MIFS

Alan Tovey is the principal of Tecnicom which is an independent consultancy. He has very wide industry experience and started his career with the Concrete Section of British Railways Western Region. He then joined a major precast manufacturer where he was involved with both design and construction of a variety of concrete structures. He gained municipal engineering experience and practical design experience of loadbearing brickwork and blockwork with a London Borough Authority. He joined the Cement and Concrete Association in 1973 and subsequently worked with the British Cement Association before setting up his consultancy in 1993. He has extensive experience of work on British Standards and Codes of Practice. His current activities include assessments, expert witness work and technical development activities.

Dr Anton Fried
BSc(Eng), MSc, PhD, FBMS

Anton Fried was educated in Zimbabwe and is a graduate of the University of Cape Town in South Africa. He studied for his MSc at Surrey University and worked for his PhD at South Bank Polytechnic. He has extensive experience of the Construction Industry starting with the Engineering Corps of the Rhodesian Army and then a consulting practice in South Africa. After completing his PhD he joined Sandbergs and gained extensive experience of investigation and testing. In 1990 he became a Senior Lecturer at Kingston University where he has developed and extended his strong research interest in Concrete and Masonry. He has published over 40 papers dealing with various aspects of these topics.

Acknowledgements

The publisher and authors would like to acknowledge the kind permission of the British Standards Institution to reproduce the following Figures within this publication:

- Figures 3.1, 3.2, 3.4 and 3.5 from BS 6073: Part 1: 1981
- Figure 3.11 from BS 4551: 1980
- Table 8.1 and Figures 8.6, 8.7, 8.8, 8.21, 10.7, 10.8, 10.11, 10.12 and 10.13 from BS 5628: Part 1: 1978 Figures 8.25, 8.26, 8.27, 8.29, 8.30, 13.1 and 13.2 from CP 121: Part 1: 1973
- Figures 14.5 and 14.6 from CP 3: Ch.III: 1972
- Figure 15.1 from BS 476: Part 8: 1972

The publisher and authors would like to thank the Department of the Environment, Transportation and the Regions, and the British Masonry Society for permission to use BMS Special Publication Number 1 *Eurocode for Masonry, ENV 1996-1-1: Guidance and Worked Examples* as Chapter 20 of this book. This publication was originally produced by a British Masonry Society working party chaired by B.A. Haseltine and comprising the following people:

- Dr. N. Crook
- Dr. G.J. Edgell
- C.A. Fudge
- G.T. Harding
- J.C. Haynes
- J.E. Long
- Prof. M.E. Phipps
- Prof. J.J. Roberts
- A.L. Taylor
- P. Watt

The authors would like to thank their friends and colleagues in all parts of the masonry industry for their help and encouragement in completing the book, and particular thanks are due to Prof. Bill Cranston and Prof. Andrew Beeby for their contributions to the First Edition. Thanks are also due to Prof. John Roberts, Dr. Geoff Edgell and Alan Rathbone, the original authors of *The Handbook to BS 5628: Part 2*, by the same publisher, which is no longer in print. Material from this book has been updated and incorporated in Chapter 9.

Notations

A	horizontal cross sectional area		e_x	ccentricity at top of a wall *or* eccentricity in x direction
A_A	area of Part A ⎫			
A_B	area of Part B ⎬ Figure 7.26		e_y	eccentricity in y direction
A_{ar}	actual area of roof lights		F_s	solid area divided by total area of slotted block
A_{aw}	actual area of windows		F_v	voided area divided by total area of slotted block
A_c	area of compressive zone		f	a stress
A_e	area of the element		f_o	characteristic anchorage bond strength between mortar or concrete infill and steel
A_f	area of opaque exposed floor (if any)			
A_m	cross sectional area of masonry		f_e	intrinsic unit strength (adjusted for aspect ratio)
A_r	area of opaque roof			
A_{roof}	total area of roof including rooflights		f'_c	uniaxial compressive strength
A_s	cross sectional area of primary reinforcing steel		f_d	design strength of a material
			f_f	characteristic compressive strength of masonry in bending
A_v	face area of void per block (mm²)			
A_{wall}	total area of wall including windows		f_g	block strength based on gross area
A_w	effective area of wall		f_{horiz}	applied horizontal stress
A_{wo}	area of opaque wall		f_k	characteristic compressive strength of masonry
a	deflection		f_{kb}	characteristic flexural strength (tension) of masonry when failure parallel to bed joint
a_r	reduced span (yield-line analysis)			
b	width of section		f_{kp}	characteristic flexural strength (tension) of masonry when failure perpendicular to bed joint
b_{eff}	effective breadth of opening (Figure 7.25)			
b_p	breadth of pier			
b_r	reduced span (yield-line analysis)		f_{kx}	characteristic flexural strength (tension) of masonry
C	internal compressive force			
C_f	coefficient of friction		f_m	block strength based on nett area
C_x	conductance of solid section		f_{self}	selfweight stress
C_y	conductance of voided section		f_{vert}	applied vertical stress
d	effective depth of tension reinforcement		f_v	characteristic shear strength of masonry
d_n	constant depending on sample size		f_w	applied vertical stress per unit area
E_i	modulus of elasticity of the part		f_y	characteristic tensile strength of reinforcing steel
e	eccentricity of load		G	specified strength of blocks
e_a	additional eccentricity due to deflection in walls		G_k	characteristic dead load
			g_A	design vertical load per unit area
e_d	drying shrinkage		g_d	design vertical dead load per unit area
e_{lat}	eccentricity at mid-height resulting from lateral load		H	overall height
			h	clear distance between lateral supports (usually height)
e_m	the larger of e_x or e_t			
e_{max}	numerically larger eccentricity		h_{eff}	effective height of wall or column
e_{min}	numerically smaller eccentricity		h_f	thickness of flange
e_t	total design eccentricity in the mid-height region of a wall		h_L	clear height of wall to point of application of lateral load
e_{temp}	temperature contraction			
e_u	ultimate tensile strain			

h_p	panel height	
h_w	window height	
I	second moment of area of total section	
I_A	second moment of area of Part A	
I_B	second moment of area of Part B	
I_v	second moment of area of linking beam	
I_i	second moment of area of the ith wall	
i	individual specimen result from the sample	
L	length or span	
L_c	contact length	
L_s	the reduction in length	
l	span	
l_g	distance between centroids of two parts of the wall	
M	bending moment due to design load	
M_d	design moment of resistance	
M_u	ultimate moment of resistance	
M_{uc}	ultimate moment of resistance when controlled by concrete	
M_{us}	ultimate moment of resistance when controlled by reinforcement	
M_w	total wind moment	
M_{wi}	wind moment carried by the ith wall	
M_y	moment in the y direction	
m	design moment per unit length	
N	design axial load on section under consideration	
N_u	ultimate axial load	
N_v	number of voids across thickness of slotted block	
n	axial load per unit length of wall available to resist an arch thrust or sample size	
n_y	vertical load per unit length of wall	
P	force	
P_m	ratio of average stress to maximum stress	
P_w	wind force carried by wall	
$\mathbf{p}$	wind load	
Q	heat flow rate through an element	
Q_g	actual heat flow through glazing	
$Q_{g\,max}$	maximum permissible heat flow through glazing	
Q_k	characteristic imposed load	
Q_{max}	maximum imposed load	
$Q_{o\,actual}$	heat flow through opaque elements of structure	
$Q_{o\,max}$	maximum permissible heat flow through opaque elements	
q	a lateral load	
q_{lat}	design lateral strength	
R	hermal resistance of solid material or factor for degree of restraint	
R_a	thermal resistance of ventilated air space	
R_{ave}	total average thermal resistance of the construction	
R_{ci}	average thermal resistance to centre line of cavity	
R_e	equivalent thermal resistance of block	
R_k	sum of known thermal resistances	
R_{max}	maximum permissible percentage of rooflight glazing	
R_{si}	inside surface resistance (thermal)	

R_{su}	outside surface resistance (thermal)	
R_v	thermal resistance of unventilated air space	
R_x	thermal resistance of solid section	
R_y	thermal resistance of voided section	
r	radius of curvature	
S	first moment of area of the section of wall to one side of the section considered about the neutral axis of the total section	
s	standard deviation calculated from sample data	
T_i	inside temperature	
T_o	outside temperature	
t	overall thickness of a wall or column	
t_1	thickness of first leaf of a cavity wall	
t_2	thickness of second leaf of a cavity wall	
t_b	block thickness	
t_{eff}	effective thickness of wall or column	
t_f	thickness of the flange	
t_p	thickness of the pier	
t_v	thickness of void in slotted block	
U	thermal transmittance value	
U_{ar}	thermal transmittance value of rooflights	
U_{aw}	thermal transmittance value of windows	
U_f	thermal transmittance value of floor	
U_t	thermal transmittance value of roof	
U_w	thermal transmittance value of wall	
v	shear force due to design loads	
V_h	horizontal shear force acting on length of a wall	
V_v	vertical shear stress on section considered	
v	shear stress due to design loads	
v_c	characteristic shear stress	
v_h	design shear stress	
W	applied load	
W_k	characteristic wind load	
W_{max}	maximum permissible percentage of glazing	
W_u	ultimate capacity of the section	
W_w	design wind load	
w	range within a sample of size n	
w_d	density of wall	
X_i	each individual result in an example	
X_n	average value of samples	
x	a dimension	
Y (or Y_n)	formulae simplification factors (for Chapter 8)	
y	a dimension	
Z	section modulus	
z	lever arm	
α	bending moment coefficient for laterally loaded panel	
β	capacity reduction factor for walls allowing for effects of slenderness and eccentricity	
γ_c	classification factor	
γ_f	partial safety factor for load	
γ_m	partial safety factor for material	
γ_{mm}	partial safety factor for blockwork or partial safety factor for compressive strength of masonry	
γ_{ms}	partial safety factor for strength of steel	
γ_{mv}	partial safety factor for material in shear	
δ	central deflection under lateral load or difference between average and actual result	

θ   rotation of joint *or* an angle

κ   a constant *or* coefficient of variation

λ   thermal conductivity

μ   orthogonal ratio or true mean value

ρ $\dfrac{A_s}{bd}$ or $\dfrac{A_s}{bt}$ as appropriate

σ   true standard deviation

φ   degree of fixity at support

Chapter 1
Introduction

Concrete has a long history. It has been used since ancient times and was known to the Ancient Egyptians and even earlier civilizations. The oldest known concrete so far discovered dates back to 5600 BC and came to light during excavations on the banks of the River Danube at Lepenski Vir in Yugoslavia[1.1]. The modern, remarkable, developments with new structures, new techniques of handling concrete and even new kinds of concrete have all taken place within a comparatively short time and, in fact, 1974 was only the 150th anniversary of the patent for the manufacture of the first Portland cement.

The first all-concrete house was built in 1835 in Kent. The only part of this house which was not constructed in concrete was the suspended first floor as to build this successfully would have required reinforcement, the introduction of which was not pioneered until some years later. The first concrete blocks were made in the United Kingdom in about 1850 by Joseph Gibbs. The blocks were hollow with moulded faces which imitated the dressed stone of that period. The process, which provided a cheap alternative to dressed stone, was patented by Gibbs. It was not until about 1910, coinciding with the significant growth in the production of cement, that the concrete block industry became properly established. Major growth took place between 1918 and 1939 with the establishment of many small block manufacturers throughout the country. It was in 1918, with thousands of troops coming home from Europe, that the then Prime Minister, David Lloyd George, began his house building scheme of *homes fit for heroes*. These concrete block houses were built entirely of concrete and were the first in this country to be built in metric units. At this time the development of the industry was still based on a low price product, with the clinker from coal-fired power stations providing much of the aggregate used.

After the Second World War, as steel was in short supply, architects were obliged to make use of reinforced and prestressed concrete. The demand for concrete blocks also began to increase and both solid and hollow blocks became widely accepted for all purposes. The resumption of house building in the 1950s and 1960s and the rebuilding of cities demolished after the war

brought emphasis to the important role to be played by concrete blocks – particularly in the use of lightweight blocks for the inner leaf of cavity walls. The low cost, lighter weight and ease of handling of these blocks ensured economy in terms of time and cost of construction. This in turn brought about the introduction of autoclaved aerated concrete blocks (aircrete) which, at that time, were probably appreciated more for their operational advantages on site than for their thermal insulation properties. Concurrent with developments in the United Kingdom, many were also taking place in the United States of America, and it is here that the evolution of high quality facing concrete masonry, and machines associated with its production, can be discovered. Perhaps what was more important was the realization by American engineers of the potential for high rise concrete masonry. The seismic problems, associated with certain areas of the country, having also fostered the development of reinforced masonry.

Within the last three decades many developments have taken place in the uses of concrete masonry. A vast range of products are now available or can be made to order. The increased use of concrete blocks and bricks, at the expense of many other masonry materials, is such that concrete masonry can be claimed to be the major masonry material in the United Kingdom (Figure 1.1). The relative proportions of the types of block and brick used are illustrated in Figure 1.2. It is apparent that the growth in the market share of autoclaved aerated concrete blocks has taken place at the expense of some of the market held by lightweight aggregate blocks. The proportionate trend in the share of insulating blocks used does not appear to have been affected by changes in the thermal regulations. At present lightweight aggregate blocks account for about 22% of the market with dense and autoclaved aerated blocks accounting for about 46% and 32% respectively.

The renewed interest in all forms of masonry as a structural material has partly been brought about the introduction of BS 5628[1.2], a Limit State Code, which enables the designer to take a more rational approach toward the use of concrete masonry. The purpose of this handbook is to cover the design of concrete masonry,

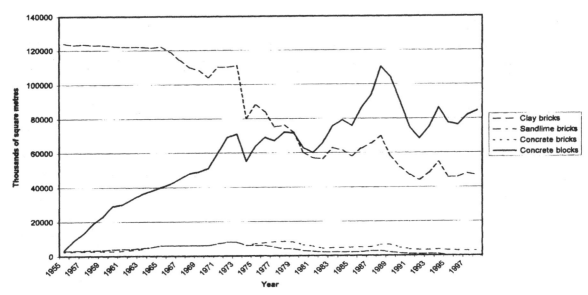

Figure 1.1 Deliveries of bricks and blocks in the UK since 1955

including many aspects not covered by the various codes of practice, as well as to provide an insight into those other aspects of masonry performance to which the designer may need to give consideration.

Another factor which has encouraged the wider use of concrete masonry has been the scarcity of ideal building sites for dwellings and buildings in general. This has also widened the experience and expertise of the engineer involved in that the foundations and superstructure of even the most modest buildings have needed careful examination. The inherent need for further information on the engineering properties of masonry has thus been uncovered. The introduction of design requirements for lateral loading in the event of an accident has led to substantial amounts of work on the lateral load

resistance of masonry, further widening the scope and ability of the engineer to effectively design with a basic understanding of the proposed masonry materials. There is no doubt that the attention recently paid to the need for conservation of energy has advanced the use of concrete masonry construction. Although Part F of the Building Regulations[1.3], when introduced in January 1975, was essentially a measure to reduce condensation, to achieve this an improvement in the resistance offered by buildings to the passage of heat through the various elements had to be effected. The use of thermal insulating blocks was therefore encouraged, and the introduction of Part FF of the Building Regulations in July 1979 gave further support to the use of blocks of this type. Familiarity with the larger size of concrete blocks compared with bricks

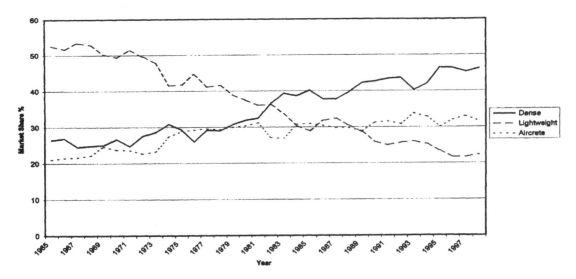

Figure 1.2 Relative proportions of block types delivered since 1965

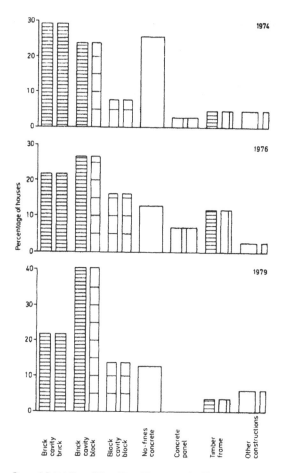

Figure 1.3 Walling of Scottish public sector dwellings
(adapted from BRE Current Paper 60/78 by kind permission of Building Research Establishment, together with private communication)

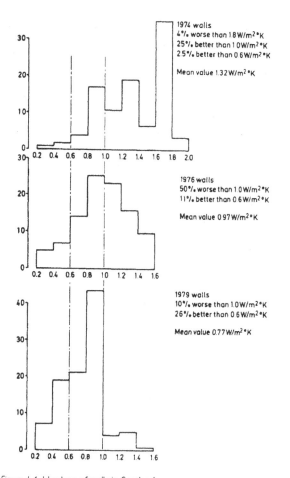

Figure 1.4 U-values of walls in Scotland
(adapted from BRE Current Paper 60/78 by kind permission of Building Research Establishment, together with private communication)

has led to a realization of the economies possible with this material, while the best concrete brick and block facing materials are also favourably comparable to any other materials. Advances still continue to be made in the scope and imagination with which masonry can be employed and while there is still room for such innovation it has to be admitted that there is a bright outlook for such an old and well tried material.

Increasingly the tendency to try to codify every aspect of design brings with it the danger that the engineer will lose sight of the overall consideration and application of engineering principles. This potential problem is evident, for example, in the way in which domestic scale techniques of building have been extended to larger buildings without sufficient consideration being given to the fact that the larger scale will have an effect on the overall stability of the building. It is hoped, therefore, that this handbook will provide a useful and practical guide to the engineer and other professionals. At the same time this handbook could not attempt to replace the training and experience of the qualified engineer. The authors have sought to classify and quantify their combined experiences in the use of concrete masonry,

introducing the results of the extensive programme of research which has been undertaken by the Cement and Concrete Association (now the British Cement Association). In addition extensive use has been made of the results of work carried out elsewhere in the United Kingdom and overseas, particularly in the United States, Australia and New Zealand.

The implementation of the Building Regulations is an interesting aspect of the work of an engineer. Although he may show to the satisfaction of the Local Authority that his proposed building complies in every way with the relevant regulations, the structure, when complete, may substantially differ from that proposed. The Building Research Station at East Kilbride in Scotland has carried out a survey of public sector housing in Scotland for compliance with the thermal regulations, the results of which are illustrated in Figures 1.3 and 1.4. Quite apart from casting an interesting insight into the relative popularity of various construction methods, it is apparent that many buildings actually under construction, as recorded by BRE staff visiting the sites, did not comply with the Building Regulations. It is also interesting to speculate just how many other aspects of

3

masonry performance, quite aside from thermal performance, are denigrated at the construction stage. It is comforting to note that except for the situation when there is vigorous engineering supervision on site, the global factor of safety for masonry at something approaching 4.5 has been retained in BS 5628.

It is currently considered necessary to pay much more attention to the overall integrity of buildings and their resistance to abnormal load. The explanation for this relates in part to the trend to codify methods of design into discrete packages thereby reducing the tendency to consider the behaviour of the building as a whole. Of course, this approach to overall stability was stimulated by the events at Ronan Point where a gas explosion blew a panel out of the side of a high rise block of flats resulting in the progressive collapse of the complete corner of the building, and subsequent attention being paid to the potential problems of progressive collapse. The fifth amendment to the Building Regulations proposed the following approach to reduce the risk of this form of collapse:

(1) The local resistance method whereby the structure is designed to have the required strength to resist abnormal loadings so that hazards will not cause any local failure.

(2) The alternative path method whereby the structure is examined to determine the effect of the removal of any single vertical or horizontal element within each storey, base or span, and ensure the ability of the remaining elements to redistribute and support the load.

The local resistance method is generally difficult to implement because of the cost implications of the approach, although there are a few masonry structures to which the method could be said to apply. The alternative path solution is the one which would seem most likely to commend itself to engineers. The philosophy of the alternative path approach is that any damage which occurs as a result of progressive collapse should be localized so that sufficient bridging is able to take place to ensure the integrity of the structure as a whole. Clearly there needs to be sufficient tying and localized redundancy within the building to confine the extent of the damage.

Finally, it should be noted that the focus of the production of design guidance and the supporting standards is now very much pan European. The final chapters of this book are provided to give the reader an insight into the developments in the Eurocodes and CEN standards.

The background has, therefore, been set for the *Concrete Masonry Designer's Handbook*. The layout and arrangement of the subsequent chapters is classified in the contents list. A number of design examples have been provided as well as design charts. Feedback on the experience of designers in the use of these and of other contents of the handbook would be of interest to the authors, as would suggestions on further areas of development or improvements to the text.

References

1.1 STANLEY, C C. Highlights in the history of concrete. Cement and Concrete Association, Slough, 1979. Publication No 97.408.

1.2 BRITISH STANDARDS INSTITUTION. BS 5628: Part 1: 1978 *Code of Practice for the structural use of masonry*. Part 1: *Unreinforced masonry*. BSI, London. pp 40.

1.3 The Building Regulations 1976. HMSO, London.

Chapter 2
Principles of limit state design

2.1 Introduction

Building construction in common with much else in this modern world, has become complex and sophisticated. Early types of construction provided minimal shelter from wind and rain whereas today's contrasting construction provides an internal environment where every aspect of climate and lighting can be under fingertip control. With this progress has come the clearer definition and understanding of performance requirements. It is against this background that the ideas of limit state design have developed. It must be said at the outset that the introduction of limit state philosophy into Codes has caused considerable confusion and a general feeling that design has become even more complicated. In fact, limit state philosophy is nothing new, it is merely a method of formalizing what has always been done in design. Limit state philosophy is nothing more than a statement that, in design, all the possible ways in which a structure can become unfit for its desired use should be considered and, if necessary, designed against explicitly.

Limit state design was originally conceived for use in reinforced or prestressed concrete design and its operation can probably be seen most clearly by considering reinforced concrete. It is perfectly possible to design a beam or slab which is adequately strong but which, under service conditions, will deflect excessively. It is also possible to design a beam which will not deflect seriously under service loads but, nevertheless, will not be strong enough. Furthermore, it is possible to design members which are both sufficiently strong and sufficiently stiff but which will be disfigured by unsightly cracking. Design to satisfy one criterion (e.g. strength) thus does not guarantee that other aspects of performance will be satisfied. These other aspects have to be carefully considered.

CP 110: 1970[2.1], the first Limit State Code, set down criteria which had to be met for each aspect of performance, and design procedures were given for satisfying these criteria. Equations were given for strength; span/effective depth ratios were given to ensure that deflections were not excessive and bar spacing rules and stress limitations were given to ensure that cracking is not excessive. It can thus be seen that limit state design is not new, since most of these provisions have appeared in previous Codes, the only difference is a more formal statement of the design objectives.

2.2 Limit states appropriate to masonry

For unreinforced masonry Clause 19 of BS 5628: Part 1: 1992[2.2] states that 'The design of loadbearing masonry members should be primarily to ensure an adequate margin of safety against the ultimate limit state being reached'. This arises because the margin of safety specified is such that the loads in walls under service conditions are lower (in the order of one quarter or one fifth) than those which will cause collapse assuming the construction is built to specification. In tests on concrete block masonry walls to collapse severe deformation or cracking does not develop until loads of around 80% of ultimate are attained. In practice far more severe deformations can arise from shrinkage and temperature effects than will ever arise from loading. While shrinkage and expansion cracking due to temperature change can cause damage which affects the serviceability, they are not dealt with in limit state terms, but by detailing rules given BS 5628: Part 3: 1985[2.3] which are referred to in *Chapter 16*.

It should be emphasized that although this part of the Code does not specify any serviceability criteria such as acceptable crack widths or acceptable movements at expansion or contraction joints, the designer is not thereby excused from considering these matters. As an example, it may be necessary in tall structures to consider differential movement between the vertical walls which are carrying the vertical loads and those walls which function as partitions. These and other similar problems must be treated on their individual merits.

With reinforced and prestressed masonry, Clause 16 of BS 5628: Part 2: 1995[2.4] states that 'the design should provide an adequate margin of safety against the ultimate limit state. This is achieved by ensuring that the design strength is greater than or equal to the design load. The design should be such that serviceability limit state criteria are met. Consideration should be given to the limit states of deflection and cracking and others where appropriate e.g. fatigue.'

Limitations on deflection are specified in clause 16.2.2 of BS 5628: Part 2: 1995[2.4] but as with unreinforced masonry, the effects of temperature, creep, shrinkage and moisture movement are dealt with using rules in clause 20 of BS 5628: Part 3: 1985[2.3].

2.3 Partial safety factor format

The limit state method is normally associated in Codes with a partial safety factor format. This is a new departure but appears an eminently logical step forward. The idea is that instead of a global safety factor, different partial safety factors are associated with different types of loading and different materials. These factors are chosen to reflect the uncertainty with which the particular parameter can be assessed. Thus, dead load, which should be relatively accurately determinable, has a lower load factor associated with it than does a live load, which is less certain. Similarly, in reinforced concrete, concrete has a larger partial factor than steel since its quality is likely to be more variable. This approach gives results which accord with common sense; structures supporting dominantly their own self weight can be designed for lower overall safety factors than those carrying a large proportion of ill defined imposed loads; members where the strength depends mainly on concrete (e.g. columns) are built to higher safety factors than those where the strength depends dominantly on the reinforcement (e.g. lightly reinforced beams). In masonry, there is only one material – masonry. This limits the partial factors and makes life simpler. The factors are applied to the *characteristic strengths* of the material to give *design strengths*. Characteristic strengths are the specified strengths of materials (i.e. the characteristic strength of a 5 N/mm² block is 5 N/mm²). Characteristic loads are those given in design regulations.

In the Code there are two types of partial safety factor:

γ_m allows for variability of materials and workmanship and is applied to strengths

Thus, design strength is given by:

$$f_d = \frac{f_k}{\gamma_m}$$

According to BS: 5628: Part 1: 1992[2.2] the partial safety factor (γ_m) for unreinforced masonry can vary from 3.5 to 2.5. For reinforced and prestressed masonry, BS 5628: Part 2: 1995[2.4] provides more details. When assessing the direct or compressive strength of masonry, the partial safety factor varies from 2.0 to 2.3 but the value of 2.0 is specified for the shear strength of masonry. A value of 1.5 is required when the bond between reinforcement and concrete infill or mortar is considered.

γ_f allows for variability of loading, and is applied to characteristic loads to produce design loads.

Thus the design load is given by

design load = γ_p × characteristic load

Values for various combinations of load are given in Clause 22 of BS 5628 Part 1: 1992[2.2] and Clause 20.2 of BS 5268: Part 2: 1995[2.4].

The factors for γ_m are consistent with those used in BS 8110: Part 1: 1997[2.5], which has replaced CP 110: 1970[2.1]. This has the advantage that where reinforced concrete elements form part of the structure, loadings calculated for the design of those elements can be used directly in the masonry design. The γ_m factors were originally based on an earlier code[2.6] but changes to the existing values of γ_m are envisaged when EC6[2.7] is introduced.

2.4 References

2.1 BRITISH STANDARDS INSTITUTION. CP 110: 1970 *The structural use of concrete.*
2.2 BRITISH STANDARDS INSTITUTION. BS 5628: Part 1: 1992. *Code of Practice for the structural use of masonry. Part 1: Unreinforced masonry.*
2.3 BRITISH STANDARDS INSTITUTION. BS 5268: Part 3: 1985. *Code of Practice for the structural use of masonry. Part 3: Materials and Components. Design and Workmanship.*
2.4 BRITISH STANDARDS INSTITUTION. BS 5268: Part 2: 1995: *Code of Practice for the structural use of masonry: Part 2: Structural use of reinforced and prestressed masonry.*
2.5 BRITISH STANDARDS INSTITUTION. BS 8110: Part 1: 1997. *Code of practice for concrete.*
2.6 BRITISH STANDARDS INSTITUTION. CP 111: 1970 *Structural recommendations for loadbearing walls.*
2.7 Comité Européan de Normalisation, Eurocode 6: Design of masonry structures. Part 1-1; *General rules for buildings. Rules for reinforced and unreinforced masonry.* ENV 1996-1-1, 1995.

Chapter 3
Materials

3.1 Introduction

This chapter covers the general specification of concrete masonry units as given in BS 6073[3.1], together with their physical properties. More comprehensive information on thermal insulation, sound insulation, rain resistance and fire resistance is given in *Chapters 12 to 15*.

The specification of concrete masonry units is covered by BS 6073: 1981[3.1]. This Standard replaced the two previous Standards, BS 2028, 1364: 1968[3.2] *Precast concrete blocks* and BS 1180: 1972[3.3] *Concrete bricks and fixing bricks*. As well as being simplified, the Standard has been separated into two parts – Part 1: *Specification for precast masonry units* and Part 2: *Methods of specifying precast masonry units* (see *Chapter 17*).

The current definitions as given in BS 6073 are repeated below:

Masonry unit
A block or brick.

Block
A masonry unit which, when used in its normal aspect, exceeds the length or width or height specified for bricks (see *Bricks* below), subject to the proviso that its work size does not exceed 650 mm in any dimension, and its height when used in its normal aspect does not exceed its length or six times its thickness.

Brick
A masonry unit not exceeding 337.5 mm in length, 225 mm in thickness, (thickness is termed *width* in BS 3921[3.15]) or 112.5 mm in height.

Types of block
The presence of transverse slots to facilitate cutting or the filling of holes or cavities with non-structural insulant shall not alter the definitions given below:

Solid block
A block which contains no formed holes or cavities other than those inherent in the material.

Note: An autoclaved aerated concrete block is a solid block under this definition.

Cellular block
A block which has one or more formed holes or cavities which do not wholly pass through the block.

Hollow block
A block which has one or more formed holes or cavities which pass through the block.

Types of brick
Solid brick
One in which small holes passing through, or nearly through, the brick do not exceed 25% of its volume, or in which frogs (depressions in the bed faces of a brick) do not exceed 20% of its volume. For the purposes of this definition small holes are defined as being less than 20 mm wide or less than 500 mm^2 in area. Up to three larger holes, not exceeding 3250 mm^2 each may be incorporated as aids to handling, within the total of 25%.

Perforated brick
One in which small holes (as defined above) passing through the brick exceed 25% of its volume. Up to three larger holes, not exceeding 3250 mm^2 each, may be incorporated as aids to handling.

Hollow brick
One in which holes passing through the brick exceed 25% of its volume and the holes are not small, as defined above.

Cellular brick
One in which the holes closed at one end exceed 20% of the volume of the brick.

7

Sizes

Co-ordinating size

The size of a co-ordinating space allocated to a masonry unit, including allowances for joints and tolerances.

Work size

The size of a masonry unit specified for its manufacture, to which its actual size should conform within specified permissible deviations.

Compressive strength

The average value of the crushing strengths of ten masonry units tested in accordance with Appendix B of BS 6073.

3.1.1 Strength

3.1.1.1 Concrete blocks

It should be noted that in BS 6073 only blocks equal to or greater than 75 mm are tested for compressive strength. Blocks less than 75 mm are only tested for transverse strength and are primarily intended for non-loadbearing partitions. BS 6073: Part 1 states that the average crushing strength of 75 mm or greater blocks shall be not less than 2.8 N/mm² and that the corresponding lowest crushing strength of any individual block in the test sample of ten, shall not be less than 80% of this minimum permissible average crushing strength.

A similar criteria applies to blocks of greater strength except that the strength specified shall apply in place of the 2.8 N/mm² minimum permissible average crushing strength quoted above. Concrete blocks are generally available with mean compressive strengths ranging from 2.8 to 20 N/mm², although blocks of greater strength can be produced. The average and corresponding lowest individual strengths for the strength ranges quoted in BS 5628[3.1] are given in Table 3.1. For blocks of thickness less than 75 mm the average transverse strength of the sample shall not be less than 0.65 N/mm².

3.1.1.2 Concrete bricks

In the case of concrete bricks, the average compressive strength shall not be less than 7.0 N/mm² and the corresponding coefficient for variation for the samples shall not exceed 20%. The physical requirements for the strength categories in BS 5628 are given in Table 3.2.

3.1.2 Dimensions and tolerances

3.1.2.1 Concrete blocks

The typical range of work size of concrete blocks is given in Table 3.3, although blocks of entirely non-standard dimensions or design may be produced. The maximum deviation on the sizes of units are as follows:

length   + 3 mm and – 5 mm
height   + 3 mm and – 5 mm

Table 3.1 Compressive strength of blocks corresponding to BS 5628 strength categories

Block strength	Minimum compressive strength N/mm²	
	Average of ten blocks	Lowest individual block
2.8	2.8	2.24
3.5	3.5	2.8
5.0	5.0	4.0
7.0	7.0	5.6
10.0	10.0	8.0
15.0	15.0	12.0
20.0	21.0	16.8
35.0	35.0	28.0

Table 3.2 Physical requirements for concrete bricks

Physical property	Compressive strength category					
	7.0	10.0	15.0	20.0	30.0	40.0
Compressive strength (wet) Average of ten bricks not less than (N/mm²)	7.0	10.0	15.0	20.0	30.0	40.0
Coefficient of varients of compressive strength not to exceed (%)	20	20	20	20	20	20
Drying shrinkage not to exceed	0.06	0.04	0.04	0.04	0.04	0.04

thickness   + 2 mm and – 2 mm average
               + 4 mm and – 4 mm at any individual point.

In practice, since most blocks are produced in accurate steel moulds, it should be recognized that a consignment of blocks is likely to comprise of units all of a similar size, although within the tolerances specified. This can have an effect on joint width, especially for facing work.

3.1.2.2 Concrete bricks

Concrete brick sizes are typically manufactured to the work sizes shown in Table 3.4. The maximum dimensional deviation for bricks is as follows:

length   + 4 mm and – 2 mm
height   + 2 mm and – 2 mm
thickness   + 2 mm and – 2 mm

3.1.3 Frost resistance

3.1.3.1 Concrete blocks

The concrete block is inherently durable and suitable for a variety of exposure conditions. The British Standard BS 5628: Part 3[3.5] gives the following recommendations as to the minimum quality of blocks and mortar for durability (Table 3.5).

The Code permits the use of blocks below ground level damp-proof course, as described in Table 3.5 providing that there is no likelihood of sulphate attack. If there is a

Table 3.3 Work sizes of blocks

Length × height mm mm	Thickness mm														
	60	75	90	100	115	125	140	150	175	190	200	215	220	225	250
390 × 190	×	×	×	×	×		×	×		×	×				
440 × 140	×	×	×	×			×	×		×	×			×	
440 × 190	×	×	×	×			×	×		×		×	×		
440 × 215	×	×	×	×	×	×	×	×	×	×	×	×	×	×	×
440 × 290	×	×	×	×			×	×		×	×	×			
590 × 140		×	×	×			×	×		×	×	×			
590 × 190		×	×	×			×	×		×	×	×			
590 × 215		×	×	×		×	×	×	×		×	×		×	×

Notes:
1. To obtain the co-ordinating size of a masonry unit, add the nominal joint width, which is normally 10 mm, to the length and height of the unit given in the Table. (The thickness remains unchanged.)
2. Other work sizes are available and in use. No single manufacturer necessarily produces the complete range of work sizes shown.

Table 3.4 Work sizes of bricks

Length × height mm mm	Thickness mm	
	90	103
290 × 90	×	
215 × 65		×
190 × 90	×	
190 × 65	×	

Notes:
1. To obtain the co-ordinating size of a masonry unit, add the nominal joint width, which is normally 10 mm, to the length and height of the unit given in the Table. (The thickness remains unchanged.)
2. Other work sizes are available and in use. No single manufacturer necessarily produces the complete range of work sizes shown.

possibility of such attack, additional precautions may be necessary.

3.1.3.2 Concrete bricks

The frost resistance of a given combination of category of brick and mortar designation will be adequate if the requirements in Table 3.5, for a given use, are complied with.

3.1.4 Drying shrinkage and wetting expansion

3.1.4.1 Concrete blocks

The average value of the drying shrinkage of the sample should not exceed 0.06% except for autoclaved aerated concrete blocks, for which the maximum permissible value shall be 0.09%.

3.1.4.2 Concrete bricks

The maximum permitted drying shrinkage for concrete bricks is indicated in Table 3.2. It should be noted that if bricks are to be used under permanently damp conditions the drying shrinkage specification is of no significance.

3.1.5 Impact resistance

The most recent proposals on performance criteria for impact are those contained in the Code of Practice for design of non-load bearing external vertical enclosures of buildings[3.6] in which an attempt has been made to provide a sensible framework for the evaluation of the performance of walls liable to suffer impact damage, although it is clear that there is a lack of information in some areas.

Two distinct criteria must be satisfied:

(1) Walling should be capable of withstanding hard and soft body impacts applied or transferred to either of its faces during normal use without sustaining damage and without deterioration of its performance. There should be no significant irreversible deformation, visually unacceptable indentation marks or irreparable damage resulting from such impact.
(2) The walling, if subjected to more severe accidental impact should not be penetrated or become dislodged from its supporting structure. Any fracture resulting from such impact should not produce debris which may be a hazard to occupants or to people outside the building.

The code covers the use of walling materials and considers both the impact resistance required for the retention of performance, including appearance, and that needed for maintaining the safety of persons. The latter takes the form of larger soft body impacts for lateral stability and is applicable to very lightweight constructions. It is extremely unlikely that any masonry construction will fail to have adequate resistance in this situation.

In considering the performance levels required, the proposals recognise six levels of exposure:

A. Readily accessible to public and others with little incentive to exercise care. Prone to vandalism and abnormally rough use.
B. Readily accessible to public and others with little incentive to exercise care. Chances of accident occurring and of misuse.

Table 3.5 Minimum quality of concrete units and mortars for durability

Situation	Element of construction		Minimum quality of units[1]				Minimum quality of mortar[1]	
			Calcium silicate bricks (Class)	Concrete bricks (Category)	Concrete blocks (Type) Thickness mm	Density kg/m³	When there is no risk of frost during construction	When freezing may occur during construction
a	Inner leaf of cavity walls and internal walls	Unplastered	2	7.0	Any	Any	(iv)	(iii) or plasticized[2] (iv)
		Plastered	1	7.0	Any	Any	(v)	(iii) or plasticized[2] (iv)
b	Backing to external solid walls	Unplastered	2	7.0	Any	Any	(iv)	(iii) or plasticized[1] (iv)
		Plastered	1	7.0	Any	Any	(iv)	(iii) or plasticized[2] (iv)
c	External walls including the outer leaf of cavity walls and facing to solid construction	Above damp-proof course near to ground level	2	15.0	≥ 75	Any	(iv)	(iii)
		Below damp-proof course but more than 150 mm above finished ground level	2	15.0	≥ 75	Any	(iii)	(iii)
		Within 150 mm of ground level or below ground	3	20.0	≥ 75	≥ 1500	(iii)[5]	(iii)[5]
d	External freestanding walls		3	15.0	≥ 75	≥ 1500	(iii)	(iii)
e	Parapets	Rendered	3	20.0	≥ 75	Any	(iv)	(iii)
		Unrendered	3	20.0	≥ 75	≥ 1500	(iii)	(iii)
f	Sills and copings of bricks		4	30.0	--	--	(ii)	(ii)
g	Earth retaining walls[3,8]		4	30.0	≥ 75	≥ 1500	(ii)[5]	(ii)[5]

Notes:
(1) The designation of mortars is given in Table 3.11. Loading requirements or other factors may necessitate the use of a higher designation.
(2) Concrete bricks should not be used in contact with ground from which there is a danger of sulphate attack unless they are protected or have been made specifically for this purpose.
(3) Bricks of a lower strength category may be used if the supplier can provide direct evidence that they are suitable in the given location.
(4) Where sulphates are present in the ground water, the use of sulphate-resisting cement for the mortar may be necessary.
(5) An effective and continuous damp-proof course should be provided at the top of the wall as well as just above ground level.
(6) It is essential that rendering is on one side only.
(7) Walls should be backfilled with free draining material.

Table 3.6 Test impacts for retention of performance of exterior wall surfaces

Wall category	Test impact energy for impactor shown		
	H1	H2	S1
	Nm	Nm	Nm
A	(see note 1)		
B		10	120
C	6		120
D	(see note 2)		
E	6		
F	3		

Notes:
1. No test impact values are given for category A walls. In each case the type and severity of vandlism needs to be carefully assessed and appropriate impact values determined.
2. With category D walls the risk of impact is minimal and impact test values are therefore not appropriate.

C. Accessible primarily to those with some incentive to exercise care. Some chance of accident occurring and of misuse.

C. Only accessible, but not near a common route, to those with high incentive to exercise care. Small chance of accident occurring or of misuse.

E. Above zone of normal impacts from people but liable to impacts from thrown or kicked objects.

F. Above zone of normal impacts from people and not liable to impacts from thrown or kicked objects.

Recommendations for category (b) and (c) conditions are provided in Table 3.6. Two types of impactor are used in the tests:

(a) *Hard body* A steel ball, 50 mm in diameter and weighing 0.5 kg is dropped vertically onto the weakest part of the test construction which is supported horizontally. This is known as type H1. Alternatively a 62.5 mm diameter steel ball weighing 1.0 kg may be used – this being type H2.

(b) *Soft body* A canvas bag of spherical/conical shape and 400 mm diameter, filled with 3 mm diameter glass spheres and weighing 50 kg its swung on a cord at least 3 m long against the weakest part of the specimen. This is test type S1.

In practice testing with a soft body impactor is unlikely to damage a masonry construction. Tests on fairly soft unrendered aerated concrete blocks with the hard body impactor indicate that even at the highest energy levels specified in Table 3.6 blocks are unlikely to suffer visually unacceptable damage. The addition of a render coat further improves the performance of these blocks. The impact resistance of concrete bricks is very good and they are generally suitable for use in areas likely to be vandalized. Particular attention should, however, be paid to the impact resistance of tiling and other vertical cladding systems.

3.1.6 Density

BS 6073: Part 2 gives methods of measurement for both block density (gross density) i.e. oven dry mass ÷ gross volume, and also concrete density (nett density), i.e. oven. dry mass ÷ concrete volume. The block density is used, for example, in determining loads on a structure and handling requirements. The concrete density can be used to assess the thermal conductivity of the material and hence the thermal resistance of the block.

3.1.7 Fire resistance

Concrete blocks and bricks have inherently good fire resistance. In the case of dwellings most concrete masonry constructions are capable of providing a degree of fire resistance far in excess of the notional required period of not less than half-an-hour. The fire resistance of different types of units depends upon the class of aggregate employed and the presence of hollows or slots. The notional fire resistance periods for masonry walls are given in *Chapter 15*. Being non-combustible, concrete masonry walling does not produce smoke or toxic gasses but consideration must be given to the performance of any added insulants or finishes.

3.1.8 Thermal insulation

There are number of ways of producing concrete blocks with good thermal insulation properties, such as aerated blocks, lightweight aggregate blocks, multi-slotted lightweight aggregate blocks and foam-filled blocks of lightweight or dense aggregate. It is simpler, therefore, to consider the thermal resistance required of the block. If a thermal transmittance (U value) of 1.0 W/m^2 °C is required when a typical brick-cavity-block wall is to be employed with dense plaster internally, then a block with resistance of 0.491 m^2 °C/W or larger will be needed.

While *it is* comparatively easy to calculate the resistance of a solid block given the material coefficient of thermal conductivity, special assumptions need to be made for multi-slotted blocks, and this is dealt with in further detail in *Chapter 12*.

Historically, it has been usual to relate the thermal conductivity of a given concrete to the density of the material – the Jacob curve. Whilst it appears that this relationship gives a reasonable prediction of thermal conductivity based on a knowledge of the material density, some types of units, e.g. blocks made with a foamslag aggregate, appear to differ considerably from the standard curve and have better thermal conductivity values than predicted by density.

In general terms, thermal insulating blocks will have a density less than 1400 kg/m^3, except in the case of foam-filled dense blocks. The lowest density unit likely to be encountered is 475 kg/m^3 for the lower density autoclaved aerated concrete blocks.

3.1.9 Acoustic performance

Concrete masonry walls in general, including those comprising two leaves of the lighter thermal insulating blocks, have sufficient mass to provide adequate levels of sound insulation, both between adjoining properties and between the property and an outside sound source. The use of blocks for internal partitions (provided joints are well filled with mortar) can also provide good levels of sound insulation between adjoining rooms where problems often arise with very lightweight partitions. Typical examples of walls able to satisfy the sound insulation requirements are shown in *Chapter 14*. These include usual 'deemed-to-satisfy' constructions, although some manufacturers are able to offer constructions which are lighter than those indicated while still complying with the regulations.

It is possible that the inner leaf of a cavity wall will be required to assist the separating wall in attaining the necessary level of sound insulation, in which case it is advisable to consult the manufacturer concerning the wide variety of blocks available.

Generally speaking, the sound insulation will be a consideration in selecting the density of the units employed, as the greater the mass built into the wall the greater its ability to resist sound waves. It is necessary to check on proprietary products, however, since even quite dense units may be sufficiently open textured to allow fairly direct transmission of sound. Conversely, the high proportion of closed cells in autoclaved aerated blocks appear to give these units a high resistance to sound transmission. Where a sound absorbent material is required, open textured units have the ability to provide a large reduction in reflected sound levels. Special blocks designed to absorb sound in set frequency ranges can be made available by a few manufacturers.

In practice, the in situ sound performance of a wall is very dependent on workmanship although a survey in Belgium has indicated that 50% of sound problems occurred as a result of faults in the conception of the building.

3.1.10 Water absorption

Unlike clay brickwork there are no test procedures or requirements for measurement of water absorption of concrete blocks or bricks so that this parameter is not referred to in determining any other aspect of performance.

3.1.11 Resistance to rain

There is no simple correlation between a standard unit test for permeability or porosity and the performance of 2 m square test panels tested in a rain rig in accordance with BS 4315[3.7]. Open textured facing blocks of very similar appearance and identical densities, but different manufacture, may perform very differently in terms of the amount of water passing through the block when tested in a rain penetration rig. Reference should be made to local experience when assessing a particular block for resistance to rain or a wall test could be commissioned, although the latter may prove fairly expensive and there are not many suitable rigs available in the UK.

Rendering or other forms of surface finish will generally provide a wall with very good resistance to rain penetration. In the case of very dense concrete blocks and bricks where large amounts of 'run-off' of water down the face of the wall may be expected to occur, the additional water will tend to be blown through points of weakness in the wall, such as poorly filled mortar joints.

The rain penetration resistance of various wall constructions is discussed in *Chapter 13*.

3.1.12 Bond to mortar

As the suction applied by concrete blocks is generally less than that applied by clay bricks it is not necessary to wet blocks before laying. If problems are encountered, for example with some aerated blocks, because of high suction (particularly in hot weather) then a proprietary water retaining admixture may be considered which might comprise of methyl cellulose, although others are available.

As a general recommendation the best compromise between performance and workability will be given by a 1:1:6 cement:lime:sand mortar, although a 1:2:9 cement:lime:sand mortar may be used for low strength lightweight aggregate blocks and for aerated blocks where there is no risk of freezing during construction. The alternative plasticized or masonry mortar may also be considered. If a strong mortar is employed particularly with lighter units, the shrinkage between the block and mortar may reduce the mortar bond strength. The mortar consistence will need to be adjusted by the block layer to suit both the weight and suction of the block in use. Dense concrete blocks, especially if hollow, exert more pressure on the mortar than light-weight blocks and a stiffer mortar may be required.

3.1.13 Ease of finishing

Most blocks are rendered without any problem, but the general rule applicable to the mortar used for rendering is that it should be weaker than the block to which it is applied. A 1:1:6 cement:lime:sand render is normally suitable for lightweight aggregate blocks while a 1:2:9 render is more suitable for aerated blocks. For dense blocks there is little point in using a render stronger than a 1:1:6 except in the case of special applications. The mix proportions of some proprietary render systems do differ and advice should be sought from the manufacturer. The texture and suction of blocks will also vary considerably and reference should be made to the specific recommendations applicable to the block in use.

Autoclaved aerated blocks are likely to have a fairly high suction compared to other types of blocks which may prove a problem, particularly in hot weather. Although it is common practice to wet these blocks on site before rendering, this is often done to excess and may give rise to cracking due to movement of the wall. It is recommended that either a polymer:cement:sand bonding coat is applied before rendering, or a water retaining admixture, such as methyl cellulose, is added to the render mortar. Some manufacturers of aerated concrete blocks add a water repellent during the manufacture of the units to limit subsequent suction. A polymer:cement:sand coat can be used to bond to very smooth dense blocks or bricks and other types of unit may also benefit if treated with a bonding coat before subsequent rendering. An extra key for the render may also be obtained by raking back the mortar joints.

Most blocks will provide a good base for paint but due consideration must be given to the manufacturer's recommendations. In particular, the alkali present in the block may attack some paints and this would require the use of an alkali resistant primer on the surface. Selection of an open textured block usually provides a very attractive finish when painted and allows the wall to 'breathe'. In general paint for use with blockwork should be vapour permeable.

3.1.14 Ease of providing fixings

There is a wide range of proprietary fixings available for use after construction of a wall. Some fixings are built-in during construction, but these are unlikely to vary in performance with different types of blocks unless the loads imposed are so high that local crushing failure occurs – an unlikely event in practice because most fixings are comparatively lightly loaded.

The ease with which small fixings may be provided will depend largely on the ease with which the block in question may be drilled. Some very dense open textured blocks may require the use of special drill heads and, as the dense aggregate may tend to deflect the drill, an undersized drill head may be required to avoid too large a hole. Aerated blocks are easily drilled but because of the friable nature of the material one of the proprietary fixings specifically for this type of block may be required. Cut nails may be used during construction for making fixings in autoclaved aerated concrete (see also *Section 3.10*).

3.1.15 Resistance to chemical attack

There is no simple means of assessing the resistance of a block to sulphate attack and reference should be made to the past experience of the manufacturer or specifier. Open textured units are likely to be more vulnerable to chemical attack than dense hydraulically pressed concrete units. Finishing specifications are available which will provide protection to blockwork against chemical attack, e.g. protection to silage retaining walls from attack by acid. However, in the case of the effect of particular chemicals on the durability of a given block, reference should be made to the manufacturer.

3.2 Standard tests for blocks

3.2.1 Strength

3.2.1.1 Concrete blocks

The standard method of determining block strength is to test ten blocks capped with mortar. Strength tests to BS 6073 should be carried out on whole blocks. The old provision of BS 2028, 1364: 1968 for testing sawn part blocks was unsatisfactory, particularly where the blocks which contained a void or other feature were used which resulted in an asymmetric specimen on sawing.

Preparation of specimens

A smooth surface is required for a capping plate, for which machined steel plates or plate glass are employed, which is covered with a suitable release agent. The ten blocks must be immersed for at least 16 hours in water at a temperature between 10–25 °C and allowed to drain for about 30 minutes under damp sacking or similar material before the capping procedure is carried out.

The mortar for capping each block should consist of one part by weight of rapid-hardening Portland cement to one part by weight of sand complying with the requirements of grading zones 2 or 3 of BS 882: 1201[3.8]. From the sand, any material retained on a No. 7 fine mesh normal or special test sieve should have been rejected as specified in Part 1 of BS 410[3.9]. Water is added to produce a mortar having a consistence value of not less than 6 mm and not more than 9 mm when measured by the test for the determination of consistence by the dropping ball method specified on page 22. Prisms or cubes should be made from the mortar and tested for compressive strength in accordance with the procedures given in BS 1881[3.10]. Prisms should be cured using plain tap water and the compressive strength determined by a strength test carried out near the ends of the unbroken prisms. The results should be used to determine the age at which the mortar reaches an average compressive strength of at least 28 N/mm², which will normally be two to four days.

The mortar is placed as a uniform layer 5 mm thick on the capping plate. One bed face of the block is pressed into the mortar so that the vertical axis of the specimen is perpendicular to the plane of the plate. The verticality of the blocks is checked by using a level against each of the four vertical faces of the block, making allowance for any taper of the block sides. It is important to ensure that the mortar bed is at least 3 mm over the whole area and that any cavity in the bed face, which is normally filled when the blocks are laid in the wall, is completely filled with mortar. Surplus mortar is trimmed off flush with the sides of the block and the specimen is covered with cloth or polythene which is kept damp. The specimen should be left undisturbed for at least 16 hours before being carefully removed from the capping plate. If the bed is free from defects the second bed face should be capped using the same procedure. After checking, the complete specimens should be placed in water at a temperature of 20 ± 2 °C until the mortar used for both test faces has attained the required minimum strength.

Determination of compressive strength

Each specimen is removed from the water and allowed to drain for about 30 minutes under damp sacking or similar material and is tested while still in a wet condition. The specimen is placed centrally in the machine (see details of testing machine below) and the load increased continually at a rate of 5 ± 0.5 N/mm² for blocks of specified strength less or equal to 7 N/mm², or 10 ± 1 N/mm² for blocks of specified strength greater than 7 N/mm². The rate of application of load should be maintained as far as possible right up to failure.

The maximum load in Newtons carried by each specimen during test is recorded. The maximum load divided by the gross area of the specimen in mm² is taken as the compressive strength of the block in N/mm² and is reported to the nearest 0.05 N/mm² for blocks with a specified strength less than 7 N/mm² and to the nearest 0.1 N/mm² for blocks with a specified strength greater than 7 N/mm². The mean of the compressive strength of ten blocks is taken as the average compressive strength of the sample and is reported to the same accuracy as the individual results.

The testing machine

A testing machine of sufficient capacity for the test and equipped with a means of providing the rate of loading specified and with a pacing device is used. The capacity of the machine shall be such that the expected ultimate load on a specimen is greater than one-fifth of the machine scale range. The machine shall comply, as regards accuracy, with the requirements of BS 1610[3.11].

The testing machine shall be equipped with two permanent ferrous bearing platens which shall normally be as large as the bedding faces of the specimen being tested. Where the permanent platens of the testing machine are not as large as the specimen to be tested, auxiliary bearing platens having dimensions not less than that of the specimen shall be used. These shall not be fixed to the permanent platens, but shall be brought to bear in intimate contact, care being taken to exclude dirt from the interfaces.

The upper machine platen shall be able to align freely with the specimen as contact is made but the platens shall be restrained by friction or other means from tilting with respect to each other during loading. The lower compression platen shall be a plain, non-tilting bearing block. The auxiliary platen that will bear on the upper surface of the specimen shall be attached loosely to the testing machine by flexible wire or chain, to prevent it falling if the specimen collapses suddenly underload.

The testing face of the main platen and both faces of each auxiliary platen shall be hardened and shall have:

(1) A flatness tolerance of 0.05 mm.
(2) A parallelism tolerance for one face of each platen with respect to the other face as datum of 0.10 mm.
(3) A surface texture not greater than 3.2 μm CLA, measured in accordance with BS 1134[3.12].

The testing faces, where case-hardened, shall have a diamond pyramid hardness number of at least 600. Where the platens are through-hardened, the steel shall have a minimum specification of EN 26Y. The permanent platens shall be solid and not less than 50 mm thick, unless blocks of specified compressive strength of 7 N/mm² or more are to be tested, in which case the platens shall be not less than 75 mm thick. Auxiliary platens shall be solid and have a thickness not less than two thirds the amount by which they overhang the permanent platens unless blocks of specified compressive strength of 7 N/mm² or more are to be tested, in which case the thickness shall not be less than the overhang. In no case, however, shall the auxiliary platen overhang the permanent platen by more than 75 mm.

Alternative rapid control test

An alternative to the use of mortar capping is the use of fibre board. The strength indicated in this test varies from that achieved using board capping (see page 15) and the simplified test procedure is generally only used by manufacturers as a rapid control test.

The procedure for preparing the blocks for test is simply to remove fins or small pieces of aggregate proud of the surface by means of a carborundum stone. The specimens are then immersed in water at a temperature between 10–25 °C for at least 16 hours and are stored under damp sacking as described for the mortar capped test. Two new pieces of 12 mm insulation board, complying with BS 1142[3.13] and 10 mm larger than the bed faces of the specimens, are used as caps. The specimens are tested in the machine in a similar manner to the mortar capped specimens. In general, the source of manufacture of the capping board will not affect the indicated block strength.

Determination of transverse breaking load

This procedure is used for relatively thin non-loadbearing partition blocks. Five blocks are selected at random and immersed for at least 16 hours in water at a temperature of between 10–25 °C. Each block is then covered with damp material and allowed to drain for 30–45 minutes before testing. The testing frame is shown in Figure 3.1. Each block is placed centrally on the support bearers with the bedding faces perpendicular to the plane of the bearers. Load is applied so that the extreme fibre stress increases at a rate of approximately 1.5 N/mm² per minute. Appropriate rates of loading are shown in Table 3.7. The maximum load carried by the block during the test is recorded to the nearest 25 N and the transverse strength determined from the expression $f = \dfrac{570P}{bt^2}$. The mean of the five determined strength loads is taken as the

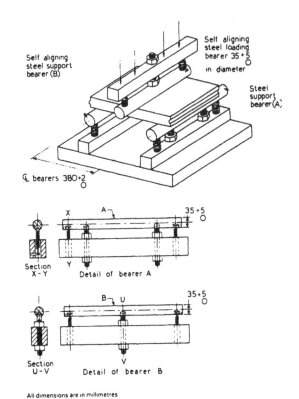

Figure 3.1 Apparatus for transverse test

Table 3.7 Appropriate rates of loading

Height (work size) mm	Thickness (work size) mm	Approximate rate of loading N/mm
190	60	1800
	75	2800
215	60	2000
	75	3200
290	60	2700
	75	4300

transverse breaking strength of the sample and is reported to the nearest 0.05 N/mm².

Limitations of the standard test for block strength

The clauses in BS 6073 relating to the performance of the testing machine are insufficient to describe whether it will perform adequately in practice. Although the accuracy of load scale indication is specified, this does not ensure that the machine is of sufficient stiffness, that the ball seating moves to accommodate the specimen and subsequently locks under load and that the specimen fails correctly even if slightly misplaced. It is recommended that consideration be given to a more comprehensive machine performance specification, details of which have been published[3.1,3.2].

Table 3.8 Relationships between some of the test specimens

Test specimens compared	Ratio
Mortar cap, block dry / Mortar cap, block wet	1.15
Board cap, block dry / Board cap, block wet	1.14
Board cap prism / Mortar cap prism	0.90
Board cap, block wet / Mortar cap, block wet	0.80

The relationship between the strengths of blocks, wet and dry, and between mortar and board capped specimens have been investigated[3.14] and the results are summarized in Table 3.8. Although these relationships are obtained from tests on a wide range of blocks, it is possible that the ratio between particular conditions might be slightly different for special materials or blocks. It must be remembered that because of the high variability of most concrete blocks a large number of specimens need to be tested to predict confidently the ratio between two forms of test.

One further point to note is the use of prism specimens which stems largely from testing small brick size units. Two block high specimens will not indicate significant differences in either mortar strength or joint thickness for most types of block as they affect the axial strength of a wall and their use is not, therefore, recommended[3.14].

3.2.1.2 Concrete bricks

The compressive strength of concrete bricks is found by testing a sample of ten units. The British Standard covering the testing of concrete bricks, BS 6073, includes a detailed sampling procedure to ensure that bricks selected are representative of the stack. The average overall dimensions of each brick, including the length and width of the frog if present, is measured to the nearest 1 mm and in the case of units with no frog or indentation, the area of the small bed face is calculated. For bricks containing a frog the gross area of the bed face in which the frog lies is calculated and the net area of application of the load is derived as the gross area minus the area of the frog. If the frog or indentation is so ill-defined that measurement is impracticable the bricks should be tested in accordance with the clause of BS 3921[3.15] which applies to frogged bricks laid frog upwards. Where both bed faces contain frogs, the net area of the smaller bed face should be used and for bricks having holes or perforations the gross area of the smaller bed face is calculated.

Before testing the bricks must be immersed in water at a temperature of 20 ± 5 °C for approximately 16 hours. Each brick is taken from the water and placed on the platen of the testing machine with the bed faces perpendicular to the direction of application of load. The testing machine must meet the requirements for accuracy of BS 1610[3.11] at the maximum load expected. To ensure uniform load the brick must be placed between plywood sheets of 2.4 to 4.8 mm thick, and the plywood must extend beyond the face of the bricks. Fresh plywood caps should be used for each brick. Load is applied at a rate of 200 ± 20 kN/minute for a frogged brick and 400 ± 40 kN/minute for a brick with no frog, until the brick has failed. The rate of loading may be doubled until half the anticipated maximum load has been applied. The compressive strength of each brick is taken as the maximum load divided by the area of the bed face. The strength of each brick is expressed in MN/m^2 to the nearest 0.5 MN/m^2 and the average of the ten bricks reported to the nearest 0.5 MN/m^2.

A procedure is included in BS 6073: 1981 for calculating the coefficient of variation of compressive strength from the individual values.

3.2.2 Shrinkage

3.2.2.1 Concrete blocks

The test procedures for determining drying shrinkage given in the current Standard, also in BS 1881: Part 5[3.10], have been found to have certain drawbacks with respect to consistency and repeatability. The following are the current test requirements.

Preparation of specimens

Four whole blocks are taken at random to determine drying shrinkage and wetting expansion and a specimen is sawn from each of the blocks. The length of each specimen must be not less than 150 mm and not more than 300 mm, the cross section of the specimen must have one dimension of 50 ± 3 mm and the second not less than 50 mm, and the centre area of the rectangle enclosing the cross section of the specimen must be solid. Reference points in the form of stainless steel balls 6.3 to 6.5 mm in diameter must be fixed to each end of the specimen such that they lie at the central axis. The balls may usually be fixed with neat rapid-hardening Portland cement and the specimens stored in moist air for at least one day to allow the cement to harden. The steel balls should ideally be located in holes, 2 mm deep, drilled in the ends of the specimen so that a hemispherical bearing is provided, but with some hard aggregates this may be difficult to achieve, in which case simple roughening of the ends of the blocks would have to suffice. The reference points should be cleaned and coated with lubricating grease, and the specimens stored in water at room temperature for a minimum of four days and a maximum of seven, the water being maintained at a temperature of 20 ± 1 °C for the final four hours. Each specimen is then measured to determine the distance over the steel balls to the nearest 1 mm and the length is taken as this distance minus 13 mm.

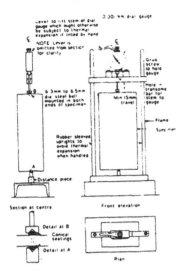

Figure 3.2 Typical measuring apparatus for drying shrinkage and wetting expansion

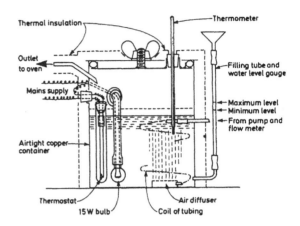

Figure 3.3 Air-conditioning equipment for supplying saturated air at controlled temperature to conditioning oven

Procedure for testing

The specimens are removed from the water and the grease is immediately wiped from the steel balls. The length of each specimen is measured* on apparatus similar to that shown in Figure 3.2. The specimen is placed between the recessed seating of the frame and the recessed end of the gauge and is rotated, the minimum reading of numerator gauge being recorded. By reversing the specimen and repeating the reading, an average of two readings may be taken as the original wet measurement. The calibration of the apparatus is checked using a reference rod.

The specimens are dried in an oven, ensuring that there is free access of air to all surfaces. (It is important not to place additional wet specimens in an oven containing partly dried specimens.) The specimens are removed at the specified time and placed in a cooling cabinet.

Ideally measurements should be carried out in a controlled temperature of 20 ± 2 °C at a time when the specimen temperature in the cabinet does not change by more than 0.5 °C in half an hour and is within 3 °C of the measured room temperature. The length of each specimen is measured as soon as possible after removal from the cabinet, a correction of 0.001% being applied for each °C difference in temperature from 20 °C and another correction made if the reading of length of the standard reference rod has changed. The cycle of drying, cooling, measuring and correcting is continued until the specified consistency and length are attained.

Readings are taken at daily intervals each side of seven and fourteen day increments from the time the specimens are taken out of the water. The average of each pair is

*Full details of this equipment are available in BS 6073. A detailed description of the oven is contained in BS 6073. Essentially the oven is designed to maintain a RH of 17% at a temperature of 55 °C. Saturated air is provided by means of the apparatus shown in Figure 3.3 and two air changes per hour take place.

calculated as the seven and fourteen day values respectively. For at least two of the three specimens A, B and C the difference between the original wet measurement and the seven day value is not less than 90% of the difference between the fourteen day value and the original wet measurement and the third is not less than 85%, the dry measurement is regarded as the fourteen day value provided the test for consistency is satisfied. If it is not satisfied, specimen D is substituted accordingly, and if the above requirement for completion of the dry measurement is then satisfied by the three specimens so obtained, take the fourteen day mean value as the dry measurement. If the criteria for completion of drying are not satisfied all the specimens are returned to the oven and readings taken at 20 and 22 days. Provided the test for consistency is satisfied the mean of the results from specimens A, B and C is regarded as the dry measurement. If the test for consistency is not satisfied substitute the result from specimen D and recalculate. Calculate the drying shrinkage for each specimen as the difference between the original wet measurement and the dry measurement expressed as a percentage of the length.

It is now necessary to measure the wetting expansion and the specimens must be immersed in water for four days, the water being maintained at 20 ± 1 °C for the final few hours. The length of each specimen is then recorded and the wetting expansion is calculated as the difference between the dry measurement and the final wet measurement expressed as a percentage of the length.

Calculation of results

If the values for drying shrinkage or wetting expansion obtained with any one of the three specimens A, B and C differ from the average value for the same three specimens by more than 25% then the excessive shrinkage value of the individual specimen is discarded and the value obtained with specimen D substituted. The average value of drying shrinkage and wetting expansion is reported to the nearest 0.005%. If the value of D is also found to differ from the average for the second set of three specimens by more than 25% the result of the test is to be regarded as invalid.

Limitations of the test procedure

The regime of curing and testing at 55 °C at a relative humidity of 17% is unrealistic in terms of being more severe than the worst conditions likely to be encountered by blockwork in practice. Furthermore, it is apparent that no limits are set to the age or initial curing applied to specimens, and drying shrinkage may change significantly because of further hydration of the cement or of atmospheric carbonation over a period of a few weeks. The rank applied to different types of blocks when tested to the above regime may differ from that observed in practice. This test procedure is, therefore, likely to be reviewed with the object of producing more practicable results.

3.2.2.2 Concrete bricks

Preparation of specimens

For the determination of drying shrinkage four whole bricks are taken at random. A depression not more than 2 mm deep is drilled or cut into the centre of each end of all four specimens although if the nature of the brick makes this difficult the surface should just be roughened. A 6.3 to 6.5 mm diameter stainless steel ball is cemented with neat rapid-hardening Portland cement into each depression, and is wiped dry and coated with lubricating grease to prevent corrosion.

The specimens marked A, B, C and D are stored in moist air for at least one day. They are then completely immersed in water at room temperature for a minimum of four days and a maximum of seven, the water is maintained at a temperature of 20 ± 1 °C for the final four hours. Each specimen is measured to determine the distance over the steel balls to the nearest 1 mm and this distance minus 13 mm is taken as the length. The equipment and subsequent method of test is similar to that used for concrete blocks.

3.2.3 Determination of block density, concrete volume and net area of hollow blocks

3.2.3.1 Determination of density

Measurement of volume of cavities:

(1) Place the blocks on a thin sheet of foam rubber or other resilient material with the open ends of the cavities uppermost.
(2) Close any cavities at the ends of the block by clamping flat sheets of 13 mm insulating board to the ends of the block without distortion. Ignore the effects of tongues or grooves.
(3) Fill a one litre glass measuring cylinder accurately with dry sand which has been graded between a 300 m BS test sieve and a 600 m BS test sieve, both sieves complying with BS 410[3.9].

(4) Fill the cavities with the sand by pouring from the cylinder, refilling if required, keeping the cylinder lip within 25 mm of the top of the cavity and pouring steadily and striking off level.
(5) Return to the cylinder any sand struck off and note, in ml, the total volume of sand used to the nearest 50 ml. Convert this volume to the equivalent volume in mm^3 of the cavities to the nearest 250 mm^3.
(6) Calculate the gross volume of the block to the nearest 250 mm^3 by multiplying the average thickness by the specified length and height of the block. (Ignore formed protrusions and indentations.)
(7) Express the volume of cavities in each block as a percentage of the gross volume of the block.
(8) Record, to the nearest 5%, the greatest volume of cavity detected.

3.2.3.2 Determination of concrete volume

(1) Remove all random flashings with carborundum stone.
(2) Measure to the nearest 1 mm using callipers and rule the dimensions of formed indentations and protrusions on the external faces and ends of the block.
(3) Calculate the algebraic sum of the volume of all indentations and protrusions to the nearest 250 mm^3. (Treat volume of indentation as negative and volume of protrusion as positive.)
(4) Calculate the concrete volume, to the nearest 250 mm^3, using the following equation:

Concrete volume = (gross volume of block)
  − (volume of cavities and voids)
  + (algebraic sum of volume of indentations and protrusions)

3.2.3.3 Determination of block density and concrete density

(1) Dry three blocks for at least 16 hours in a ventilated oven having the temperature controlled at 105 ± 5 °C.
(2) Cool the blocks to ambient temperature and weigh.
(3) Repeat (1) and (2) until the mass lost in one cycle does not exceed 0.05 kg.
(4) Calculate the block density and the concrete density by using the following formulae:

Block density in kg/m^3 = oven dry mass in kg ÷ gross volume in m^3

Concrete density in kg/m^2 = oven dry mass in kg ÷ concrete volume in m^3.

(5) Record the mean densities of the three blocks to the nearest 10 kg/m^3.

3.2.3.4 Determination of net area of hollow blocks

(1) Obtain the mean height from six height measurements.
(2) Calculate the net area using the following equation:

Net area = concrete volume ÷ mean height.

(a) Four positions for checking length of whole blocks

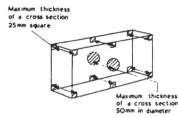

(b) Six positions for checking height of whole blocks

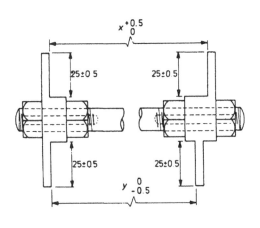

Maximum thickness of a cross section 25 mm square

Maximum thickness of a cross section 50 mm in diameter

(c) Seven measurements of thickness

Figure 3.4 Measuring points for block dimensions

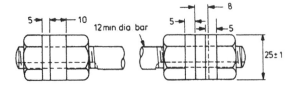

Figure 3.5 GO/NOT GO gauges for checking length and height of concrete blocks

Notes:
1. Keys are used for keeping fittings at both ends in the same plane
2. Fittings may be made from the solid as shown or made up from separate pieces
3. The two measuring dimensions are to be accurate to 0.5 mm

NOTE: The net area of hollow blocks is required for assessing the characteristic compressive strength of walls of hollow concrete blocks, filled with in situ concrete (see Chapter 9).

3.2.4 Dimensions

To determine the dimensions of a batch of concrete blocks it is usual to take ten whole blocks from the consignment at random. Any fins may be removed with a carborundum stone before checking the dimensions at the points shown in Figure 3.4. There are four positions at which the length is checked, six positions for height and seven measurements of thickness. As a routine control of length and height GO/NOT GO gauges of the sort shown in Figure 3.5 may be used but callipers and a rule must be used for the thickness.

When checking compliance with dimensional tolerances it is necessary to calculate to the nearest 1 mm the average of the seven measurements of thickness of each block. If the dimensions are being employed to calculate the block density, the gross volume of the block is calculated to the nearest 50,000 mm^3 from the average thickness and the specified length and height of the block. When the dimensions are being used to work out the area for compressive strength determinations, the average thickness at the top and bottom bed faces are calculated to the nearest 1 mm and the gross area is calculated to the nearest 500 mm^2 as the smaller of the two average thicknesses multiplied by the specified length of block.

3.2.5 Moisture content

There is no standard procedure for determining the moisture content of concrete masonry, but it is suggested that a procedure similar to that indicated in BS 1881: Part 5: 1970[3.10] is adopted.

The dry density may be determined by placing the weighed block or block sample, in a drying oven at a temperature of 105 ± 5 °C. This may take up to three days for very dense concrete, after which the specimen should be removed from the oven and immediately transferred to a dessicator or dry airtight vessel. Once the temperature has fallen below 60 °C it may be reweighed and the moisture content by weight then deduced. As with a concrete cube, the saturated weight may be determined by weighing the specimen under water and in air before drying in the oven.

3.2.6 Thermal insulation

In the United Kingdom the thermal insulation value of a concrete block is determined by measuring the thermal conductivity in accordance with BS 874: 1973[3.16]. *Methods for determining thermal insulating properties, with definitions of thermal insulating terms.* This Standard allows a variety of procedures but aspects relating to building materials are considered here.

The basic specimens required for test are two 305 × 305 × 50 mm prisms from the block. If the face size of block is smaller than the above, sections must be sawn from the block, glued and the faces ground to an accurate thickness. A heating element is then placed between the prisms, with thermocouples on either side of each prism either on the surface or buried. The thermal conductivity of the material is calculated by applying a known input of electrical energy and measuring the surface temperatures. The apparatus can be a plain hot plate with surrounding layers of insulation to prevent edge losses, or a guarded hot plate.

While materials as delivered can vary, the variation in results between laboratories testing the same type of block appears to be due to differences in the testing procedure. Among the observations made by Spooner[3.17] on the results of a 'round robin' thermal conductivity test were the following:

(1) A comparison of results taken from plain and guarded hot plate equipment recorded no difference for aerated concrete specimens but 7% difference for lightweight concrete.

(2) Good contact must be maintained between the thermocouples and the specimen surfaces and for thermal conductivities above 0.5 W/m K the thermocouples must be cemented into grooves in the surface of the specimens.

3.3 Quality control for blocks

3.3.1 Introduction

The Code of Practice, BS 5628, introduces the concept of quality control such that the value of γ_m should be commensurate with the degree of control exercised during the manufacture of the structural units, the site supervision and the quality of the mortar used during construction. Two levels of control are recognized for manufacture:

Normal category

This category should be assumed when the requirements for compressive strength in the appropriate British Standard are met, but the requirements for the special category below are not.

Special category

This may be assumed where the manufacturer:

(1) agrees to supply consignments of structural units to a specified strength limit, referred to as the *acceptance limit* for compressive strength, such that the average compressive strength of a sample of structural units, taken from any consignment and tested in accordance with the appropriate British Standard specification has a probability of not more than 2½% being below the acceptance limit;

(2) operates a quality control scheme, the results of which can demonstrate to the satisfaction of the purchaser that the *acceptance limit* is consistently met in practice and the probability of failing to meet the limit is never greater than that stated above.

The British Standards Institution runs a quality control scheme whereby a registered certification trade mark (kitemark) and legend 'Approved to British Standard' may be marked on the blocks by manufacturers licensed under the scheme. The mark indicates that the units have been produced to comply with the requirements of the British Standard under a system of supervision, control and testing operated during manufacture and including periodical inspection at the manufacturers' works.

The lack of correlation between unit and wall strength for small units such as bricks has led to the concept of testing small assemblages or prisms of masonry. The use of prisms has also been applied to larger units such as concrete blocks, but there is little purpose in using more than single blocks, as unit tests correlate well with wall tests on blockwork panels, and prism tests in the case of blocks do not even reflect large changes in mortar strength, joint thickness or workmanship[3.14]. The use of these tests for site control of masonry construction is, therefore, not recommended, and block, mortar and workmanship quality is better controlled by means of separate tests.

3.3.2 Performance of the testing machine

It is now widely accepted that such features as the stiffness of the testing machine frame, hydraulic system and the behaviour of the ball seat under load can all reduce the indicated strength of the specimen and thus affect the recorded strength of concrete cubes. These aspects are quite separate from the accuracy of load scale indication of the machine. Although the effect of poor machine performance on indicated block strengths has not been well researched, a limited number of tests carried out by The Cement and Concrete Association have indicated strengths 10% lower than those obtained on a good machine. For concrete cube machines, for which many more test results are available, differences of up to 30% have been recorded. It is important to ensure that testing is carried out in accordance with the correct British Standard for the particular product and on a machine performing satisfactorily. Many manufacturers of testing machines offered 'special' versions of standard concrete cube testing machines for block testing only – possibly the only concession to block testing being the provision of a larger platen. BS 6073 makes specific recommendations about platen thickness and such like, which should be followed. Purpose designed block testing machines are now available and providing these are regularly checked should provide good results. The Foote prooving device is one method of checking the performance of block testing machines, and is reported by Roberts[3.18] who has also published a testing machine specification[3.19] from which appropriate clauses can be extracted.

3.4 Mortar strength

Mortar fulfils a multi-purpose role in a masonry assemblage. In addition to providing a level bed so that the load is evenly spread over the bearing area of the units, the mortar bonds units together to help them resist lateral forces, allows the control of alignment and plumb, and prevents the ingress of rain through the joints.

3.4.1 Properties relevant to mortar selection

Mortar should be very workable and hence readily and economically used by the mason, it must offer adequate durability but should not be stronger than the units so that movements can be accommodated. Table 3.5 indicates the minimum quality of unit and mortar for a given exposure condition.

Traditionally, lime mortars were employed for masonry construction, typical proportions being 1:3 lime:sand. Although lime mortars offer excellent workability characteristics they rely on the loss of water and carbonation to slowly gain strength and this constraint on the rate of construction has lead to the widespread use of cement mortars. Portland cement mortar quickly gains strength allowing rapid construction. Lime is often added to the mix to improve workability producing the so called cement:lime:sand (compo) mortar. Masonry mortar consists of a mixture of Portland cement with a very fine mineral filler and an air-entraining agent but should not be overmixed because an excessive amount of air could be entrained. Air-entrained or plasticized mortar is produced by using a plasticizer to replace the role of the lime with the cement:sand:mix, whereby the air bubbles increase the volume of the binder paste and fill the voids in the sand.

In Table 3.9 comparison is made of the properties of different types of mortar and their approximate equivalence. For most general purpose concrete masonry construction with conventional detailing the lowest grade of mortar practicable should be employed. Thus for aerated concrete blocks a 1:2:9 cement:lime:sand mortar will be appropriate, providing:

(1) the sand grading is good;
(2) there is no risk of frost attack during setting;
(3) the surface finish to be applied as a decorative coat does not demand a stronger background.

For most general purpose concrete masonry construction a 1:1:6 cement:lime:sand mortar is suitable. High strength loadbearing masonry and reinforced masonry generally require a 1:¼:3 or 1:½:4½ cement:lime:sand mortar. Table 3.10 indicates the lime:sand mixes required for specified cement:lime:sand mortars. Retarding agents can be added to delay the set and prolong the working life of the mortar, but cement mortars should otherwise be used within two hours of mixing. It should be noted that when retarded materials set the strength is no weaker than that obtained without the use of a retarder. Table 3.10 also indicates the order of strength that may be aimed at for control purposes. In practice the strength achieved by a given mortar will depend on the quality and grading of the sand as well as the cement and water content. The bond strength of the mortar to the units is often far more important than the compressive strength

Table 3.9 Lime:sand mixes required for specified cement:lime:sand mortars

Specified cement: lime: sand mortar	Lime: sand mix	Gauging of cement with lime:sand mix
Proportions by volume	Proportions by volume	Proportions by volume
1:½:4–4½	1:8–9	1:4–4½
1:1:5–6	1:5–6	1:5–6
1:2:8–9	1:4–4½	1:8–9
1:3:10–12	1:3½:4	1:10–12

which is not a prime requirement. Given a reasonable sand grading the bond strength achieved will depend on the workmanship of the masonry and the type and condition of the units at the time of laying, together with the subsequent curing regime. Sands for use with mortar should comply with BS 1199, 1200: 1976[3.20].

3.4.2 Quality control for mortars

Specified quality control of mortars on site is generally confined to large jobs where loadbearing masonry is involved and construction is taking place under the special category indicated in BS 5628: Part 1[3.4]. It is prudent to specify effective sampling and test procedures for the mortar on any large site, particularly where, for reasons of strength or appearance, consistent batching and use is critical.

Cement for mortars should comply with the requirements of one of the following Standards: BS 12:[3.21] *Portland cement*; BS 146: Part 2: 1973[3.22] *Portland blast furnace cements*; BS 4027: 1980[3.23] *Sulfate-resisting Portland cement* or BS 5224: 1975[3.24] *Masonry cement*.

Lime should be hydrated non-hydraulic (high calcium) or semi-hydraulic (calcium) conforming to BS 890: 1972[3.25] *Building limes*. Sands for mortar should comply with the grading requirements of Table 2 of BS 1200[3.20] and with the general requirements of that Standard in respect of its limitations on deleterious substances.

Ragsdale and Birt[3.26], have conducted an extensive survey of the availability, usage and compliance with specification requirements of building sands. It is apparent that sands available in the UK for use in building mortars vary considerably in their properties and characteristics and that those which conform fully and consistently to the current British Standards are unobtainable from local sources in many parts of the country. As a result Ragsdale and Birt found that many sands which did not comply with the British Standards could satisfactorily be used and that the factors of prime importance were cleanliness and grading, rather than compliance with the appropriate British Standards gradings. Figures 3.6 and 3.7 reproduced from this survey indicate the specified grading limits for masonry mortar sands and for plastering sands respectively. The British Standard gradings have subsequently been widened.

20

Table 3.10 Requirements for mortar (adapted from BS 5628: Part 1: 1978)

	Mortar designation	Type of mortar						Mean compressive strength at 28 days *B.nn²)	
		Cement:lime:sand		Air-entraining mixes					
				Masonry cement: sand		Cement:sand with plasticizers			
		by volume	by weight (mass)	by volume	by weight (mass)	by volume	by weight (mass)	Preliminary	Site tests
Increasing strength and → Increasing ability to accommodate movements ↓	(i)	1:0–¼:3	1:0–¼:3½	–	–	–	–	16.0	11.0
	(ii)	1:½:4–4½	1:¼:5	1:2½–3½	1:3½	1:3–4	1:4	6.5	4.5
	(iii)	1:1:5–6	1:½:6½	1:4–5	1:5½	1:5–6½	1:6½	3.6	2.5
	(iv)	1:2:8–9	1:1:10	1:5½–6½	1:7½	1:7–8	1:9	1.5	1.0

Direction of change in properties is shown by the arrows

Increasing resistance to frost attack during construction →

Improvement in adhesion and consequent resistance to rain penetration ←

Notes:
1. Where mortar of a given compressive strength is required by the designer, the mix proportions should be determined from tests conforming to the recommendations of Appendix A of BS 5628: 1978.
2. The different types of mortar that comprise any one designation are approximately equivalent in strength and do not generally differ greatly in their other properties. Some general differences between types of mortar are indicated by the arrows at the bottom of the Table, but these differences can be reduced (see CP 121).
3. 'Lime' refers to non-hydraulic or semi-hydraulic lime. The proportions are based on dry hydrated lime.
4. Proportions by volume. The proportions of dry hydrated lime may be increased by up to 50% to improve workability. The range of sand contents is to allow for the effects of the differences in garding upon the properties of the mortar. Generally, the higher value is for sand that is well graded and the lower for coarse or uniformly fine sand. The designer should clearly indicate which proportions are required for the particular sand being used. Where no specific instructions are given, it will be assumed that the designer has satisfied himself that the sand will have no significant effect on the mortar and that it may be batched within the given range to achieve workability.
5. Proportions by weight (mass). The proportions of dry hydrated lime may be increased by up to 25% to improve workability. The proportions for sand given are mean values for each mortar designation. The proportions by mass may alternatively be determined by the designer by measuring the actual bulk densities of the constituents of a volume-batched mix known to produce satisfactory results.
6. An air-entrained admixture may, at the discretion of the designer, be added to a lime:sand mix to improve its early frost resistance.

Pigments should comply with the requirements of BS 1014: 1975[3.11] and generally should not exceed 10% by weight of the cement, except in the case of carbon black which should not exceed 3% by weight of the cement. Normally adding such high percentages would not, in any case, be economic.

The mixing and use of mortars should be in accordance with the recommendations of BS 5628: Part 3[3.5] and BS 5390 1976[3.28]. Ready mixed lime:sand for mortar should comply with the requirements of BS 4721: 1981[3.29].

Preliminary tests
Six weeks prior to building the masonry, the strengths of the grades of mortar to be used should be determined in the laboratory with materials taken from the sources which are to supply the site. Six specimens should be produced of one of the following types: 75 mm cubes, 100 mm cubes or 100 mm × 25 mm × 25 mm prisms. The mortar should have a consistence corresponding to a 10 mm penetration of the dropping ball* (without suction method) and should be cured hydraulically and tested in compression in accordance with the procedures given in BS 4551: 1980[3.30]. The type of specimen to be used on the Site should be identical to that used for these preliminary tests.

Interpolation of test results
The average compressive strengths for the various grades of mortar are shown in Table 3.10. If desired half of the specimens may be tested at seven days, the results of which will normally give an indication of the strength to be expected at 28 days. For the mortars in Table 3.10 the strengths at seven days will approximate to two thirds of the strengths at 28 days provided that the mortars are based on Portland cement with no additive to retard or accelerate the rate of hardening. If the average of these seven day strengths equals or exceeds two thirds of the appropriate weight in the Table the requirements are likely to be satisfied, but if the average strength is less, the designer may choose to wait for 28 day strengths or to have the test repeated on a more suitable sand.

Site tests
Six 100 mm × 25 mm × 25 mm prisms or four cubes should be prepared on Site for every 150 m² of wall of one grade of mortar, or for every storey of the building,

*Note that a mortar of this consistence is not workable enough for most purposes.

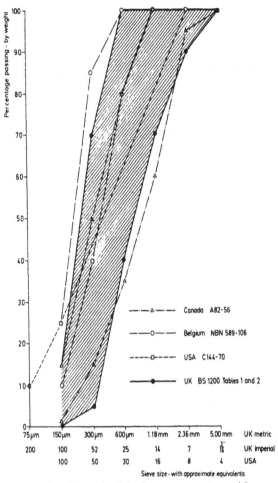

Figure 3.6 Specified grading limits for masonry mortar sands*

*It should be noted that these are the gardings that applied at the time the survey was conducted.

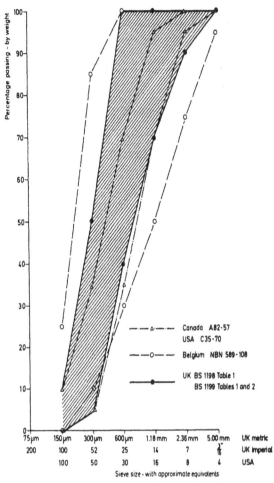

Figure 3.7 Specified grading limits for plastering sands*

*It should be noted that these are the gardings that applied at the time the survey was conducted.

whichever is the more frequent. Specimens should be stored and tested in accordance with BS 4551[3.30]. If required, half the site specimens may be tested at seven days. The average strength should exceed two thirds of the appropriate 28 day strength in Table 3.10. When the Site samples are tested at the age of 28 days, the results will be deemed to pass if the average strength of three 100 mm × 25 mm × 25 mm prisms or two cubes exceeds the appropriate Site values given in Table 3.10.*

3.5 Standard tests for mortars

3.5.1 Consistence by dropping ball

Flow, stiffening time and strength properties of a mortar will all be affected by the consistence which may be determined by the dropping ball test. Since some other

test procedures require a standard consistence of 10 ± 0.5 mm penetration of the dropping ball, preliminary tests in accordance with this Clause may have to be made for adjustment of the water content.

Suitable apparatus for carrying out the test, as shown in Figure 3.8, consists of a dropping mechanism for a methyl methacrylate ball. This release mechanism must be such that it does not impart any appreciable spin, friction or acceleration to the ball other than that due to gravity (Figure 3.9). The polished ball has a diameter of 25 ± 0.1 mm, weighs 9.8 ± 0.1 g and is dropped into the mortar contained in a brass mould of 100 mm internal diameter and 25 mm internal length. The apparatus includes a device for measuring the depth of penetration of the ball in mm to an accuracy of 0.1 mm, which also needs to be capable of measuring any fall in the surface level of the mortar for consistence retentivity. The brass mould is filled by pushing the mortar with the end of a palette knife in about ten stages. When the mortar is slightly above the rim the surface is levelled with the top

---

*For retarded mortars the retardation time should be added to the 7 or 28 days.

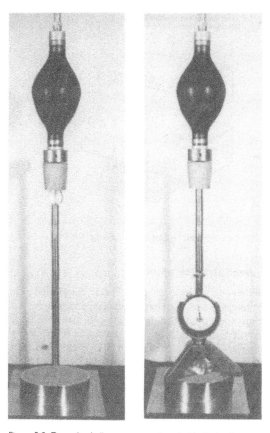

Figure 3.8 Dropping ball apparatus and a suitable device for measuring penetration

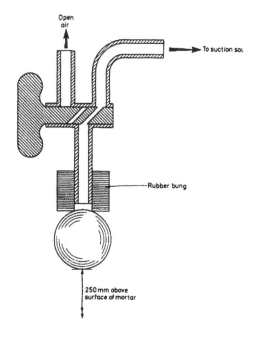

Figure 3.9 Diagrammatic representation of ball release

of the mould using the palette knife, which is held at about 45° and moved once across the mould with a sawing motion to strike off excess material, and then at a slightly flatter angle to trowel the surface.

The methyl methacrylate ball is allowed to drop from a height of 250 mm and must land within 12 mm of the centre of the surface of the mortar. The penetration of the ball is measured and recorded to the nearest 0.1 mm. The test is repeated twice with two other samples of the mortar, after cleaning the apparatus, and the average of the three penetrations is recorded to the nearest 0.1 mm as the consistence.

3.5.2 Consistence retentivity and water retentivity

The retention of consistence and of water in mortars is of considerable practical importance, especially if the mortar is to be applied on materials of high suction. The consistence of any particular mortar will depend upon the water content and the content of entrained air. Thus, measuring the water retained under a standard test condition is not always adequate to describe the performance of some mortars, and the degree to which the mortar retains its consistence is the more useful measure for general use. In this test the properties of the mortar are measured before and after standard suction treatment. The final consistence value, expressed as a percentage of the original value, is termed the 'consistence retentivity'. The weight of water retained after suction, expressed as a percentage of the original water content, is termed the 'water retentivity'.

For this test, the mould used for the dropping ball test is filled and the average penetration is determined. The depression left by the ball is filled with mortar and struck off level. The surface of the mortar is then covered with two pieces of white cotton gauze (two circles, 110 mm in diameter or two squares of 110 mm wide) and eight circles of filter paper are placed on top. Extra thick white filter paper of weight 200 g/m² and 110 mm in diameter should be used on top of which a non-porous plate 110 mm in diameter should be placed loaded with a 2 Kg weight. After two minutes the weight, filter paper and cotton gauze are removed and the gauze discarded.

The fall in the level of the mortar is then measured and a single dropping ball test is carried out after suction. The apparent penetration of the ball is corrected by subtracting from it any measured fall in the level of the mortar so that the corrected penetration of the ball after suction, expressed as a percentage of the average penetration before suction, is the consistence retentivity. This procedure is repeated with a second sample of mortar, and the average of the two retentivity values to the nearest 5% is taken as the consistence retentivity of the mortar.

To measure the water retentivity the mould in a dry condition and light circles of filter paper are weighed. The mould is filled as for the dropping ball test and, with all mortar removed from the outside, the mould and its contents are weighed. The mortar is then subjected to suction as described earlier and the filter paper weighed to the nearest 0.05 g. The weight of water originally present in the mould is calculated from the weight of mortar in the mould and the moisture content of the mortar. The weight of water retained by the mortar after suction (i.e. the weight of water originally present in the mould minus the weight of water absorbed by the filter paper), expressed as a percentage of the weight of water originally present in the mould full of mortar, is the water retentivity. The procedure is repeated with a second sample of mortar and the average of the two retentivity values to the nearest 1% is taken as the water retentivity of the mortar.

3.5.3 Determination of the compressive strength of mortar

The standard cube for testing mortar is 100 mm but 75 mm moulds may also be used, or prisms 100 mm × 25 mm × 25 mm.* The mould must be substantial in construction and comply with the requirements of BS 1881: Part 5 1970[3.10]. Particular attention is paid to the dimensional tolerances of the mould.

A representative sample of the mortar should be taken and the test specimens made as soon as practicable. The mould should be filled in layers approximately 50 mm deep and each layer should be compacted by hand or by vibration. Once the top layer has been compacted the surface of the concrete should be finished level with the top of the trowel. Ideally vibration should be used to ensure compaction but a standard compacting bar may be used to complete compaction by hand.

Specimens made in the laboratory should be stored in a place free from vibration, in moist air of at least 90% relative humidity at a temperature of 20 ± 2 °C for 16–24 hours. Either a moist curing room or damp matting covered by polythene should be used. Demoulding can take place after 24 hours, and specimens should then be stored in a tank of clean water at a temperature of 20 ± 5 °C until tested. Specimens made on site should be stored under damp matting covered with polythene at a temperature of 20 ± 5 °C for 16–24 hours. After demoulding, usually 24 hours, the specimens should be stored in clean water contained in a tank maintained at a temperature of 20 ± 2 °C until they are removed for testing. The specimens should be tested in a concrete cube testing machine of Grade A in terms of BS 1610[3.11] and the performance aspects of the machine should be verified by regular reference testing. Load should be applied to the specimens at a rate of 15 N/mm² per minute and the maximum load failure recorded.

It should be noted that BS 4551: 1970[3.30] also contains a flexural test for mortar using the 100 mm 25 mm × 25 mm prism but it is unlikely that the designer will require this information for the mortar in isolation.

---

* The new CEN prism size is 160 × 40 mm.

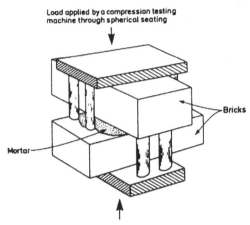

Figure 3.10 Crossed brick couplet test for mortar bond

3.5.4 Bond strength of mortar

A number of techniques have been developed for determining the bond strength of mortar to units but no direct methods are incorporated into a British Standard. The method of testing for the flexural strength of a masonry assemblage specified in BS 5628: Part 1[3.4], (Appendix A2) relies on testing panels 1.2 m to 1.8 m long by 2.4 m to 2.7 m high, where the bond of the mortar to the units plays an important part in the overall performance of such a panel.

A cross-brick couplet test was developed for incorporation in ASTM specifications[3.31] involving the use of a testing machine fitted with a special jig applying such tensile force that the unit eventually parts, thus giving a measure of the bond at the mortar. A diagram of the apparatus is shown in Figure 3.10.

3.5.5 Stiffening rate

This test involves an arbitrary procedure for determining the time taken after mixing for the resistance offered by the mortar to the penetration of a metal rod 30 mm³ in cross sectional area to reach 1.0, 1.5 or 2.0 N/mm². The higher value is suitable for higher cement content mortars when testing can be completed within the day, whereas the lower values are more appropriate for weaker mortars. This is a comparative test procedure which may, for example, be used to investigate the effect of an admixture. Alternatively a procedure may be used to compare the time to give an agreed resistance to penetration with that of a control mix.

The time of stiffening is taken from the time water or cement is added to the mix. Mortar should be prepared to a standard consistence of 10 ± 0.5 mm dropping ball penetration. A mould approximately 75 mm diameter and 50–100 mm high should be filled in ten layers and tapped on the bench four times after each increment. The top surface should be struck level and the whole operation completed within 15–20 minutes of commencement of stiffening time. The specimens should be stored in air at a temperature of 20 ± 2 °C and a

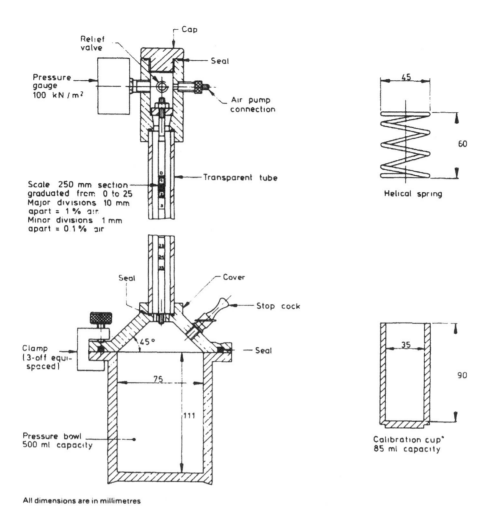

Figure 3.11 Apparatus for measurement of air content (pressure method)

relative humidity greater than 90%. The resistance to penetration is measured after two hours, and then at hourly intervals until the value is greater than half the required resistance. Thereafter the resistance should be measured at half hourly intervals.

To carry out the measurement the mould is placed on a platform scale under the penetration rod, which is held in a drill stand independent of the platform. The level of the stand is used to lower the rod slowly into the mortar until the loose washer* just touches the surface at which point the reading of the scale in kg is noted. The resistance to penetration in N/mm² can then be calculated.

*The brass rod is usually 65 mm long and 6.175 ± 0.025 mm in diameter. The end which penetrates the mortar has a reduced diameter of 5 mm for a length of 40 mm. A loosely fitting brass marker (washer) with an external diameter of 20 mm rests on the stop formed at the diameter change point.

3.5.6 Air content of freshly mixed mortars

The air content of mortars can be determined by the density method or the pressure method. The density method probably gives more reproducible results, but requires a knowledge of the densities of the constituents and the mix proportions (including the water content) by weight. The pressure method does not require this information but involves the use of special equipment.

To determine the air content using the density method a thick walled brass measure is first weighed and then filled with mortar in ten increments, the measure being tapped against the bench four times after each increment. Filling is carried out in such a way that little excess has to be struck off with a palette knife, and the measure is wiped clean on the outside and weighed. It is then possible to calculate the density of mortar and, from a knowledge of the capacity of the measure, to determine the air content.

Using the pressure method the apparatus shown in Figure 3.11 must be employed. The bowl of the equipment is filled with mortar, filter paper wetted and placed on the surface of the mortar, and the head unit clamped into position. The glass tube is lowered through the gauge glass until the end is just above the filter paper. Water is poured slowly through this tube as the tube is slowly withdrawn until the water is up to the zero mark in the gauge glass. With the tube withdrawn, the assembly is tilted and rotated while being struck with a mallet to remove air bubbles. Water is again brought to the zero mark and the top opening capped. The required pressure, as determined during calibration, is then applied. The bowl is again tapped with the mallet and the pressure brought to calibration pressure. The air content may then be read. The pressure is reduced slightly and then returned to the calibration level, and the air content is read again. The cap is then removed and a check made to ensure that the water level returns to the zero mark. The value of air content is taken as the average of both readings.

3.5.7 Test for hardened mortars

3.5.7.1 Sampling of hardened mortars

In a number of circumstances it may be considered necessary to carry out tests on hardened mortars. The following notes serve only to give guidance on the more important factors to be considered.

Sufficient samples of mortar should be taken from regularly spaced positions of those parts of the construction under consideration* to enable a representative picture of the mortar in the construction to be built up. A minimum of 100 g of material made up of not less than three equal increments will be required. In cases where only chemical analysis is required, and not sand grading, the sample may be obtained by drilling with a masonry drill as an alternative to cutting out work. Some care needs to be taken in evaluating the range of volume proportions determined by chemical analysis, and reference should be made to BS 4551: 1970[3.30] for further details.

3.5.7.2 Analysis of hardened mortars

Before any tests can be carried out it is necessary to separate the various layers of the construction and remove any contaminating materials, such as paint, gypsum plasters, spatter-dash, grime, brick and block. The main sample should be lightly crushed and, if necessary dried at 105 ± 5 °C to remove all free moisture.

It is not within the scope of this book to detail all the test procedures for the chemical analysis of mortars and reference should be made to BS 4551: Part 1: 1970. Sufficient information has, however, been included so that the reader is aware of the range of tests available.

Loss on ignition
This test is only required when either it is suspected that organic ingredients are present in a substantial proportion, or where the composition is required on an anhydrous carbonate free basis for more accurate assessment of cement:lime:sand ratios with carbonate free sands. Lime in an old fully carbonated mortar, for example, will have a loss on ignition of 44 parts by weight for each 56 parts of calcium oxide, as against about 20 parts for each 56 parts of calcium oxide, in fresh, dry, hydrated lime.

Sand content, sand grading, clay and silt content
This test provides information on the suitability of the sand used in the mortar.

Soluble silica, mixed oxides, iron oxide, calcium oxide, magnesium oxide and chloride
It is sometimes necessary for these tests to be carried out, particularly to differentiate between ordinary Portland cement and rapid-hardening Portland cement because both show a ratio of calcium oxide to soluble silica of approximately three. In the case of sulphate-resisting Portland cement the ratio of aluminium oxide to iron oxide will be 1.0 or less, but tests will also need to be carried out on a sample of the sand used because the amount of acid soluble aluminium and iron oxides may vary considerably. White Portland cement is characterized by ratios of aluminium to iron oxide of the order of 20. High alumina cement will produce mortars (unless calcium carbonate has been added) with a ratio of calcium oxide to the aluminium oxide of approximately 1.0.

Mixes based on lime usually have calcium oxide to silica ratios much higher than 3. If a magnesium lime has been employed the magnesium oxide content of the mortar will usually be higher than 5% (i.e. higher than Portland cement, high calcium lime or hydraulic lime).

Sulphur trioxide content
Where the sulphur trioxide content is higher than might be expected for the use of Portland cement (sulphur trioxide > 3%) and lime (sulphur trioxide > 1%) the following courses should be considered:

(1) the use of calcium sulphate plaster;
(2) attack of the mortar by sulphates from extraneous courses;
(3) the use of super sulphated cement.

Proportions by volume
Most measurements will generally indicate the relative proportions of each constituent of a mortar by weight. To find out the relative proportions by volume the following relationship should be used:

Amount of constituent by volume =

$$\frac{\text{\% of the constituent by weight}}{\text{bulk density of constituent}}$$

---

*It is important to store the sample in an air-tight container.

3.6 Wall ties

3.6.1 Durability

The purpose of a wall tie is to restrain together two leaves of masonry. The two common types of tie available are the metal strip tie with fish tails at both ends and a twist in the centre (the fish tail tie) and the wire type which is known as the butterfly because of its shape. Wall ties are made in a range of materials from hot dipped galvanized steel* (sometimes coated in bitumen) to stainless steel, copper, or copper alloy. Plastic wall ties are also available but are not accepted by some local authorities and are not generally recommended by the authors. In some areas all cavity wall ties are required to be of a non-ferrous metal to ensure long term durability and this is advocated as a sensible precaution for all large masonry constructions. The specification of ties is controlled by BS 1243[3.32].

3.6.2 Flexibility

Where the inner and outer leaves of the building are constructed from different materials, such as clay brick and concrete brick, differential movement has been known to lead to cracking when very stiff wall ties have been employed. Standard sizes of fish tail ties are 150 mm and 200 mm in length, 20 mm in width and thicknesses of 3.0 and 2.5 mm. Butterfly ties are of 12 SWG wire and 150, 200 and 300 mm in length. The latter will clearly impose less lateral restraint when differential thermal and moisture movements are expected to be a problem and should thus be employed in these situations. In practice butterfly or double triangle ties are preferred for constructions using blockwork.

3.6.3 Location of ties

In the case of cavity walls, ties should be embedded in the mortar during construction so that at least 50 mm of the tie is embedded in the mortar of each leaf, and should be evenly spaced at a rate of not less than 2.5 per square metre. In non-loadbearing masonry they should normally be placed at intervals of no more than 900 mm horizontally and no more than 450 mm vertically. In loadbearing masonry the spacing should be in accordance with Table 3.11. Additional ties must be provided around openings so that there is one tie for each 300 mm height of the opening.

When lateral loads on walls need to be calculated, reference should be made to Table 3.12 for guidance on working loads in ties.

Table 3.11 Spacing of ties

Minimum leaf thickness (one or both)	Cavity width	*Spacing of ties horizontally	vertically	Number of ties per square metre
mm	mm	mm	mm	
75	50–75	450	450	4.9
90 or more	50–75	900	450	2.5
90 or more	75–100	750	450	3.0
90 or more	100–150	450	450	4.9

*This spacing may be varied provided that the number of ties per unit area is maintained.

3.7 Damp-proof courses

Materials for damp-proof courses should comply with the requirements of one of the following British Standards: BS 743: 1970[3.33], BS 6925: 1988[3.34] or BS 6577: 1985[3.35]. Where there is no British Standard for a particular material, the manufacturer should provide evidence as to its suitability for the intended purpose and conditions of use.

A damp-proof course should be provided in a building to prevent the entry of water from an external source into the building or between parts of the structure and should be bedded in mortar and protected from damage during construction. Consideration must be given to the provision of a damp-proof course at the following points:

(1) where the floor is below ground level;
(2) 150 mm above ground level in external walls;
(3) under sills, jambs and over openings;
(4) in parapets;
(5) chimneys and other special details.

Stepped damp-proof courses at openings should extend beyond the end of the lintel by at least 100 mm. All horizontal damp-proof courses should protrude 10 mm from the external face of the wall and be turned downwards. Vertical damp-proof courses should be of adequate width and be fixed so as to separate the inner and outer leaves of the wall.

In loadbearing masonry, care must be taken to use a material which will not creep or otherwise spread under load for the damp-proof course. The type of damp-proof course provided can effect the restraint conditions to be assumed for design, for example, if engineering bricks are employed a wall might be assumed to be continuously supported in a situation where the use of a flexible damp-proof course would mean the assumption of simple support. On site rolls of flexible damp-proof course should be stored on a level surface protected from heat and other damage.

3.8 Reinforcement

Reinforcing steel should conform to the requirements of one of the following British Standards: BS 4449: 1988[3.36]; BS 970: Part 4: 1970[3.37], BS 1449: 1956[3.38]; BS 4482: 1985[3.39] or BS 4483: 1969[3.40]. Reinforcement may need to be galvanized according to the requirements of BS 729:

---

*It should be noted that some methods of manufacture may result in butterfly ties which are likely to be less durable than the corresponding strip tie made from galvanized steel.
It must be remembered that factors such as sound performance may be greatly influenced by the type of tie employed.

Table 3.12 Working loads in ties

Specification	Working loads in ties engaged in dovetail slots set in structural concrete	
	Tension	Shear
	N	N
Dovetail slot types of ties		
(1) Galvanized steel or stainless steel fishtail anchors 3 mm thick in 1.25 mm galvanized steel slots 150 mm long. Slots set in structural concrete	1000	1200
(2) Galvanized or stainless steel fishtail anchors 2 mm thick in 2 mm galvanized steel slots 150 mm long. Slots set in structural concrete	700	1100
(3) Copper fishtail anchors 3 mm thick in 1.25 mm copper slots 150 mm long. Slots set in structural concrete	850	1000

	Working loads in ties embedded in mortar of different designations*			
	Tension			Shear
	(i)	(iii)	(iv)	(i), (ii) or (iii)
	N	N	N	N
Cavity wall ties				
(1) Wire butterfly				
(a) Zinc coated or stainless steel 3.25 mm diameter	650	650	500	550
(b) Copper 2.65 mm diameter	500	500	450	500
(2) Strip fishtail, 3 mm × 10 mm; zinc coated; copper, bronze or stainless steel	1400	1000	700	900

*Designations of mortar (i) to (iv) as in Table 3.11. Unless stress grading is carried out the above table of safe working loads may not be used when the percentage of colouring is in excess of the following:

(1) 10% by weight of the cement if pigments complying with the requirements of BS 1014 are used.
(2) 2% to 3% by weight of the cement if carbon black is used.

Table 3.13 Strength of reinforcement

Designation	Nominal size	Specified characteristic strength
	mm	mm
Hot rolled steel grade 250 (BS 4449)	All sizes	250
Hot rolled steel grade 460/425 (BS 4449)	Up to and including 16 Over 16	460 425
Cold worked steel grade 460/425 (BS 4461)	Up to and including 16 Over 16	460 425
Hard drawn steel wire and fabric (BS 4482 and BS 4483)	Up to and including 12	485

1971[3.41] or otherwise protected from corrosion. The characteristic tensile strengths of reinforcement are given in the appropriate British Standards and are quoted in Table 3.13. Bed joint reinforcement should be of the tram line type so that it does not deform under tensile load. The effective diameter should be between 3 and 5 mm.

Reference should be made to *Chapter 9* dealing with reinforced masonry for further information on detailing.

3.9 Infill concrete and grout for reinforced masonry

Cement for use with infill concrete (grout) should comply with the requirements of one of the following British

Standards: BS 12: 1989[3.21] *Portland cements*; BS 146: Part 2: 1991[3.22] *Portland blast furnace cement* or BS 4027: 1991[3.23] *Specification for sulphate-resisting Portland cement*. Masonry cement and high alumina cement should not be used. In some circumstances a small proportion of lime may be added to the mix and this should be non-hydraulic (high calcium) or semi-hydraulic (calcium) complying to BS 890: 1972[3.25] *Building limes*. Sands and coarse aggregates should comply with BS 882: 1992[3.8].

Concrete infill should be one of the following mixes:

Cement : lime (optional) : Sand : 10 mm aggregate

(1)   1   0–¼   3   –
(2)   1   0–¼   3   2
       (Proportioned by volume of dry materials)

(3) A prescribed or designed mix of grade 25 or better in accordance with BS 5328 with a nominal maximum size of aggregate of 10 mm.

(4) A prescribed or designed mix of grade 25 or better in accordance with BS 5328[3.42] with a nominal maximum size of aggregate of 20 mm.

Mix (1) must be liquid and should be used for grouting internal joints not completely filled by mortar at the time of laying. Mixes (2), (3) and (4) should have a slump between 75 mm and 175 mm. Mixes (2) and (3) should be used for filling spaces with a minimum dimension of not less than 50 mm and mix (4) may be used for filling spaces with a minimum dimension of 100 mm.

Superplasticizers may be employed to produce high workability concrete of low water/cement ratio but should only be employed under supervision and on the authority of the engineer in charge. Expansive additives are also available to reduce the shrinkage of the infilling concrete within the masonry. This obviates the need for 'topping up' of cores and recompaction after the initial slump of the material.

The strength of the infilling concrete or grout should be checked by making 100 mm cubes in accordance with BS 1881[3.10], the details of which are essentially covered in *Section 3.5.3*. The characteristic strength should not fall below that specified or, alternatively, other forms of acceptance criteria must be proposed. The sampling rate should be set with respect to the method of mixing and the method of placing. Generally speaking a rate of sampling comparable to that for the mortar is appropriate.

3.10 Mechanical fixings for attaching components to masonry

The majority of fixings in masonry comprise of plugs placed in drilled holes to receive screws, bolts or nails, mechanical anchorages such as sockets or channels, or fired or precision fixed pins and nails. The performance requirement for the fixing can be summarized as:

(1) it must be strong enough to support the load (whether direct tension, shear or both);
(2) the fixing must be durable so that it will last the life of the building or at least the component, and will not deteriorate causing stains, etc;
(3) it must be economic.

It is beyond the scope of this book to consider in detail the various ways in which a fixing can be made, but the following general principles apply to most types of fixing.

In the case of drilled fixings it is necessary to ensure that the drilled hole is of the correct size for the plug and the side should be perpendicular to the surface of the wall. Certain forms of dense aggregate open textured block tend to deflect the tip of the drill, which can result in an oversized hole being produced. A variety of plugs may be inserted into the drilled hole, including fibre, metal and plastic and in the case of irregular holes a fibre plugging compound may be employed. Plugs are usually specified by British Screw Gauge, but there is an increasing tendency to produce fixings capable of taking a range of screw sizes, in which case it is important to note that the most efficient fixing in terms of strength performance will be obtained when the largest screw size is used.

The action of inserting the screw expands and distorts the plug to such an extent that a good mechanical bond is achieved between the screw, fixing and substrate. Fibre plugs and some metal type plugs will tend to be pushed too far into holes that have been over-drilled, whereas plastics and some forms of metal plug have a tip which prevents them disappearing into an oversize hole.

Special fixings may be used with aerated and other very light blocks, which usually incorporate some degree of expansion to provide mechanical bond. Cut nails can also be driven into the block to carry component loads.

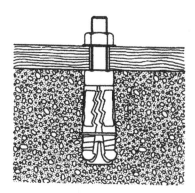

Figure 3.12 Single cone anchor

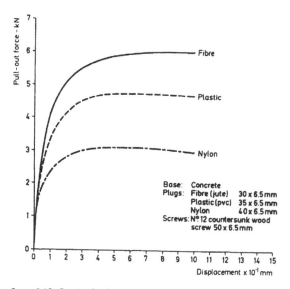

Figure 3.13 Graph of pull-out force and relative displacement

Heavier fixing devices may be required for some purposes, and these may be of the single cone anchor form (Figure 3.12) which expand when the bolt is tightened, the retraction stud bolt type whereby a wedge expands on retraction of the stud, or the self drilling type where a plug is hammered into the anchor to cause expansion and the collar proud of the surface is snapped off. In addition to various other forms of expansion fixing there are a number of chemical fixings which rely on the breaking of a capsule of chemicals to initiate setting.

Most manufacturers will supply detailed information of the loadbearing performance of their products. For general information only, some pull out results are presented to indicate how a plug type fixing will perform under load. Figure 3.13 shows tests on the three types of plug shown in Figure 3.14. Note that an initial slip takes place leading to a bond type failure in most cases.

29

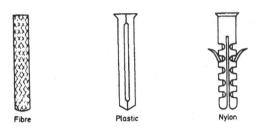

Figure 3.14 Typical plugs

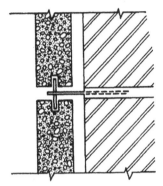

Figure 3.15 Dowelled tie holding precast concrete unit to masonry wall

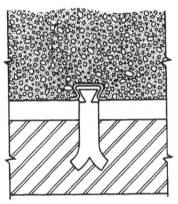

Figure 3.16 Masonry tied to a column by means of a metal tie held in a doevtail slot

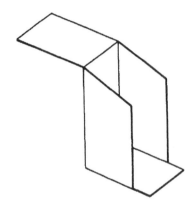

Figure 3.17 A joist hanger

Figure 3.18 Wall plate holding down strap

3.11 Mechanical fixing for tying structural elements

The most common form of mechanical fixing used in masonry is the wall tie, and the specifications and standards covering standard wall ties are discussed under that heading in *Section 3.6*. There are, in addition, a number of specialized ties, such as fishtail ties with lips or dowels (Figure 3.15) used to fix stone or precast concrete to masonry. Cramps may be used with some types of coping stone to locate adjacent stones.

Cast in slots or bolt connections are sometimes provided in concrete panels so that adjoining masonry can be tied in. The dovetail slot, as shown in Figure 3.16, overcomes one of the main problems of building masonry to the frame accurately, in that the ability of the tie to move up and down the slot provides some freedom. Specialist fixings are also available to tie brick slips to beams and such like.

Various forms of loadbearing devices, such as angle corbels, are available to restrain heavy structural members. Built-in devices such as joist hangers (Figure 3.17) or wall plate holding down straps (Figure 3.18) are very widely used. These devices should be designed and used in accordance with the manufacturers instructions, with due regard for the local stress concentration as detailed in *Chapter 10*.

3.12 Method of test for structural fixings

The method of test for structural fixings in concrete and masonry is covered in BS 5080: Part 1: 1993[3.43]. The procedure consists mainly of a means of applying a tensile force to a fixing installed in a solid base material. Figure 3.19 is a photograph of a test rig capable of

Figure 3.19 Equipment suitable for performing a tensile test on a fixing

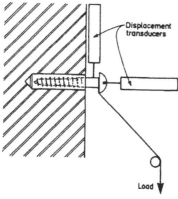

Figure 3.20 Diagrammatic representation of equipment used to measure the creep of a fixing

carrying out such a test. The Standard gives information on the use of standard specimens as well as set procedures for the installation of the fixing. During the test, load is applied in discrete increments and measurements are taken of both the load and the movement of the fixing as shown in Figure 3.20. A graph is then drawn of load against relative movement. It may be necessary to test a number of samples of each type of fixing for a given substrate so that some idea of variation can be obtained. It is common practice to apply, quite large safety factors, often a factor of 5, for pull-out loads.

It may well be that some fixings are required to operate with a high shear load rather than a direct pull-out load, while in other situations it may be necessary to provide a test rig which will enable both forms of loading to be applied simultaneously.

In the case of some chemical and plastic forms of fixing long term creep under load may need to be investigated.

3.13 References

3.1 BRITISH STANDARDS INSTITUTION. BS 6073: 1981 *Precast concrete masonry units.* Part 1: *Specification for precast masonry units* (pp 12) and Part 2: *Method for specifying precast masonry units* (pp 8). BSI, London.

3.2 BRITISH STANDARDS INSTITUTION. BS 2028, 1364: 1968 *Precast concrete blocks.* BSI, London. pp 44.

3.3 BRITISH STANDARDS INSTITUTION. BS 1180: 1972 *Concrete bricks and fixing bricks.* BSI, London. pp 20.

3.4 BRITISH STANDARDS INSTITUTION. BS 5628: Part 1: 1992 *Use of masonry.* Part 1: *Structural use of unreinforced masonry.* BSI, London, pp 62.

3.5 BRITISH STANDARDS INSTITUTION. BS 5628: Part 3: 1985 *Use of masonry* Part 3: *Materials and components, design and workmanship.* BSI, London. pp 103.

3.6 BRITISH STANDARDS INSTITUTION. BS 8200: 1985 *Code of practice for design of non-load bering external vertical enclosures of buildings.* BSI, London. pp 73.

3.7 BRITISH STANDARDS INSTITUTION. BS 4315: Part 2: 1970 *Methods of test for resistance to air and water penetration.* Part 2: *Permeable walling constructions (water penetration).* BSI, London. pp 16.

3.8 BRITISH STANDARDS INSTITUTION. BS 882: 1992 *Aggregates from natural sources for concrete.* BSI, London. pp 12.

3.9 BRITISH STANDARDS INSTITUTION. BS 410: 1986 Amd 1 *Test sieves* (AMD 8255 July 15 1994). BSI, London. pp 23.

3.10 BRITISH STANDARDS INSTITUTION. BS 1881: Part 5: 1970 Amd 2 *Methods of testing concrete.* Part 5: *Methods of testing hardened concrete other than strength* (AMD 6267 March 30 1990). BSI, London. pp 27.

3.11 BRITISH STANDARDS INSTITUTION. BS 1610: 1992 *Materials testing machines and forces verification equipment.* Part 1: 1992 *Specification for the grading of the forces applied by materials testing machines when used in the compression mode* (pp 13) and Part 2: 1985 *Grading of equipment used for the verification of the forces applied by materials testing machines* (AMD 6174 October 31 1989) (pp 14) and Part 3: 1990 *Grading of the forces applied by deadweight and lever creep testing machines* (pp 13). BSI, London.

3.12 BRITISH STANDARDS INSTITUTION. BS 1134: Part 1: 1988 *Method for the assessment of surface texture.* Part 1: *Method and instrumentation.* BSI, London. pp 31.

3.13 BRITISH STANDARDS INSTITUTION. BS 1142: 1989 Amd 2 *Fibre building boards* (AMD 7776 July 15 1993). BSI, London. pp 44.

3.14 ROBERTS, J J. The effect of different test procedures upon the indicated strength of concrete blocks in compression. *Magazine of Concrete Research*, Vol 25, No 83, June 1973. pp 87–98.

3.15 BRITISH STANDARDS INSTITUTION. BS 3921: 1985 *Clay bricks.* BSI, London. pp 22.

3.16 BRITISH STANDARDS INSTITUTION. BS 874: 1973 Amd 3 *Determining insulating properties with definitions of thermal insulating terms* (AMD 5228 December 31 1986). SI, London. pp 47.

3.17 SPOONER, D C. Results of a round robin thermal conductivity test organized on behalf of the British Standards Institution. *Magazine of Concrete Research*, Vol 32, No 111, June 1980. pp 117–122.

3.18 ROBERTS, J J. The performance of concrete block testing machines as assessed by the Foote proving device. *Cement and Concrete Association*, London, January 1973. Publication No 42.478. pp 5.

3.19 ROBERTS, J J. Specification for a machine for testing concrete blocks in compression. *Cement and Concrete Association*, London, October 1974. Publication. No 42.499. pp 12.

3.20 BRITISH STANDARDS INSTITUTION. BS 1198, 1199, 1200: 1976 *Building sands from natural sources*. BSI, London. pp 8.

3.21 BRITISH STANDARDS INSTITUTION. BS 12: 1989 *Portland cements*. BSI, London. pp 9.

3.22 BRITISH STANDARDS INSTITUTION. BS 146: 1991 *Portland blastfurnace cements*. BSI, London. pp 19.

3.23 BRITISH STANDARDS INSTITUTION. BS 4027: 1991 *Sulfate-resisting Portland cement*. BSI, London. pp 17.

3.24 BRITISH STANDARDS INSTITUTION. BS 5224: 1976 *Specification for masonry cement*. BSI, London. pp 11.

3.25 BRITISH STANDARDS INSTITUTION. BS 890: 1972 *Building limes*. BSI, London. pp 32.

3.26 RAGSDALE, L A, AND BIRT, J C. Building sands: availability, usage and compliance with specification requirements. *CIRIA Report 59*. London, June 1976. pp 30.

3.27 BRITISH STANDARDS INSTITUTION. BS 1014: 1975 *Pigments for Portland cement and Portland cement products*. BSI, London. pp 12.

3.28 BRITISH STANDARDS INSTITUTION. BS 5390: 1976 *Code of Practice for stone masonry*. BSI, London. pp 48.

3.29 BRITISH STANDARDS INSTITUTION. BS 4721: 1981 Amd 1 *Ready-mixed building mortars*. BSI, London. pp 15.

3.30 BRITISH STANDARDS INSTITUTION. BS 4551: 1980 *Methods of testing mortars, screeds and plasters*. BSI, London. pp 35.

3.31 AMERICAN SOCIETY FOR TESTING AND MATERIALS. ASTM: E 149–66 *Tentative method of test for the bond strength of mortar to masonry units*. Philadelphia, 1966. pp 10.

3.32 BRITISH STANDARDS INSTITUTION. BS 1243: 1978 *Specification for metal ties for cavity wall construction*. BSI, London. pp 8.

3.33 BRITISH STANDARDS INSTITUTION. BS 743: 1970 Amd 4 *Materials for damp proof courses. Metric units* (AMD 6579 January 31 1991). BSI, London. pp 26.

3.34 BRITISH STANDARDS INSTITUTION. BS 6925: 1988 Amd 1 *Mastic asphalt for building and civil engineering (limestone aggregate)* (AMD 7150 July 15 1992). BSI, London. pp 12

3.35 BRITISH STANDARDS INSTITUTION. BS 6577: 1985 *Mastic asphalt for building (natural rock asphalt aggregate)*. BSI, London. pp 13.

3.36 BRITISH STANDARDS INSTITUTION. BS 4449: 1988 *Carbon steel bars for the reinforcement of concrete*. BSI, London. pp 16.

3.37 BRITISH STANDARDS INSTITUTION. BS 970: Part 4: 1970 Amd 4 *Wrought steels in the form of blooms, billets, bars and forgings. Part 4: valve steels*. BSI, London. pp 26.

3.38 BRITISH STANDARDS INSTITUTION. BS 1449: 1991 *Steel plate, sheet and strip.* Part 1: 1991 *Carbon and carbon-manganese plate, sheet and strip* Sections 1.1–1 15 (pp 144) and Part 2: 1983 *Specification for stainless and heat-resisting steel plate, sheet and strip* (AMD 6646 April 30 1991) (pp 18). BSI, London.

3.39 BRITISH STANDARDS INSTITUTION. BS 4482: 1985 *Cold reduced steel wire for the reinforcement of concrete*. BSI, London. pp 12

3.40 BRITISH STANDARDS INSTITUTION. BS 4483: 1985 *Steel fabric for the reinforcement of concrete*. BSI, London. pp 12.

3.41 BRITISH STANDARDS INSTITUTION. BS 729: 1971 *Hot dip galvanized coatings on iron and steel articles*. BSI, London. pp 16.

3.42 BRITISH STANDARDS INSTITUTION. BS 5328: 1991 Concrete. Part 1: 1991 *Guide to specifying concrete* (pp 26) and Part 2: 1991 *Methods for specifying concrete mixes* (pp 23) and Part 3: 1990 *Procedures to be used in producing and transporting concrete* (pp 12) and Part 4: 1990 *Procedures to be used in sampling, testing and assessing compliance of concrete* (pp 12). BSI, London.

3.43 BRITISH STANDARDS INSTITUTION. BS 5080: Part 1 1993 *Structural fixings in concrete and masonry.* Part 1: *Method of test for tensile loading*. BSI, London. pp 15.

Chapter 4
Types of block and brick

4.1 Introduction

When categorizing the types of block and brick available, a number of factors need to be considered, the influences of which are discussed under the following four headings:

(1) methods of manufacture
(2) appearance
(3) performance
(4) ease of use

The possible compromises which may need to be made with respect, for example, to producing a facing block with good thermal insulation properties, are covered to some extent in each section, but it should be recognized that the above categories must widely overlap.

It is not within the scope of this book to go into details of materials used for manufacturing concrete blocks. The reader should, however, be aware that in addition to various dense aggregates, materials such as pulverized fuel ash, pumice, clinker, foamed slag, expanded clay, expanded shale and sawdust are employed.

4.2 Methods of manufacture

4.2.1 Concrete blocks (Aircrete)

Concrete blocks in the United Kingdom are generally manufactured by the following processes:

(1) a foaming process in the case of autoclaved aerated concrete
(2) a mobile machine called an egglayer
(3) a static machine

4.2.1.1 Autoclaved aerated blocks

Aerated concrete blocks are manufactured under controlled factory conditions. The constituent materials are usually cement and sand, although in many cases pulverized fuel ash is used as a replacement for some or all of the sand. The raw materials are mixed with a foaming agent and discharged into large steel moulds. These moulds are usually wheeled through the various stages of manufacture on a track or rails. The 'cake' in the mould rises due to the action of the foaming agent until it is completely filled with the aerated material, and is then cut by wires to give the desired unit size before being placed in an autoclave – a high pressure steam curing chamber – where curing takes place at pressures up to 15 atmosphere.

This method of production is only used for making very lightweight blocks, typically between 475–750 kg/m³, which are primarily produced for their good thermal insulation properties. Two particular advantages in this are:

(1) it is comparatively easy to alter the position of the wires and produce non-standard block sizes given a substantial order.
(2) the dimensional use of autoclaves stabilizes the blocks at a comparatively early stage and enables them to be used direct from the factory.*

Although autoclaved aerated blocks are sometimes used fair-faced and painted, it is not possible to produce a facing block using this process. To overcome this problem some manufacturers have in the past bonded their blocks to a 10–25 mm facing to produce a faced aerated block. This technique of manufacture of facing units also necessitates the provision of a number of special units to maintain the appearance of the corners, etc. and is not currently used in the U.K.

It should also be noted that it is only possible to manufacture solid blocks in aerated concrete and that it is not possible to produce blocks stronger than 7 N/mm² without greatly increasing the density.

4.2.1.2 The egglayer

This appropriately named machine may be used to produce a very wide range of dense and lightweight blocks. Essentially the plant comprises a central batching

---

*Note, however, that the building of hot blocks into a wall is not necessarily desirable.

and mixing facility, tipping bucket truck, the egglayer machine and a large flat concrete apron. The mix is discharged from the mixer into the bucket of the truck and transferred to the hopper of the egglayer. The egglayer moves in small steps along the concrete apron and at each step stamps out blocks directly onto the surface of the concrete. The blocks are produced by drawing some of the mix contained in the hopper into a steel mould where it is vibrated and compacted. The design and control of the mix has to be such that the blocks are free standing directly from the mould without subsequent slumping. The blocks are then left to air cure although in the summer they may be wetted down or covered.

The obvious limitations to this method of production are:

(1) the large area of concrete apron required means that in bad weather most manufacturers can only afford to cover a small part of the apron;
(2) the apron needs to be well laid and regularly maintained to ensure good dimensional accuracy;
(3) the open air curing means that the blocks are subject to the vagaries of the weather, although in practice manufacturers adjust the design of the mix to compensate for the prevailing weather conditions;
(4) although there are some exceptions most egglayers are not suitable for producing good quality facing units or very high strength blocks.

In spite of these limitations the egglayer produces very economic dense and lightweight blocks. Recent developments for greater production include the double drop machine, which produces two layers of blocks, and the production of blocks on end, rather than laid, on the bed face.

To produce blocks of different sizes or configurations, it is necessary to change the mould in both the egglayer and the static machine discussed in the next section. Whilst manufacturers do have a range of moulds available they are very expensive, and custom made sizes, faces and configurations are only likely to be economic for large orders. Because of the wear that will occur on moulds, block tolerances from a given mould will inevitably change from minus to plus over a large manufacturing cycle. As is also the case with the static machine, the vertical casting method makes the height of the block the most difficult dimension to control.

4.2.1.3 The static machine

The static machine is the most consistent method of producing good quality dense and lightweight blocks. A central automatically controlled batching plant is used to feed the mixing plant which directly links with the block machine or machines. Blocks are stamped out by a combination of vibration and compaction onto steel or wooden pallets. The effect of the vibration and compaction effort on the mix is usually much greater than can be applied by an egglayer so that it is possible to produce very dense, high strength blocks using this technique. Once the blocks have been produced on the pallet they move along a conveyor system and are stored on a large metal frame which, when full, is carried by a transfer car to a curing chamber. In the majority of cases low pressure steam curing is employed, but in some cases burner curing or autoclaving is used. When curing has been completed the frames are removed from the curing chambers by the transfer cars, the blocks are automatically cubed ready for despatch and the pallets returned into the system.

Low pressure curing gives the blocks high early strength. In the case of autoclaved blocks, however, the crystalline structure of the cementitious products is changed and offers a number of advantages:

(1) shrinkage is reduced;
(2) greater uniformity of colour is possible – blocks which are autoclaved tend, in general, to be lighter in colour;
(3) autoclaving can utilize the pozzolanic properties of some of the mix constituents.

Autoclaving is, however, an expensive process both in terms of capital plant costs and in energy costs.

The type of mould box required can vary between machine manufacturers but the observations made in the previous section regarding costs and use are equally applicable to this method of manufacture.

Both egglayers and static machines may be used to produce solid, cellular or hollow blocks with densities from 800–2100 kg/m^3 and strengths of 2.8–35 N/mm^2 although the higher densities and strengths are more likely to be produced by static machines.

4.2.2 Concrete bricks

Concrete bricks may be produced on the static machine as previously described or by hydraulic pressing. This technique involves the use of very dry mixes which are pressed into shape by high loads on a continuous production basis. Using this method it is possible to produce high strength, durable units. Most of the types of blocks described in the following sections can also be made available as concrete bricks.

4.3 Appearance

4.3.1 Plain facing blocks

It will already be clear that not only the method of manufacture, but also the type of aggregate, will greatly affect the appearance of a block. In dense facing blocks, local aggregates will invariably be used (indeed the availability of local materials would have played a large part in the siting of the plant), for example, a limestone aggregate might be used. The manufacturer generally has the option of changing the grading of the aggregate, which may be continuous or gap graded, employing different fine aggregates, employing an admixture, varying the degree of compaction and employing a pigment. In addition, the surface of the finished block

Figure 4.1 Split faced concrete bricks which can be either flat or frogged

may be grit or sand blasted to further change the appearance. By using a combination of the possible variables a large manufacturer will have available a range of open textured facing blocks which could be made in a range of strengths and sizes solid, cellular or hollow. Examples of typical dense aggregate facing blocks of the type described are shown in Figure 4.1. Because of the high cost of the natural material, plain facing blocks can be produced specifically to replace local stone as a building material.

In addition to dense aggregates, lightweight aggregates may be used to produce plain facing blocks. Units made from graded wood particles are also available but the very lightweight blocks are not generally produced as homogeneous facing units.

4.3.2 Split face blocks

Split face blocks are made by splitting a dense block (often a hydraulically pressed block) so that the aggregate is split and exposed. The appearance of this type of block will clearly depend upon the aggregate employed and its size. The heavy emphasis of the aggregate tends to make this type of block suitable in areas where stone is a traditional building material. The technique can be used to produce very attractive concrete bricks as shown in Figure 4.1.

4.3.3 Tooled face blocks

The faces of dense blocks may be tooled to expose the aggregate but, unlike split face units, this operation is often carried out so that the face of the unit has a more rounded appearance. Blocks and bricks produced using this technique tend to be relatively expensive but are widely employed in certain areas of the country.

4.3.4 Profiled blocks

A profiled face can be obtained during casting by simply adapting the mould box although it may be necessary to change the mix design when making this type of block. The portion of the profiled face proud of the main face of the block should not be counted when working out the thickness of the unit for design purposes.

4.3.5 Slump blocks

The slump or 'Adobe' block is produced by using a comparatively wet fine aggregate mix so that after demoulding the face slumps or bulges in a fairly random manner. Although not widely produced or used in the United Kingdom this type of block is popular in some parts of the world. The random and variable nature of the unit places less emphasis on the quality of the blocklaying.

4.3.6 Exposed aggregate blocks

Some manufacturers can offer blocks with exposed aggregate faces. This is achieved by bonding a prepared face to an aerated or lightweight block or by casting the face wet on the back-up block. Traditional methods of washing to expose the aggregate are still used, as is abrasive blasting. The aggregates used in production of this type of unit need to be selected with great care.

4.3.7 Scored face blocks

With this type of block alterations are made to the mould producing 'dummy' mortar joints on the faces so that when the blocks are laid with an appropriate colour of mortar the wall appears to be built from smaller units.

4.3.8 Screen wall blocks

These blocks are not generally used for designed load-bearing purposes but are included for completeness. They are made from dense concrete and are produced in a number of designs of perforation so that attractive screen walls can be produced.

4.3.9 Coloured units

To achieve a coloured finish on the unit pigments may be incorporated at the mixing stage or the colour may be imparted by the careful selection and bleaching of natural aggregates. Some forms of pigment can, however, be prone to fading which could lead to problems, particularly where a building is subject to differential weathering. There may also be problems with batch to batch variations of fresh blocks, so that it is wise to intermix different batches of facing blocks and bricks to ensure random rather than localized variations.

Figure 4.2 Thermal insulating blocks

4.4 Performance

This section deals with performance as it affects the type of block available in general terms only. More detailed information on particular properties is given elsewhere in the book.

Good thermal insulation has been of greatest significance in promoting the use of lightweight concrete blocks, but other properties have also been exploited for specific markets.

4.4.1 Thermal insulating blocks

Thermal insulating blocks typically have resistances between 0.4 and 2.60 m^2K/W. This category includes all aircrete (autoclaved aerated blocks) some solid and cellular lightweight aggregate blocks and aggregate blocks which are insulation filled or which incorporate an insulation layer bonded to one or more block faces. A range of thermal insulating blocks is shown in Figure 4.2.

4.4.2 Separating wall blocks

Some manufacturers offer special dense blocks specifically for producing solid separating walls meeting the statutory requirements described in Chapter 16. Proprietary lifting devices are also available for use in carrying and laying.

4.5 Ease of use

4.5.1 Hollow blocks

These have been defined in *Chapter 3* and an example is shown in Figure 4.3. Hollow blocks are generally made of dense aggregates and are particularly useful in the following circumstances:

(1) for producing single leaf walls 150–225 mm thick for which dense solid blocks might be too heavy to lay rapidly;
(2) for producing reinforced masonry by incorporating reinforcing steel in the core and filling with in situ concrete;
(3) where it is required to incorporate services within the thickness of the wall.

4.5.2 Cellular blocks

These blocks are made with dense and lightweight aggregates so that the face is similar to a corresponding solid block. The main advantage is that this type of block is lighter than the corresponding solid block and the voids can present a convenient hand hold. One end of the voids should be effectively sealed to prevent mortar droppings falling into the voids when that end is used to support the bed joint mortar. An example is shown in Figure 4.3.

Some manufacturers include blocks designed for easy cutting as a standard proportion of each delivery. These incorporate special slots which enable the unit to be split into a half or part block with the aid of a bolster.

4.5.3 Specials

Special shaped units are produced to make maintenance of bonding patterns easier particularly in facing work or as cavity closers. Other blocks such as lintel or beam blocks are designed to provide permanent form-work to in situ lintels and beam courses cut within the wall. Examples of some specials are shown in Figures 4.4, 4.5, 4.6 and 4.7. The high cost of moulds and the need to interrupt manufacturing runs to produce a small number of units tend to make production of special blocks expensive.

Figure 4.3 Solid, cellular and hollow blocks

Figure 4.4 Block with spade cuts to facilitate cutting (left), standrad block (centre) and half block (right)

Figure 4.5 Cavity closure and Quoin blocks

37

Figure 4.6 Lintel and bond beam unit

Figure 4.7 Tongue and groove blocks

Chapter 5
Types of wall

5.1 Introduction

This chapter deals with the types of wall which may be encountered in terms of bonding pattern, construction and finish. The chapter gives a ready definition of each wall type and where the contents overlap with other sections of the manual, reference is made to those sections for further information.

Particular attention should be paid to the validity of design information available for the type of construction under consideration and the bond pattern to be employed. For example, in BS 5628: Part 1: 1992[5.1], masonry is defined as 'an assemblage of structural units, either laid in situ or constructed in prefabricated panels, in which the structural units are bonded and solidly put together with mortar or grout'. This definition clearly eliminates dry stacked walls held together by glass fibre reinforced render or ground blocks held together by a thin bed adhesive, from being considered as masonry. An alternative example is the definition contained in BS 5628 of a single leaf wall, namely, 'a wall of bricks or blocks laid to overlap in one or more directions and set solidly in mortar'. By this definition a stack bonded wall may not be considered as masonry.

In the following sections the types of bonding pattern are first considered and then the various wall types are explained.

5.2 Types of bonding pattern

5.2.1 Running bond

This bonding pattern is used for the majority of concrete masonry constructed in the United Kingdom, and is popular for the ease with which the wall can be laid, the good strength properties and pleasant appearance of the wall. A wall constructed in this way is clearly stronger in flexure perpendicular to the bed joint than parallel to the bed face, because of the 'toothed' effect of the masonry, and it is this weakness which determines the orthogonal ratio of the wall, important when the design of unreinforced laterally loaded panels is under consideration. Research has shown[5.2] that a wall constructed in

running bond but with the perpend joints left unfilled can carry the same axial load as a properly built wall, but improperly filled joints will undoubtedly impair the lateral load resistance of such a wall and lead to problems in meeting bond, thermal and fire resistance requirements.

In Figure 5.1 the results of American tests[5.3] on walls built to a range of bonding patterns, and tested both axially and laterally, are presented together with the relative strengths referenced to the performance of a running bond wall. Whilst these tests have in some cases been outdated by more recent research programmes, they do at least give a working basis for comparison.

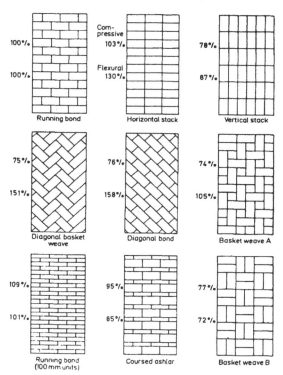

Figure 5.1 Relative strength in compression and flexure on the vertical span of patterned concrete masonry (average values for walls with two types of mortar)

39

5.2.2 Other bonding patterns

There are a number of potential bonding patterns, some of which differ little from the running bond pattern in terms of wall performance. Perhaps the smallest variation from the conventional running bond is the use of quarter bond and running with half block units. It is important to remember that a change in the bonding pattern may well result in the necessity for additional specials, and a masonry saw on site will be particularly useful. There are, of course, big differences in the use of brick size units, for which there are many traditional intricate bonding patterns, and concrete blocks, where the size and 'scale' effect tends to limit the number of options available. Accepting this, and the penalty of cost and laying rate which may prove apparent when an unusual bonding is employed, it is necessary to consider the structural implications of changes. Perhaps the most extreme example occurs with the use of stack bonded walls, where there is no interlock between the units. Even if such a wall is essentially non-loadbearing it will almost certainly prove necessary to incorporate bed joint reinforcement in order to prevent cracking due to movement.

5.3 Single leaf wall

The definition of a single leaf wall contained in BS 5628 is 'a wall of bricks or blocks laid to overlap in one or more directions and set solidly in mortar'. For many years the single leaf or 'solid' wall comprised of a 9 in. brick wall, i.e. made from 4½ in. thick bricks. When, in the period between the World Wars, the cavity wall came to be more widely used in an effort to reduce the incidence of rain penetration, the use of the solid wall declined so much that in England and Wales it was rarely used, although it enjoys some use in Scotland where it was still encompassed by the Scottish Building Regulations[5.4] and in many other parts of the world where it has continued to be widely used such as the USA and Europe. Climatic differences as well as indigenous building resources greatly influence methods of construction. In Germany, for example, cavity walling would be considered uneconomic in the South of the country whereas in the North, with natural clay resources and a wetter climate, cavity walls are more common.

The main factors which have to be considered with a single leaf wall are as follows:

(1) prevention of rain penetration – to this end the wall will have to be well constructed and the provision of a protective coating, such as rendering, may be necessary;
(2) timber joists may need to be specially treated if they are to be built into the wall – otherwise joist hangers may be used with due allowance for the eccentricity of the load on the wall;
(3) to meet current and proposed thermal regulations, the use of added insulants may need to be considered.

One system developed in Scotland is the 'posted' solid wall, as shown in Figure 5.2, which employs a single leaf

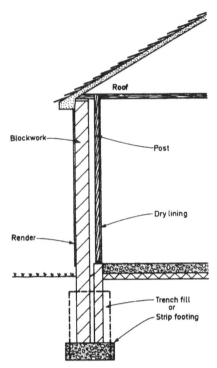

Figure 5.2 Single leaf posted construction

of 150–250 mm of insulating block, usually rendered externally. A completely separate inner lining of plasterboard, 50 mm or more away from the inner face of the masonry, is supported by means of a series of timber posts spanning from roof to floor. Traditionally such posts would be supported on sleeper walls but more recently solid floors have been employed with one and two storey dwellings. External insulation systems also provide a useful means of protecting the outside of single leaf constructions because such systems are generally very effective in preventing rain penetration and enable simple construction details to be employed without the risk of cold bridging. An example is shown in Figure 5.3.

There is no doubt that single leaf walling is very competitive in terms of overall construction costs compared to other constructions, and given the changes occurring in thermal insulation requirements the use of solid walling may receive a substantial boost. A newer development in the use of solid blockwork is the use of glass reinforced cement render to bond together dry-stacked blocks.

5.4 Cavity walls

The original concept of providing a cavity wall is basically quite sound. Even using permeable walling materials, separated by a gap of perhaps 50 mm, the ingress of water through a well built wall should be eliminated. Although water might pass through the outer leaf of the wall this would subsequently run down the back face of the outer leaf and be discharged from the

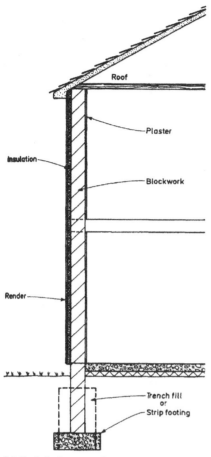

Figure 5.3 Single leaf – externally insulated

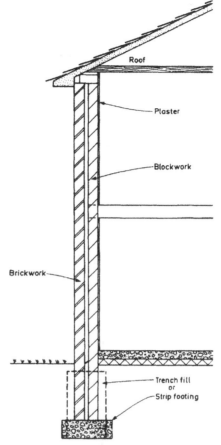

Figure 5.4 Brick – cavity – block

bottom of the cavity – providing no bridge was encountered which enabled the water to pass to the inner leaf of the construction.

Since the introduction of the cavity wall the incidence of rain penetration has indeed been reduced and problems can now be traced to:

(1) the desire to use walling materials which are more permeable than those traditionally used;
(2) poor workmanship;
(3) poor understanding of design – especially the difficulty in executing such features as stepped cavity trays, etc.

The growth in the use of cavity walls has also resulted in structural design changes. The bearing of timber joists on the inner leaf is a sensible precaution in overcoming the risk of moisture reaching the wood, and this has lead to the situation where the inner leaf is usually loadbearing and the outer leaf is not. To stop the non-loadbearing leaf from 'waving in the wind' it has been considered essential to tie the two leaves together, although the long term efficiency of this must be doubted when the durability of the ties is considered. The tying of the two leaves has also given rise to the consideration of. composite action between the two leaves from a design viewpoint.

A factor which has had a profound influence upon construction of cavity walls has been the necessity for better thermal insulation as a health measure, to prevent undue condensation or as an energy saving measure. This has lead to the use of lower density blocks in the inner leaf to increase the thermal insulation as shown in Figure 5.4 and as a direct result, blocks with a density as low as 475 kg/m³ may be employed. These blocks may be used in such a way that other aspects of performance, such as sound resistance, are not necessarily impaired, but they are usually of low strengths between 2.8 to 5 N/mm². The shape factor, however, is more favourable for blocks than for brick shaped units and using cross-wall construction, it is possible to produce two storey constructions with precast concrete floors using blocks only 4.35 N/mm² in strength, as shown in Figure 5.5. To overcome the more stringent thermal resistances now required of walls by the Building Regulations there has been a tendency to construct cavity walls with two leaves of lightweight block, as shown in Figure 5.6, and to build with a thicker inner leaf.

Although the width of the cavity between two leaves is generally considered as 50 mm, this figure is in no way sacrosanct. For a large blockwork building using two 100 mm leaves it might be very useful to incorporate a

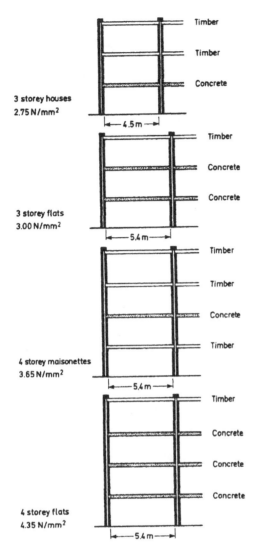

3 storey houses
2.75 N/mm²

Timber

Timber

Concrete

— 4.5 m —

3 storey flats
3.00 N/mm²

Timber

Concrete

Concrete

— 5.4 m —

4 storey maisonettes
3.65 N/mm²

Timber

Timber

Concrete

Timber

— 5.4 m —

4 storey flats
4.35 N/mm²

Timber

Concrete

Concrete

Concrete

— 5.4 m —

Figure 5.5 Structural performance: calculated required concrete block strengths for various loadbearing wall constructions up to four storeys

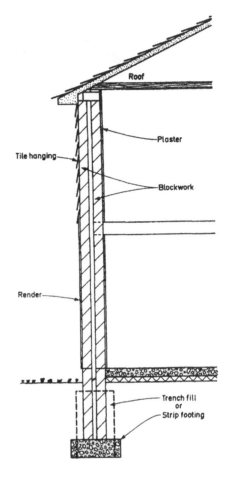

Figure 5.6 Block – cavity – block

100 mm cavity so that the whole design is based on the 300 mm module. Increasingly encroachments are being made into the cavity space to incorporate added insulation and this is discussed more fully in Chapter 12.

5.5 Buttress

A buttress is a portion of wall thicker than the main run of wall of which it is an integral part, as shown in Figure 5.7, usually projecting on one side of the wall only, but in some cases projecting on both sides. Generally, a buttress is provided either to provide local support under a structural member or reduce the eccentricity of loading at that point, or to provide lateral stability for the wall. In addition to providing a stiffer bending element buttresses also provide end fixity for walls spanning between them. Occasionally buttresses are provided

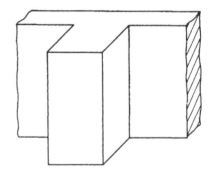

Figure 5.7 A buttress

solely as architectural features. They may be constructed by toothing masonry into the bonding pattern of the main wall or by building the buttress separately and using substantial metal ties to ensure composite action.

5.6 Infill panels

Where masonry panels are used in framed buildings as non-loadbearing panels, the main aspects of design are

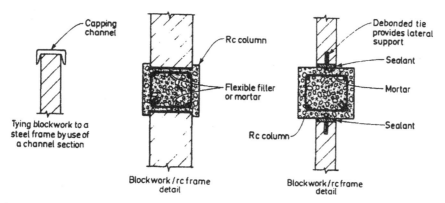

Capping
channel

Rc column

Flexible filler
or mortar

Tying blockwork to a
steel frame by use of
a channel section

Debonded tie
provides lateral
support

Sealant

Mortar

Sealant

Rc column

Blockwork/rc frame
detail

Blockwork/rc frame
detail

Figure 5.8 Typical panel/frame details

that the panel should have adequate resistance to lateral load and service loads, it should be effectively tied to the frame and should not cause distress as a result of long term movements.

The use of loadbearing material to provide panels within a frame of a different material, although considered structurally inefficient, for some types of building may still be economically viable.

There are a number of techniques for tying in panels to the frame, and a few examples of steel and concrete frames are shown in Figure 5.8. It must be remembered that concrete masonry walls tend to undergo shrinkage and a wall/frame detail is often a good place to introduce a movement joint, as discussed in *Chapter 16*.

5.7 Gravity retaining walls

Gravity retaining walls rely upon the weight of masonry built into the wall and are, therefore, if of any great height, massive in construction and costly to build. For retaining walls over 4–5 m high it will generally be cheaper to use reinforced concrete, but for walls up to 3–4 m high reinforced masonry may be a more economic solution.

Design of gravity retaining walls is usually carried out using the 'classic' theory with no tension being assumed in the masonry, and as such no further consideration is given to this aspect in this book. It should be remembered, however, that only blocks suitable for use below ground should be built into a retaining wall. Where the ground is subject to high levels of sulphates, reference should be made to the manufacturer as to the suitability of the blocks for use in the prevailing conditions. The durability of the blocks may be enhanced by tanking or otherwise providing a protective coating which will reduce the effect of aggressive conditions.

5.8 Separating (or party) walls

Separating walls between dwellings are primarily designed to reduce sound transmission and the spread of fire, although they may also serve as loadbearing walls.

Sound reduction is satisfied by traditionally solid walls but lighter building materials, when built into a cavity wall with a 50 mm or 75 mm cavity and plastered, may also prove to be perfectly satisfactory in service. This is covered in detail in *Chapter 14*.

5.9 Fair faced walls

Building a fair faced wall with concrete blocks or bricks requires attention to detail over and above that required for normal walls. Careful planning is necessary at the design stage in terms of module, layout of openings, sizing of components, etc., to obviate unnecessary cutting of blocks and breaking of the bonding pattern. The appearance of masonry openings may be maintained by using reinforced masonry lintels, or certain forms of pressed steel lintels.

The use of special blocks at corners and reveals will probably be necessary and any cutting on site should be carried out with a mechanical saw. An important point to be considered is that it is very much more difficult to build a wall fair faced on both sides than fair faced on just one side.

The mortar joint is an important feature of all facing work and must be given due attention. As the mortar joints need to be regular in appearance and properly finished, the block will need to be dimensionally accurate, more so for ordinary blocks (this is as important with bricks) and the arrisses clean and sharp. The colour of the mortar should be selected to enhance the appearance of the units, and to avoid colour variations due to batching or re-tamping, a central batching facility should be considered or pre-bagged materials or ready mixed mortar employed.

There are a number of tools for finishing joints and various types of joint profile that may be employed. Generally, a lightly tooled or flush finish is to be preferred, both from the point of view of preventing water ingress and of helping to ensure the long term durability of the materials used. It should also be remembered that a deeply recessed joint may affect the loadbearing performance of the wall.

When cavity trays are incorporated into the wall weep holes should be provided to allow water to drain away, and consideration should be given to the effect of water

Figure 5.9 Fair-faced blocks used for flats

weeping out of such holes on the appearance of the building. Clearly, all materials to be used need to be carefully handled and stored on site. Figures 5.9, 5.10 and 5.11 show examples of the use of fair faced blockwork.

5.10 Rendered masonry walls

Rendering a wall greatly increases the resistance of the wall to rain penetration and reduces the rate of air infiltration as well as providing an attractive finish. There are, however, a number of guidelines which should be followed.

In general, render should be weaker than the block or brick to which it is to be applied. Thus, a 1:1:6 cement: lime:sand render would be suitable for lightweight aggregate blocks and most bricks, while a 1:2:9 cement:lime:sand render is more suitable for some types of aerated blocks. For two or three coat work it is recommended that each successive coat should be weaker or thinner than the coat preceding it. Newly applied rendering, including stipple and spatter dash coats, should be kept damp for the first three days. The second coat should be delayed until the previous coat has had time to harden. Specifications generally require a delay period of seven days between coats and this should enable most types of cement:lime:sand render to harden sufficiently, even under cold conditions. A shorter period, depending upon prevailing conditions, may be suitable for weaker 1:2:9 1:1:6 renders. In some cases the

Figure 5.10 Fair-faced blocks used externally

masonry and subsequent rendering coats may need to be dampened to reduce suction but free water should never be left on the surface. Some blocks may form a very high suction background, particularly in hot weather, and an admixture such as a methylcellulose may be useful to reduce the loss of water from the mix. Renderings should not be applied to frost bound walls or during frosty conditions.

Most types of lightweight blocks provide sufficient key so that raked joints may not be required. Rough struck joints will usually be adequate, but in view of the variety of blocks available the manufacturer should be consulted at an early stage regarding specification of render and the preparation of the background.

Render is available in a range of types, colours and textures, some examples of which are discussed in the publication *External Rendering*[5.6]. An example of rendered housing is illustrated in Figure 5.12.

Figure 5.11 Fair-faced blocks used internally

Figure 5.12 Rendered housing

5.11 Diaphragm or cellular walls

Two parallel leaves of masonry joined by masonry cross ribs so that I or box sections are formed are referred to as diaphragm walls, as shown in Figure 5.13, which are very useful and economic for single storey buildings 4–5 m high such as sports halls and warehouses. Cellular walls are usually designed to span vertically as propped cantilevers and are considered in *Chapter 7*.

This type of construction can incorporate McAlloy bars within the cavity so that a moderate prestress may be applied to the masonry, and can prove very economic if adequate provision is made to protect the prestressing bar.

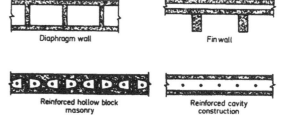

Figure 5.13 Selection of wall types

5.12 Fin walls

A fin wall is similar in appearance to a buttressed wall and consists essentially of a cavity wall stiffened with fins to act as a series of T sections as shown in Figure 5.13. As with the diaphragm wall, it is particularly applicable to 4–5 m single storey structures such as sports halls.

5.13 Reinforced masonry

The simplest and most economic form of reinforced masonry construction is the use of hollow blocks, the cores of which are filled after construction with reinforcing steel and in situ concrete. This method of construction is described fully in *Chapter 9*. The incorporation of reinforcement does not greatly alter the axial load carrying capability of the wall but significantly alters the lateral load performance. It can be useful to reinforce locally at door joints, lintels, etc.

In addition to hollow blocks it is possible to reinforce walls using special bonding patterns or filled cavity construction. Special bonding patterns tend to be expensive to execute, but filled reinforced cavity construction can be viable under certain circumstances. A comparison of the two techniques is shown in Figure 5.13.

Reinforced masonry is particularly useful as a technique for tying buildings together, for assisting performance of tall walls subjected to lateral loading and for retaining walls. An example of a high rise block in the USA is shown in Figure 5.14.

Figure 5.14 High rise reinforced blockwork in the USA

5.14 Veneered and faced walls

A veneered wall comprises of a facing unit attached to a backing (usually loadbearing) unit, but not bonded to such an extent that common action occurs under loads. There are a number of fixing and anchorage devices designed to facilitate this technique, but it is not widely used in the United Kingdom. Unlike a veneered wall, the facing of a faced wall is attached to the backing and bonded so that common action results under load.

5.15 Collar jointed walls

This wall comprises two parallel single leaf walls not further than 25 mm apart with the space between them filled with mortar and the walls tied together so that common action occurs under load. Construction of this type is useful for building an essentially 'solid' wall with brick size units.

5.16 Thin jointed walls

Thin jointed walls are masonry walls laid with a mortar joint thickness of 3 mm or less. To achieve this blocks of high dimensional accuracy are required and special thin joint mortars need to be used. The main advantages of the technique are that walls can be built with great accuracy and are relatively 'dry'.

5.17 References

5.1 BRITISH STANDARDS INSTITUTION. BS 5628: Part 1: 1992 *Code of Practice for the structural use of masonry.* Part 1: *Unreinforced masonry.* BSI, London.

5.2 READ, J B. AND CLEMENTS S.W. The strength of concrete block walls. Phase II: Under uniaxial loading. Cement and Concrete Association, London, 1972. Publication No 42.473. pp 17.

5.3 HEDSTROM, R O. Load tests on patterned concrete masonry walls. *Journal of the American Concrete Institute*, Vol 32, No 10, April 1961. pp 1265–1286.

5.4 The Building Standards (Scotland Consolidation) Regulations 1971. HMSO, London.

5.5 The Building Regulations 1976. HMSO, London.

5.6 MONKS, W L. AND WARD, F. External rendering. Cement and Concrete Association, London, 1980. Publication No 47.102. pp 32.

Chapter 6
Stability, robustness and overall layout

BS 5628: Part 1, in common with BS 8110, contains provisions relating to stability and robustness, which arise from the *Report of the inquiry into the collapse of flats at Ronan Point*[6.1]. There are four recommendations applicable to all masonry buildings and some special recommendations for structures of five storeys and over.

The four general recommendations are:

(1) a layout should be chosen for the structure to ensure 'a robust and stable design';
(2) the structure must be capable of resisting a horizontal force equal to 1.5% of the total characteristic dead load above the level being considered;
(3) adequate connections should be made between walls and floors and between walls and roofs;
(4) in regard to accidental forces, there should be 'a reasonable probability' that the structure 'will not collapse catastrophically under the effect of misuse or accident'. The Code goes on to say that 'no structure can be expected to be resistant to the excessive loads or forces that could arise due to an extreme cause, but it should not be damaged to an extent disproportionate to the original cause'.

Under accidental forces it also states 'furthermore, owing to the nature of a particular occupancy or use of a structure (e.g. floor mill, chemical plant, etc.) it may be necessary in the design concept or a design appraisal to consider the effect of an accident, there is an acceptable probability of the structure remaining after the event, even if in a damaged condition'. Furthermore if there is the possibility of vehicles running into and damaging or removing vital loadbearing members of the structure in the ground floor, the provision of bollards, retaining banks, etc. should be considered.

The special recommendations for buildings of five storeys and over effectively spell out ways in which condition (4) above, can be satisfied for such buildings. The recommendations give three options which can be followed to limit accidental damage. The first of these (*Section 6.5.1*) requires that each structural element be capable *either* of resisting a design load of 34 kN/m² applied from any direction or of being removed without leading to collapse of a *significant portion of the structure*. The second and third options involve the introduction of ties into the structure. As well as the aforementioned recommendations which apply to the completed structure, the Code also emphasises the need to consider stability during construction, both of individual walls and of the structure as a whole.

The following Sections contain remarks with regard to each of the recommendations in turn.

6.1 Category 1 and Category 2 buildings

It is recognised that it is impracticable to prevent all accidental damage and its consequences e.g. a gas explosion in a two storey house is likely to cause considerable damage, collapse and even fatality. It is, however, essential to limit damage and to take into account the consequence of damage and collapse, which typically increases with increasing height, size and occupancy of the building. BS 5628: Part 1 caters for this by making recommendations for two categories of buildings. Category 1 (four storey and below), Category 2 (five storey and over). These building categories and the options available to the designer are given in Table 6.1 and dealt with in more detail in the following clauses to this chapter.

6.2 Layout

Low and medium rise masonry structures are almost always inherently stable, firstly because walls are naturally strong in resisting lateral forces applied in their plane and secondly because it is usual for such a building to contain walls arranged in different directions. To ensure stability it is necessary to provide sufficient walls to resist lateral and torsional movements. Figure 6.1 illustrates layouts of walls which would ensure stability. It should be noted that wind loads normally applied to the face of a building have to be transmitted via the floors to the walls providing lateral resistance.

Figure 6.2 shows arrangements of walls which are not symmetrical thus giving rise to torsional moments. The

Table 6.1 Detailed accidental damage recommendations

Building type	Design recommendations		
Category 1 All buildings of four storeys and below	Plan form and construction to provide robustness, interaction of components and containment of spread of damage (see clause 20, BS 5628: Part 1)		
Category 2 All buildings of five storeys and above	Plan form and construction to provide robustness, interaction of components and containment of spread of damage (see clause 20, BS 5628: Part 1)		
	Additional detailed recommendations for category 2		
	Option (1)	Option (2)	Option (3)
	Vertical and horizontal elements, unless protected, proved removable, one at a time, without causing collapse	Horizontal ties Peripheral, internal and column or wall in accordance with clause 37.3 and table 13, BS 5628: Part 1	Horizontal ties Peripheral, internal and column or wall in accordance with clause 37.3 and table 13, BS 5628: Part 1
		Vertical ties None or ineffective Vertical elements, unless protected, proved removable, one at a time, without causing collapse	Vertical ties In accordance with clause 37.4 and table 14, BS 5628: Part 1

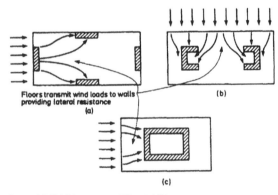

Figure 6.1 Wall layouts providing stability

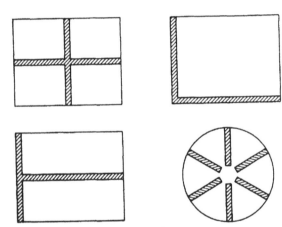

Figure 6.3 Layouts unable to resist torsion

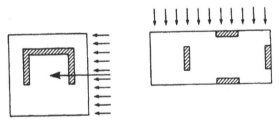

Figure 6.2 Wall layouts introducing torsional response

possibility of a torsional deformation under a gust loading applied to part of the structure should not be forgotten. Figure 6.3 shows a building stable against uniform lateral pressure but not against an eccentric gust.

In most instances the layout of a masonry building will be far more complex than is shown in Figures 6.1 to 6.3, with many more loadbearing walls continuous from foundation to roof. There may also be internal walls which, even if of lower strength blocks and perforated by door openings, can still contribute to lateral load resistance and stability. An accurate analysis of response to lateral load is not normally possible, so that simplifying assumptions, ignoring the action of particular walls or parts of walls, are justified and will be on the safe side.

Where shear walls intersect, as shown in Figure 6.1 (b) and (c), it is necessary to limit the length of wall which can be assumed to act as a flange. This is analogous to the limitations on flange width applied to T and L beams of reinforced and prestressed concrete. American practice[6.2, 6.3] limits the width taken into account in the case of T and I sections to be one-sixth of the total wall height above the level being considered but with the outstand on either side not exceeding six times the thickness of the intersected walls. In the case of L and Z sections, the outstand is limited to one-sixteenth of the

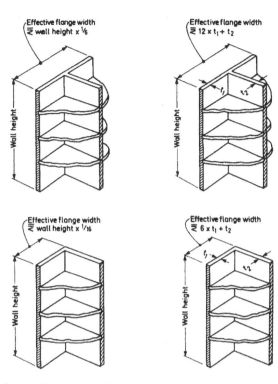

Figure 6.4 Intersecting walls

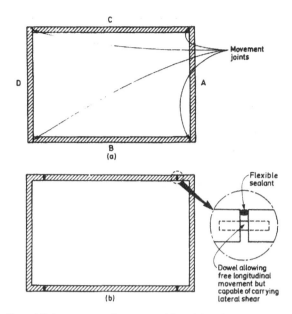

Figure 6.5 Arrangement of movement joints giving :a) a potentially unstable structure (b) an alternative more stable structure

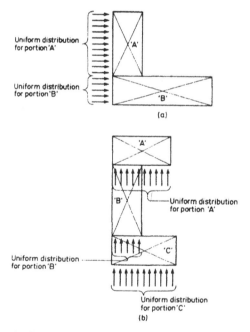

Figure 6.6 Distribution of a 1.5% horizontal load for buildings of non-rectangular plan form

total wall height above the level being considered or to six times the thickness of the intersected wall, whichever is the less. Figure 6.4 shows how the provisions apply to T and L walls. For detailed analysis of such walls, refer to *Chapter 5*.

Temperature and shrinkage movements also require consideration when deciding on a layout for walls. A layout such as that illustrated in Figure 6.1(b) will be susceptible to this problem. As has already been said, walls are naturally strong in resisting lateral forces and will be equally strong in resisting lateral movements. Shear walls acting in the same plane should, if possible, not be located further apart in that plane than 10 m. If this recommendation is exceeded it is possible that severe cracking will develop in one or other of the shear walls or in the connecting floor or roof.

The provision of movement joints also requires special care (see *Chapter 16*). Figure 6.5(a) illustrates a layout actually used on a single storey building which collapsed catastrophically when wall A was sucked out by wind. Walls B and C, being left supported at one end only, then fell over with the roof, taking wall D with them. A less convenient but stable alternative is shown (Figure 6.5(b)). Another alternative would have been to have provided ties at roof level between walls A and D.

6.3 1.5% horizontal load

In Clause 20.1, BS 5628: Part 1, is stated that the horizontal load should be uniformly distributed. This is

to avoid stress concentrations which would arise in analysis if it was considered acting at, say, the centre of gravity of the part of the building above the level being considered. It should be considered as uniformly distributed in both horizontal and vertical directions in the case of a building of rectangular plan form. In the case of a non-rectangular plan form it is suggested that the forces be calculated on convenient rectangles of the

49

plan area. Examples are shown in Figure 6.6. It will be seen that unless separate rectangles are considered, an altogether excessive load requirement could arise on projecting wings of a building. In the case of a long building considered end-on, it will be permissible to consider the load as arising at various points distributed along the length of the building, otherwise very high local stresses may arise.

It can be noted that the design for a lateral load equal to a percentage of the load of the building is a simple way of designing for the effect of earthquakes. In parts of the world prone to earthquakes, percentages much higher than 1.5 are used in design. The assessment of forces to cater for the effects of earthquakes on a European front will be covered by a Eurocode. However, the 1.5% may currently be considered as adequate to deal with the very minor earthquakes which occur in the UK.

6.4 Roof and floor connections

In Clause 20.1 it is recommended that roof and floor connections should be as in Appendix C, BS 5628: Part 1. The fixings in Appendix C are designed to provide simple resistance to lateral movement to ensure that there is no possibility of a wall buckling above a height of two storeys (see *Chapter 16*). It should be noted that more onerous tie requirements are specified for structures of five or more storeys.

6.5 Misuse or accident

A proper layout and the provision of an ability to withstand a 1.5% horizontal force will provide a considerable safeguard against excessive damage following an accidental load in a normal masonry building. Consideration should, however, be given to the excessive risks which might arise from any particular type of occupancy and to structures exposed to vehicle impact. With regard to gas explosions, studies[6.4] have shown that these accidents occur quite frequently in dwellings, although the pressures reached in the Ronan Point collapse are attained in only a handful of cases in any given year. Therefore, there can be no question of individual dwelling units being designed to resist the effects of an explosion of the severity of that at Ronan Point. The general question of what level of damage is acceptable from a given cause is a matter of judgement. Special consideration will be appropriate for buildings such as concert halls or sports centres which may contain a large number of people.

6.6 Recommendations for Category I buildings (four storey and below)

The four general recommendations given in clause 20.3, BS 5628: Part 1, and mentioned at the beginning of this chapter are aimed at the limitation of accidental damage and the preservation of structural integrity and are applicable to all building categories. No additional recommendations are made in respect to category 1 buildings.

6.7 Special recommendations – buildings of five storeys and over

At present, masonry buildings of five storeys and over are not very common in the UK. There is, however, no reason why, for some particular occupancies at least, such structures should not become as popular as they already are in certain areas of America. As structures over five storeys are likely to cover a fairly large plan area, there will usually be many loadbearing walls inserted in various directions. Such structures are naturally robust, but even so, particular requirements are specified in the Code. These arise essentially from Section 5 of Approved Document A or the alternative detailed provisions given in Section 5 of BS 5628: Part 1. The detailed provisions are given in Section 5 of BS 5628: Part 1 and, as previously mentioned, comprise of three options, each of which is discussed below.

6.7.1 Option 1

In this option it is necessary to consider each individual loadbearing element in turn and either demonstrate that it can withstand a force of 34 kN/m² applied in any direction or prove that no significant portion of the structure will collapse if it is removed.

The figure of 34 kN/m² is close to the pressure which occurred in the Ronan Point collapse which, as previously mentioned, is only likely to be reached or exceeded in a handful of incidents in a given year[6.4].

It may initially appear unreasonable to consider only one element at a time since an explosion inside a building will produce pressure on both slabs and walls. In practice, however, once the weakest element begins to fail, the burning gas in a gas explosion will begin to vent so that the load will begin to drop off. In the case of high explosives, the pressure is dependent on the distance from the centre of the explosion so that, except in extreme cases, the only severe effect will be to the elements nearest to the centre of the explosion.

The option to resist 34 kN/m² is probably only practicable for walls. Most slabs in structures will have a dead weight of only around 6 kN/m² and so will fail by being uplifted unless a lot of top steel is provided. Short span slabs may be able to withstand 34 kN/m² in a downward direction.

In the case of walls it is possible for 34 kN/m² to be resisted, provided the wall carries a substantial vertical load and is of reasonable thickness, and is also restrained between concrete floors. The formula for arch action given in Clause 37.1.1 can then be applied, which is based on the assumption of arching as illustrated in Figure 6.7.

Taking moments about C:

$$n\,(t - \delta) + q_{lat} \cdot \frac{h_a}{2} \cdot \frac{h_a}{4} = q_{lat} \cdot \frac{h_a}{2} \cdot \frac{h_a}{2}$$

Ignoring δ as being small gives:

$$nt = \frac{q_{lat}\, h_a^2}{8}$$

or
$$q_{lat} = \frac{8tn}{h_a^2}$$

Finally introducing γ_m to the equation gives:

$$q_{lat} = \frac{8 \times t \times n}{h_a^2 \, \gamma_m}$$

and letting $\gamma_m = 1.05$

$$q_{lat} = \frac{7.6tn}{h_a^2}$$

as in Clause 37.1.1 of BS 5628: Part 1.

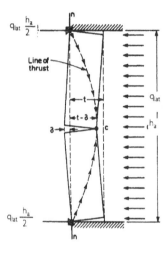

Figure 6.7 Arching failure under pre-compression load, n

It will be seen that the compressive load actually increases the capacity of arching so that the design value of *n*, the axial load/unit length of wall, is derived from an imposed load of $0.35 \, Q_k$ and a dead load of $0.9 \, G_k$. For the purpose of accidental loading γ_m is taken as 1.05. This is much lower than that used for ordinary compression in accidental loading ($0.5 \times 3.5 = 1.75$). This is because the load arising from $0.35 \, Q_k$ and $0.9 \, G_k$ is much lower than that for which the wall is designed. The actual depth of the stress block carrying the axial load at the arch hinges is usually a small fraction of the wall thickness. A shortfall in arch capacity is more likely to arise in these cases due to the wall being out-of-plumb. In practice it will be found that for walls 150–200 mm thick the compressive load will be insufficient unless the wall supports three or more storeys of construction. Thus it is concluded at this point that under *Option 1* it is necessary to consider the removal firstly of floors and roofs and secondly of walls in upper storeys.

It should be noted that for checking after removal, reduced γ_f and enhanced γ_m factors are used. The γ_f factors for dead, imposed and wind loads are given in Clause 22(d), while the γ_m factors are to be taken as half

of those specified for normal design (Clause 27.3). These reduced safety factors take account of the reduced probability of excessive imposed loads or understrength coinciding with the occurrence of very severe (and therefore unlikely) accident.

The definition of loadbearing elements which may be removed is given in Table 11 of BS 5628: Part 1. In the case of beams and columns it is their whole length which is considered to be removed.

The extent of a wall which needs to be considered to be removed is the length between lateral supports or length between a lateral support and the end of the wall. When there are no lateral supports the length to be considered to be removed is 2.25 *h*, where *h* is the height of the wall, in the case of internal walls and the full length in the case of external walls. For slabs the whole area should be considered removable unless there are partitions or other construction underneath which can form temporary supports for the reduced design load ($0.35 \, Q_k + 1.05 \, G_k$) on the remaining slab. Table 11 of BS 5628: Part 1 does not deal specifically with one way slabs, but it would seem reasonable to assume that an upper limit of 2.25 *l*, where l is the span, could be applied for internal slabs in cases where there are no substantial partitions to provide temporary supports. An acceptable lateral support is taken to include a right intersection or return wall (min 340 kg/m²) with no openings within h/2 of the supported wall. It must also have connections capable of resisting a force of 0.5 F_t/m (where F_t = 60 KN or 20 + 4 N_s, whichever is the lesser. N_s is the number of storeys.).

The horizontal force F_t is used for the requirements for other support connections and for the horizontal ties.

A lateral support can also include a pier or stiffened wall section (resisting force 1.5 F_t/m) as a connected partition (0.5 F_t/m) with an average weight not less than 150 kg/m².

It is necessary to consider the portion of the structure liable to collapse after removal and judge whether it is a 'significant portion'. Section 5 of Approved Document A provides some assistance here in that it allows removal of a portion of any one structural member provided that:

'(a) structural failure consequent on that removal would not occur within any storey other than the storey of which the portion forms part, the storey next above (if any) and the storey next below (if any); and

(b) any structural failure would be localized within each such storey'.

Section 5 of Approved Document A goes on to state that in regard to (b) the regulation may be 'deemed to be satisfied if the area within which structural failure might occur would not exceed 70 m² or 15% of the area of the storey (measured in the horizontal plane), whichever is the less'.

The matter is clearly one for the exercise of engineering judgement. It may be helpful to consider the particular case of a cross wall type of construction as in Figure 6.8. Various cases of removal and consequential collapse are indicated for removal of elements in the top two storeys,

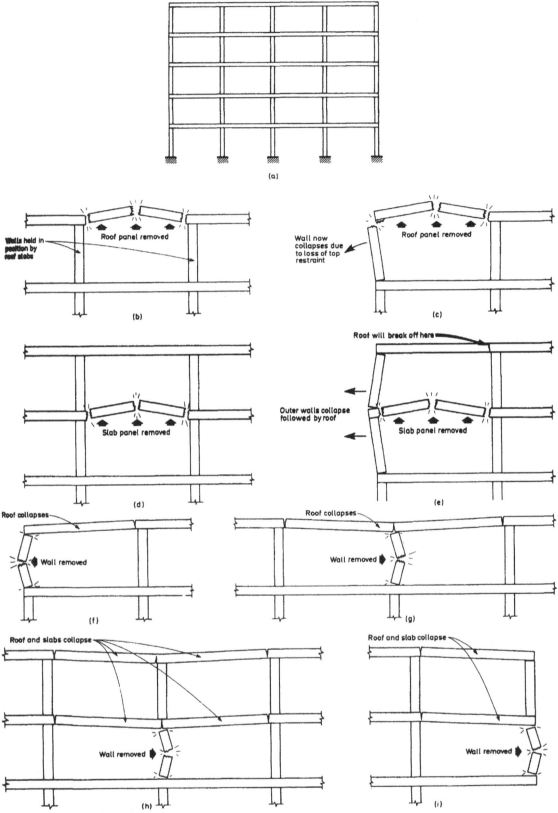

Figure 6.8 Consequence of removing elements

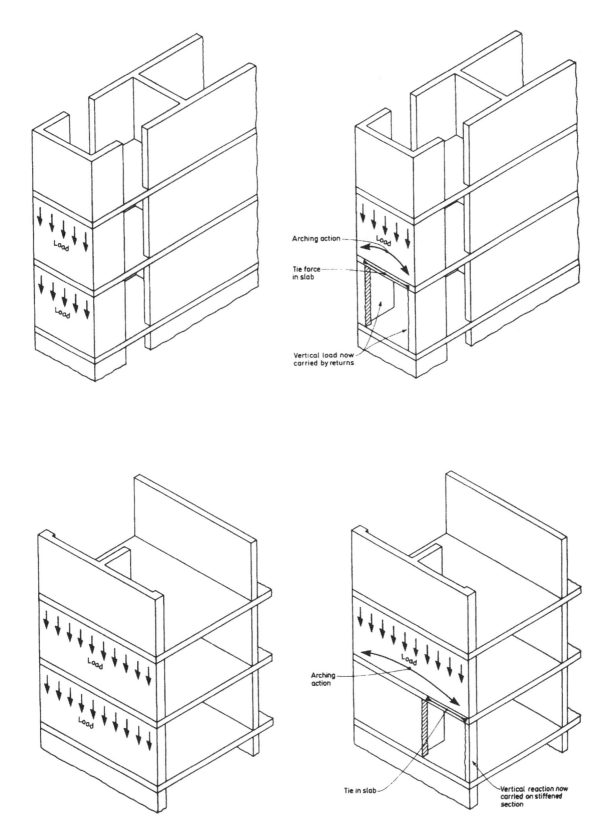

Figure 6.9 Ensuring stability following removal of an element

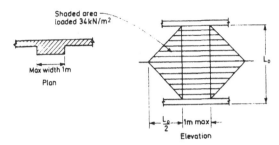

Shaded area loaded 34kN/m²

Max width 1m

Plan

$\frac{L_a}{2}$ 1m max

Elevation

L_a

Figure 6.10 Wall braced by stiffened section

all of which, in the authors' view, are acceptable as being damage which is not *disproportionate to the original cause*. Interior slab removal lower down the structure is also acceptable since it will not cause any further collapse (Figure 6.8(d)). However, exterior slab removal and wall removal in lower storeys will not be acceptable. Potential solutions in respect of removal of the external slab or wall are sketched in Figure 6.9 where the external or flank wall is provided with returns or with stiffened sections which are designed to withstand 34 kN/m². It should be noted that the latter solution involves designing the stiffened sections to carry a proportion of load transmitted from the adjoining panel (Figure 6.10). This solution can also be adopted where necessary for internal walls below the top most storeys. More sophisticated solutions than those suggested in Figure 6.9 can obviously be conceived, for instance allowing cantilevering out of portions which can lose support, but it may be preferable to use the procedures given under *Options 2* and *3* (*Sections 6.7.2* and *6.7.3*) below rather than rely on extensive calculations which will, of necessity, involve many assumptions.

6.7.2 Option 2

In this option horizontal ties are provided at all floor levels in accordance with rules specified in the Code and the vertical elements are proved to be removable, one at a time, without collapse of a significant portion of the structure.

For detailed operation of this option, reference should be made to the comments under *Option 3* for the provision of horizontal ties and those under *Option 1* for ways of dealing with vertical elements, i.e. walls.

6.7.3 Option 3

In this option both horizontal ties and vertical ties are provided at specified positions around, across and down the building.

It is worth commenting at the outset that the provision of ties in masonry buildings is not new. Many older buildings, where the masonry has deformed or where settlement has occurred, have been fitted with ties, passing to iron anchor plates which can be seen on the outer walls, so that the principle of providing ties is well established.

The primary objective may be seen as providing integrity to the structure as a whole, but there is a secondary objective in that the ties will limit the possible spread of damage. Ties in a portion of wall above a removed storey allow it to act as a tied arch. Horizontal ties acting in a catenary manner may also permit slabs to continue to carry load in a severely deformed state, thus avoiding collapse on to floors below and subsequent further damage. Where vertical ties are provided at intervals in walls, they essentially give locally protected elements; in this instance there is a close parallel in effect between *Option 3* and *Option 1*.

Horizontal ties

Options 2 and *3* require horizontal ties to be provided in the form of a peripheral tie, plus internal and external column and wall ties, and which are related to a basic horizontal tie force F_t where:

$$F_t = 60 \text{ kN or } 20 + 4 \, N_S \text{ kN}$$

where N_S is the number of storeys

The peripheral tie – is positioned internally within 1.2 m of the edge of the floor or roof or in a perimeter wall and acts somewhat like the hoops on a barrel. The tie must be anchored at re-entrant corners or changes of construction and sized to carry a force of F_t.

The internal tie – is positioned across the floor or roof in both directions and acts in a catenary/continuity manner. In the direction perpendicular to the span they must be sized to carry a force of F_t. Whilst in the direction of the span (or spans in the case of a two way span) they must be sized to carry a force of:

$$F_t \text{ or } \frac{F_t \, (G_k + Q_k)}{7.5} \times \frac{L_a}{5} \text{ whichever is the greater}$$

Where:

$G_k + Q_k$ is the average characteristic dead and imposed loads kN/m²

L_a is the greatest span in metres in the direction of the tie or 5× clear storey height, whichever is the lesser.

These ties may be uniformly distributed or concentrated/grouped in the floor or roof, or placed within walls (within 0.5 m of the floor or roof). Any such horizontal concentrated spacing should not exceed 6 metres.

The column/wall tie – is positioned so as to tie the vertical element (Corner columns should be tied in two directions) to the floor or roof slab/beam, and may be provided partly or wholly by the reinforcement used as peripheral and internal ties. These ties/connections must be sized to carry a force of:

$$2F_t \text{ or } (h + 2.5)F_t$$

(kN – in the case of columns, and kN/m – in the case of walls)

This tie connection may be provided by reinforcement, or based on either shear strength or friction, as appropriate. Wall ties may be uniformly distributed or concentrated at positions along the length of the wall. Any concentrated ties should not exceed 5 m centres or be more than 2.5 m from the end of the wall.

The provision of peripheral and internal ties is a simple matter where in situ reinforced concrete floors are used, in which instance BS 8110 specifies a minimum percentage of steel which must be provided in both directions throughout. There may be some difficulties in dealing with floors made up of precast elements, but the detailed provisions of BS 8110 should be consulted here. Manufacturers of precast floor units are aware of the problems and can advise designers on solutions appropriate to their own products.

The provision of column and wall ties can be achieved in two ways. Firstly by friction or shear arising from direct bearing of floor slabs on the wall, or secondly by arranging ties along the lines of Figures 16–18 of Appendix C to BS 5628.

The requirements for vertical ties are clearly specified in Table 14 of BS 5628: Part 1. In masonry incorporating hollow units it is probably simplest to provide ties at spacings of 1 m or so. In other cases it will be necessary to provide special masonry units every so often which provide a suitable vertical void into which ties and surrounding concrete can be placed; in this instance it would be preferable to use a larger spacing. While it is not specifically mentioned in Table 12 under *Option 3*, a statement in Clause 37.4 implies that vertical ties need not be extended to the foundation but only to a level where the wall can be proved capable of resisting 34 kN/m². This would appear sensible since the precompression load from five or six storeys will produce a wall robust enough to resist removal in nearly all cases.

6.8 References

6.1 Report of the inquiry into the collapse of flats at Ronan Point, Canning Town. HMSO, London, 1968.
6.2 Building Code requirements for concrete masonry structures. *ACI Journal Proceedings*, Vol 7, No 8, August 1978. pp 384–403.
6.3 Commentary on Building Code requirements for concrete masonry structures. *ACI Journal Proceedings*, Vol 7, No 9, September 1978. pp 460–498.
6.4 TAYLOR, N. and ALEXANDER, S J. Structural damage in buildings caused by gaseous explosions and other accidental loadings. *BRE Current Paper CP 45/74*, Building Research Establishment.

Chapter 7
Walls under vertical load

7.1 Background

7.1.1 Mortar strength and block strength

In the previous chapters, the properties of blocks and mortars have been discussed and are classified by their characteristic strengths. A plain wall is a composite structure made of these two materials and, under compressive loading, its strength would be expected to be influenced by the strengths of both materials. What sort of interaction between block and mortar strength might be expected? Two extreme possibilities can be put forward:

(a) Between horizontal joints, all loads will effectively be carried by the blocks while at the horizontal joints, all the load is carried by the mortar so that the wall strength might be expected to correspond to the strength of the weaker material;

(b) The function of the mortar joint is simply to produce a good uniform bearing between the blocks and provided the mortar is not so fluid that it could squeeze out like toothpaste, its strength is irrelevant and the wall strength will correspond to the strength of the blocks.

In fact, the second of these possibilities is the closest to the truth though the properties of the mortar may have some influence on the strength. Walls tend to be weaker than the average strength of the blocks for a number of reasons which will now be considered.

7.1.1.1 Behaviour of a thin mortar layer between stronger blocks

Assuming for the moment that both mortar and block are elastic, one influence of the mortar on the block can easily be determined. The elastic modulus of the mortar is commonly substantially less than that of the block and, as a consequence, the vertical strains under axial load are greater. This in turn implies a greater transverse dilation due to Poissons ratio. Since the stronger block will not tend to dilate so much, it will restrain the mortar which will be put into horizontal compression, thus putting the mortar into a state of triaxial compression. Very roughly,

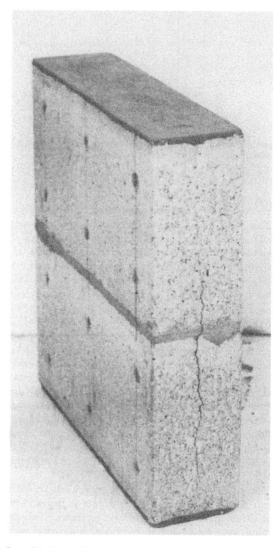

Figure 7.1 Mode of failure of couplet

concrete or mortar in triaxial compression will be able to withstand a vertical stress, f_{vert}, given by

$$f_{vert} = f'_c + 4f_{horiz}$$

where

f'_c = the uniaxial compressive strength
f_{horiz} = the applied horizontal stress

57

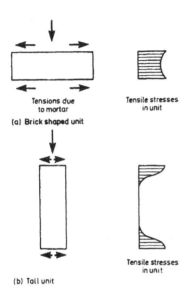

Figure 7.2 Effect of aspect ratio on tensile stress induced by mortar:
(a) brick-shaped unit (b) tall unit

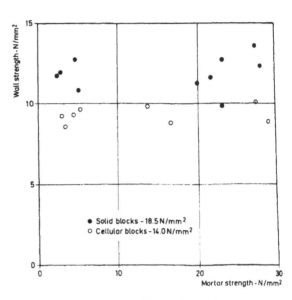

Figure 7.3 Effect of mortar strength on wall strength

The restraint from the blocks ensures that f_{horiz} can attain whatever value is required to sustain the applied vertical load and hence the mortar joint cannot fail before the block. This does not imply that the strength of a wall is totally unaffected by mortar strength since the compression induced in the mortar by the restraint provided by the block must be equilibrated by a tension in the block. While not an exact model, it is convenient to consider the failure criterion for concrete to be one of limiting tensile strain. Compressive failure occurs when the transverse tensile strain produced by Poissons ratio effects reaches the limiting value and the material cracks vertically. It can be seen in this situation that transverse tensions produced by the action of the mortar will tend to increase the transverse tensile stresses induced by the vertical loads and hence reduce the vertical load that can be carried. Inspection of the mode of failure of couplets (Figure 7.1) or short walls in axial compression show that failure is initiated by the development of a vertical split in the plane of the wall starting in the region of the mortar joint.

This argument would lead one to expect that the load carrying capacity of a wall would reduce as the ratio of unit strength to mortar strength increases. This is certainly true for brickwork but evidence indicates very little influence of mortar strength in blockwork walls. The reason for this is almost certainly the difference in aspect ratio between a brick and a block, as illustrated in Figure 7.2. It will be seen that, while the whole brick will be put into more or less uniform tension, the block will only be subjected to local stress at the ends.

Figure 7.3 shows results of tests on blockwork carried out by the Cement and Concrete Association[7.1] which indicate the minimal influence which mortar strength appears to have on blockwork. In solid blockwork, therefore, mortar can be considered to be merely a bedding material for the blocks and, as far as axial loading is concerned, its strength is largely irrelevant. As will be seen later, this may not necessarily be true for walls under eccentric loads.

7.1.1.2 Influence of aspect ratio of unit on wall strength

The suggestion that unit aspect ratio influences wall strength is incorrect apart from its effect on the influence of mortar discussed above. The effect of aspect ratio is actually upon the measured unit strength and not the wall strength. When a unit is crushed in a testing machine the platens of the machine restrain the unit from expanding and thus put the top and bottom into lateral compression. A degree of triaxial compression is induced and the unit is able to withstand greater compressive loads. If the unit is short compared with its breadth (e.g. a brick) then all the material in the unit will be triaxially compressed and a substantial increase in strength will be obtained. If the unit is high compared with its breadth ($h/b \geq 2$) then the strength enhancement will only occur at the ends and the overall ultimate strength will not be affected. The actual relationship between apparent unit strength and aspect ratio varies somewhat from one material to another, and examples are shown in Figure 7.4. Tests on units with a low aspect ratio give an over estimate of the actual material compressive strength and, as a consequence, the ratio of wall strength to unit strength is low. As the aspect ratio increases, the ratio of wall strength to unit strength increases. Thus, the ratio of wall strength to unit strength is much higher for blockwork than for brickwork, and it is occasionally stated that blockwork is more efficient than brickwork. It must be emphasized here that this is not the case, it is simply that tests on single units do not give a direct measure of uniaxial compressive strength.

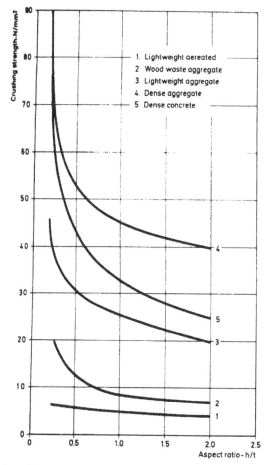

Figure 7.4 Crushing strength gagainst aspect of ratio for various blocks

7.1.1.3 Other factors

Many other factors will influence the relationship between block strength and wall strength in axial compression, such as:

Workmanship

One would not expect a poorly built wall to perform as well as a well built wall. Indications from test evidence[7.1] suggest that this is so but that for blockwork the effect may not be large.

Hollow or cellular blocks

If walls are built so that the transverse webs of the blocks do not line up vertically the strength of the wall will be substantially reduced since the webs will become ineffective as far as carrying vertical load is concerned, but this factor has been taken into account in the Code[7.2].

7.1.2 Characteristic wall strength

In design the characteristic strength of a wall is employed to define the strength of masonry, which is that of a short wall subjected to axial compression. As discussed, the relation between wall strength and unit strength is somewhat complex, depending upon the mortar, the aspect ratio of the unit and other factors. No generally accepted method exists by which this strength may be calculated from a knowledge of the unit and mortar strengths and so recourse is made to a series of empirical relationships for the various types of element. These arise from direct tests on walls and, due to the large variety of units, are liable to be rather approximate for particular cases.

7.1.3 Short walls subjected to eccentric load

Concrete in compression may be considered to be roughly plastic and capable of sustaining a stress of about 80% of the cube strength. In fact the unit strength, corrected for aspect ratio effects, will also be of about this value. Assuming a plastic stress distribution leads to the conditions shown in Figure 7.5 for the ultimate stress distribution in an eccentrically loaded block.

From this, the following equations can be derived:

$$N = 2b \left(\frac{t}{2} - e \right) f_c$$

or

$$N = (t - 2e) \, bf_c$$

where

t = block thickness
b = length of block
e = eccentricity of load
N = vertical load
f_c = unit strength (adjusted for aspect ratio)

In a wall, the ultimate stress obtainable will be rather less than f_c due to the influence of mortar strength, workmanship, etc. These will have a similar influence to that found for axially loaded walls.

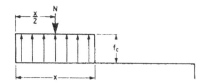

Figure 7.5 Stress in blockwork under ultimate load

7.1.4 Slender walls

So far, the strength of walls has been dealt with in situations where any deformations which occur will have a negligible influence on the strength. This, however, is by no means always the case. Consider the wall shown in Figure 7.6 which is supporting an eccentric load. The eccentric load induces a moment in the wall which will cause the wall to deflect. As a consequence, the eccentricity of the load at mid-height of the wall is not the same as the top but is increased by the amount of the deflection. This will lead to the stress conditions at mid-height being considerably more severe than would be assumed if the deflection were ignored. In the limit, this

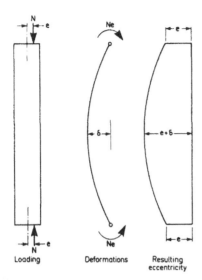

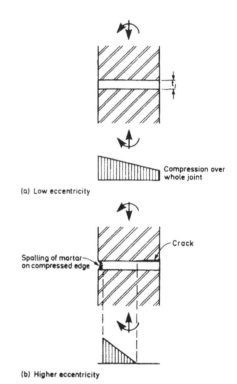

Figure 7.6 Slender wall

(a) Low eccentricity

Spalling of mortar
on compressed edge

Crack

(b) Higher eccentricity

Figure 7.8 Behaviour of joint under eccentric load: (a) low eccentricity
(b) high eccentricity

extra eccentricity can cause more deflection, resulting in more eccentricity and an instability failure can result. The normal situation, however, is simply that the increased eccentricity causes the centre section to fail at a lower load than would otherwise be the case.

To estimate the reduction in capacity of the wall, it is necessary to obtain an estimate of the deflections, for which it is necessary to know the moment-curvature relationship for a blockwork wall. Moment-rotation relationships have been reported by Cranston[7.3] for small units made up of two blocks plus a single mortar joint. Figure 7.7 shows typical results for different levels of axial load. It will be seen that as the eccentricity increases, the stiffness decreases. For the lower loads the behaviour could almost be considered to be elastic-

plastic. The basic reasons for this type of behaviour are quite easy to discern if the behaviour of the joint is considered (see Figure 7.8(a) and (b)). For low eccentricities of load (Figure 7.8(a)) the joint is completely in compression and the rotation of the joint will be given by:

$$\theta = \frac{N \, e \, t_i}{E_i \, I} \qquad (1)$$

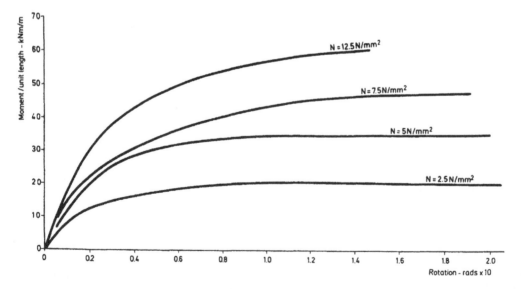

Figure 7.7 Moment rotation characteristics of blockwork

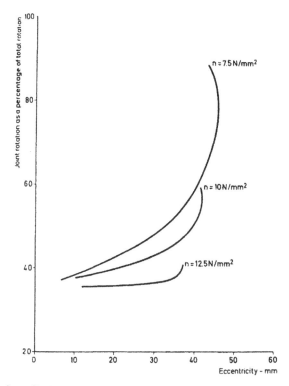

Figure 7.9 Proportion of total rotation accommodated in joints for various levels of axial load

The tensile stress that can develop between the block and the mortar is very low and cracks will develop in the joints almost as soon as any part of the joint goes into tension. Also, the mortar on the compressive face will start to spall at quite an early stage. This leaves the load to be carried on a relatively small and decreasing area of mortar. The stress and deformations in this piece of mortar become very large as eccentricity increases. As a consequence of the lower strength and stiffness of the mortar relative to the block and of the small area of mortar actually supporting the loads, a large proportion of the total rotation takes place in the joints. This is illustrated in Figure 7.9 which shows the proportion of the rotation occurring in the joint as a function of the total rotation for various situations. As the eccentricities increase or the axial load lowers, the behaviour of the joint begins to dominate the overall behaviour.

There are two modes of failure of the critical section of a wall:

(1) for relatively higher vertical loads, failure will occur in the block when the eccentricity reaches the value given by the relation below:

$$e = \tfrac{1}{2} \left(t - \frac{N}{b f_c} \right)$$

(this can be obtained from equation (1) above);

(2) for walls with relatively light vertical loads the mortar in the joint crushes. This crushing starts at

the compression face and as the deformations increase, the crushing works its way into the joint until there is insufficient sound mortar to sustain the load at the required eccentricity.

From the above it will be seen that the behaviour of very slender walls, which will inevitably be lightly loaded at failure, will be dominated by the behaviour of the mortar and the properties of the block will be largely irrelevant. This is opposite to the situation for heavily loaded short walls where the strength depends upon the properties of the block and properties of the mortar are largely irrelevant.

7.1.5 Development of theoretical equations for slender walls

Information on the basic moment-curvature behaviour of blockwork can be obtained by numerical methods, and can be used to predict the strengths of vertically loaded slender walls. However, for design purposes such an approach is impracticable and, in any case, insufficient data on basic behaviour exists. As a consequence, some alternative approach is required and it has been found that the approach used for reinforced concrete slender columns can be adapted for use on masonry walls.

In very simple terms, the derivation for a reinforced concrete member is as follows. Consider a pinned ended strut of height h. At failure, the strain in the concrete on the compression face at mid-height will be 0.0035. Assuming a balanced section, the strain in the steel at yield will be, say, about 0.002. The curvature of the critical section will thus be given by:

$$\frac{1}{r} = \frac{0.0035 + 0.002}{d} \qquad (2)$$

The central deflection will be given by the relation

$$a = \kappa \, h^2 \, \frac{1}{r} \qquad (3)$$

where κ is a constant which depends upon the shape of the curvature diagram. If the axial load on the column is N_u then this deflection under ultimate conditions will increase the central moment on the column by an amount equal to $N_u a$. The column is then designed for the initial applied moment plus this additional moment. Equation (3) above can be rearranged as follows by substituting for $\frac{1}{r}$ from (2).

$$\frac{a}{d} = \kappa' \frac{h^2}{d^2}$$

From this it can be seen that all columns with the same slenderness ratio are predicted to have the same ultimate deflection expressed as a fraction of the effective depth. Figure 7.10 shows a series of load-central deflection curves for 140 mm masonry walls with various eccentricities of load. It will be seen that despite the great variation in ultimate load obtained and the differences in form of some of the curves, all the walls are at, or close

61

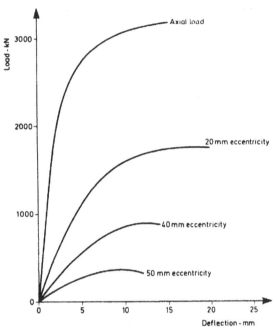

Figure 7.10 Load-deflection curves for a series of 13 course high walls

7.2 Walls under vertical load – design

7.2.1 General

The basic information required for the design of a wall to resist a vertical load (eccentric or axial) is as follows:

(1) the loading appropriate to the ultimate limit state which consists of the characteristic loads (dead, imposed, wind) and appropriate partial safety factors;
(2) a characteristic compressive strength for the particular type of masonry being adopted and an appropriate partial safety factor for reducing the characteristic strength to a design value;
(3) to assess the effects of slenderness and eccentricity of load, it is necessary to estimate the effective height and effective thickness of the wall;
(4) an assessment is needed of the effective eccentricity of the loading at the top of the wall.

Each of these aspects of design will be considered in turn and, where appropriate, the provisions of the Code described and explained.

7.2.2 Design loads and partial safety factors

Characteristic loads are multiplied by partial safety factors which take account of:

(1) possible unusual increases in load beyond the characteristic values;
(2) possible inaccuracies in assessment of load effects or unforeseen stress redistributions within the structure;
(3) variations in dimensional accuracy.

Since the characteristic load is ideally a load with a 5% chance of being exceeded during the life of a structure, it is clear that design should be carried out for a load somewhat greater than this as a 1 in 20 chance of collapse due to overload would be unacceptable. The size of this increase can be expected to depend upon the inherent variability of the type of loading considered, hence a larger factor will be expected for live loads than for dead.

In (2) above, load effects are the moments, stress axial forces, and such like, at particular points in the structure resulting from the loading; they are the results of structural analysis. This part of the safety factor is thus to allow for inaccuracies in analysis. An effect of constructional inaccuracies is also to produce errors in the assessed load effects and in the ISO Standard setting out the limit state format errors in analysis and errors due to tolerances are combined together into a partial factor γ_f [7.4]. This is also employed in the Bridge Code, BS 5400 [7.5]. In the most rigorous formulations this factor is applied to the load effects and not, as in BS 5628, to the loads.

A further function of the partial factors on the loads is to take account of the relative probability of various combinations of loading occurring. This function is not explicitly mentioned in Clause 19 but is nevertheless

to, their maximum load at the same deflection of about 10 mm. Thus, the same form of equation as that used for reinforced concrete can be expected to work reasonably well for masonry. Such an equation would have the form:

$$\frac{e_a}{t} = \kappa \left(\frac{h_{ef}}{t}\right)^2 \qquad (4)$$

where:

e_a = additional eccentricity
h_{ef} = effective height

This is the equation used in Appendix C of the Code. A value of $\frac{1}{2400}$ is used for κ and an arbitrary adjustment has been made so that there will be zero additional moment for an eccentricity of $0.05t$. This additional eccentricity has to be added to the eccentricity of the vertical loading. To achieve this, it is necessary to know the distribution of the initial and additional eccentricities up the height of the wall. For convenience, Appendix B makes the assumptions shown in Figure 7.11.

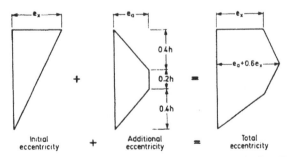

Figure 7.11 Assumed variations of eccentricity over the height of a wall

implicit in the values given for the factors in Clause 22. It is clearly less likely that maximum dead, live and wind loads will all occur simultaneously than that maximum dead and wind loads or dead and live loads will occur simultaneously. Thus for combination (3), dead, imposed and wind load, the partial factors are all lower than those adopted for combinations (1) and (2).

Two factors may be given for the dead loading because in some cases, the capacity of a wall may actually be increased by the addition of load, an example of which would be a wall subjected to a substantial wind load. The vertical load on such a wall provides a degree of prestress and increases the capacity of the wall to withstand the wind. In this case, the wall's ability to withstand the wind should be checked under the lower value of dead load $(0.9 G_k)$.

7.2.3 Characteristic compressive strength of masonry

Appendix A2 specifies a test procedure for determining the characteristic strength of masonry, from which it will be seen that the characteristic strength of a particular type of masonry is that strength obtained for a panel of the masonry 1.2 to 1.8 m wide and from 2.4 to 2.7 m in height. The load should be axial and the top and bottom of the panel are restrained against rotation. In the absence of a value obtained from such a test, the Code gives charts and tables which may be used to estimate a value appropriate to the particular units and mortar employed. Interpolation is required between these tables for particular sizes of block and so, to simplify this procedure, Table 7.1 gives values of the design ultimate strength of masonry for common types and thickness of block.

Occasionally it may be convenient to build a wall of hollow blocks and then fill the cavity with concrete. The assessment of an appropriate characteristic strength in this case is dealt with in Clause 23.1.7. This states that, provided the infill concrete is at least as strong as that in the blocks, the blocks may be treated as solid and an appropriate characteristic strength obtained using Code Tables (b) and (d). For this purpose, the characteristic strength of the concrete in the block is assessed using the net cross sectional area of the block rather than the gross area of the block as is normally used when assessing block strength.

When blocks are laid flat in a wall rather than in the normal orientation it is necessary to derive a suitable characteristic strength for the wall. This is a common form of construction in Ireland and the Irish Standard IS 325: Part 1: 1986[7.6], which is based on BS 5628 Part 1, contains in Table 2 characteristic compressive strength for masonry when the blocks are laid flat. As an alternative the recommendation produced by the Cement and Concrete Association as a result of their test programme may be used. This suggested that for 100 mm thick blocks laid flat the values for characteristic compressive strength of the masonry determined from Table 2(d) of BS 5628: Part 1 should be reduced by a factor ranging from 0.84 for 10 N/mm² blocks to 0.70 for blocks of 30 N/mm². The block strength used in conjunction with 2(d) should be for blocks tested in the normal direction.

Walls have occasionally been built in stack bond. This in fact, lies outside the scope of BS 5628 since BS 5628 Part 3[7.7] requires that the horizontal distance between joints in successive courses should not be less than a quarter of the unit length. Nevertheless, there may be reasons why the use of stack bond is desired and test work has been carried out in America and the National Masonry Association of America have developed recommendations for its use[7.8]. In principle, this report suggests the following:

(1) the vertical loadbearing capacity may be considered to be the same as for normal masonry except that concentrated loads should be considered to be carried only by the tier of masonry on which the load acts;

(2) the flexural strength across the horizontal joints can be considered to be the same as in normal masonry;

(3) the flexural strength across the vertical joints should be considered to be zero. To compensate for this, horizontal reinforcement should be provided.

The report gives recommendations as to amounts and arrangements of steel but it would seem reasonable to ensure sufficient steel in the joints to provide the same vertical flexural strength as given in Table 7.3 for the particular block used.

Hollow blocks are sometimes laid with mortar on the two outer strips of the blocks only. This is referred to as shell bedding. Clause 23.3 states that, where this is done, the appropriate value of f_k is obtained in the usual way from Tables 2(b) and (c) of the Code but that the design strength is then reduced by the ratio of the bedded area to the *net* area of the block.

7.2.4 Design strength

The design strength of masonry is obtained from the characteristic strength by dividing it by a material partial safety factor, γ_m. This factor makes allowance for variations in quality of the materials and differences between the strength of masonry constructed under site conditions and those built under laboratory conditions. Bearing this in mind, it is to be expected that a higher degree of quality control in the manufacture of the units will permit the use of a lower partial safety factor and similarly for construction control. A proportion of this safety factor must allow for the approximate nature of the tables given in the Code for determining characteristic masonry strength as a function of unit and mortar strength. Consequently, one would expect some reduction if the characteristic strength were determined directly. Hence the 10% reduction permitted when the Appendix A2 test has been used. If the special category of unit control and construction control are employed and if the characteristic strength has been obtained by test, a 36% increase in design strength can be obtained.

Table 7.1 Design ultimate strength of masonry (kN/m)

Mortar designation		Compressive strength of unit (N/mm²)							Block type
		2.8	3.5	5.0	7.0	10.5	14.0	21.0	
(a)		Block height = 215		Block thickness = 100		Partial safety factor = 3.5			
	(I)	80	100	143	194	261	325	438	solid
		80	100	143	163	176	190	222	hollow
	(II)	80	100	143	183	246	290	377	solid
		80	100	143	157	164	172	191	hollow
	(III)	80	100	143	183	239	275	342	solid
		80	100	143	154	158	162	174	hollow
	(IV)	80	100	126	160	205	241	305	solid
		80	100	126	137	141	145	155	hollow
(b)		Block height = 215		Block thickness = 140		Partial safety factor = 3.5			
	(I)	93	116	167	227	304	379	512	solid
		93	116	167	197	225	253	309	hollow
	(II)	93	116	167	214	288	339	440	solid
		93	116	167	189	211	228	266	hollow
	(III)	93	116	167	214	280	322	399	solid
		93	116	167	187	203	215	242	hollow
	(IV)	93	116	147	187	240	282	256	solid
		93	116	147	165	179	191	216	hollow
(c)		Block height = 215		Block thickness = 190		Partial safety factor = 3.5			
	(I)	105	129	187	255	342	425	574	solid
		105	129	187	232	281	329	418	hollow
	(II)	105	129	187	240	323	380	494	solid
		105	129	187	221	263	295	360	hollow
	(III)	105	129	187	240	314	261	448	solid
		105	129	187	219	255	279	327	hollow
	(IV)	105	129	165	210	269	316	400	solid
		105	129	165	193	222	246	292	hollow
(d)		Block height = 215		Block thickness = 215		Partial safety factor = 3.5			
	(I)	111	136	197	269	360	449	606	solid
		111	136	197	249	308	366	473	hollow
	(II)	111	136	197	253	340	401	521	solid
		111	136	197	237	290	329	407	hollow
	(III)	111	136	197	253	331	381	472	solid
		111	136	197	235	281	311	369	hollow
	(IV)	111	136	174	221	284	333	422	solid
		111	136	174	207	244	274	330	hollow

7.2.5 Effective height of wall

The concept of effective height appears to have been adopted almost universally for the treatment of instability effects in design. In an earlier section consideration was given to the failure by buckling of a pinned ended column as sketched in Figure 7.12(a). Analysis of a fixed ended column gives points of zero moment at the quarter points. The centre half of the column is thus behaving exactly as the pinned ended column and will fail at the same load as a pinned ended column of height *l*/2. A cantilever may be considered as half of a pinned ended strut; its strength will thus be the same as a pinned ended strut of length 2*l*. Obviously, a convenient way of dealing with these in design is to treat them all as pinned ended struts but with effective lengths chosen to given the correct failure load for the particular end conditions. The 'effective' lengths are thus:

Pinned ended strut	*l*
fixed ended strut	*l*/2
cantilever strut	2*l*

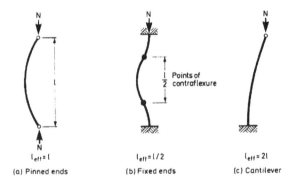

Figure 7.12 Buckling of various struts: (a) pinned ends (b) fixed ends (c) cantilever

Actual structural situations differ from these ideal structures; the tops and bottoms of columns are not

Table 7.1 continued

Mortar designation	2.8	3.5	5.0	7.0	10.5	14.0	21.0	Block type
(e)	Block height = 190		Block thickness = 90		Partial safety factor = 3.5			
(I)	72	90	129	175	235	292	394	solid
	72	90	129	147	159	171	200	hollow
(II)	72	90	129	165	222	261	339	solid
	72	90	129	141	148	155	172	hollow
(III)	72	90	129	165	215	248	308	solid
	72	90	129	139	142	146	156	hollow
(IV)	72	90	113	144	185	217	275	solid
	72	90	113	123	127	130	140	hollow
(f)	Block height = 190		Block thickness = 100		Partial safety factor = 3.5			
(I)	77	96	138	187	251	313	422	solid
	77	96	138	158	173	188	222	hollow
(II)	77	96	138	176	237	280	364	solid
	77	96	138	152	161	170	191	hollow
(III)	77	96	138	176	231	266	330	solid
	77	96	138	150	155	160	173	hollow
(IV)	77	96	121	154	198	233	294	solid
	77	96	121	133	138	143	155	hollow
(g)	Block height = 190		Block thickness = 140		Partial safety factor = 3.5			
(I)	86	107	154	210	281	350	473	solid
	86	107	154	186	217	248	309	hollow
(II)	86	107	154	197	266	313	407	solid
	86	107	154	178	203	224	266	hollow
(III)	86	107	154	197	258	297	369	solid
	86	107	154	176	196	211	241	hollow
(IV)	86	107	136	173	221	260	329	solid
	86	107	136	155	172	187	216	hollow
(h)	Block height = 190		Block thickness = 190		Partial safety factor = 3.5			
(I)	98	120	174	237	318	396	535	solid
	98	120	174	220	273	324	418	hollow
(II)	98	120	174	223	301	355	461	solid
	98	120	174	209	256	290	360	hollow
(III)	98	120	174	223	292	336	417	solid
	98	120	174	208	248	275	326	hollow
(IV)	98	120	154	195	251	295	373	solid
	98	120	154	183	216	242	291	hollow

totally rigidly restrained and their effective lengths will exceed *l*/2. A common assumption is to take 0.75 times the actual height.

A failure of slender masonry walls is not strictly by buckling in the classical sense; nevertheless, the concept of effective height remains useful. Clause 28.3.1.1 defines the effective height as either

(1) ¾ of the clear distance between lateral supports where some rotation restraint exists; or
(2) the clear distance between lateral supports where the restraint is only to lateral movement and not to rotation.

The question remains as to what constitutes a lateral support.

Consider the buckling of an external wall (see Figure 7.13). If the tie to the intermediate floor were to break, the buckling mode would become as shown in Figure 7.13(b). The effective height is doubled and the capacity of the wall is reduced catastrophically, and a tie of sufficient strength must be provided to ensure that this cannot happen. Clause 28.2.1 states that a lateral support must

be able to carry a horizontal force of 2.5% of the design vertical load on the wall in addition to any horizontal forces induced by the design loads (e.g. wind suction). This force would develop if the upper and lower walls were both 1.25% out of plumb (i.e. 40 mm in 3 m height), which should cope with the worst that is likely to arise in practice. Details of what will constitute a simple lateral support (giving no rotation restraint) and an enhanced lateral support which can be assumed to provide some restraint to rotation are given in Clauses 28.2.2.1 and 28.2.2.2 respectively.

In assessing the effective height of short lengths of wall where the length is less than or equal to four times the thickness, the rules in Clauses 28.3.1.2 or 28.3.1.3 for columns should be used rather than the provisions of 28.3.1.1. In the case of cavity walls, the thickness should be taken as the thickness of the loaded leaf.

7.2.6 Effective thickness

In the same way that it is convenient to reduce all buckling problems to equivalent pinned ended struts by

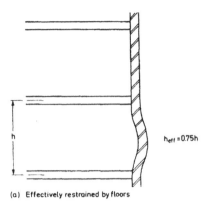

(a) Effectively restrained by floors

$h_{eff} = 0.75h$

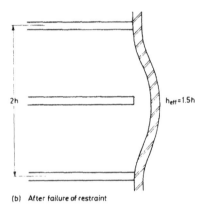

(b) After failure of restraint

$h_{eff} = 1.5h$

Figure 7.13 Buckling of external wall: (a) effectively restrained by floors (b) after failure of restraint

using effective heights, it is convenient to reduce problems to consideration of solid rectangular sections by introducing an effective thickness. For solid walls or piers, the effective thickness equals the actual thickness. Clause 28.4 gives values of effective thickness for walls stiffened by piers and cavity walls.

The precise derivation of these factors is unclear but the basic principles can be illustrated without difficulty. The buckling strength of a wall depends upon its stiffness and a convenient measure of stiffness is the moment of inertia of a section. Consider a wall stiffened with piers spaced at a centre to centre distance of six times the pier width (top line of Table 5 in the Code) as shown in Figure 7.14. If $t_p = 2t$, the moment of inertia of the section is given as $0.03 \times 6l_p \times 8t^3$. The moment of inertia of an equivalent rectangular section is given by $\frac{6 \ l_p \ t'_{ef}}{12}$.

Equating these gives $t_{ef} = 1.42 \ t$ which is (to one decimal place) the same as given in Table 5. If the same operation is carried out for $\frac{t_p}{t} = 3$, then $\frac{t_{ef}}{t} = 2$; again as given in Table 5. However, if these calculations are carried out for the case where the pier spacing is 10 or 20 times the pier width, then larger values than those given in Table 5 will be obtained. This indicates that while a pier can fully stiffen a length of wall equal to six times its width, it is

Figure 7.14 Wall with pier

not fully effective when the spacing is larger. By the time the pier spacing reaches 20, large areas of wall between the piers must be considered to be unstiffened.

The effective thickness of a cavity wall can only be obtained empirically as it will depend upon the effectiveness of the ties in inducing some composite action between the two leaves.

7.2.7 The strength of short walls under eccentric loads

Under axial loading, the design strength of a wall which is sufficiently short for there to be a negligible deflection under ultimate loads is simply the characteristic strength of the wall divided by the appropriate partial safety factor. Once the load is applied eccentrically, however, the load capacity is reduced. Appendix B of BS 5628 indicates that the reduced capacity may be calculated on the assumption that, at ultimate load a plastic distribution of stress will act over the whole compression zone. The assumptions are illustrated in Figure 7.15.

In Figure 7.15

e_p = the eccentricity of the load
n_v = the vertical load per unit length of wall
t = the thickness of the wall
f_k = the characteristic strength of the masonry
γ_m = the partial safety factor

It will be seen that the vertical load must act at the centre of the compression block and hence the neutral axis depth is given by:

$$x = t \left(1 - \frac{2e_m}{t} \right) \qquad (5)$$

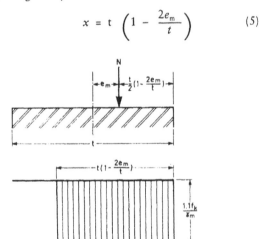

Figure 7.15 Stress block under ultimate conditions

By equilibrium of forces:

$$n_\gamma = \frac{1.1f_k}{\gamma_m} x = \frac{1.1f_k t}{\gamma_m}\left(1 - \frac{2e_m}{t}\right). \qquad (6)$$

For very small eccentricities this equation will give a value of n_γ greater than the axial capacity. This is clearly not permissible and so an upper limit on the value of the load of $\frac{f_k t}{\gamma_m}$ must be applied.

If this is substituted into equation (5) we get, for the limiting case:

$$\frac{f_k t}{\gamma_m} = \frac{1.1f_k t\left(1 - \frac{2e_m}{t}\right)}{\gamma_m} \qquad (7)$$

This gives a limiting value of $\frac{e_m}{t} = 0.045$ (the Code rounds this to 0.05). Below this value of eccentricity, the capacity of the section is unaffected by the eccentric loading. Above this eccentricity, the capacity will be reduced to that given by equation (5). Writing $n_\gamma = ß n_\gamma$ axial, it will be seen that:

$$ß = 1.1\left(\frac{1 - 2e_m}{t}\right)$$

This gives the following values for ß as a function of $\frac{e_m}{t}$.

Table 7.2 Values of ß as a function of e_m/t

Ratio	eccentricity / wall thickness	Capacity reduction factor
	0.05	1.00
	0.10	0.88
	0.15	0.77
	0.20	0.66
	0.25	0.55
	0.30	0.44
	0.35	0.33
	0.40	0.22
	0.45	0.11
	0.50	0.00

When the wall is sufficiently slender for the deflection at its centre to have a significant influence on the central eccentricity, there will be a further reduction in load capacity. The derivation of equations for estimating the increase in eccentricity due to deflection of the wall was given in *Section 7.1.5*. Appendix B gives the relationship:

$$e_a = t\left[\frac{1}{2400}\left(\frac{h_{ef}}{t_{ef}}\right)^2 - 0.015\right] \qquad (8)$$

and, as discussed in Section 7.1.5, the eccentricity in the middle region of the wall is assumed to be:

$$e_t = e_a + 0.6e_x \qquad (9)$$

where e_x is the eccentricity at the top of the wall. Clearly, if e_a is small, e_t can be smaller than e_x. If this is the case, then, obviously e_x governs. The design eccentricity e_m, is thus the greater of e_t and e_x. The capacity of the wall has to be estimated for this eccentricity.

If a value of $\frac{h_{ef}}{t}$ of 6 is used, it will be seen that e_a will be calculated to be zero. Thus, slenderness effects are assumed in the Code to be zero for slenderness ratios of 6 or less.

Equations (7), (8) and (9) may be used to derive the capacity reduction factors in Table 7 of the Code. A typical example will show how this is achieved. Assume a slenderness ratio of 12. Substituting this into (8) gives an additional eccentricity in the central region of $0.045t$. If the eccentricity at the top is $0.1t$, then (9) gives the total eccentricity in the central region as $0.105t$. This is greater than the eccentricity at the top and hence $\frac{e_m}{t} = 0.105$. For this eccentricity, equation (7) gives a value of β of 0.869. If the eccentricity at the top is $0.3t$, then the central eccentricity will be found to be $0.225t$. Since this is less than the eccentricity at the top, $\frac{e_m}{t}$ becomes the top value of 0.3. For this, equation (7) gives a β value of 0.44. These values can be seen to be the same as are given in Table 7 of the Code for the particular eccentricities and slenderness ratios chosen.

From the design point of view then, calculation of the strength of a vertically loaded wall can be achieved using the formula

$$\text{Vertical load resistance} = \frac{\beta b t f_k}{\gamma_m}$$

Columns are tackled in exactly the same way as walls except that it may be necessary to consider the effects of bending about both axes. Where the eccentricities are small about either or both axes (less than $\frac{1}{20}$ of the section size) the strength can be checked by assuming a value of ß from Table 7 of the Code on the basis of the larger ratio of eccentricity to section dimension in that direction and the slenderness ratio appropriate to the minor axis. Where both eccentricities exceed $\frac{1}{20}$ of the section dimension, the Code states that a capacity reduction factor must be calculated using the assumption made in Appendix B. The exact procedure envisaged by the Code in this latter situation is not set out but the following would appear satisfactory:

(1) for the eccentricity about each axis in turn calculate the additional eccentricity using equation (8), assess the design eccentricity as either the initial eccentricity at the top or that derived from equation (9), whichever is the greater;

(2) by trial and error, define a neutral axis position so that the centroid of the compression zone coincides with the centroid of the applied load. The vertical load capacity of the section is then given by:

$$N_u = \frac{1.1\,A_c f_k}{\gamma_m}$$

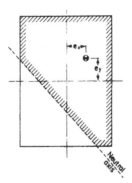

Figure 7.16 Biaxial bending of a column section

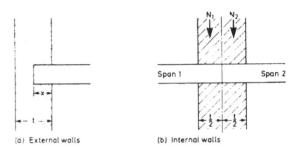

(a) External walls   (b) Internal walls

Figure 7.18 Assessment of eccentricities: (a) external walls (b) internal walls

7.2.8 Assessment of eccentricity

Practice in the assessment of eccentricity is summarized in Building Research Establishment Digest No 246[7.9]. The introductory paragraph of the section of this digest dealing with eccentricity is worth quoting in full:

'The assessment of eccentricity of vertical loading at a particular junction in a brickwork or blockwork wall is made complex by the many possible combinations of wall and floor systems. The structural behaviour of the junction will be affected by the magnitude of the loads, the rigidity of the interconnected members and the geometry of the junction. It is seldom possible to determine the eccentricity of load at a junction by calculation alone, and the designer's experience must be his guide.'

The advice given may be summarised as follows:

(1) bearing on external walls

In general eccentricity $= \dfrac{t}{2} + \dfrac{x}{3}$

where a concrete floor of short span spans the full width of the wall or the load is applied by a centred wall plate, it may be more reasonable to regard the load as axial.

(2) bearing on internal walls
   (a) span on either side not differing by more than 50%

   $$\text{eccentricity} = 0$$

   (b) spans differing by more than 50%
   Imagine the floor split along the centreline of the wall and treat each side as for an external wall (see Figure 7.18(b)). Treat each side as for (1) above to obtain eccentricities for N_1 and N_2. Combine these to obtain overall eccentricity.

(3) joints supported on hangers
   Load assumed to act 25 mm outside face of wall.

In all cases, the Code makes the assumption that the eccentricities apply at the top of the wall and that the eccentricity at the bottom is zero. If this is clearly unreasonable then the Code approach should be modified appropriately.

where

N_u = ultimate axial load
A_c = area of compression zone

(see Figure 7.16).

Provided the section is rectangular, this problem can be solved by the use of the design chart in Figure 7.17. To use the chart, calculate the ratio of the eccentricity to the section dimension in the minor axis and major axis bending directions. Using these two values, obtain a capacity reduction factor β from Figure 7.17. The vertical load capacity is then given by:

$$N_u = \beta \, \frac{btf_k}{\gamma_m}$$

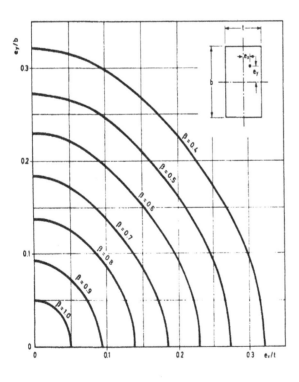

Figure 7.17 Design chart for biaxial bending of masonry piers

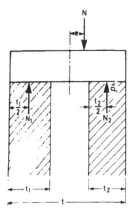

Figure 7.19 Cavity wall with both leaves loaded

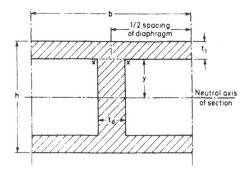

Figure 7.20 Diaphragm wall

7.2.9 Cavity walls with both leaves loaded

If, at the level where the load is applied, both leaves are spanned by some form of spreader which carries the applied load, each leaf may be designed to carry an axial load of magnitude such that, acting together, they equilibrate the applied load. This is illustrated in Figure 7.19.

From this Figure,

$$N_e = N_2 \left(\frac{t}{2} - \frac{t_2}{2} \right) - N_1 \left(\frac{t}{2} - \frac{t_1}{2} \right)$$

$$N = N_1 + N_2$$

These equations may be solved for N_1 and N_2.

Failure will occur as soon as one or other leaf reaches its capacity. This will commonly be the inner leaf of an external wall.

A modification to this procedure is required as soon as the line of action of the load passes outside the centroid of one of the leaves. When this occurs, the load is clearly carried by only the one leaf at an appropriate eccentricity.

7.2.10 Diaphragm Walls

In a diaphragm wall the two leaves of a double leaf wall are effectively connected by diaphragms which permit the whole wall to act as a cellular section. This is distinct from the normal cavity wall where the connection, via wall ties, does not provide sufficient connection to allow the two leaves to act compositely in resisting load.

Clearly, the fundamental requirement for a double leaf wall to behave as a diaphragm or cellular wall is that diaphragms should be provided at sufficiently close spacings and should be adequately bonded into the two leaves. Provided this is done, the stresses in the wall may be estimated on the basis of the properties of the whole section. If the spacing of the diaphragms is too large, areas of the wall between the diaphragms will become ineffective but, provided the spacing of the diaphragms is not greater than about twelve times the width of the thinner leaf, it can be assumed that the whole wall is effective. In the event that a wider spacing becomes

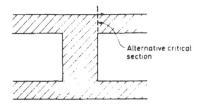

Figure 7.21 Alternative critical section for shear

essential, then a reasonable approach would seem to be to assume that any part of the wall at a distance further than six times the leaf thickness from the centre line of the nearest diaphragm is incapable of resisting load.

The vertical shear stress at the junction between the diaphragms and the leaves of the wall may be calculated using the standard elastic formulae. Thus, the vertical shear stress is given by:

$$v_v = \frac{V_h S}{It} \qquad (10)$$

where

v_v = vertical shear stress on section considered
V_h = the horizontal shear force acting on the length of wall equal to the spacing of the diaphragms
I = second moment of area of total section
t = thickness of section on which shear is being checked
S = first moment of area of the section of wall to one side of the section considered about the neutral axis of the total section.

(See Figure 7.20 for clarification.)

Thus, for the section sketched in Figure 7.20, the stress across the junction between the diaphragm and the leaf is given by:

$$v = \frac{V_h b t_1 \ (h - t_1)}{2 \ It_d}$$

If the diaphragm is thick compared with the thickness of the leaves, the vertical shear should be checked in the leaf rather than at the joint (see Figure 7.21). Obviously, this is only likely to be critical where the diaphragm is more than twice the thickness of the leaves.

It is suggested that the vertical shear force in the walls should be limited to $\frac{0.7}{\gamma_m}$ N/mm². This is twice the value given in Clause 25 for shear across a bed joint with mortar classes (i), (ii) or (iii) and, in the absence of better information, seems a reasonable value.

Assuming that the diaphragms are not bonded into the leaves, the vertical shear across the joint between them should be carried by the provision of ties.

In the first edition of this handbook one of two approaches was suggested to design the wall to resist vertical and horizontal loads. Firstly, it can be designed entirely elastically so that the tensile stress does not exceed half the design flexural stress and the compressive stress does not exceed 0.8 of the design compressive stress $\left(0.8\dfrac{f_k}{\gamma_m}\right)$. Thus:

$$\frac{N_{u\,max}}{A_c} - \frac{N_{u\,max}ey_1}{I} \leq 0.8\frac{f_k}{\gamma_m} \tag{11}$$

$$\frac{N_{u\,min}}{A_c} - \frac{N_{u\,min}ey_2}{I} \geq \tfrac{1}{2}f_t \tag{12}$$

where

y = distance from the neutral axis to the face of the wall

A_c = cross-sectional area of wall

If walls are designed in this way, it may be assumed that deflections under the design loads are negligible and thus there will be no slenderness effects.

Secondly, they can be designed as cracked sections and thus unable to carry tensile stresses. In this case, the vertical load may be assumed to be resisted by a stress block as shown in Appendix B of BS 5628. Slenderness effects have to be allowed for and a capacity reduction factor can be estimated using equation (1) from Appendix B and equation (2) modified to:

$$e_t = 0.6\,e_{xt} + e_a + e_{lat}$$

where e_{lat} is the eccentricity at mid height resulting from the lateral loading. Equation (3) will require re-definition as follows (see Figure 7.22).

(i) by moments about section centroid:

$$(b - t_d)\left(y_1 - \frac{t_1}{2}\right)t_1 + xt_d\left(y_1 - \frac{x}{2}\right)$$
$$+ \left[\tfrac{1}{2}(t_2 - h + x)(b - t_d)(2y_2 - t_2 - h + x)\right]$$
$$= e[(b - t_d)t_1 + t_dx + [(t - h + x)(b - t_d)]]$$

Solve equation (1) for x. If $x < y_2 - t_2$ (i.e. x is not within leaf) delete terms in square brackets and solve again for x. If $x < t_1$ (i.e. compression zone is entirely in one leaf) design as single wall of thickness t. Having found x, the area of the compression zone A_c is found and the vertical load capacity is given as:

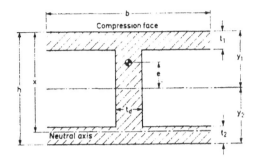

Figure 7.22 Definitions used in ultimate load equations

$$n_y = \frac{1.1\,A_c f_k}{\gamma_m}$$

Using Equation (1) in Appendix B, the effective thickness of the wall may be taken as the overall wall thickness.

Subsequently a number of design approaches have been promulgated by Curtin et al[7.10, 7.11, 7.12], Phipps et al[7.13,7.14], Beck [7.15], and the Institution of Structural Engineers[7.16]. An amendment has now been prepared for BS 5628: Part 1 which clarifies the approach to be taken when designing in accordance with this code.

The amendments to BS 5628: Part 1: 1992 were agreed by the responsible British Standards Institution Committee B/525/6/WG100 and finalized on 23 November 1999. The planned amendments to Clause 36.9.1, which deals with Propped Cantilever Wall Design, are as follows:

(a) The design flexural tensile stress in the wall (other than in the vicinity of the lower support) should not exceed the appropriate value of design flexural strength (clauses 24 and 27). At the lower support the design moment of resistance due to gravity forces should be assessed using factors of safety of 0.9 on the dead load and 1.4 on the wind load.

(b) The wall should be stable when its strength is derived from gravity forces only (i.e. when the flexural tensile strength of the masonry is ignored) with a factor of safety of 1.0 on the dead and wind loads.

(c) The design stress in the compressive zone of the wall should not exceed the design compressive strength.

7.2.11 Non-standard eccentricities of load

As stated earlier, the assumption has been made in the derivation of the capacity reduction factors that the eccentricity used in design is present at the top of the wall reducing to zero at the bottom. In practice, the eccentricity at the bottom of the wall will commonly be in the opposite direction to that at the top. As far as consideration of slenderness effects are concerned, it is the eccentricity in the central region of the wall which is required and thus to assume opposite eccentricities at top and bottom will lead to a smaller central eccentricity and a larger load capacity. Since the precise behaviour of the

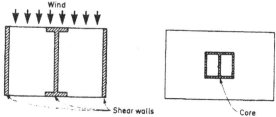

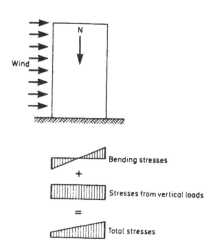

Figure 7.23 Shear walls or cores

Figure 7.24 Stresses in a shear wall

bottom of walls is difficult to define, it would be inadvisable to assume that the eccentricities changed sign unless the wall was constructed in such a way as to make this inevitable. If, however, the nature of the structure was such that the eccentricity at the top and bottom were in the same direction, then it would be advisable to make allowances for this since the capacity of the wall will be reduced. The required modification to the design procedure is to change equation (2) in Appendix B from:

$$e_t = 0.6\,e_x + e_a$$

to:

$$e_t = e_a + e_{min} + 0.6\,(e_{max} - e_{min})$$

where

e_{max} is the numerically larger eccentricity
e_{min} is the numerically smaller eccentricity
e_{max} and e_{min} have the same sign if they are in the same direction

Suitable values of β may now be calculated using equations (1) (3) and (4) from Appendix B together with the modified equation (2).

7.2.12 Walls with eccentricities in their own plane (shear walls)

The overall stability of masonry structures subjected to wind loads will commonly be provided by walls or groups of walls acting as 'shear walls' or 'cores' (see Figures 7.23 and 7.24).

In principle these act as cantilevers in resisting the horizontal forces shown in Figure 7.25.

In most cases stress can be checked on this basis.

$$\text{Stress at base of wall} = \frac{N_y}{A} \pm \frac{M_w}{I}\,y$$

where

N_y = vertical load
A = cross sectional area of wall
M_w = total wind moment
I = moment of inertia of wall section
y = distance form centroid to point considered

This approach will work for single walls or groups of connected walls. Where wind loads are resisted by a series of separate walls, the total wind forces will be carried by the various walls in proportion to their relative stiffness. Thus the wind moment carried by a particular wall is given by:

$$M_{wi} = M_w\,\frac{I_i}{\Sigma I}$$

M_{wi} = wind moment carried by $i\,th$ wall
M_w = total wind moment
I_i = moment of inertia or $i\,th$ wall
ΣI = sum of moments of inertia of all walls assumed to be resisting wind moments

Possible difficulties can arise where a wall is pierced by a series of openings so that it may not be clear whether it can be considered as a single unit or whether it should be considered as separate walls. Consider the wall illustrated in Figure 7.26. Clearly, the actual conditions will lie between those illustrated in Figure 7.25 (b) and (c). If the openings are relatively small, assumption (b) will be closest while, if they are larger (a) will be more appropriate. For walls with a single line of openings, Pierce and Matthews[7.17] give a simplified design approach which should be applicable to masonry structures. The wind moments are considered in two parts

$\beta_1 M_w$ is carried by the two parts of the wall acting independently (as in Figure 7.25(c))
$(1 - \beta_1)M_w$ is assumed to be carried by the wall acting as fully connected wall (see Figure 7.25(b))

The stress can then be found by summing the stress obtained from the two calculations. β_1 is given by the following relationship

$$\beta_1 = \frac{2}{\alpha H}\left(\frac{\alpha H - 1}{\alpha H}\right)$$

where

$$\alpha = \sqrt{\frac{12 I_b}{h b_{eff}^3}\left(\frac{I_g}{I_A + I_B} + \frac{1}{A_A} + \frac{1}{A_B}\right)}$$

H = overall height
I_b = moment of inertia of linking beam

71

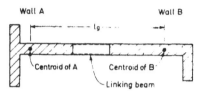

Figure 7.26 Pierced shear wall – definitions

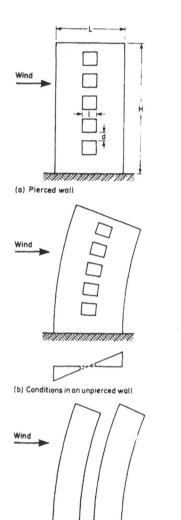

(a) Pierced wall

(b) Conditions in an unpierced wall

(c) Conditions in wall if openings produce a complete break

Figure 7.25 Pierced shear walls: (a) pierced wall (b) conditions in an unpierced wall (c) conditions of wall if openings produce a complete break

h = storey height

b_{eff} = effective breadth of opening = $l + d$ in Figure 7.25(a)

l_g = distance between centroids of the two parts of the wall (designated as A and B)

I_A, I_B respectively the moments of inertia of parts A and B

A_A, A_B respectively the areas of parts A and B

If $\alpha H > 16$ then the wall should be considered fully coupled (i.e. $\beta_1 = 0$).

If $\alpha H < 4$ the two halves should be considered to act independently.

Calculation of ß for $\alpha H = 4$ does not give $\beta_1 = 1$ which this would suggest, but Pierce and Matthews say at this end of the range the result is so sensitive to minor changes in the parameters that it is advisable to take $\beta_1 = 1$ for $\alpha H < 4$.

The calculation outlined above is tedious and a further simplification has been made. Figure 7.27 gives values of ß$_1$, as a function the parameters $\frac{d}{h} \times \frac{l}{L}$ and $\frac{h}{L}$. This Figure is an approximation to the approach given above but is probably adequate for most design purposes.

The maximum and minimum vertical stress can now be assessed as:

$$f = \frac{P}{A_A + A_B} \pm \frac{ß_1 My_1}{I_A \text{ or } I_B} \pm \frac{(1 - ß_1)\, My_2}{I'}$$

I' is the moment of inertia of A and B acting together

Having found the maximum stress, this can be compared with the design stress, making due allowance for the capacity reduction factors in Table 9. In assessing the eccentricity for use with Table 9, the forces resulting from the wind can be assumed to be acting axially. Appropriate assumptions should be made about the eccentricities of loads applied by floors supported by the wall.

The problem which remains is the stress condition which may arise in the connecting beams in pierced shear walls. Pearce and Matthews give the following formula for the maximum shear force carried by a connecting beam:

$$Q_{max} = \beta_2 P_w \, \frac{h}{\mu l_g} \tag{13}$$

where

$$\mu = 1 + \frac{(I_A + I_B)\,(A_A + A_B)}{A_A\, A_B\, l_g^2}$$

P_w = wind force carried by wall and
β_2 is obtained from Table 7.3.

Table 7.3 Values of coefficient β_2 as a function of β_1

β_1	β_2
.1	.79
.15	.71
.2	.54
.25	.58
.3	.51
.35	.44
.4	.40

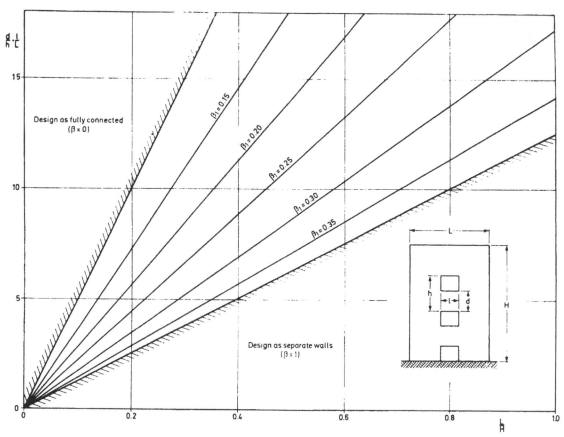

Figure 7.27 Design chart for shear walls

In most cases it will be sufficiently accurate to take μ as equal to 1.15.

The shear stress in the connecting beam should not exceed $\dfrac{0.2 \sqrt{f_k}}{\gamma m}$

7.3 Examples

7.3.1 Combined vertical and lateral load

BS 5628: Clause 36.8 gives two possible approaches to check the lateral load capacity of a wall carrying significant vertical loads. The most direct method is to use the equation given in the clauses:

$$q_{lat} = \frac{4\,tn}{h^2_a}$$

where

q_{lat} = design lateral strength per unit area
n = axial load per unit length of wall
t = wall thickness
h_a = clear height of wall

This equation is based on arch action and the derivation can be seen to be approximately as follows:

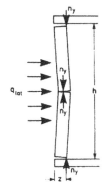

taking moments about the centre of the wall,

$$\frac{q_{lat}\,h^2_a}{8} = nz$$

z is the distance between the lines of action of the vertical load at the centre and at the ends.

Assuming z to be roughly equal to t and introducing a safety factor of z gives:

$$q_{lat} = \frac{4\,tn}{h^2_a}$$

The second approach is by consideration of the effective eccentricity due to the lateral load and any other eccentricity using Clause 32 and Appendix B. No indication of how this should be attempted is given so there may be ways other than the approach suggested below.

73

There are three sources of eccentricity and the distribution of eccentricity up the height of the wall needs to be assessed for each.

(i) Vertical loads

The eccentricity at the top of the wall is assessed in the normal way and is assumed to reduce linearly from this value to zero at the foot of the wall, i.e.

(ii) Additional moments due to slenderness effects

e_a can be calculated from equation 1 in Appendix B. The distribution of this eccentricity up the wall is open to interpretation. Appendix B states that the value may be taken as zero at top and bottom and e_a over the middle fifth of the height. However, this is derived from consideration of vertical load only. Consideration of buckling behaviour suggests that it would be more realistic to assume zero at points of contraflexure, giving the distribution sketched below:

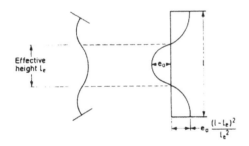

(iii) Eccentricity due to lateral load

The problem here is to assess the distribution of moment over the height of the wall. Provided there is some reasonable degree of axial load acting it seems unreasonable to assume the wall to be pinned top and bottom; the ends will have some moment capacity. A reasonable assumption for most cases would seem to be to assume the distribution as follows:

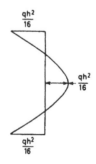

The eccentricity can be found by dividing these moments by the axial load, i.e.

$$e_q = \frac{q\,h^2}{16n}$$

The total eccentricities can now be obtained by addition. The additional eccentricity due to slenderness effects is applied in the direction which will produce the worst effect.

The wall strength can now be checked by applying equation 3 in Appendix B to the section with the greatest eccentricity.

7.3.2 Combined vertical and horizontal load

For convenience this example is based on the same structure as that in *Problem* 2 on vertical load. Also, so that the vertical loads previously assessed can be used, the example will consider the wall between ground and first floor, though in practice this might not necessarily be the most severe case and other floors would also have to be checked.

PROBLEM

Check that 10.5 N/mm² blocks are satisfactory for the wall between ground floor and first floor if the characteristic wind load is 0.55 kN/m² over the whole height of the building.

SOLUTION

To check that 10.5 N/mm² blocks are satisfactory for the wall between ground and first floor is the characteristic wind load is 0.55 kN/m² over the whole height of the building, the structure and loading are the same as used in *Problem* 2 on vertical loads *(Section 7.3.3)*.

Clause 36.8 gives two methods of carrying out the design:

(a) by using the formula provided in Clause 36.8, or
(b) by consideration of effective eccentricities and the principles of Clause 32 and Appendix B.

Both methods will be used in this example.

METHOD (A)

$$q_{lat} = \frac{4 \times t \times n}{h^2_a}$$

using kN and m units

t = 140 mm = 0.14 m
h = 2.9 m

The design vertical load for this load case, $(0.9\,G_k + 1.4\,W_k) = 0.9 \times 76.4 = 68.8$ kN/m

$$\therefore q_{lat} = \frac{4 \times 0.14 \times 68.8}{2.9^2} = 4.58 \text{ kN/m}^2$$

the required capacity = $1.4 \times 0.55 = 0.77$ kN/m²
∴ design is OK.

As before, the design vertical load is 68.8 kN/m.

The connection with the slabs above and below will provide some moment restraint, therefore assume the distribution of lateral moment is as sketched below:

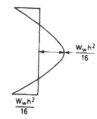

$$\frac{W_w h^2}{16} = \frac{1.4 \times 0.55 \times 2.9^2}{16} = 0.40 \text{ kN/m}$$

This moment corresponds to an eccentricity of:

$$\frac{0.4 \times 10^6}{68.8 \times 10^2} \text{ mm} = 5.88 \text{ mm}$$

The eccentricity of the load at the top of the wall can be assessed as follows:

(i) load from second, third floors and roof, axial
(ii) load from first floor at eccentricity of $\frac{t}{6}$ by moments

centre of wall,

$$\frac{0.9 \times 12 \times 140}{6} = 0.9 \times 57.15 \times e_t$$

$\therefore e_t = 4.9 \text{ mm}$
(57.15 load of roof + two floors + three heights of wall)

The eccentricity of the vertical load at the base of the wall can be taken at zero.

Additional moments

$$\frac{l_e}{t} = 14.18$$

hence, from equation 1 in Appendix B, the additional eccentricity developing due to slenderness effect is:

$$e_a = 140 \left[\frac{1}{2400} (14.18)^2 - 0.015 \right]$$

$= 9.63 \text{ mm}$

The Code does not define how the eccentricity varies up a wall in bending with some fixity but a reasonable assumption is as follows:

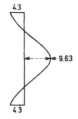

The total eccentricity can now be obtained by adding the additional eccentricities to the initial eccentricity and the eccentricity due to lateral load:

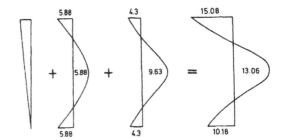

Hence maximum eccentricity = 15.08
Vertical load capacity
$\dfrac{f_k}{\gamma_m}$ for 10.5 N/mm² block = 1.84 N/mm²

$$= 1.1 \left(1 - \frac{2 \times 15.08}{140} \right) 140 \times 1.84$$

$= 222 \text{ kN/m}$

This exceeds 68.8 kN/m $\therefore$ wall is OK.
By inspection, load case (c) will be OK.

7.3.3 Vertical load

PROBLEM 1

Find the characteristic strength of masonry built from 5 N/mm² solid blocks 190 mm high by 140 mm thick using 1:1:6 mortar. What design load will an axially loaded short wall (i.e. no slenderness effects) made from this masonry carry per metre at the ultimate limit state?

PROBLEM 2

The sketch below illustrates part of a four storey building of loadbearing masonry. The floors and roof spanning onto the inner leaf of the wall shown have a span of 6 m. Carry out calculations for the load case of vertical loads only (load case (a) in Clause 22 of BS 5628) for the wall between ground and first floor. What strength of block will be needed?

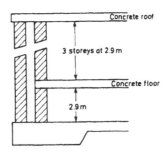

Loadings

Roof	Dead load	=	3.8 kN/m²
	Live load	=	0.75 kN/m²
Floors	Dead load	=	4.0 kN/m²
	Live load	=	3.5 kN/m²

Weight of inner skin of masonry assume 2.5 kN/m²
Assume 90 mm thick outer skin
   140 mm thick inner skin – block height 190 mm
   50 mm cavity
   designation (iii) mortar

PROBLEM 3

A precast floor unit is supported on a 140 mm wide block wall as sketched below. The wall is constructed of 140 mm thick modular blocks of 5 N/mm² strength. The wall is 2.9 m high and the beams on the wall at 0.75 m centres. If the design ultimate end reaction from the beams is 20.18 kN, check that the local bearing stresses are satisfactory.

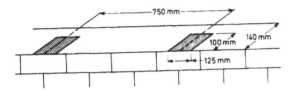

PROBLEM 4

The 12 m high structure sketched below is subjected to a wind load of 0.5 kN/m² over the whole height. Assuming that the walls are built of 190 mm thick solid blocks and that the characteristic dead and live loads at the base of the wall are as given below, check that a 10.5 N/mm² block will be satisfactory.

Loads in kN/m	Wall A	Wall B	Wall C
Dead	76.4	152.0	76.4
Imposed	33.8	67.5	33.8

Wind force coefficient = 1.0 (long face), 0.875 (short face)
Storey height 2.9 m

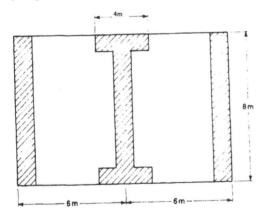

SOLUTION 1

Find the characteristic strength of masonry made of 5 N/mm² solid blocks 190 mm high by 140 mm thick using 1:1:6 mortar. What design load will a low wall made from this masonry carry per metre?

 To find the characteristic strength, f_k, it is necessary to interpolate between Tables 2(b) and 2(d) in BS 5628. A 1:1:6 mortar is a Type (iii) mortar as defined in Table 7.1.

The height to thickness ratio of the block is

$$\frac{190}{140} = 1.36$$

Table 2(b) is for blocks with $\frac{h}{t} = 0.6$

Table 2(d) is for blocks with $\frac{h}{t} \geq 2.0$

The characteristic strengths are given for those $\frac{h}{t}$ ratios as:

$$\frac{h}{t} = 0.6, \qquad f_k = 2.5$$

$$\frac{h}{t} \geq 2.0, \qquad f_k = 5.0$$

Hence, by linear interpolation, f_k for $\frac{h}{t} = 1.36$ can be found to be:

$$2.5 + \frac{(1.36 - 0.6)}{(2 - 0.6)} \ (5 - 2.5) = 3.86 \ \text{N/mm}^2$$

Assuming normal control on site and in the manufacture of the blocks, the partial safety factor should be 3.5 (From Table 4). The design ultimate strength is thus:

$$\frac{3.86}{3.5} = 1.10 \ \text{N/mm}^2$$

The load capacity of a metre of wall is thus:

$$\frac{1.10 \times 1000 \times 140}{1000} \ \text{kN} = 154 \ \text{kN/m}$$

A similar calculation can be done for the common sizes of block and the standard specified compressive strengths. This has been done in Table 7.4. In all following examples, the figures from Table 7.4 will be used as the basic data. It will be seen that this Table gives a value of 154 kN/m for a 140 mm thick 5 N/mm² modular block.

Table 7.4 BS 5268 Design strengths of blockwork (kN/m)
1:1:6 mortar
$\gamma_m = 3.5$

Type	Thickness	Characteristic block strength N/mm²				
		3	5	10.5	21	28
Metric	100	86	143	239	342	414
blocks	140	100	167	280	399	483
215 mm high	215	118	197	331	472	572
Modular blocks	90	77	129	214	308	372
	140	92	154	258	369	446
190 mm high	190	104	174	292	417	505

(i.e. this is the strength of an axially loaded wall where there are no effects due to slenderness)

SOLUTION 2

Design wall for bottom floor of the four storey building sketched below. The roof and floors span 6 m onto the wall.

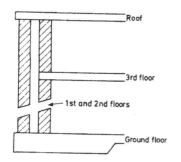

Loadings

Roof	Dead	3.78 kN/m²
	Live	0.75 kN/m²
Floors	Dead	4.0 kN/m²
	Live	3.5 kN/m²

Weight of inner skin of masonry – assume 2.5 kN/m²
Assume 90 mm thick outer skin
        140 mm thick inner skin – block height 190 mm
        50 mm cavity
        designation (iii) mortar.
Assess loading to ground floor level

Dead		
Roof	½ × 6 × 3.8 =	11.4
Three floors	3 × ½ × 6 × 4.0 =	36.0
Four storey heights of wall	4 × 2.9 × 2.5 =	29.0
Total kN/m		76.4

Live		
Roof	½ × 6 × 0.75 =	2.25
Three floors	3 × ½ × 6 × 3.5 =	31.50
Four storey heights of wall		0
Total kN/m		33.75

Design ultimate load
$= 1.4 \times G_k + 1.6 \, Q_k$
$= 1.4 \times 76.4 + 1.6 \times 33.75$
$= 161$ kN/m

Eccentricity of load

Building Research Station Digest No. 246 gives guidance on the assessment of eccentricity. For an end wall with a concrete floor supported over the whole thickness of the leaf, *BRS Digest 61* suggests that if the ratio of span to wall thickness is less than 30, the eccentricity of the load from the floor may be taken as zero but where the ratio of span to wall thickness exceeds 30, a value of ⅙ of the wall thickness will be more appropriate. In this case, $\frac{L}{t} = \frac{6000}{6} = 43$, hence $\frac{t}{6}$ is used. The load from the first floor is therefore taken at this eccentricity while the loads from the floors above are assumed to be axial.

Ultimate load from first floor

$$= 1.4 \times 12 + 1.6 \times 10.5 = 33.6$$

hence eccentricity of total load

$$= \frac{33.6 \times 1/6}{161} = 0.035 \, t$$

This is less than $0.05t$ which is the minimum design value.

Slenderness ratio

Effective thickness of cavity wall $= ⅔ (90 + 140)$
                                         $= 153.33$
effective height $= 0.75 \times 2900$ $= 2175$

$$\frac{h_{ef}}{t_{ef}} = \frac{2175}{153.3} = 14.18$$

∴ From Table 7 of BS 5628 ß = 0.89

Design equation

$$N = \frac{ß \cdot f_k \cdot t}{\gamma m} \qquad f_k = \frac{N \gamma_m}{\beta t}$$

Assume normal category of control of manufacture and construction

$$\gamma_m = 3.5$$

$$f_k = \frac{161 \times 10^3 \times 3.5}{0.89 \times 140 \times 10^3} = 4.52 \text{ N/mm}^2$$

Unit selection

Consider 190 × 140 mm solid unit with a designation (iii) mortar.
    Use Tables 2(b) and 2(d)

$$\text{Aspect ratio} = \frac{190}{140} = 1.36$$

Try 7.0 N/mm² unit

$$f_k = 3.2 + \frac{(6.4 - 3.2)}{1.4} \times (1.36 - 0.6)$$

$$f_k = 4.94 \text{ N/mm}^2 \text{ Therefore O.K.}$$

Use unit as selected.

SOLUTION 3
Concentrated load under bearing.*

   A precast floor unit is supported on a 140 mm wide block wall as sketched below. The wall is constructed from 5 N/mm² modular blocks. The wall is 2.9 m high. The beams are at 0.75 m centres.

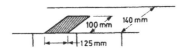

The end reactions from the beams are:
Dead 5.5 kN
Live 7.8 kN
The design ultimate reaction is thus:

$$5.5 \times 1.4 + 7.8 \times 1.6 = 20.18 \text{ kN}$$

*See *Chapter 10.*

The bearing stress is thus:

$$\frac{20.18 \times 1000}{125 \times 100} = 1.61 \text{ N/mm}^2$$

Inspection of Figure 4 in BS 5628 indicates that the bearing is Type 2. The local design stress is therefore $1.5 \frac{f_k}{\gamma_m}$. From Example 1, $\frac{f_k}{\gamma_m}$ for a 140 mm thick block of 5 N/mm^2 is 1.10 N/mm^2. The permissible bearing stress is thus 1.65 N/mm^2. The stress immediately below the bearing is thus OK.

A check should also be done to ensure that the stress at 0.4 times the wall height below the bearing is less than $\beta \frac{f_k}{\gamma_m}$ assuming a 45° spread of load. 0.4×2.9 m = 1.07 m.

However, since the beams are at 0.75 m intervals, the stresses will be uniform at a level of 0.75 m below the joint. A check on the whole wall for $1.6 Q_K + 1.4 G_K$ will have ensured that this stress is OK, no further check is therefore needed.

SOLUTION 4
Simple shear wall

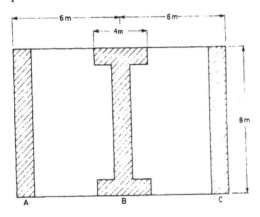

Structure is 12 m high, walls are of 140 mm thick solid blocks. Loading from floors and roof as in *Example 2*. The wind load may be assumed to be 0.5 kN/m^2 over the whole height.
Wind load on N face = $1.0 \times 12 \times 12 \times 0.5 = 72$ kN
Wind load on E face = $0.875 \times 8 \times 12 \times 0.5 = 42$ kN
These forces at 6 m from the ground.
Loads at base of walls (kN/m)

	A	B	C
Dead	76.4	152	76.4
Imposed	33.8	67.5	33.8

Consider wind from E or W. The only wall which will provide significant resistance is wall B.
Calculate second moment of area of B about N–S axis

$$= \frac{7.6 \times 0.19^3}{12} + \frac{2 \times 0.19 \times 4^3}{12} \text{ m}^4$$

$$= 0.004 + 2.026 = 2.03 \text{ m}^4$$

Consider load case (b):

0.9 G_k or 1.4 G_k + greater of 1.4 W_k or 0.015 G_k
1.4 W_k = 1.4 × 42 = 58.8 kN
0.015 G_k = 8 × (152 + 2 × 76.4) × 0.015
= 36.6 kN
Design horizontal load = 58.8 kN
Design vertical load = 0.9 × 152 × 8 = 1094 kN
or = 1.4 × 152 × 8 = 1702 kN

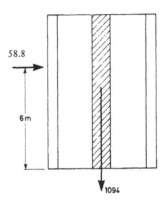

Length of wall = (7.6 + 4 + 4) = 15.6 m
∴ load on wall per metre = $\frac{1094}{15.6}$ = 70 kN/m

or $\frac{1702}{15.6}$ = 109 kN/m

Stress at outer edges = $\pm \frac{M_y}{I} + \frac{N}{A}$

$$= \pm \frac{4}{2} \cdot \frac{58.8 \times 6}{2.03} + \left(\frac{70}{0.19} \text{ or } \frac{109}{0.19} \right)$$

= ± 348 + (368 or 574)
= 922 Kn/m^2 maximum
= 175 kN/m of wall

Slenderness ratio of wall = $\frac{2900 \times 0.75}{190}$ = 11.45

Eccentricity of load from floor onto returns can be assumed to be zero.
Hence β = 0.94
Hence required design capacity in kN/m

$$= \frac{175}{0.94} = 186 \text{ kN/m}$$

Table 7.4 shows that this can be achieved using a 10.5 N/mm^2 block.
Load case (c) should also be considered
Horizontal load = 1.2 × 42 = 50.4 kN
Vertical load = 1.2 × (152 + 67.5) × 8 = 2107 kN
similarly as before:

$$\frac{N}{A} + \frac{M_y}{I} = \frac{2107}{15.6 \times 0.19} + \frac{50.4 \times 6 \times 4}{2.03 \times 2} \text{ kN/m}^2$$

= 711 ± 298
= 1009 kN/m^2 maximum
= 192 kN/m of wall
As before, β = 0.94
∴ Design resistance = 204 kN/m

This is more critical than load case (b) but still within the capacity of a 10.5 N/mm² block which will take 292 kN/m (see Table 7.4).

Now consider wind from the other direction. This will be much less critical and is being done purely to illustrate the principal of splitting the wind forces between the three walls.

Calculate moments of inertia of walls about E–W axis.
Walls A and C

$$I = \frac{0.19 \times 8^3}{12} = 8.11 \text{ m}^4$$

Wall B

$$I = \frac{0.19 \times 8^3}{12} + 2 \times 0.19 \times 3.8 \times 3.9^2$$

$$= 30.07 \text{ m}^4$$

Wind force is shared between walls in proportion to their stiffness. For load case (c) the design force is 1.2 × 72 = 86.4 kN

Wall A takes $\frac{86.4 \times 8.11}{2 \times 8.11 \times 30.07} = 15$ kN

Wall C will also take 15 kN

Wall B will take 56.4 kN

The capacity of the walls is checked as for bending in the other phase. Note that the walls should also be checked for the horizontal shear at the base and wall B should be checked for vertical sheer at the junction of the 'web' and the 'flange'.

7.3.4 Strength of biaxially bent pier

PROBLEM AND SOLUTION 5
This is not a problem which is likely to arise very often in practice. Should such a problem arise, however, this example illustrates how it may be tackled within the intentions of the Code.

Check whether the pier having the cross section drawn below can withstand a vertical design load of 130 kN at an eccentricity in one direction of 125 mm and of 90 mm in the direction at right angles. The pier is hollow and is constructed of 5 Nmm² blocks in 1:1:6 mortar for which the value of $\frac{f_k}{\gamma_m}$ is 1.1 N/mm².

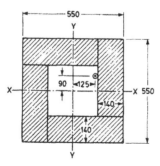

This problem can be solved by trial and error by guessing a neutral axis position and then estimating the position of the centroid of the area of the cross section in compression. When a neutral axis position has been found such that the centroid of the compressed area coincides with the centroid of the load, the axial load capacity can be calculated from the equation:

$$N = 1.1 \frac{f_k}{\gamma_m} A_c$$

where A_c = area of masonry in compression

If N > 130 kN, the section is satisfactory. The problem is probably best solved by drawing out the section to scale. Only two tries will be set out in detail here.

Try 1

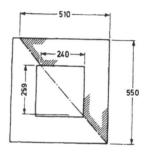

Area of compression zone $= \frac{510 \times 550}{2} - \frac{240 \times 259}{2}$

$$= 140,250 - 31,080$$
$$= 109,170 \text{ mm}^2$$

Moments about Y–Y

$$140,250 \times \left(\frac{550}{2} - \frac{510}{3} \right)$$

$$- 31,080 \left(\frac{270}{2} - \frac{240}{3} \right) = 109,170 \bar{x}$$

hence $\bar{x} = 119$
Moments about X–X

$$140,250 \times \left(\frac{550}{2} - \frac{550}{3} \right)$$

$$- 31,080 \left(\frac{270}{2} - \frac{259}{3} \right) = 109,170 \bar{y}$$

hence $\bar{y} = 104$
$\bar{x} = 125$, $\bar{y} = 90$ is required.

It is necessary to increase $\bar{x}$ very slightly and decrease $\bar{y}$ by about 14%. Moving the neutral axis to the right without rotation will increase $\bar{x}$, clockwise rotation of the neutral axis line will decrease $\bar{y}$.

Try 2
Area of compression zone

$$= 550 \times 40 + \frac{450 \times 550}{2} - 270 \times 20 - \frac{215 \times 270}{2}$$

$$= 22,000 + 123,750 - 5400 - 29,029$$
$$= 111,325 \text{ mm}^2$$

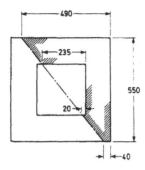

Moments about Y–Y

$$22{,}000 \times \left(\frac{550}{2} - \frac{40}{13} \right)$$

$$+ \ 123{,}750 \left(\frac{5500}{2} - 40 - \frac{450}{3} \right)$$

$$-5400 \qquad \left(\frac{270}{2} - \frac{20}{2} \right)$$

$$- \ 29{,}029 \qquad \left(\frac{270}{2} - \frac{215}{3} \right) = 111{,}325 \ \bar{x}$$

hence $\bar{x} = 122.3$

Moments about X–X

$$123{,}750 \times \left(\frac{550}{2} - \frac{550}{3} \right)$$

$$- \ 29{,}029 \qquad \left(\frac{270}{2} - \frac{270}{3} \right) = 111{,}325 \ \bar{y}$$

hence $\bar{y} = 90.16$

This is close enough to 125 and 90.
Vertical load capacity with this eccentricity

$$= \frac{1.1 \times 1.1 \times 111{,}325}{10{,}000} \ \text{kN}$$

= 134.7 kN

This is greater than 130 therefore design is satisfactory.

7.4 Structural design of low-rise buildings

In order to provide safe designs for housing without the need for calculations of loading and strength criteria BS 8103: Part 2: 1996 *Structural design of low-rise buildings*. Part 2. *Code of practice for masonry walls for housing*[7.18] was produced. This code sets a number of limitations in various clauses such that if the building is constructed within these limitations no detailed calculations are required. The overall stability of the building still needs to be achieved and Part 1 of the code[7.19] covers stability aspects.

7.5 References

7.1 READ, J B, and CLEMENTS, S W. The strength of concrete block walls. Phase III: Effects of workmanship, mortar strength and bond pattern. Cement and Concrete Association, London, 1977. Publication No 42.518. pp 10.

7.2 BRITISH STANDARDS INSTITUTION. BS 5628: Part 1: 1995. *Code of Practice for Use of Masonry*. Part 1. *Structural use of unreinforced masonry*. London.

7.3 CRANSTON, W B, and ROBERTS, J J. The structural behaviour of concrete masonry – reinforced and unreinforced. *The Structural Engineer*, Vol 54, No 11, November 1976. pp 423–436.

7.4 INTERNATIONAL STANDARDS ORGANIZATION. General principles for the verification of safety of structures. ISO 2394, February 1973.

7.5 BRITISH STANDARDS INSTITUTION. BS 5400–2: 1978. *Steel, concrete and complete bridges specification for loads*. London.

7.6 NATIONAL STANDARDS AUTHORITY OF IRELAND. I.S. 325: Part 1: 1986 *Code of Practice for use of masonry*. Part 1: *Structural Use of Unrefined Masonry*. Dublin.

7.7 BRITISH STANDARDS INSTITUTION. BS 5628: Part 3: 1985. *British Standard Code of Practice for Use of Masonry*. Part 3. *Materials and components, design and workmanship*. London.

7.8 NATIONAL CONCRETE MASONRY ASSOCIATION OF AMERICA, VIRGINIA. Technical Report No 37, 1952.

7.9 BRITISH RESEARCH ESTABLISHMENT. Strength of brickwork and blockwork walls: design for vertical load. HMSO, 1981, BRE Digest 246.

7.10 CURTIN, W.G., SHAW, G., BECK, J.K. and BRAY, W.A. Design of Brick Fin Walls in Tall Single-Storey Buildings. Brick Development Association. Woodside, 1980 p.32.

7.11 CURTIN, W.G., SHAW, G., BECK, J.K. and BRAY, W.A. Design of Brick Diaphragm Walls. Brick Development Association, Woodside, 1982, p. 41.

7.12 CURTIN, W.G., SHAW, G., and BECK, J.K. Structural Masonry Designers' Manual. Oxford, Blackwell Scientific. 1987.

7.13 PHIPPS, M.E. and MONTAGUE, T.I. The design of concrete blockwork diaphragm walls. Leicester, Aggregate Concrete Block Association, pp. 18.

7.14 PHIPPS, M.E. and MONTAGUE, T.I. The design of prestressed concrete blockwork diaphragm walls. Leicester. Aggregate Concrete Block Association, pp.18.

7.15 SHAW, G. and BECK, J.K. Design of concrete masonry diaphragm walls. Concrete Society Working Part Technical Report No. 27. CS53.053. 1985. pp.29.

7.16 INSTITUTION OF STRUCTURAL ENGINEERS. Manual for the design of plain masonry in building structures. London, 1997, p. 77.

7.17 PEARCE, D J, and MATTHEWS, D D. *An appraisal of the design of shear walls in box frame structures*. Property Services Agency, London, 1972.

7.18 BRITISH STANDARDS INSTITUTION. BS 8103: Part 2: 1996. *Structural design of low-rise buildings* Part 2. *Code of Practice for masonry walls for housing*. London.

7.19 BRITISH STANDARDS INSTITUTION. BS 8103: Part 1: 1986 *Structural design of low-rise buildings* Part 1. *Code of Practice for stability, site investigation, foundations and ground floor slabs for housing*. London.

Chapter 8
Lateral loading on plain masonry (including walls with bed joint reinforcement)

8.1 Background

Chapter 7 deals with the design of walls subject to vertical loads and walls subject to both vertical and lateral loads. There are, however, many walls, for instance those used for cladding steel or concrete portal framed buildings, subjected to predominantly lateral loads with the vertical load being very small or limited to self weight. Thus it is often necessary to check the adequacy of a non-vertical loadbearing wall to resist wind pressure. This need for some specific design information for laterally loaded walls was brought sharply into focus when the deemed-to-satisfy *Schedule 7* of the *1972 Building Regulations*[8.1] introduced restrictions limiting its application to residential buildings of not more than three storeys and certain small single storey structures in areas where the design wind speed does not exceed 44 m/sec. Previously *Schedule 7* covered a wider range of structures.

The application of CP 111[8.2] to the design of walls to resist wind pressure caused difficulties to engineers over the years despite the introduction of an amendment in 1970 specifying permissible tensile stresses in the bed and perpend joints, because the scope of the Code remained unamended and could be interpreted as applying only to walls with significant vertical load. There was also a somewhat ambiguous statement that in general no reliance should be placed on the tensile strength of masonry in calculations but that the engineers could, at their discretion, design for tensile stresses. To complicate matters further the designers, having already taken responsibility to allow for tensile stresses, had very little published information to enable them to evaluate the bending moments in the panel. In practice, however, many designs were carried out using the permissible tensile values given in CP 111. Bending moments being either arbitrarily assessed or determined by using, for example, the bending moment coefficients available for two way spanning reinforced concrete slabs. It must be said that few failures of such designs have occurred, and those which have, have been due to inadequate edge fixing of the panels or because no structural calculations

were carried out at all. To partially remedy the situation some simple rules for assessing the size of a wall subject to wind pressures were introduced in 1975 as an amendment to CP 121[8.3] The guidance was amended in 1985 to include panels with openings and may now be found in BS 5628: Part 3[8.17]. This simplified approach is summarised in *Section 8.5* of this chapter, but is limited in application. It became clear when drafting the limit state code for masonry that the engineer still required more comprehensive design methods properly supported by research evidence.

Some work was carried out by Isaacs[8.5] as early as 1948 and by Fishburn[8.8] in 1961. Several other papers on the strength of walls in lateral loading and the effect of the bond between the mortar and the masonry have been written, the most notable UK ones being referred to in the following sections. However, there was still insufficient data available in the early 1970s to put forward a positive design procedure. As a result, considerable research was undertaken in the UK from around 1973 with the result that for the first time a definite calculation procedure was included in BS 5628[8.4], the current limit state version of CP 111[8.2]. The majority of the research[8.6, 8.7] was conducted by the British Ceramic Research Association for clay brickwork and by the Concrete Block Association for concrete blockwork. In addition, some special sponsored work on flexural strength, supervised by the Building Research Establishment has been carried out at the British Ceramic Research Association at the request of the Property Services Agency on both brick and concrete block masonry. This work was extended to deal with calcium silicate bricks and concrete bricks.

So far reference has only been made to the stability of walls as determined by the flexural strength of the masonry. As well as this method BS 5628 permits an approach based on arching. Most of the information in this Chapter is devoted to the flexural strength method although some guidance is given on the arching method.

Table 8.1 Characteristic flexural strength of masonry f_{kx} N/mm²

Mortar designation	Plane of failure parallel to bed joints			Plane of failure perpendicular to bed joints		
	(i)	(ii) & (iii)	(iv)	(i)	(ii) & (iii)	(iv)
Clay bricks having a water absorption						
less than 7%	0.7	0.5	0.4	2.0	1.5	1.2
between 7–12%	0.5	0.4	0.35	1.5	1.1	1.0
over 12%	0.4	0.3	0.25	1.1	0.9	0.8
Calcium silicate		0.3	0.2		0.9	0.6
Concrete bricks		0.3	0.2		0.9	0.6
Concrete blocks (solid or hollow) of compressive strength in N/mm²:						
2.8 ⎫ used in walls of					0.40	0.4
3.5 ⎬ thickness* up to		0.25	0.2		0.45	0.4
7.0 ⎭ 100 mm					0.60	0.5
2.8 ⎫ used in walls of					0.25	0.2
3.5 ⎬ thickness* up to		0.15	0.1		0.25	0.2
7.0 ⎭ 250 mm					0.35	0.3
10.5 ⎫ used in walls of					0.75	0.6
14.0 ⎬ any thickness*		0.25	0.2		0.90†	0.7†
and over ⎭						

Note linear interpolation between entries is permitted for:
(a) concrete block walls of thickness between 100 mm and 250 mm;
(b) concrete blocks of compressive strength between 2.8 N/mm² and 7.0 N/mm² in a wall of given thickness.
*The thickness should be taken to be the thickness of the wall, for a single-leaf wall, or the thickness of the leaf, for a cavity wall.
†When used with flexural strength in parallel direction, assume the orthogonal ratio, μ = 0.3.

8.1.1 Method 1 – Flexural strength approach

Determining the lateral load capacity of a wall panel based on attainment of ultimate tensile stresses in the bed joint and perpend joints.

The problem of determining the lateral load capacity of a masonry wall revolves around two basic points: (a) the flexural strength of the masonry and (b) the distribution of bending moments induced in the panel.

8.1.1.1 Characteristic flexural strengths

The research work referred to above demonstrates that items which have the major influence on the flexural strength of masonry are: (i) the composition and strength of the mortar, (ii) the initial flow of the mortar, (iii) the curing conditions, and (iv) the properties of the masonry units. Of these items it is reasonable to expect the safety factors and permitted stresses to accommodate (ii) and (iii). With regard to item (i), this is covered by allowing different stresses for different strength mortars.

With regards to item (iv), the flexural strength of concrete block masonry was found to be broadly related to the compressive strength of the units. With clay bricks it was found in contrast that the water absorption of the unit was a major parameter. Thus in Table 8.1 (taken from BS 5628) the characteristic flexural strength of clay bricks relates to water absorption and that of concrete blocks to compressive strength. The strength and water absorption of calcium silicate and concrete bricks are fairly constant and therefore only one stress is given.

It will also be seen in Table 8.1 that the flexural strength of blocks of compressive strength between 2.8 N/m² and 70 N/m² is affected by the unit thickness. The ultimate flexural moment of resistance thus does not always increase in proportion to the increase in section modules. A linear interpolation between entries in Table 8.1 is permitted for blocks of compressive strength between 2.8 N/mm² and 7.0 N/mm² in wall thickness between 100 mm and 250 mm. Table 8.2 gives this interpolation for a range of common block sizes. The characteristic strength values given in table 8.1 may be used for the categories of brick, block or mortar shown. Alternatively, tests may be carried out in accordance with A.3 of BS 5628: Part 1.

Table 8.2 Characteristic strength, capacity and orthogonal ratios for general thickness blocks

| Thickness (mm) | 2.80 | | | 3.50 | | | 7.00 | | | 2.8–7.0 | 2.8 | 3.5 | 7 |
| | f_{kp} | M | | f_{kp} | M | | f_{kp} | M | | f_{kb} | μ (0.60) | μ (0.55) | μ (0.40) |
		$\gamma_m = 3.1$	$\gamma_m = 3.5$		$\gamma_m = 3.1$	$\gamma_m = 3.5$		$\gamma_m = 3.1$	$\gamma_m = 3.5$				
100	0.400	0.215	0.190	0.450	0.242	0.214	0.600	0.323	0.286	0.250	0.53	0.56	0.42
110	0.390	0.254	0.225	0.437	0.284	0.252	0.583	0.379	0.336	0.243	0.62	0.56	0.42
115	0.385	0.274	0.242	0.430	0.306	0.271	0.575	0.409	0.362	0.240	0.62	0.56	0.42
125	0.375	0.315	0.279	0.417	0.350	0.310	0.558	0.469	0.415	0.233	0.62	0.56	0.42
140	0.360	0.379	0.336	0.397	0.418	0.370	0.533	0.562	0.498	0.223	0.62	0.56	0.42
150	0.350	0.423	0.375	0.383	0.464	0.411	0.517	0.625	0.554	0.217	0.62	0.57	0.42
175	0.325	0.535	0.474	0.350	0.576	0.510	0.475	0.782	0.693	0.200	0.62	0.57	0.42
190	0.310	0.602	0.533	0.330	0.640	0.567	0.450	0.873	0.774	0.190	0.61	0.58	0.42
200	0.300	0.645	0.571	0.317	0.681	0.603	0.433	0.932	0.825	0.183	0.61	0.58	0.42
215	0.285	0.708	0.627	0.297	0.737	0.653	0.408	1.105	0.899	0.173	0.61	0.58	0.42
220	0.280	0.729	0.645	0.290	0.755	0.668	0.400	1.041	0.922	0.170	0.61	0.59	0.43
225	0.275	0.748	0.663	0.283	0.771	0.683	0.392	1.066	0.944	0.167	0.61	0.59	0.43
250	0.250	0.840	0.744	0.250	0.840	0.744	0.350	1.176	1.042	0.150	0.60	0.60	0.43

Compressive strength of units N/mm²

Note: The bracketed values of orthogonal ratio at the head of the end columns are simple lower bound values that may simplify calculation

8.1.1.2 Behaviour of wall panels

The research work on flexural strength was mainly carried out on prisms of masonry or on small wallettes subject to bending either across the bed joints or across the perpend joints. Quite a number of tests[8.6, 8.7, 8.18] on actual wall panels have also been carried out, covering clay, calcium silicate and concrete units. The failure loads were found to be different depending upon the flexural strength of the materials, but the actual mode of failure was broadly similar for panels of similar size and having the same support conditions. Patterns of cracking observed on the inside faces for typical walls are shown in Figure 8.1. The failure patterns are reasonably similar to what would be expected from reinforced concrete slabs analysed in accordance with yield line theory. Yield line theory was, of course, developed for use with reinforced concrete in which it is assumed that the element has sufficient ductility to enable several lines to develop with constant moment until there are sufficient to cause failure. It is clear that a masonry wall cannot truly behave in the same manner since it is a brittle material and is thus not capable of holding its moment across a line after a crack has formed. Despite this, however, yield line theory has been found to be a reasonable method for predicting the capacity of walls. The coefficients given in Table 9 of the Code and the tables to this Chapter (*Section 8.5*) were derived by yield line theory. For many designs it will be a simple matter of using the coefficients given but there will be occasions where it may be desirable to refer directly to the basic yield line equation as, for example, in the case of a wall of irregular shape or one with openings.

There are a number of text books and publications on the subject of yield line analysis but the most commonly available are those by Johansen[8.9, 8.15], Jones[8.10], Jones and Wood[8.11] and the publication by Comité Européan du Béton[8.12]. More recent equations are given in reference

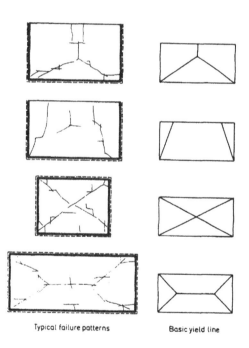

Typical failure patterns   Basic yield line

Figure 8.1 Patterns of cracking observed on the inside faces for typical walls

8.13 which includes for a panel supported on one vertical edge and its base. This reference also contains a series of graphs for determining the strength of lateral leaded panels. The derivation of yield line formula is outside the scope of this handbook as indeed would a comprehensive review of formula for various panel shapes be.

Reference should, therefore, be made to the above documents when detailed information is required, but for convenience, formula relating to the three standard panel shapes as given in the Code are given below.. Further information with regard to panels of irregular shape and with openings is given in *Sections 8.2.6* and *8.2.7*. An alternative ultimate load analysis using a fracture line approach has also been proposed by Sinha[8.13, 8.14].

A comprehensive thesis dealing with laterally loaded masonry walls both unreinforced and reinforced with bed joint reinforcement has been prepared by Cajdert[8.16] of Sweden. The thesis covers work carried out in Sweden and other countries including the UK.

PANEL WITH FREE TOP EDGE

Failure pattern (i)

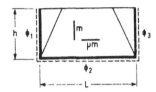

$$\alpha = \frac{1}{6\mu} \left(\frac{h}{L}\right)^2 \left(\frac{1}{[(3 + \phi_2) \div Y^2] - 1} \right)$$

$$\text{where } Y = \sqrt{\left(\left[\frac{\sqrt{1 + \phi_1} + \sqrt{1 + \phi_3}}{\sqrt{\left[\frac{\mu(3 + \phi_2)}{\left(\frac{h}{L}\right)^2}\right]}}\right]^2 + 3\right) - \left(\frac{\sqrt{1 + \phi_1} + \sqrt{1 + \phi_3}}{\sqrt{\left[\frac{\mu(3 + \phi_2)}{\left(\frac{h}{L}\right)^2}\right]}}\right)}$$

Failure pattern (ii)

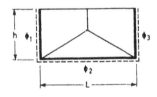

$$\alpha = \frac{1}{6Y_{13}^2} \left[\sqrt{\left(3 + \frac{\mu Y_2^2}{\left(\frac{h}{L}\right)^2 Y_{13}^2}\right)} - \frac{\sqrt{\mu} Y_2}{\left(\frac{h}{L}\right) Y_{13}}\right]^2$$

$$\text{where } Y_2 = \sqrt{1 + \phi_2} \quad Y_{13} = \sqrt{1 + \phi_1} + \sqrt{1 + \phi_3}$$

The perpend moment of resistance required so that a uniform lateral load, W_k, can be carried is given by:

$$m = \alpha W_k \lambda_f L^2$$

where

α is the bending moment coefficient greater of above equations

W_k is the characteristic wind load

λ_f is the partial safety factor for load

In both yield line equations

μ = orthogonal ratio

= $\dfrac{\text{moment of resistance on bed joint}}{\text{moment of resistance perpendicular to bed joint}}$

ϕ = degree of fixity at support

= $\dfrac{\text{moment of resistance at support}}{\text{moment of resistance of span}}$

i.e. 0 for pinned (simple) support

1 for fixed support

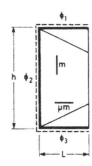

$$\alpha = \frac{1}{6} \left(\frac{1}{[(3 + \phi_2) \div Y^2] - 1} \right)$$

$$\text{where } Y = \sqrt{\left(\left[\frac{\sqrt{1 + \phi_1} + \sqrt{1 + \phi_3}}{\sqrt{\left[\frac{\left(\frac{h}{L}\right)^2 (3 + \phi_2)}{\mu} \right]}} \right]^2 + 3 \right) - \left(\frac{\sqrt{1 + \phi_1} + \sqrt{1 + \phi_3}}{\sqrt{\left[\frac{\left(\frac{h}{L}\right)^2 (3 + \phi_2)}{\mu} \right]}} \right)}$$

Failure pattern (ii)

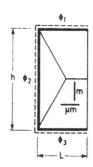

$$\alpha = \frac{\left(\frac{h}{L}\right)^2}{6\mu \, Y_{13}^2} \left[\sqrt{\left(3 + \frac{\left(\frac{h}{L}\right)^2 Y_2^2}{\mu \, Y_{13}^2} \right)} - \frac{\sqrt{\frac{1}{\mu}} \left(\frac{h}{L}\right) Y_2}{Y_{13}} \right]^2$$

$$\text{where } Y_2 = \sqrt{1 + \phi_2} \quad Y_{13} = \sqrt{1 + \phi_1} + \sqrt{1 + \phi_3}$$

Again

$$m = \alpha \, W_k \, \lambda_f \, L^2 \qquad\qquad W_k, \lambda_f, \text{ and } \phi \text{ as before}$$

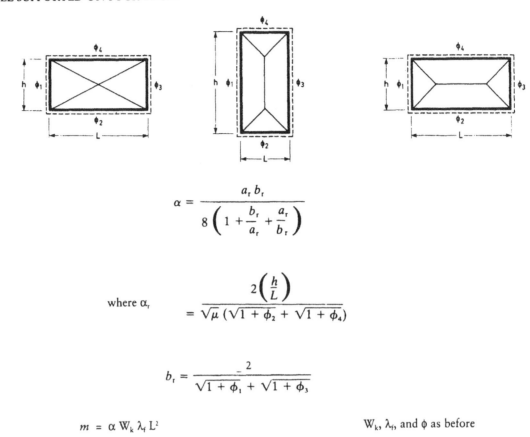

$$\alpha = \frac{a_r b_r}{8\left(1 + \dfrac{b_r}{a_r} + \dfrac{a_r}{b_r}\right)}$$

$$\text{where } \alpha_r = \frac{2\left(\dfrac{h}{L}\right)}{\sqrt{\mu}\,(\sqrt{1 + \phi_2} + \sqrt{1 + \phi_4})}$$

$$b_r = \frac{2}{\sqrt{1 + \phi_1} + \sqrt{1 + \phi_3}}$$

$$m = \alpha\,W_k\,\lambda_f\,L^2 \qquad\qquad W_k,\ \lambda_f,\ \text{and } \phi \text{ as before}$$

8.1.2 Method 2 – Arching approach

The alternative method given in BS 5628 for the design of walls subjected to lateral loads is, as mentioned previously, based on the assumption that an arch thrust can be developed in the plane of a wall which is built solidly between rigid supports. The concept of this design approach may be illustrated by considering the member shown in Figure 8.2 which illustrates a horizontally spanning wall being subjected to a lateral load q_{lat}, the central deflection under this load being δ.

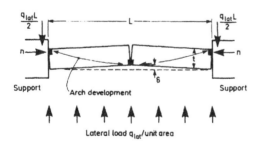

Lateral load q_{lat}/unit area

From Figure 8.2, and by reference to Figure 8.3 which shows the compressive blocks at supports and mid span, the internal compressive force n is equal to the permitted compressive stress multiplied by the depth of the compressive block. Thus $n = fx$.

This force is being applied internally over a level arm a. The internal moment of resistance

$$\begin{aligned}
Mi &= na \\
&= f_x \left(t - \frac{x}{2} - \frac{x}{2} - \delta \right) \\
&= f_x\,(t - x - \delta)
\end{aligned}$$

The applied moment over the span L gives:

$$M = \frac{q_{lat}\,L^2}{8}$$

Figure 8.2

Figure 8.3

Since for equilibrium the internal and external moments must balance, then:

$$\frac{q_{lat}\,L^2}{8} = (f_x\ (t - x - \delta)$$

$$\text{i.e.}\quad q_{lat} = 8 f x\ \frac{(t - x - \delta)}{L^2}$$

Now from BS 5628 the characteristic strength of masonry under these conditions may be taken as $1.5\,f_k$, and the depth of the compressive block is assumed to be $0.1\,t$. The ultimate design load q_{lat} predicted using the simplified and conservative approach with a partial safety factor for masonry of γ_m is determined as follows:

$$q_{lat} = \frac{8 \times 1.5\,f_k \times 0.1\,t\ (t - 0.1\,t - \delta)}{\gamma_m\,L^2}$$

$$= \frac{f_k}{\gamma_m} \times \frac{1.2\,t\ (0.9\,t - \delta)}{L^2}$$

When deflection, δ, equals 0 then the ultimate design load tends towards

$$q_{lat} = \frac{f_k}{\gamma_m}\left(\frac{t}{L}\right)^2\ 1.08$$

In BS 5628, Clause 36.4, we find for simplicity that in a panel of $L/t < 25$ the deflection can be ignored and that the design load may be found from the equation:

$$q_{lat} = \frac{f_k}{\gamma_m}\left(\frac{t}{L}\right)^2$$

It is reasonable to consider that the 1.08 figure was rounded down to 1.0 since some small deflections even at $L/t < 25$ may occur. An alternative and perhaps better way is to simply state that the equation given in Clause 36.4 of BS 5628: Part 1 will be conservative providing $\delta \leq 0.067t$ since putting this value into the above equation results in $q_{lat} = f_k/\gamma_m\ (t/L)^2$.

The statement made in the opening paragraph of this section with regard to rigid supports and the subsequent illustration given in Figure 8.3 seem to suggest that the arch approach can only be made when a panel spans between two columns. This is not necessarily the case since the Code also makes reference to walls built continuously past the support. In this instance the arch approach can be applied to a section of wall providing that the wall which continues past the support has sufficient strength to resist the internal arch force n. Obviously even the column or member providing the support, as in the case of a wall as shown in Figure 8.2, must also be capable of resisting the arch thrust. The support itself must also be able to resist the applied lateral load with negligible deflection.

The formula for q_{lat} given in BS 5628 and reproduced above may be used provided that the length/thickness is not greater than 25 and provided that there is no appreciable reduction in the length of the wall due to moisture and thermal movements. The Code gives no

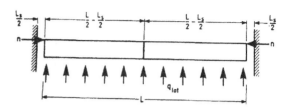

Figure 8.4 Effect of shortening

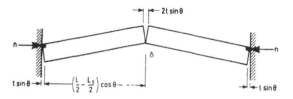

Figure 8.5 Effect of shortening

guidance on determination of the deflection in the wall or allowance for the effects of movement. These aspects have been considered and the following approach is suggested. Shortening will arise due to:

(a) the axial load or thrust in the arch
(b) shrinkage movements
(c) thermal movements

The deflection of the supports must of course also be small. To put the matter into perspective, it is of interest to look at the effects of shortening. Consider Figures 8.4 and 8.5 which show a wall in which shortening of L_s has taken place. It is required to assess the deflection, δ. Assuming that θ is small (an assumption that is checked later), then

$$t \sin \theta = t\,\theta$$

As θ and L_s are small

$$\left(\frac{L}{2} - \frac{L_s}{2}\right)\cos\theta = \frac{L}{2} - \frac{L_s}{2} \cong \frac{L}{2}$$

It follows then that

$$L_s = 4\,t\,\theta = \frac{4\,t\,\delta}{\dfrac{L}{2}} = \frac{8\,t\,\delta}{L}$$

and hence

$$\delta = \frac{L_s\,L}{8\,t}$$

It can also be useful to be able to calculate the actual arch thrust per unit depth of wall. The application of statics gives

$$n = \frac{q_{lat}\,L^2}{8\,(0.9\,t - \delta)}$$

Two practical examples will now be considered.

Example 1

Consider a wall with a span of 4.75 m and a thickness of 190 mm (i.e. span/depth 25). The effects of shortening due to a shrinkage strain of 400×10^{-6} and a shortening under axial load of 100×10^{-6}, are to be assessed.

$$L_s = 4.75 \times 5 \times 10^{-6} \times 10^{-3} = 2.4 \text{ mm}$$

$$\therefore \delta = \frac{2.4 \times 4.75 \times 10^{-3}}{8 \times 190} = 7.5 \text{ mm} = 0.039\,t$$

The strain used in the example is tending towards the maximum which would normally be expected in practice and yet δ is considerably less than $0.067\,t$ and therefore indicates why the effect of deflection in members with a span/depth ratio not greater than 25 may be neglected.

$$\text{Now } \theta = \frac{7.5}{2375} = 0.00316 \text{ rad.}$$

and $\cos \theta = 0.999995$

This is equivalent to a further shortening strain of 5×10^{-6} which is negligible in comparison to the basic 500×10^{-6} originally considered. Thus the capacity of the arch may be determined from formula

$$q_{lat} = \frac{f_k}{\gamma_m} \times \frac{1.2\,t\,(0.9\,t - \delta)}{L^2}$$

For a unit with a characteristic strength of 3.5 N/mm^2 and adopting $\gamma_m = 3.5$, this gives:

$$q_{lat} = \frac{3.5 \times 1.2 \times 0.19\,(0.9 \times 0.19 - 0.0075) \times 10^{-6}}{3.5 \times 4.75^2}$$

$$= 1.65 \text{ kN/m}^2$$

If the effect of the deflection had not been determined, the q_{lat} from Clause 36.4 would have been:

$$q_{lat} = \frac{3.5 \times 10^{-6}}{3.5} \left(\frac{0.19}{4.75} \right)^2 = 1.6 \text{ Kn/m}^2$$

Example 2

Consider a wall which spans 5.00 m and is 100 mm thick and resisting an ultimate load q_{lat} of 0.6 kN/m². Estimate the deflection arising from an axial shortening due to thrust and shrinkage of 400×10^{-6}. Check the design assuming $f_k = 6.4 \text{ N/mm}^2$.

$$L_s = 400 \times 5.0 \times 10^{-6} = 2 \text{ mm}$$

$$\delta = \frac{L_s L}{8\,t} = \frac{2 \times 5000}{800} = 12.5 \text{ mm}$$

$$\theta = \frac{12.5}{2500} = 0.05 \text{ radians}$$

$\therefore \cos \theta = 0.9999875$

This is equivalent to a further shortening strain of 12.5×10^{-6}, which is negligible. Thus the capacity of q_{lat}

$$= \frac{7 \times 1.2 \times 100\,(90 - 12.5) \times 10^3}{3.5 \times 5000^2}$$

$$= 0.74 \text{ kN/m}^2 > 0.6 \text{ kN/m}^2$$

therefore wall will be acceptable.

The actual thrust developed at ultimate

$$= \frac{q_{lat}\,L^2}{8\,(0.9\,t - \delta)}$$

$$= \frac{0.6 \times 52 \times 10^3}{8\,(90 - 12.5)}$$

$$= 24.2 \text{ kN/m depth of wall.}$$

To check that the support is acceptable it is simply a matter of determining the deflection of the support member carrying the thrust, in this case 24.2 kN/m, and adding this additional δ into the q_{lat} formula. If q_{lat} is still greater than the applied lateral load then the supports are adequate. If not, the support must be stiffer.

8.2 Design to BS 5628: Part 1

The general design procedure of BS 5628 is interactive in that the following approach is required:

(1) make initial assumption of support conditions;
(2) make assumptions as to strength of unit required;
(3) determine orthogonal ratio and hence bending moment coefficient ;
(4) determine flexural strength required;
(5) check flexural strength of chosen masonry;
(6) if too low return to either (1) or (2) and modify.

In addition, a similar approach is required to check for shear, slenderness and ties to supports. These various factors are examined in greater detail as follows.

8.2.1 Support conditions

The support conditions have to be assessed first. The Code covers most of the common uses (Figure 8.6). A free edge is easily identified but some judgement is necessary in deciding between simply supported or fixed. In the examples given in the Code reference is made to 'fixity at the discretion of the engineer'. The engineer appears to have two alternatives, to consider the support as either pinned or fixed. It would be simple to suggest that if in doubt the worst condition should be assumed. Unfortunately, this wording tends to be rather restrictive and it would appear better to suggest that the worst *practical* condition should be assumed. This would then allow an engineer and a checking authority to take a more reasonable line to a particular problem. For example, if in a certain design it is clear that the support is not a pinned condition, but there is doubt about the support being fully fixed then it would seem reasonable to allow for partial fixity. Obviously, if serious doubts do exist then the worst condition, pinned, must be assumed. If doubts exist about a pinned condition then it would certainly be better to consider that as a free edge. The text to the illustration (Figure 8.6) in the handbook showing support conditions has deliberately been reworded to bring the readers attention to the fact that only partial fixity may exist. Reference may also be made to *Section 8.6* which includes guidance on the assessment

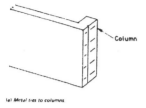
(a) Metal ties to columns.

(b) Bonded return walls.

(c) Metal ties to columns or unbonded return walls.

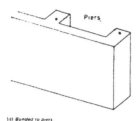
(d) Bonded to piers.

Figure 8.6

(a)   metal ties to column – normally simple support but possible degree of fixity with rigid supports and stiff well anchored ties
(b)   bonded return walls – degree of fixity at the discretion of the designer, the shorter the return the less the fixity – very short returns could reach free edge condition
(c)   metal ties to return walls – normally simple support but possible degree of fixity with stiff wall anchored ties
(d)   bonded to piers – intermediate pier, degree of fixity at discretion of engineer. End pier, simple support or free edge where pier insufficient to carry reaction.

of support conditions when adopting the simplified procedure given in BS 5628: Part 3 (8.17).

8.2.1.1 Accommodation for movement

It is appropriate while considering the vertical support condition to draw the readers attention to the question of accommodation of movements within the wall. Provision for movement may necessitate the inclusion of movement joints within the wall which may alter the initial assessment of the support conditions. This matter will also need to be reconsidered when the actual panel size is being determined to ensure that the designed length is not greater than the acceptable length between movement joints otherwise cracking may result.

8.2.1.2 Effect of tied piers

It is usual with concrete blockwork to use tied piers rather than fully bonded piers. This, however, will make little difference to the support condition since end piers are generally only taken as simple supports. Intermediate piers are often taken as fixed supports but since it is the continuing wall rather than the pier which provides the fixity it again makes little difference.

8.2.2 Design of solid single leaf walls

Having considered and assessed the support conditions the next stage is to determine the bending moments

which occur in the panel as a result of the applied lateral load and hence calculate the required thickness or characteristic flexural strength.

The general design procedure may be split into four basic forms: (a) free standing walls, (b) vertically spanning walls, (c) horizontally spanning walls, and (d) walls which span in two directions – these include both three-sided and four-sided supported walls.

8.2.2.1 Free standing walls

From Clause 36.5.2 it may be seen that the design moment of resistance of a free standing wall subject to horizontal forces is given by:

$$m = W_k \, \gamma_f \, \frac{h^2}{2} + Q_k \, \gamma_f \, h_L$$

where

m is the design moment per unit length
W_k is the characteristic lateral load
γ_f is the partial safety factor for loads
h is the clear height of wall or pier above restraint
Q_k is the characteristic imposed load
h_L is the vertical distance between the point of application of the horizontal load, Q_k, and the lateral restraint.

In the case of wind load only:

$$m = W_k \, \gamma_f \frac{h^2}{2}$$

8.2.2.2 Vertically spanning walls

In accordance with Clause 36.4.2 it will be found that the design bending moment per unit length of a wall spanning vertically may be written as:

$$m = \alpha \, W_k \, \gamma_f \, h^2$$

where:

- m is the design moment per unit length
- W_k is the characteristic lateral load
- γ_f is the partial safety factor for loads
- h is the height between horizontal supports
- α is the bending moment coefficient

For walls which have simple supports at both top and bottom we find that the bending moment coefficient is taken as for any other member which has a uniformly distributed load over a simple span, i.e.

$$\alpha = 0.125$$

thus $m = 0.125 \, W_k \, \gamma_f \, h^2$

or $\dfrac{W_k \, \gamma_f \, h^2}{8}$ as given in the Code.

The general approach has been explained in this way rather than giving the Code formula directly since the general formula $m = W_k \, \gamma_f \, \chi^2$ can be used to apply to all walls.

Vertically spanning walls will often have simple supports top and bottom as may be the case in single storey structures. However, the Code does indicate that allowance may be made, even in walls which span vertically, for any fixity which occurs at the supports. In this type of member the bending moment may be determined from the following formula:

$$m = \frac{W_k \, \gamma_f \, h^2}{2} \left[\frac{1}{\sqrt{1 \times \phi_1} + \sqrt{1 \times \phi_2}} \right]^2$$

where

- m is the design bending moment per unit length
- ϕ_1 and ϕ_2 is the degree of fixity at each support
- W_k is the characteristic wind load
- γ_f is the partial safety factor for loads
- h is the distance between supports

Some attention, however, needs to be given to the values of fixity taken at the supports since the presence of horizontal damp-proof courses may limit the amount of flexural strength which can be developed. This is particularly important with walls spanning vertically only since total reliance is made on the flexural strength of the bed joints.

Obviously where the moments due to the self weight stresses and any stress due to light vertical loads at the support under consideration is greater than the moment which would be produced by the characteristic strength of the bed joint. Adopting the enhancement given in 36.4.2 of the Code then full fixity might be achieved when $g_d \geq f_{kb}/\gamma_m$. Where the stress is less than $f_{kb} \gamma_m$ then the fixity will be at least $g_d / (f_{kb}/\gamma_m)$. (See Section 8.2.2.4 with regards to definition of f_{kb}.) Further comment on the effect of dpc's is given in Section 8.6.

It may be argued that this approach does not allow for any direct flexural strength given by the bed joint. However, this may be justified since some redistribution of bending moments over that found by normal elastic analysis of moments in beams has already been used to produce the yield line moments. If direct flexural strengths are allowed for, it may be more appropriate to determine the moments by normal elastic analysis with no allowance for redistribution. Where appreciable vertical loads are in existence then it will be advisable to design as indicated in Chapter 7.

Where piers are incorporated in a vertical spanning wall the load carried by the pier should be assessed from normal structural principles – for example, in a wall with piers the walls may be considered as spanning horizontally between the pier and the piers designed as spanning vertically carrying the load exerted on it by the span of the wall, the section modulus of the pier being assessed as recommended in Clause 36.4.3 of BS 5628. When the piers are closely spaced (flange width equal to distance between piers) then the wall will tend to act as a stiffened plate and thus the full section modulus may be adopted in assessing the moment of resistance of the wall.

8.2.2.3 Horizontally spanning walls

There is no section in the Code dealing directly with the design of walls assumed to span horizontally. However, in the text to Table 9 it is indicated that a panel having h/l greater than 1.75 will tend to span horizontally, thus it may be taken that it is acceptable to design a panel on the assumption of only spanning horizontally. When the aspect ratio is less than 1.75 it will be more economical to design as a two way span.

By reference to Clause 36.4.2 and to the general design approach adopted earlier, the design bending moment per unit height for a wall spanning horizontally may be taken as:

$$M = \alpha \, W_k \, \gamma_f \, L^2$$

in which W_k and γ_f are as before,

L is the length between vertical supports and α is again the bending moment coefficient for which a span between simple supports may be taken as 0.125.

Thus

$$m = 0.125 \, W_k \, \gamma_f \, L^2$$

$$\frac{W_k \, \gamma_f \, L^2}{8}$$

It is again logical that account may be taken for any fixity that occurs at the supports but since there is no direct reference made to partially fixed edges as there is with vertically spanning walls it may be assumed that full fixity can be taken in cases where this is justified. Therefore, the bending moment coefficient for a panel with fully fixed supports can be expected to reach 0.063, giving:

$$m = 0.063 \, W_k \, \gamma_f \, L^2$$

$$\frac{W_k \, \gamma_f \, L^2}{16}$$

8.2.2.4 Walls spanning in two directions

This type of wall includes panels with one free edge such as the U shape and C shape panels as they are referred to as well as panels which are supported on all four sides.

The general expression for the required design bending moment per unit height of these types of panels is given in Clause 36.4.2 as:

$$m = \alpha \, W_k \, \gamma_f \, L^2$$

The symbols α, W_k and γ_f are as previously defined. However, some confusion did occur in respect to the panel length 'L'.

The coefficients given in Table 9 of BS 5628 and in the tables given in *Section 8.4* of this handbook for all panels (panels A to L, Table 9) are given for use with the horizontal length L, and not as given in the previous edition of the Code, the length between supports. It is important to note that L is the horizontal distance between vertical supports (A–I) or the distance between the vertical support and the free edge in the case of C shaped panels (J–L).

With this type of panel α depends on:

(a) support conditions – free, pinned, fixed;
(b) aspect ratio $\frac{h}{L}$
(c) orthogonal ratio μ

Support conditions

To establish the bending moment coefficient, α, account must first be taken of the support conditions as explained in detail in *Section 8.2.1*.

Aspect ratio

Another factor which influences the bending moments in a panel is its aspect ratio, which is simply the ratio of the height between supports divided by the length of the panel either between supports or between a vertical support and the free edge h/l. Where the aspect ratio is between 0.3 and 1.75 the panel will tend to span in two directions. When h/l is less than 0.3 the Code indicates that the wall will tend to cantilever towards the free edge. When h/l is greater than 1.75 the wall will tend to span one way across the shortest dimension.

Orthogonal ratio

The orthogonal ratio is defined as being the ratio of the flexural strength of the masonry when failure is parallel to the bed joints to that when failure is perpendicular to the bed joints. The characteristic flexural strength of masonry is given in Table 3 of the Code and is given the general symbol f_{kx}. For convenience, however, and to avoid confusion as to which strength is used under which conditions, the following terms will be used in the following text and examples:

f_{kb} is the characteristic flexural strength when failure is parallel to the bed joints (see Figure 8.7)
f_{kp} is the characteristic flexural strength when failure is perpendicular to the bed joints (see Figure 8.8)

The orthogonal ratio

$$\mu = \frac{M_{ub}}{M_{up}} = \frac{\left(\dfrac{f_{kb}}{\gamma_m}\right) Z_b}{\left(\dfrac{f_{kp}}{\gamma_m}\right) Z_p}$$

In the general case of a wall of constant thickness $Z_b = Z_p$.

Thus for general cases the orthogonal ratio, $\mu = \dfrac{f_{kb}}{f_{kp}}$

When a wall is stiffened by piers it will be necessary to determine the orthogonal ratio as:

$$\mu = \frac{M_{ub}}{M_{up}}$$

For the moment of resistance of a wall with piers see *Section 8.2.2.5*.

Clause 36.4.2 indicates that the orthogonal ratio may be calculated allowing for any vertical load that acts so as to increase the flexural strength along the bed joint. Generally, it is the self weight of the wall which is to be considered, but to take accurate account of the effect of self weight is difficult since the effective stress on the bed joints, and thus the orthogonal ratio, varies within the height of the wall. An acceptable simplification is to determine the self weight stress at mid-height of the panel and to use this with the characteristic flexural strength, f_{kb}, to determine a constant orthogonal ratio.

As the flexural strength in Table 3 of BS 5628 is given in characteristic terms and the self weight stresses in ultimate design terms, the former may be modified by the partial safety factor for material before adding to any design vertical dead load per unit area (g_d) so acting. Thus the modification to allow for self weight gives:

$$\mu = \frac{\dfrac{f_{kb}}{\gamma_m} + f_{self}}{\dfrac{f_{kp}}{\gamma_m}}$$

where

f_{self} is the design self weight stress at mid-height
γ_m is the partial safety factor for materials

which can be simplified to:

$$\mu = \frac{f_{kb} + f_{self} \, \gamma_m}{f_{kp}}$$

or as given in the Code

$$\frac{f_{kx} + \gamma_m \, g_d}{f_{kp}}$$

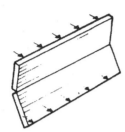

Figure 8.7 Plane of failure parallel to bed joints

Figure 8.8 Plane of failure perpendicular to bed joints

where g_d is the design vertical dead load per unit area. In examining the above equation it will be readily apparent why f_{kb} and f_{kp} have been adopted.

Adopting the suggestion of using the self weight stress at mid-height of the panel and allowing for a partial safety factor for load of 0.9 gives, for single storey panel:

$$f_{self} = 0.5\ h\ W_d\ 9.81 \times 10^{-6} \times 0.9$$
$$= 4.4 \times 10^{-6}\ h\ W_d$$

where

h is the panel height (m)
W_d is the density of wall (kg/m³)

8.2.2.5 Moment of resistance of panel

Having determined required m, a check has to be made that the masonry can be of that strength.

The ultimate moment of resistance of panels subject to bending, when assuming elastic distribution, is given in Clause 36.4.3 of the Code as:

$$M_u = \frac{f_{kx}}{\gamma_m} Z$$

for panels spanning vertically this is more conveniently written as:

$$M_u = \frac{f_{kb}}{\gamma_m} Z \text{ or } M_u = \frac{f_{kb}}{\gamma_m} + g_d$$

where self weight and small vertical loads act on the wall.

For panels spanning either horizontally or in two directions this is again more conveniently written as:

$$M_u = \frac{f_{kp}}{\gamma_m} Z$$

where

γ_m is the partial safety factor for materials
Z is the section modulus

With panels spanning in two directions it is only necessary to determine M_u in the direction parallel to the perpend joint. This is because the coefficients given in Table 9 of BS 5628 have been derived to give moments in this direction. Moments and hence stresses along the bed joints do not need to be determined since these are automatically allowed for by using the orthogonal ratio in the formula used to determine coefficients. For a plain wall spanning vertically $Z = \frac{bt^2}{6}$.

In assessing the section modulus of a wall containing piers (Figure 8.9), the outstanding length of flange from the face of the pier should be taken as (a) 4 × thickness of wall forming the flange when the flange is unrestrained, or (b) 6 × thickness of wall forming the flange when the flange is continuous, but in no case more than half the distance between the piers. Obviously the appropriate section modulus must be determined depending on which face is in tension. Allowance for the section modulus of a wall with piers only applies when assessing the bending stresses in a vertically spanning element.

Where the wall is spanning horizontally or in two directions, the section modulus should be taken in a vertical plane between any piers, i.e.

$$Z = \frac{ht^2}{6}$$

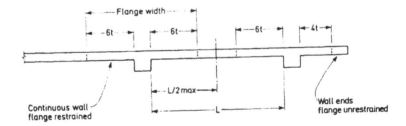

Figure 8.9

92

8.2.2.6 Shear strength of panel

The characteristic shear strength of masonry, f_v, is given in Clause 25 of the Code* as: $0.35 + 0.6\ g_A \le f_v < 1.75$ N/mm² except for mortars of designation (iv) where the maximum shear strength is reduced to $0.15 + 0.6\ g_A$ and must not be greater than 1.4 N/mm². g_A = design vertical load per unit area.

The design for shear stress, vh, along the bed joints must be controlled so that:

$$v_h \le \frac{f_v}{\gamma_{mv}}$$

where

γ_{mv} is the partial safety factor for material strength in shear
= 2.5 for general use.

Taking the design shear stress, v_h, being equal to the design wind force, $\gamma_f W_k$, acting on the wall times the effective area of wall A_w on which it is assumed to act divided by the cross sectional area resisting the shear force A

That is:

$$v_h = \frac{\gamma_f\ W_k\ A_w}{A}$$

thus

$$\frac{f_v\ A}{\gamma_{mv}} \ge \gamma_f\ W_k\ A_w$$

If yield line analysis were adopted it would be possible to determine the total reactions on each panel support. However, as has been explained, this method of analysis is not strictly applicable to a brittle material and although it can assist in determining the moments in a wall it does not seem appropriate to go to the necessary lengths to determine reactions. It will generally be acceptable to consider the edge reactions to be due to an area bounded by a line drawn at 45° to form the supported corners and for such to be uniformly distributed along the supports (Figure 8.10).

8.2.2.7 Slenderness limits

The control on slenderness limits or limiting dimensions is given by Clause 36 which states that the dimensions of a laterally loaded panel should be limited as follows:

(1) panel supported on three edges
   (a) two or more sides continuous: height × length equal to 1500 t_{ef}^2 or less

---

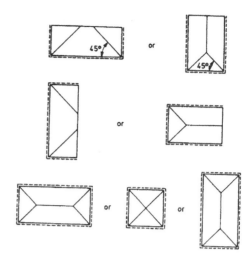

Figure 8.10

(b) all other cases: height × length equal to 1350 t_{ef}^2 or less
(2) panel supported on four edges
   (a) three or more sides continuous: height × length equal to 2250 t_{ef}^2 or less
   (b) all other cases: height × length equal to 2025 t_{ef}^2 or less
(3) panel simply supported at top and bottom height equal to 40 t_{ef} or less
(4) free standing wall height equal to 12 t_{ef} or less

In cases (1) and (2) no dimension should exceed 50 times the effective thickness, t_{ef}.

No direct reference is made to panels which are assumed to span horizontally but it may be deduced from (1)(a) and (b) that where the distance to the free edge is less than 30 t_{ef} the horizontal span for both simple and fixed supports should not exceed 50 t_{ef}. Where the distance to the free edge exceeds 30 t_{ef} the horizontal span should be limited to 45 t_{ef} for single supports and 50 t_{ef} for fixed supports. In the case of panels spanning vertically with partial fixity at the supports the limit of 40 t_{ef} should still be assumed to be applied. The effect of these controls may readily be seen by reference to Figure 8.11. The symbol t_{ef} is the effective thickness as given in Figure 8.12.

8.2.2.8 Summary of design requirements

The detailed information contained in the last section may be summarized into an overall design procedure as follows:

Noting that
h is the panel height
L is the panel length
W_k is the characteristic wind load and
γ_f is the partial safety factor for loads

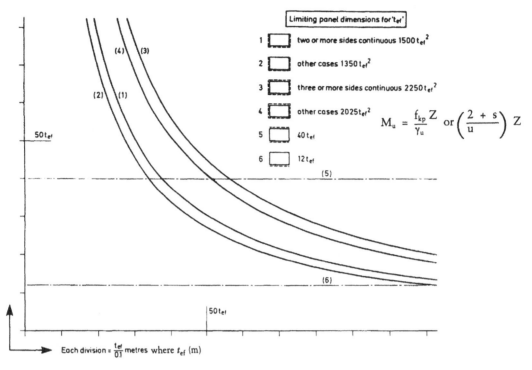

Limiting panel dimensions for 't_{ef}'

1		two or more sides continuous $1500\,t_{ef}^2$
2		other cases $1350\,t_{ef}^2$
3		three or more sides continuous $2250\,t_{ef}^2$
4		other cases $2025\,t_{ef}^2$
5		$40\,t_{ef}$
6		$12\,t_{ef}$

$$M_u = \frac{f_{kp}}{\gamma_u}Z \quad \text{or} \quad \left(\frac{2+s}{u}\right)Z$$

Each division $= \dfrac{t_{ef}}{0.1}$ metres where t_{ef} (m)

Figure 8.11

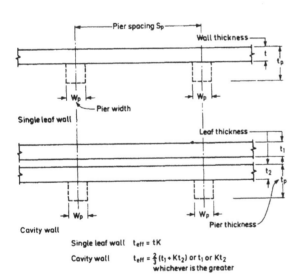

Single leaf wall $\quad t_{eff} = tK$

Cavity wall $\quad t_{eff} = \frac{2}{3}(t_1 + Kt_2)$ or t_1 or Kt_2 whichever is the greater

Values of K			
$\dfrac{S_p}{W_p}$	\multicolumn{3}{c}{$\dfrac{t_p}{t}$ or $\dfrac{t_p}{t_2}$}
	1 or NO piers	2	3
6	1.0	1.4	2.0
10	1.0	1.2	1.4
20 or NO piers	1.0	1.0	1.0

Figure 8.12

Walls spanning vertically

$$m = \alpha\,\gamma_f\,W_k\,h^2$$

where $\alpha = 0.125$ simple supports (for fixed supports see *Section 8.2.2.2*)

$$M_u = \frac{f_{kp}}{\gamma_m}Z \quad \text{or} \quad \left(\frac{f_{kb}}{\gamma_m} + f_{gd}\right)Z$$

when only self weight acts $g_d = f_{self}$

Walls spanning horizontally

$$m = \alpha\,\gamma_f\,W_k\,L^2$$

where $\alpha = 0.125$ simple support (0.063 fixed support – see *Section 8.2.2.3*)

$$M_u = \frac{f_{kp}}{\gamma_m}Z$$

Walls spanning two directions

$$m = \alpha\,\gamma_f\,W_k\,L^2$$

where α depends on (i) support conditions, (ii) orthogonal ratio (μ), (iii) aspect ratio $\dfrac{h}{l}$

This may be found from Table 9 (BS 5628), the tables in *Section 8.5*, or in appropriate cases by yield line analysis (8.1.1.2).

$$M_u = \frac{f_{kp}}{\gamma_m}Z$$

94

(1) In the case of blocks 2.8 and less than 10.5 N/mm², it tends to be less easily solvable because the flexural strength and orthogonal ratio changes according to thickness. Thus it will generally be easier to select a unit of a given strength and thickness and try against the applied moment ($M_u \geq m$). A simplification can, however, be made by adopting a lower bound orthogonal ratio of 0.6, 0.55 or 0.40 for block strengths of 2.8, 3.5 and 7.0 respectively in order to determine the bending moment coefficient α and hence the applied moment. The required thickness appropriate for the block strength can then be determined from Table 8.2.

In the case of concrete bricks or blocks of strength 10.5 N/mm² and greater the thickness is again determined so that $M_u \geq m$ but since

$$m = \alpha\, \gamma_f\, W_k\, L^2 \text{ and } M_u = \frac{f_{kp}\, b\, t^2}{6\gamma_m}$$

thickness required:

$$t = \sqrt{\frac{6\gamma_m\, \alpha\, \gamma_f\, W_k\, L^2}{f_{kp}\, b}}$$

(2) the shear stress should be limited so that

$$v_h \leq \frac{f_v}{\gamma_{mv}}$$

$$\leq \frac{f_v}{2.5}$$

(3) the panel dimensions must be checked against the limiting dimension requirements of Clause 36.3.

8.2.3 Design of double leaf (cavity) walls

Section 8.2.2 dealt with the basic design of laterally loaded walls as applicable to single leaf walls in general. With cavity walls the same basic procedure is adopted but the contribution of two leaves must be considered. Before the introduction of BS 5628 many engineers would have assessed the strength of a cavity wall to flexural loads by proportioning the applied loads to the two leaves in accordance with their stiffness. This approach was perhaps reasonable, although in practice the difference in the stiffness of the two leaves was not particularly marked since either the leaves were of similar thickness, or, where a difference in thickness did exist, the thicker leaf invariably had a lower modulus of elasticity. There is also a tendency for outer leaves to be more continuous than inner leaves. Table 9 of BS 5628: Part 1 is derived from yield-line analysis which relates to strength rather than stiffness.

These are some possible explanations to why the Code allows the design moment of resistance of a cavity wall to be taken as the sum of the design moments of resistance of the two leaves, except when the wall ties are not capable of transmitting the full force, in which case the contribution of the appropriate leaf should be limited accordingly (Figure 8.13). It would be unlikely for an

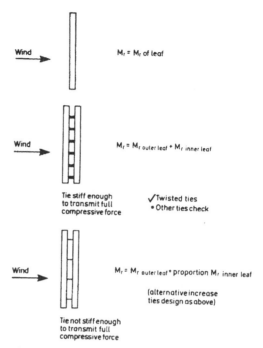

Figure 8.13 Moment of resistance of single leaf and cavity walls

engineer to actually design in the latter case and no doubt it would only be used in instances where a check is to be made on an already completed wall.

With regard to the ties, the Code indicates that vertical twist type ties will always provide sufficient strength when placed at normal centres (Clause 29.1.4 of the Code), but butterfly or double triangle ties should be checked. In addition, recent tests indicate the possibility of some composite action with the stiffer ties. An important point here is for engineers not to take the simple way out and specify vertical twist ties at all times, since the greater stiffness of these ties may cause problems of cracking due to the differential movement between the two leaves, particularly if the leaves are built of different materials such as fired clay bricks and concrete blocks. It will, therefore, often be desirable to retain the more flexible type of ties and carry out a simple check on the compression in the ties (see Chapter 3 with regard to durability of ties). In fact, from Clause 36.2 it is given that the characteristic strength of a double triangle or wire butterfly tie, in a cavity not wider than 75 mm, may be taken as 1.25 kN. Since, by reference to Clause 29.1.4, ties must be provided at not less than 2.5/m² it may readily be deduced, taking the partial safety factor for ties $\gamma_m = 3.0$ as given by Clause 27.5, that this type of tie will be quite acceptable where the design wind load required to be carried by the inner leaf does not exceed: $2.5 \times 1.25/3 = 1.04$ kN/m² and hence where the design wind pressure exceeds this figure ties will be required at not less than:

$$\frac{3.0\ W_k\ \gamma_f}{1.25}$$

i.e. $2.4\ W_k\ \gamma_f$ ties per m².

The slenderness limits applicable to cavity walls are the same as for single leaf walls except the effective thickness as explained in the previous section is modified as for any loadbearing cavity wall.

8.2.4 Design with small vertical precompression

Where a panel carries a small vertical load in addition to its self weight, the characteristic flexural strength may be modified in a similar manner to that explained in *Section 8.2.2.4* for accommodation of the self weight. Thus in designing a wall that spans vertically or in two directions the characteristic flexural strength may be modified to $f_{kb}/\gamma_m + g_d$, where g_d includes both self weight and dead load, and used to determine a modified orthogonal ratio or the strength of a vertically spanning member. As the vertical load gradually increases the panel will behave as a loadbearing element subject to lateral pressure and no longer controlled by flexural stresses. No precise value can be given but it is likely that it would be preferable to design as a loadbearing wall when the applied vertical design stress exceeds 0.1 N/mm^2.

8.2.5 Design with substantial vertical load

The design of this type of wall would be carried out by effectively modifying the applied vertical load and the appropriate capacity reduction factors as explained in *Chapter 7*.

8.2.6 Design of irregular shaped panels

Generally, most masonry walls will be rectangular. However, on occasions it is necessary to evaluate a trapezoidal shaped wall such as the gable wall to a mono pitched structure or the familiar gable wall to a conventionally pitched structure. With the first of these walls, as shown in Figure 8.14, the panel may be designed assuming an aspect ratio based on the mean height of the wall, i.e. $(h_1 + h_2)$ $0.5/L$ with three or four sided support being taken as appropriate. In the case of a vertical free edge, two cases arise as shown in Figure 8.15 and discussed below.

Since the vertical free edge will tend to dominate the moments induced in the panels, it will be conservative in case (1) to design to an assumed aspect ratio based on the mean height of the wall, i.e. $(h_1 + h_2)$ $0.5/L$.

Case (2) suggests a more critical situation than the first and thus a different aspect ratio seems justified. It would appear unnecessary to design for the full height of the unsupported edge and thus a reasonable compromise would be to design for an aspect ratio between the height in case (1) and the height, h, which gives 0.75 $(h_1 + h_2)/L$. This height may also be reasonable to apply to the case where both vertical edges are free.

With the conventional shaped gable, as shown in Figure 8.16, three distinct support conditions may occur which will influence the way in which design should be tackled, as shown in Figure 8.17.

In case (1) the panel may be designed on an assumed rectangular panel supported on four sides with an aspect ratio equal to 0.5 $(h_1 + h_2)/L$. The degree of fixity being determined as usual, the top support generally being assumed as a simple support. In case (2) the panel supported in this fashion will need two designs. Firstly, to design the main body of the wall as being supported on four sides with an aspect ratio equal to h/L, appropriate allowance being made for any fixity that occurs along the supports. Secondly, to design the triangular section as a cantilever from the top support. In the particular case of support to the sloping portion as well as a top horizontal support, the general design of the main section will suffice, unless the pitch is very steep, in which case it would seem appropriate to design the triangular section as a panel spanning between two opposite edges having a height of $(h_1 - h)/2$ or alternatively, yield line theory could be used.

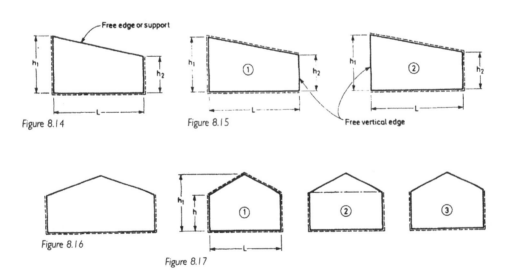

Figure 8.14   Figure 8.15   Figure 8.16   Figure 8.17

The type of panel in case (3) should be tackled directly by yield line theory, but the following simple approach may be used to give an indication of the thickness required. The suggestion is to tackle the wall in two parts, the top triangular section being designed as a cantilever. Since, due to the triangular section, some concentration of stress will occur towards the centre of the wall and as the horizontal support to the cantilever is the underlying masonry and thus undefined, it is suggested that this section could reasonably be designed as a rectangular section of height equal to two thirds of the height from eaves to apex, i.e. $2(h_1 - h)/3$ which would give an effective bending moment coefficient of 0.22 instead of 0.17 for a full triangular section. This approach would also seem appropriate for the cantilever section of case (2). The main wall section may then be designed as a panel supported on three sides with a line load acting on the free edge, the line load being taken as the reaction from the cantilever section. In effect this will give a result similar to that of a panel with a full length window of height equal to twice cantilever span, i.e. $h_w = 4h_1 - h)/3$ (see *Section 8.2.7*), again allowance being made for any fixity to the supports.

Cases (2) and (3) are not particularly desirable situations and every effort should be made to provide support to the wall along the sloping portion of the roof. However, in certain situations, as with a large studio roof window, support may not be possible, in which case the approach given here may be considered.

8.2.7 Design of panels with openings and line loads

The influence of openings on a laterally loaded masonry wall is very difficult to determine in specific terms since very few, if any, results are available on this type of wall. Appendix D of the Code gives some guidance and introduces the concept of using yield line analysis to determine the bending moments. It has already been stated that the current analytical methods are not particularly applicable for use with a brittle material with discrete planes of weakness, such as masonry and it is important, therefore, to emphasize the final determination of thickness which will require some engineering judgement rather than relying solely on single mathematical answers. It is not possible at present to give a definitive mathematical solution to panels with openings or additional line loads. The information given in this section should, however, provide the designer with sufficient guidance to provide a more confident answer.

It is convenient to start with the reference to the frame surround to the opening made in Appendix D of the Code. Most frames will possess a reasonable amount of strength such that they may either have sufficient strength to replace that lost by the area of the opening (particularly when the opening is small) or the strength of the frame will tend to transfer the loading from the opening towards the corners of the frame. This latter point could in many cases have the effect of a reduction of the bending moment in the panel from that which would otherwise be obtained by consideration of the

opening to apply a full uniformly distributed load along the unsupported edge.

There are several analytical methods which could be used to give an indication of the effect of openings in masonry walls but yield line analysis appears the most adaptable, is well documented and has, therefore, been adopted as the basic mathematical approach used in the following part of this section. The most useful reference for this type of problem is that written by Johansen[8.9].

8.2.7.1 Line loads on panels

Before dealing with the actual case of openings within the panel, such as windows and doors, it will be useful to consider the situation where a load is applied along the entire length of an unsupported edge as shown in Figure 8.18, since it should be noted that the bending moment coefficients as given in Table 9 of BS 5628 for U and C shaped panels (panels A, B, C, D, J, K and H – Table 9) do not allow for any load on the unsupported edge. The solution to this problem will generally be either, (a) to simply consider the load on the free edge to be carried by a given band of wall adjacent to the opening, or (b) to determine a new yield line analysis with a line load included. The first approach needs little explanation and is simply a matter of assuming a width of wall over which the loading is considered to be distributed. Many widths have been suggested, some as low as 300 mm. However, by consideration of the distribution of line loads on slabs as given by Johansen[8.9] (Section 1.4) and the normal distribution of edge loads as given in BS 8110, it would be reasonable to consider the additional load from the opening to be carried by a band of masonry $0.25\,L$ wide, the bending moments being taken as suggested in Section 8.1 for walls spanning vertically or horizontally as appropriate. The second approach is more complex in that it is necessary for a yield line analysis to be carried out. For certain panel shapes it is possible to use standard formulae such as that given by Johansen. The panel supported on three sides with a uniformly distributed load and line load on the free edge is usefully given in Section 1.2 of Johansen. In fact, by using the approach given by Johansen, a simple modification to the coefficients given in the Code can be proposed as follows.

Johansen indicates that a slab with a uniformly distributed load and a line load may be converted to the case of a slab with a uniformly distributed load only by use of certain transformations. The transformations given by Johansen basically modify the aspect ratio of the slab and effectively increase the uniformly distributed load applied. The equations given by Johansen (Section

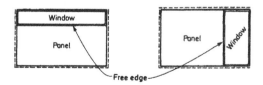

Figure 8.18

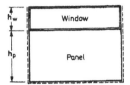

Figure 8.19

Table 8.3

$\dfrac{h_w}{h_p}$	Modification to load $(1 + 2\beta)$	Modification to aspect ratio $\dfrac{(1 + 3\beta)}{(1 + 2\beta)}$
0.5	1.5	1.17
0.4	1.4	1.14
0.3	1.3	1.12
0.2	1.2	1.08
0.1	1.1	1.04

1.2) are in terms of a variable line load in two directions and a variable uniformly distributed load. In the case of the wall shown in Figure 8.18, it is only necessary to consider the line load along the free edge and, since the line load will thus be a product of the applied uniformly distributed load, the equations may be simplified in terms of only one variable load, i.e. the wind load.

Taking the example in Figure 8.19 and defining the window height as h_w and the panel heights as h_p, the transformations or modifications to the panel dimensions are controlled by an expression

$$\sqrt{\frac{1 + 3\beta}{1 + 2\beta}}$$

for height and

$$\sqrt{\frac{1 + 2\beta}{1 + 3\beta}}$$

for length, which in effect increases the aspect ratio $\dfrac{h}{L}$ by

$$\frac{1 + 3\beta}{1 + 2\beta} \quad \text{where } \beta = \frac{\text{total line load}}{\text{total uniformly distributed load}}$$

The total line load for the panel is shown as $h_w/2 \times L \times p$ (p is the wind load). The total uniformly distributed load is $h_p \times L \times p$. Hence $\beta = 0.5\, h_w/h_p$.

The transformation or modification to the applied distributed load is controlled by $1\beta + 2\beta$ where β is as indicated previously.

The modifications to the aspect ratio and applied uniformly distributed load, i.e. that which is required to convert a panel with uniformly distributed and line load to a panel with uniformly distributed load only, may be tabulated in terms of the ratio of the window height to the panel height as shown in Table 8.3.

By inspection of the bending moment coefficients given in Table 9 of the Code it will be found that for small values of $\dfrac{h_w}{h_p}$ the modification of the aspect ratio in Table 8.2 does not have a very significant effect. The transformation also allows for modification to the fixity of the base but by considering the worst case when the fixity = 0 (i.e. simple support), this term can conveniently be neglected, thus leaving only the modification to load and aspect ratio to be considered. However, by ignoring the modification to the aspect ratio, which is reasonable on the grounds that its effect will often be small and that the argument of the stiffness of the frame can be used to counter this, it can be suggested that the effect of a panel with a line load can be catered for by simply increasing the applied uniformly distributed load by the factor $1 + 2\beta$. This can be rewritten as $[1 + h_w/h_p]$. Thus for panels with line loads it can be suggested that the design moment is given by $m = \alpha W_k\, \gamma_f\, [1 + h_w/h_p]\, L^2$ for U shaped panels or $m = \alpha\, W_k\, \gamma_f\, [1 + L_w/L_p]\, L^2$ for C shaped panels where α is taken from Table 9 of the Code, from the equations given in 8.1.1.2 or other appropriate formula. An alternative and perhaps easier way is to suggest that the coefficient α should be multiplied by $[1 + h_w/h_p]$ or $[1 + L_w/L_p]$, as appropriate.

In effect this approach suggests that the bending moment coefficient should be increased by a factor equal to the window height divided by the panel height, i.e. if the window is 10% of the height of the panel, increase the coefficient by 10%. The transformation for the alternative yield line pattern can be tackled in a similar way. This tends to give lower modification factors than for the previous case and therefore indicates that the simple modification still applies. Due to the simplification made and because the coefficients given in Table 9 of the Code may be from a different basic formula to that given by Johansen, the simple modification may not be technically correct but gives good agreement with a full rigorous analysis. When it is remembered that yield line analysis is only an approximate solution to masonry walls, the simple suggestion for dealing with a panel with a line load appears quite reasonable.

A rigorous analysis of the panel shown in Figure 8.20 gives a bending moment coefficient in the region of 0.09. The basic coefficient from Table 9 for $h/L = 2.4\,3.6 = 0.68$ and orthogonal ratio $\mu = 0.35$ gives $\alpha = 0.074$.

Now height of window = 0.6 m and height of panel 2.4 m, therefore $h_w/h_p = 0.6/2.4 = 0.25$. Therefore increase coefficient by 25%, i.e. $0.074 \times 1.25 = 0.093$. This gives good agreement with the rigorous analysis.

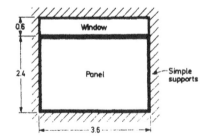

Figure 8.20

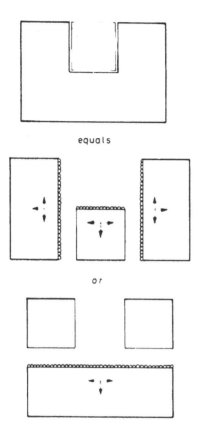

equals

or

Figure 8.21

8.2.7.2 Panels with openings

It is mentioned in the Code that small openings, due in part to the strength of the surrounds, will have little effect on the strength of the wall. The Code gives no guidance as to what a small opening is. An opening of only 5% each way of the panel is small but the effect of the opening will depend on exactly how the openings are distributed within the panel. A small slit completely up one edge will have a negligible area but will convert a panel from four edge support to three edge support. When openings become larger one basic suggestion made in the Code is to divide the panel into sub panels and then to design each part in accordance with the rules given in Section 3.6 of the Code. This example given in the Code is illustrated in Figure 8.21. Unfortunately, it is difficult to see how this can be done since the panels as shown have line loads on the free edge which, as explained in previous sections, are not calculated for in the Code. However, by using methods suggested in the previous section this can now be done.

The alternative suggestion made in the Code, Appendix D, is to design the panel using a recognized method of obtaining bending moments in flat plates, e.g. finite element or yield line. Since the use of yield line was made in the previous section dealing with line loads on the free edge, it will be continued within this section.

Section 1.15 of Johansen will be found to be invaluable for openings within panels since several commonly occurring situations have been dealt with. In the previous section a simplification for dealing with the problem was made using Johansen's work as a basis. In the case of panels with openings, however, Johansen himself gives a simple method for dealing with the problem, suggesting in Section 1.24 that slabs with small holes may be dealt with by simply reducing the fixity of the supports to an amount equal to the length of the opening divided by the length of the side. The approach used by Johansen can be made selective by determining the basic yield line pattern and modifying for any openings which interfere with a yield line. However, adoption of the basic transformation approach and modification of the fixity of all supported edges in proportion to the total length of openings adjacent to each side, can provide the designer with a method for assessing the strength of panels with openings. The question of 'what is small' must be considered and indeed some limitation as to the disposition of the openings must be made. Since this method is not an attempt to replace a rigorous analysis but purely the provision of a simple indication or rule of thumb which may be used to assess the required panel thickness, a limit of openings up to 25% of the panel area is suggested since this conveniently reduces the fixity by 50%, i.e. it reduces a panel with fully fixed edges to half way between this and a simple support condition. In addition the maximum length of any opening should not exceed half the panel dimension. Where the dimensions or area of openings are in excess of these limits the tendency will be to reduce the panel to a different form, for example, a four sided support may well reduce to a panel with three supports and a three sided support may reduce to a two sided support or a dangerous condition where the panel is supported on two adjacent edges. Where a panel has initially simple supports the transformation will reduce the fixity to effectively a negative condition.

The determination of the bending moment or coefficient may be made using the formula given in *Section 8.1.1.2*. For convenience the coefficients given in the Code are repeated. The simplifications for dealing with panels with openings can be demonstrated by considering the panel shown in Figure 8.22.

Taking the orthogonal ratio of the basic panel to be 0.5 and the aspect ratio of h/L 3/6 = 0.5 the bending moment coefficient for a wall with fixed supports from Table 9.1 of Code is given as 0.014. The corresponding coefficient for a simple support condition from Table 9A is 0.028. Since the original fixity was 1.0 reducing the fixity of the supports to $\phi (1 - \frac{h_w}{h_p})$ and $\phi (1 - \frac{L_w}{L_p})$

gives a partial fixity of 0.75 which, by linear interpolation, suggests a bending moment coefficient for the panel shown in Figure 8.22 of 0.014 + 0.014 × 0.25, i.e. 0.0175, say 0.018.

Where the opening is such that the modifications give a different change in fixity to adjacent sides it will be necessary to determine a new bending moment coefficient

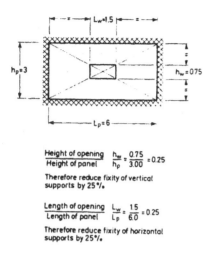

Height of opening / Height of panel $\dfrac{h_w}{h_p} = \dfrac{0.75}{3.00} = 0.25$

Therefore reduce fixity of vertical supports by 25%.

Length of opening / Length of panel $\dfrac{L_w}{L_p} = \dfrac{1.5}{6.0} = 0.25$

Therefore reduce fixity of horizontal supports by 25%.

Figure 8.22

by use of the formula given in *Section 8.1.1.2* with modified fixity to the supports, or a rigorous analysis of the type given by Johansen. In the case of a panel with a line load and openings this may be tackled by first adjusting for modified fixity to take care of the openings and then modifying the coefficients so obtained by the method suggested in the previous section.

8.2.7.3 Final assessment

It has already been mentioned several times that any method of analysis, either rigorous analysis or the simplification given, should not automatically be taken to give an absolute answer. It is essential to give the matter some thought before reaching a final design solution. One way to tackle the general subject of laterally loaded panels with either line load and/or openings would perhaps be to assess initially what thickness or strength of unit would be required without openings, then to carry out either a rigorous design or the simple design approach given in this Section to assess the thickness or strength required for the panel with openings. In some situations it may be necessary to consider what thickness would be required if the panel were idealized into separate elements as suggested by the Code. Having done this and with consideration to the fixity of the frame to the opening, the engineer should be able to confidently select the units for the wall. In many cases the engineer will also be able to draw on previous experience.

8.2.8 Influence of hollow blocks

In Table 3 of BS 5628 particular reference is now made to the flexural strength of hollow blocks. Since no restriction is made in the Code with regard to the use of hollow or cellular blocks in a laterally loaded wall it can be taken that solid, hollow and cellular blocks should be treated in the same way. This has been demonstrated by

tests and the following is also included as a reasoned explanation.

As mentioned in *Section 8.1* the test results give some correlation for concrete blocks between the unit strength and the flexural strength. This aspect is not uncommon as there is also a good relationship between the tensile and cube strength of normal concrete. Since the unit strength of blocks is assessed on the gross area (see *Chapter 3*) it follows that the concrete strength or nett compressive strength of a hollow block will be higher than its quoted unit strength. If the relationship between the compressive strength and the flexural strength is correct it may be found that the flexural strength of a hollow block will be greater than for the solid block, when bending occurs perpendicular to the bed joints. Hence for hollow blocks it could be argued that the characteristic flexural strength of hollow blocks could be obtained from Table 3 of BS 5628 by using the nett strength of the unit instead of the normally quoted gross strength. This will give an increase in the flexural strength of some 35% for the typical hollow block.

The presence of voids, although possibly increasing the flexural strength, will reduce the section modulus of the wall. If a vertical section through a wall containing hollow blocks is considered to be depicted by the section shown in Figure 8.23, with a mean void width of $0.6\ t$, the section modulus is given as $(1 - 0.6^2)\ bt^2/6 = 0.64\ bt^2/6$, i.e. some 36% less than a solid block. The conclusion drawn from this is that the reduction in the section modulus is counteracted by the possible increase in flexural strength of the unit and that a hollow block could be treated as if it were solid when bending occurs perpendicular to the bed joints.

A slightly different approach is necessary when considering failure parallel to the bed joint since the flexural strength in this direction is controlled by the adhesion between the mortar and the unit rather than by the tensile or flexural strength of the concrete within the unit. It will be noted from Table 3 (BS 5628) that only one value of the characteristic flexural strength along the bed joint is given, i.e. 0.25 N/mm², (0.20 N/mm² in the case of mortar designation (iv) for all block strengths. Although some increase in the flexural strength with respect to unit strength was indicated from research the

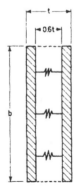

Figure 8.23

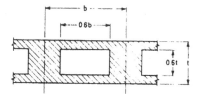

Figure 8.24

amount was relatively small and was ignored. Again this work was conducted on solid units but should be reasonable for hollow blocks since the stresses quoted are basically quite low (with $\gamma_m = 3.5$, $\gamma_f = 1.2$, the bed joint stresses are effectively slightly less than those given in CP 111).

A horizontal section through a wall built with hollow blocks may be represented by Figure 8.24 which gives the section modulus as $(1 - 0.6 \times 0.6^2)\, bt^2/6$ i.e. $0.784\, bt^2/6$.

Rather than determining the section modulus of the particular block being used, the maximum reduced section modulus, as given above, could be adopted which would effectively mean the same as treating the block as solid with flexural strengths from Table 3 of BS 5628 reduced by 0.8. However, as hollow blocks are likely to be better bedded than solid blocks, and as there is some test evidence to indicate that bed joint stresses tend to increase with the unit strength, it can be justifiable to ignore the reduction in the section modulus and just treat hollow blocks as solid with flexural strengths directly from Table 3.

8.2.9 Influence of damp-proof courses

The introduction of a damp-proof course in a wall subject to flexure may alter the degree of fixity that occurs at the base. Clause 36.4.2 indicates that the bending moment coefficient, $\mu\alpha$, at a damp-proof course may be taken as for an edge over which their full continuity exists when there is sufficient vertical load on the damp-proof course to ensure that its flexural strength is not exceeded.

The application of this is not particularly clear and a number of procedures are possible.

Firstly by considering an uncracked section and stresses; the flexural strength f_{kb}/γ_m will not be exceeded where a vertical stress of the same value is applied.

This gives

$$\frac{P}{A} - \frac{f_{kb}}{\gamma_m} = 0$$

Thus with a stress of $P/A = f_{kb}/\gamma_m$ acting, a moment resulting in a flexural stress of f_{kb}/γ_m could be carried before tension develops. By this method the degree of fixity, adapting a linear relationship, may then be written as:

$$\phi = \left[\frac{\text{design vertical load per unit area}}{\left(\dfrac{f_{kb}}{\gamma_m} \right)} \right]$$

or where the vertical load gd has been used to determine the orthogonal ratio:

$$\phi = \left[\frac{\text{design vertical load per unit area}}{\left(\dfrac{f_{kb}}{\gamma_m} \right) + g_d} \right]$$

The alternative approach which is appropriate for a cracked section is to define fixity as the ratio of the moment of resistance of the panel at the support (base) to the moment of resistance at midspan (midheight).

$$\phi = \frac{M_d \text{ support}}{M_d \text{ span}}$$

The moment of resistance at the support is the product of the design vertical load gd from the centroid of the compressive stress block.

$$\frac{n_w}{2} \left[t - \frac{n_w \gamma_m}{f_k} \right] \quad \text{(clause 36.5.2)}.$$

Relating this to just the self weight stress gives:

$$M_{d\,support} = (g_d\, bt/2)\,(t - g_d\, \gamma_m\, t/f_k)$$
$$= 3 g_d\, (1 - g_d\, \gamma_m\, f_k)\, Z$$

This takes into account the characteristic strength of the material but for a wall carrying self weight only, the eccentricity may, for simplicity, be taken as $0.45\, t$.

The moment, when also introducing the partial factor for loads, is then:

$$M_{d\,support} = 0.45\, h\, \rho t^2\, \gamma_f$$

This value was determined from an extensive range of tests carried out by Anderson[8.19].

Now the moment of resistance in the span at the design level is taken as:

$$M_{d\,span} = \left(\frac{f_{kb}}{\gamma_m} \right) Z$$

Where the vertical load gd has been used to determine the orthogonal ratio, the characteristic moment then becomes:

$$M_{d\,span} = \left(\frac{f_{kb}}{\gamma_m} + g_d \right) Z$$

However as stresses in excess of f_k/γ_m develop the effect will be to reduce the degree of fixity since the moment at the base is a constant resulting from the cracked section. This will result in a lowering of the bending moment coefficient and hence the failure strength would be less than that expected by setting γ_m to one. Now in order to sustain the same fixity up to failure the support moment would need to be compared with the characteristic moment in the span. This then gives

$$M_{d\,span} = (f_{kb})\, Z$$

Similarly, where the vertical load g_d has been used to determine the orthogonal ratio, the characteristic moment then becomes:

$$M_{d\ span} = (f_{kb} + g_d)\ Z$$

Now when

h is the height of the panel (m)
t is the thickness of the wall (m)
ρ is the density of the masonry (kg/m³)

and where the vertical load g_d has been used to determine the orthogonal ratio:

The degree of fixity is then

$$\phi = \frac{0.45\ h\ \rho\ t^2\ \gamma_f\ 9.81 \times 10^{-6}}{\left((f_{kb} + g_d)\dfrac{t^2}{6}\right)}$$

Where

$g_d = 0.5\ h\ \rho\ \gamma_f\ 9.81 \times 10^{-3}$
    $= 4.4\ \rho\ h \times 10^{-6}$

Adopting $\gamma_f = 0.9$, the most onerous condition, the equation may be simplified to:

$$\phi = \frac{0.24\ h\ \rho \times 10^{-4}}{(f_{kb} + 4.4\ h\ \rho \times 10^{-3})}$$

The term gd ($4.4h\ \rho \times 10^{-3}$) may be omitted when the vertical load (self weight) has not been used for determining the orthogonal ratio. The advantage of the above expression is that it is independent of f_k and simplifies tabulation.

The degree of fixity, when gd is used for the orthogonal ratio, resulting from the last equation is given in Table 8.4 and in Table 8.5 when it is not. When = 1 or greater fully fixed, when < 1 partially fixed. The bending moment coefficient may then be obtained by linear interpolation from Table 8.7 (Table 9 BS 5628) or from a yield line analysis (e.g. the equations in 8.1.1.2).

Table 8.4 Fixity at dpc due to self weight – when gd used for othogonal ratio

	f_{kb} >	0.15	0.17	0.19	0.21	0.22	0.24	0.25
hp	gd	fixity	fixity	fixity	fixity	fixity	fixity	fixity
0	0	0.00	0.00	0.00	0.00	0.00	0.00	0.00
500	0.0022	0.08	0.07	0.06	0.06	0.05	0.05	0.05
1000	0.0044	0.16	0.14	0.12	0.11	0.11	0.10	0.09
1500	0.0066	0.23	0.20	0.18	0.17	0.16	0.15	0.14
2000	0.0088	0.30	0.27	0.24	0.22	0.21	0.19	0.19
2500	0.011	0.37	0.33	0.30	0.27	0.26	0.24	0.23
3000	0.0132	0.44	0.39	0.35	0.32	0.31	0.28	0.27
3500	0.0154	0.51	0.45	0.41	0.37	0.36	0.33	0.32
4000	0.0176	0.57	0.51	0.46	0.42	0.40	0.37	0.36
4500	0.0198	0.64	0.57	0.51	0.47	0.45	0.42	0.40
5000	0.022	0.70	0.63	0.57	0.52	0.50	0.46	0.44
5500	0.0242	0.76	0.68	0.62	0.56	0.54	0.50	0.48
6000	0.0264	0.82	0.73	0.67	0.61	0.58	0.54	0.52
6500	0.0286	0.87	0.79	0.71	0.65	0.63	0.58	0.56
7000	0.0308	0.93	0.84	0.76	0.70	0.67	0.62	0.60
7500	0.033	0.98	0.89	0.81	0.74	0.71	0.66	0.64
8000	0.0352	1.04	0.94	0.85	0.78	0.75	0.70	0.67
8500	0.0374	1.09	0.98	0.90	0.82	0.79	0.74	0.71
9000	0.0396	1.14	1.03	0.94	0.87	0.83	0.77	0.75
9500	0.0418	1.19	1.08	0.98	0.91	0.87	0.81	0.78
10000	0.044	1.24	1.12	1.03	0.94	0.91	0.85	0.82

Table 8.5 Fixity at dpc due to self weight – when gd not used for orthogonal ratio

	f_{kb}	0.15	0.17	0.19	0.21	0.22	0.24	0.25
hp	gd	fixity	fixity	fixity	fixity	fixity	fixity	fixity
0	0	0.00	0.00	0.00	0.00	0.00	0.00	0.00
500	0	0.08	0.07	0.06	0.06	0.05	0.05	0.05
1000	0	0.16	0.14	0.13	0.11	0.11	0.10	0.10
1500	0	0.24	0.21	0.19	0.17	0.16	0.15	0.14
2000	0	0.32	0.28	0.25	0.23	0.22	0.20	0.19
2500	0	0.40	0.35	0.32	0.29	0.27	0.25	0.24
3000	0	0.48	0.42	0.38	0.34	0.33	0.30	0.29
3500	0	0.56	0.49	0.44	0.40	0.38	0.35	0.34
4000	0	0.64	0.56	0.51	0.46	0.44	0.40	0.38
4500	0	0.72	0.64	0.57	0.51	0.49	0.45	0.43
5000	0	0.80	0.71	0.63	0.57	0.55	0.50	0.48
5500	0	0.88	0.78	0.69	0.63	0.60	0.55	0.53
6000	0	0.96	0.85	0.76	0.69	0.65	0.60	0.58
6500	0	1.04	0.92	0.82	0.74	0.71	0.65	0.62
7000	0	1.12	0.99	0.88	0.80	0.76	0.70	0.67
7500	0	1.20	1.06	0.95	0.86	0.82	0.75	0.72
8000	0	1.28	1.13	1.01	0.91	0.87	0.80	0.77
8500	0	1.36	1.20	1.07	0.97	0.93	0.85	0.82
9000	0	1.44	1.27	1.14	1.03	0.98	0.90	0.86
9500	0	1.52	1.34	1.20	1.09	1.04	0.95	0.91
10000	0	1.60	1.41	1.26	1.14	1.09	1.00	0.96

8.3 Design examples to BS 5628: Part 1

In the following examples it should be noted that the values for γ_m and γ_f are arbitrary values and in practice these values will need to be adjusted to suit the actual category of manufacturing and construction control being used. Also, in instances where the panel is providing stability to the structure, then γ_f of 1.4 will be required.

8.3.1 Laterally loaded masonry walls: Basic design example

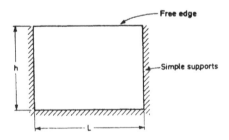

Determine the required thickness of a single leaf wall supported as shown above using the following criteria:

Characteristic wind load   = 0.45 kN/m²
Height of wall to free edge   = 4.5 m
Length of wall between restraints   = 4.5 m
Concrete blocks (solid) strength   = 7.0 N/mm²
Normal category construction control } = γ_m = 3.1
Special category manufacturing control }
Panel *not* providing stability to structure = γ_f = 1.2
Mortar designation   = (iii)

8.3.1.1 Flexural design

The design bending moment per unit height of the wall is given by the following expression (Clause 36.4.2, BS 5628: Part 1):

$$m = \alpha \, W_k \, \gamma_f \, L^2$$

The bending moment coefficient depends on: (i) orthogonal ratio, (ii) aspect ratio, h/L, (iii) support conditions.

(i) Try thickness of 190 mm, 7.0 N/mm² block
Determine f_{kx}
for 7 N/mm² at 100 (f_{kb})   = 0.25
for 7 N/mm² at 250 (f_{kb})   = 0.15
Therefore by linear interpolation at 190 mm (f_{kb}) = 0.19
for 7 N/mm² at 100 (f_{kb})   = 0.6
for 7 N/mm² at 240 (f_{kb})   = 0.35
Therefore at 190 mm (f_{kb})   = 0.45

$$\text{Orthogonal ratio} = \frac{f_{kb}}{f_{kp}} = \frac{0.19}{0.45} = 0.42$$

(ii) Aspect ratio $\dfrac{h}{L} = \dfrac{4.5}{4.5} = 1.0$

(iii) Support conditions – simple
From BS 5628 Table 9A (Table 8.6.1)
for $\dfrac{h}{L} = 1.0$

$\alpha = 0.083$ with $\mu = 0.5$
$\alpha = 0.087$ with $\mu = 0.4$
Therefore by linear interpolation
$\alpha = 0.0862$ when $\mu = 0.42$
Since $W_k = 0.45$ kN/m² $\gamma_f = 1.2$ L = 4.5
then applied design moment per unit height
$m = 0.0862 \times 0.45 \times 1.2 \times 4.5^2$
    $= 0.943$ kNm/m
The design moment of resistance is given as (Clause 36.4.3)

$$M = \frac{f_{kx}}{\gamma_m} Z$$

For a panel bending in two directions

$$= \frac{f_{kx}}{\gamma_m} Z$$

Now, since $f_{kp} = 0.45$ and $t = 190$

$$\text{then } M = \frac{0.45 \times 190^2 \times 10^{-3}}{3.1 \times 6}$$

$$= 0.873 \text{ kNm/m} \ (< 0.943)$$

This is too small and the moment capacity needs to be increased.
Try 215 $f_{kb} = 0.173$, $f_{kp} = 0.408$

Therefore $\mu = \dfrac{0.173}{0.408} = 0.42$

and α = 0.0862

Hence m = 0.943 kNm/m

The design moment of resistance

$$M = \frac{0.408 \times 215^2 \times 10^{-3}}{3.1 \times 6}$$

= 1.015 kNm/m

This is greater than the applied design bending moment and therefore adequate. Therefore use block of thickness 215 mm (7 N/mm²).

Rather than using the above iterative approach the thickness could alternatively be determined by using the simple method described in 8.2.2.8.
With the simplified method the thickness of a 7 N/mm² is determined as follows:
Let $\mu = 0.4$ and $f_{kp} = 0.6$
Since $h/L = 1.0$, then $\alpha = 0.087$ (BS 5628 Table 9A) (Table 8.6.1) and as $W_k = 0.45$, $\gamma_f = 1.2$, L = 4.5
$M = 0.087 \times 0.45 \times 1.2 \times 4.5^2 = 0.951$ kNm/m
Then from Table 8.2 (page 83) the required thickness is 215 mm ($m = 1.015$ kNm/m, $\gamma_m = 3.1$)
Maximum characteristic wind for this thickness

$$m = M$$

$$0.0862 \, W_k \times 1.2 \times 4.5^2 = 1.015$$

$$\therefore \ W_k = \frac{1.015}{0.0862 \times 1.2 \times 4.5^2}$$

= 0.48 kN/m²

8.3.1.2 Slenderness limits

Panel simply supported on more than one side, therefore, from Clause 36.3, $h \times L \leq 1350 \, t_{ef}^2$.
   Since wall is single leaf $t_{ef} = t$
therefore:

$$t \geq \sqrt{\frac{h \times L}{1350}} \geq \sqrt{\frac{4.5 \times 4.5 \times 10^6}{1350}} \geq 123 \text{ mm}$$

In addition no dimension shall exceed 50 t_{ef}. Since h and L both = 4.5 m, then t $\geq \dfrac{4.5 \times 10^3}{50} \leq 90$ mm.

8.3.1.3 Design for shear

Consider the wind load to be distributed to the supports as shown below:

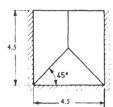

Then total load to support = $\gamma_f \times W_k \times$ loaded area

Thus the total shear along base

$$= \frac{1.2 \times 0.45 \times (4.5 \times 2.25)}{2} = 2.734 \text{ kN}$$

Assuming that this load is uniformly distributed along base, the design shear force per metre run

$$= \frac{2.734}{4.5} = 0.608 \text{ kN}$$

The design shear stress (v_h) is therefore

$$\frac{0.608 \times 10^3}{215 \times 1000} = 0.0028 \text{ N/mm}^2$$

The characteristic shear strength (f_v) from Clause 25 = $0.35 + 0.6 \, g_A$.

Since the self weight is to be ignored, characteristic strength, f_v, = 0.35 N/m². A lower value may be required with the presence of a damp-proof course (see *Section 8.2.2.6*).

The design shear stress (v_h) must be limited so that

$$v_h \leq \frac{f_v}{\gamma_{mv}}$$

In this case v_h = 0.028 N/mm² and

$$\frac{f_v}{\gamma_{mv}} = 0.35 = 0.14 \text{ N/mm}^2.$$

Therefore shear resistance is adequate along base.

Total shear to each vertical support = 1.2 × 0.45 × 2.25 (2.25 + 4.5) = 4.10 kN. Again, considering load to be uniformly distributed along support, then design shear force per metre run = 4.10/4.5 = 0.91 kN.

Using 2 mm thick anchors into dovetail slots in column, characteristic strength of each tie = 4.5 kN (Table 8, BS 5628). Placing ties at 900 mm centres and taking the partial safety factor for material as 3.5. The design load resistance per metre run of wall = 4.5/3.5 × 1000/900 = 1.43 kN. This is greater than the design shear force and therefore adequate.

8.3.2 Example I

PROBLEM

A single leaf wall to a warehouse, as shown below, has to be designed to withstand a characteristic wind load of 0.6 kN/m². Architectural requirements stipulate a maximum wall thickness of 140 mm. The panel does not provide stability to the structure. Both construction and manufacturing control are to be normal category (γ_m = 3.5).

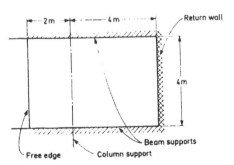

SOLUTION
Design for flexure
(1) *Right hand portion of panel 4 m long by 4 m high.* This panel is simply supported top and bottom. Consider return wall and support at column to give fixed vertical edges, i.e.

Try using 125 mm, 7 N/mm² block.

Determine f_{kx} by linear interpolation from Table 3, BS 5628.

Alternatively by reference to Table 8.2, page 83
f_{kx} (f_{kb}) = 0.233 and
  (f_{kp}) = 0.558
Orthogonal ratio

$$= \frac{f_{kb}}{f_{kp}} = \frac{0.233}{0.558} = 0.42 \text{ ignoring self weight}$$

Aspect ratio of panel $\frac{h}{L} = \frac{4}{4} = 1.0$

Using support condition as above, bending moment coefficient from BS 5628 Table 9G (Table 8.4.13) = 0.0376.

Now, W_k = 0.60 kN/m² γ_f = 1.2 L = 4 m
Therefore, applied design moment per unit height
m = 0.0376 × 0.60 × 1.2 × 4²
   = 0.433 kN m/m

The design moment of resistance $M_d = \frac{f_{kp}}{\gamma_m} Z$

$$= \frac{0.558}{3.5} \times \frac{1000 \times 125^2 \times 10^{-6}}{6}$$

= 0.415 kN/m/m. This is just too low.
Try 140 mm, f_{kb} = 0.223 f_{kp} = 0.533 (Table 8.2).
Since μ = 0.42, α and m remain unaltered.

$$M_d = \frac{0.533}{3.5} \times \frac{1000 \times 140^2 \times 10^{-6}}{6}$$

= 0.498 kNm/m
Alternatively the thickness of 140 mm could have been determined directly from Table 8.2.

Slenderness limits
Panel has less than three sides continuous, therefore Clause 36.3(b)(2) controls. $h \times L \leq 2025 \, t_{ef}^2$
$t_{ef} = t$ since single leaf wall. Therefore

$$t \sqrt{\frac{4 \times 4 \times 10^6}{2025}} \geq 89 \text{ mm}$$

Check 50 t_{ef} requirement

$$t \geq \frac{4 \times 10^3}{50} = 80 \text{ mm}$$

Loading condition controls and which would be met by a panel thickness of 140 mm

(2) *Left hand portion of panel 2 m long by 4 m high.* This portion of the panel is simply supported top and bottom. One vertical edge is free, the other is to be considered as fixed, i.e.

Again, using 7 N/mm² solid block.

Thus, as before, $\mu = 0.42$ (ignoring self weight).

Aspect ratio of panel $\dfrac{h}{L} = \dfrac{4}{4} = 2.0$

Although this is outside the extent of Table 9 (BS 5628) it does not mean that the panel cannot be designed. However, it is necessary to consider the panel behaviour more closely since there may be a tendency for the panel to span in one direction or for an alternative yield line pattern to develop. Note 2 to Table 9 (BS 5628) relates to a panel with a free top edge. In the case of a panel with a free vertical edge it is reasonable to consider the panel cantilevering from the vertical support at a ratio h/L of 1/0.3, i.e. around 3.0.

In addition the bending moment coefficient for a horizontal cantilever span would be of the order of 0.5 γ_f $W_k L^2$, i.e. a coefficient of 0.5. The coefficient in Table 9, Figure K, for $\mu = 0.42$ and $h/L = 1.75$ is in the order of 0.16. It is therefore clear that the panel under consideration with a h/L of 2.0 is still spanning significantly in two directions. In addition, since the panel is only slightly outside Table 9 the yield line pattern will remain basically the same and thus the bending moment coefficient may be determined by use of the formula given in *Section 8.1.1.2.*

For the panel given

$$\phi_1 = \phi_3 = 0 \quad \phi_2 = 1.0 \quad \mu = 0.42 \quad h = 4 \quad L = 2$$

$$\frac{h}{L} = \frac{4}{2} = 2.0$$

For failure pattern (i)

$$\alpha = \frac{1}{6}\left(\frac{1}{[(3 + \phi_2) \div Y^2] - 1}\right)$$

where $Y = \sqrt{\left(\left[\sqrt{\dfrac{\sqrt{1 + \phi_1} + \sqrt{1 + \phi_3}}{\dfrac{\left(\dfrac{h}{L}\right)^2 (3 + \phi_2)}{\mu}}}\right]^2 + 3\right) - \left(\dfrac{\sqrt{1 + \phi_1} + \sqrt{1 + \phi_3}}{\dfrac{\left(\dfrac{h}{L}\right)^2 (3 + \phi_2)}{\mu}}\right)}$

thus $Y = \sqrt{\left(\left[\sqrt{\dfrac{\sqrt{1} + \sqrt{1}}{\dfrac{2^2 \times 4}{0.42}}}\right]^2 + 3\right) - \left(\dfrac{\sqrt{1} + \sqrt{1}}{\dfrac{2^2 \times 4}{0.42}}\right)}$

$$= 1.44$$

Then $\alpha = \dfrac{1}{6} \times \dfrac{1}{\dfrac{4}{1.44^2} - 1}$

$$= 0.178$$

For failure pattern (ii)

$$\alpha = \frac{\left(\dfrac{h}{L}\right)^2}{6\mu Y_{13}{}^2}\left[\sqrt{\left(3 + \dfrac{\left(\dfrac{h}{L}\right)^2 Y_2{}^2}{\mu Y_{13}{}^2}\right)} - \dfrac{\sqrt{\dfrac{1}{\mu}}\left(\dfrac{h}{L}\right)Y_2}{Y_{13}}\right]^2$$

where $Y_{13} = \sqrt{1 + \phi_1} + \sqrt{1 + \phi_3}$ and $Y_2 = \sqrt{1 + \phi_2}$

$$\text{thus } Y_{13} = \sqrt{1} + \sqrt{1} = 2 \qquad Y_2 = \sqrt{2} \qquad \frac{h}{L} = \frac{4}{2} = 2$$

$$\text{hence } \alpha = \frac{2^2}{6 \times 0.42 \times 2^2} \left[\sqrt{\left(3 + \frac{2^2 \ (\sqrt{2})^2}{0.42 \times 2^2}\right)} - \frac{\sqrt{\frac{1}{0.42}} \ (2)\sqrt{2}}{2} \right]^2$$

$$= 0.145$$

$$\therefore \text{ Critical coefficient } = 0.178$$

$W_k = 0.60 \quad \gamma_f = 1.2 \quad L = 2$ therefore
$m = 0.178 \times 0.60 \times 1.2 \times 2^2$
$= 0.513$ kN m/m
This is greater than that provided by a 140 mm block.
Try 150 mm, $f_{kb} = 0.217$, $f_{kp} = 0.517$, thus $\mu = 0.42$
leaving $m = 0.513$ kN m/m.

$$M_d = \frac{0.517}{3.5} \times \frac{1000 \times 150^2 \times 10^{-6}}{6}$$

$$= 0.554 \text{ kN m/m (adequate)}$$

(Alternatively from Table 8.2, M = 0.554 for 150 mm block.)

Slenderness limits
The panel has simply two supported edges, therefore, Clause 36.3(1)(b) is applied, $h \times L \leq 1350 \ t_{ef}^2$

$$\therefore t \geq \sqrt{\frac{4 \times 2 \times 10^6}{1350}} \geq 77 \text{ mm}$$

50 t_{ef} requirement requires $\frac{4 \times 10^3}{50} = 80$ mm

Again loading condition controls panel size

The panel requires to be around 140 mm for the right hand portion and not less than 150 mm for the left hand portion. In this instance since both panels need to be approximately the same, the assumption that the central column support is fixed is confirmed. Had the panel sizes been significantly different then some readjustment would be required to allow for only partial fixity to the column support to the right hand panel. *Therefore, select 7 N/mm² block of thickness 150 mm.* By inspection of the results it may be possible, by taking self-weight into account, to justify a 140 mm block (see 8.3.3, for an example taking self-weight into account.)

Shear
The wind load to be assumed to be distributed to the supports as shown.

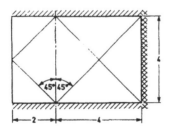

Total shear load to base (also top support)

$$= 1.2 \times 0.6 \left(\frac{4 \times 2}{2} + \frac{2 \times 2}{2}\right)$$

$$= 4.32 \text{ kN}$$

Assuming a uniform distribution the design shear force per metre run

$$= \frac{4.32}{6} = 0.72 \text{ kN}$$

The design shear stress,

$$v_h = \frac{0.72 \times 10^3}{140 \times 1000} = 0.005 \text{ N/mm}^2$$

Since self weight is to be ignored, design shear strength

$$= \frac{f_v}{\gamma_{mv}} = \frac{0.35}{2.5} = 0.14 \text{ N/mm}^2$$

Therefore, shear resistance is adequate.

Central column support
The central column support must be designed to withstand a total applied design load $= 1.2 \times 0.6 \times 2(4 \times 2)/2 = 5.76$ kN or $5.76/4 = 1.44$ kN/metre. This should be capable of taking the load without undue deflection, say span/500. Top support to be designed to resist design load 0.72 kN/metre. Connections also required to transmit forces to support.

8.3.3 Example 2

PROBLEM
This example illustrates how allowance may be made for the self weight of the wall. A wall as used in the basic design example (see 8.3.1) is to be used but allowance made for the density of the wall at 1200 kg/m³. Determine the characteristic pressure that may be carried.

Height of wall to free edge $= 4.5$ m
Length of wall between restraints $= 4.5$ m
Concrete blocks (solid) strength $= 7.0$ N/mm²
Normal category construction control $\}$ $\therefore \gamma_m = 3.1$
Special category manufacturing control $\}$
Panel *not* providing stability to structure $\therefore \gamma_f = 1.2$
Density of wall $= 1200$ kg/m³

SOLUTION
Design for flexure

$$m = \alpha \ W_k \ \gamma_f \ L^2$$

Determination of bending moment coefficient

(1) Modified orthogonal ratio

$$\mu = \frac{f_{kb} + \gamma_m\, g_d}{f_{kp}}$$

where g_d is the self weight stress at mid height of the panel. Other symbols are as indicated previously.

$g_d = 0.5\, h\, \rho\, 9.81 \times 10^{-6}\, \gamma_f$

γ_f for maximum criteria is 0.9

ρ is density of wall 1200 kg/m³

$\therefore g_d = 0.5 \times 4.5 \times 1200 \times 9.81 \times 10^{-6} \times 0.9 = 0.024$

now $f_{kb} = 0.17$ $f_{kp} = 0.41$ $\gamma_m = 3.1$

$$\therefore \mu = \frac{0.17 + 0.024 \times 3.1}{0.41} = 0.60$$

(2) Aspect ratio $h/L = 4.5/4.5 = 1.0$

(3) Support conditions – edge supports are again simple supports. Both supports have an effective restraint equivalent to that exerted the self weight of the wall. Since full wall height acts at base self weight, stress = $0.024 \times 2 = 0.048$ N/mm².

Since g_d was used to modify the orthogonal ratio then the effective fixity at the base using the stress approach is:

$$\phi = \frac{0.048}{\dfrac{0.17 + 0.024}{3.1}} = 0.61$$

Therefore effectively between a simple and fully fixed support.

There is no table that deals directly with partial supports and, therefore, the coefficient must be determined by use of yield line formula (e.g. *Concrete Masonry Designer's Handbook*) or, alternatively, an approximate solution may be obtained by linear interpolation from standard bending moment tables.

In this particular case, the condition for a fixed base with simple side supports is not contained in the Code but is given in *Section 8.5* of this handbook.

From TDH 6052

$\alpha = 0.080$ for simple base ($\phi = 0$), when $\mu = 0.60$ and $h/L = 1.0$

From Table 8.7.1

$\alpha = 0.067$ for fixed base ($\phi = 1$), when $\mu = 0.60$ and $h/L = 1.0$.

Hence, for partial fixity $\phi = 0.61$

$\alpha = 0.080 - (0.080 - 0.067) \times 0.61$
$= 0.072$

The design moment

$m = \alpha W_k \gamma_f L^2$
$= 0.072\, W_k\, 1.2 \times 4.5^2$
$= 1.75\, W_k$

and design moment of resistance is $M = \dfrac{f_{kp}}{\gamma_m} Z$

But $M = m$, and since $t = 215$ mm and $Z = bt^2/6$ then:

$$W_k \le \frac{f_{kp}\, b\, t^2}{1.75 \times \gamma m \times 6}$$

$$\le \frac{0.41 \times 1000 \times 215^2}{1.75 \times 3.1 \times 6 \times 10^6}$$

$$\le 0.58 \text{ kN/m}^2$$

Thus, by allowing for the self weight of the wall (density 1200 kg/m³) an additional 0.1 N/m² wind may be carried on the wall as shown in the basic example. This is a 21% increase.

The effective fixity adopting the moment approach (see *8.2.9*) which would be more appropriate when the dpc cannot provide tension is:

$$\phi = \frac{0.24\, h\, \rho \times 10^{-4}}{f_{kb} + 4.4\, h\, \rho \times 10^{-3}}$$

$$= \frac{0.24 \times 4.5 \times 1200 \times 10^{-4}}{0.17 + 4.4 \times 4.5 \times 1200 \times 10^{-3}}$$

$$= \frac{0.13}{0.17 + 0.024}$$

$$= 0.67$$

This suggests a slightly greater fixity than by the stress method but the effect on the maximum capacity would be minimal.

Shear

By comparison with *Example 1* shear is adequate at base. Shear to vertical supports is increased to $0.19 \times 0.50/0.45 = 1.10$ kN/metre. This is less than that provided by the ties (1.43 kN/metre), therefore adequate.

8.3.4 Example 3

PROBLEM

In this example a continuous window opening along the upper edge of a wall panel is to be designed. The loading on the window applies an effective line load on the free edge.

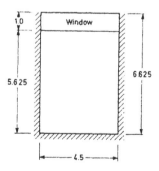

The wall is to be built with concrete bricks in designation (iii) mortar. γ_m to be taken as 3.5 and γ_f as 1.2. The characteristic wind pressure has been determined as 0.7 kN/m².

SOLUTION

Design for flexure

Adopting the approach given in *Section 8.2.7*, then for the basic bending moment coefficient:

(1) orthogonal ratio μ = 0.310.9 = 0.33
(2) aspect ratio h/L = 5.625/4.5 = 1.25
(3) support conditions – consider as simple.

Then BS 5628, Table 9A (Table 8.7.1) for h/L = 1.25 = 0.096 by linear interpolation
Height of window h_w = 1.0 m and height of panel h_p = 5.625 m
therefore h_w/h_p = 1.0/5.625 = 0.18
Modification factor for line load = 1 + 0.18 = 1.18
hence modified bending coefficient 0.096 × 1.18 = 0.113
Now W_k = 0.7 kN/m² γ_f = 1.2 L = 4.5 m
Then applied design moment per unit height
m = 0.113 × 0.7 × 1.2 × 4.5²
= 1.92 kN m/m
Design moment of resistance

$$M_d = \frac{f_{kp}}{\gamma_m} Z$$

Since with concrete bricks the orthogonal ratio remains constant and as design capacity increases in proportion to the section modules then the required thickness may be obtained from:

$$t \geq \sqrt{\frac{1.99 \times 3.5 \times 6 \times 10^6}{0.9 \times 1000}} \geq 212 \text{ mm}$$

Use 215 mm thick wall.

The example may be continued by considering what thickness would be required for the basic wall panel without the window and what thickness would be required had the panel been taken for the full 6.625 m height.

For the basic wall panel 5.625 × 4.5 m, bending moment coefficient as determined previously is 0.096 which gives a required thickness of 212 × 0.0965/0.113 = 180 mm.

Had the panel been 6.625 × 4.5 m then the aspect ratio is 6.625/4.5 = 1.47 and BS 5628, Table 9A (Table 8.7.1) for h/L say 1.5 α = 0.100. Hence thickness required = 212 × 0.100/0.113 = 188 mm. This latter value is useful in comparing to the original case with window opening since both panels are subject to a similar total wind load but in the case of the panel with openings the total length of yield line is less and hence a slightly greater panel thickness, as determined as 215 mm, appears justified.

The loading from the window could be considered to be simply carried by a strip of brickwork running parallel to the opening. Had this method been adopted then an additional loading, as illustrated below, could be assumed to be carried by the top 1.125 m of wall (i.e. 0.25 L as given in *Section 8.2.7*).

Equivalent uniformly distributed loading transferred to strip

$$= \frac{0.7 \times 0.5 \times (3.5 + 4.5)}{1.125 \times 4.5 \times 2} = 0.28 \text{ kN/m}^2$$

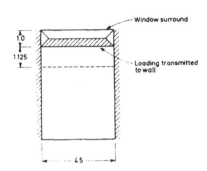

Total uniformly distributed loading on strip therefore = 0.7 + 0.28 = 0.98 kN/m². Designing strip as simply supported span m = 0.125 × 0.98 × 1.2 × 4.5² = 2.98 kN/metre, giving t required = 264 mm.

If moment were arbitrary, assessed as $\dfrac{W_k \gamma_m L^2}{10}$

then t required reduces to 236 mm.

Considering this, and that there will be some stiffness within the window frame, then a wall thickness as originally derived of 215 mm would appear adequate.

8.3.5 Example 4

PROBLEM

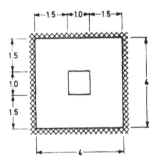

A single leaf panel 4 m long by 4 m high with a centrally placed window 1 m × 1 in has to be designed to resist a characteristic wind pressure of 0.6 kN/m². All edges are considered to be fixed.

$$\gamma_m = 3.5 \quad \gamma_f = 1.2 \quad 10.5 \text{ N/mm}^2 \text{ blocks}$$

SOLUTION

Using the method indicated in *Section 8.1.1.2* (h_w/h_p) fixity to edges is reduced by ¼ = 0.25, i.e. ϕ = 0.75. Now by use of the formula given in *Section 8.1.1.2* the bending moment coefficient may be determined as = 0.0375. An approximate solution may, however, be obtained by interpolating between E and I of Table 9 (BS 5628) – see Tables 8.7.7 and 8.7.15.

Now

$$\mu = \frac{0.25}{0.75} = 0.33 \quad \frac{h}{L} = \frac{4}{4} = 1.0$$

For the pinned edged condition panel E ($\phi = 0$) then $\alpha = 0.066$ (Table 8.7.7)

For fixed edged condition panel 1 ($\phi = 1$) then $\alpha = 0.033$ (Table 8.7.15)

For linear interpolation for all edges with fixity $\phi = 0.75$ $\alpha = 0.041$

To continue with the simple interpolated figure $m = 0.041 \times 0.60 \times 1.2 \times 4^2$

$= 0.47$ kN/m per metre

Using 10.5 N/mm² blocks then t required

$$= \sqrt{\frac{0.47 \times 3.5 \times 6 \times 10^6}{0.75 \times 1000}}$$

$= 115$ mm

Had the panel had simply supported edges top and bottom then the following would arise.

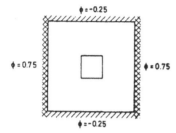

This could not be interpolated from Tables in BS 5628 and the formula would have been necessary. In this case the bending moment coefficient may be derived as:

$\alpha = 0.047$

$$m = 124 \times \sqrt{\frac{0.047}{0.038}} = 144 \text{ mm}$$

Since some strength will be obtained from self weight and even the window surround, a 140 mm thick wall would be adequate.

Referring back to *Example 2*, it is indicated that an opening of 1 m × 1 m may also have been able to be incorporated in the wall since its thickness was determined to lie between 125 mm and 140 mm but had to be increased to 140/150 mm to cater for the left hand panel.

8.3.6 Example 5

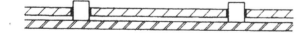

PROBLEM

A cavity wall of span 4.5 m by 3.375 m high has to be designed to withstand a wind pressure of 0.6 kN/m². The outside leaf is brickwork running past supporting columns. The inner leaf is blockwork abutting the columns with a soft control joint. The wall is to be designed assuming a simple support at its base but allowing for self weight. Taking density of brickwork as 1700 kg/m³ and blockwork as 1000 kg/m³.

SOLUTION
Outer leaf

Consider brick $f_{kb} = 0.3$ and $f_{kp} = 0.9$. Wall density 1700 kg/m³. Orthogonal ratio 0.3/0.9 = 0.33.

Consider as simply supported at base but allow for self weight of wall. Self weight stress at mid-height = $0.5 \times 3.375 \times 1700 \times 9.81 \times 10^{-6} \times 0.9 = 0.025$. Modified orthogonal ratio

$$\mu = \frac{0.3 + 3.1 \times 0.025}{0.9} = 0.42 \text{ consider as } 0.4$$

Therefore bending moment coefficient α from Table 9C, BS 5628 (Table 8.7.5) for aspect ratio 3.375/4.5 = 0.75 $\alpha = 0.044$

Consider thickness as 100 mm. Thus design moment capacity of wall

$$M_d = \frac{f_{kx}}{\gamma_m} Z = \frac{f_{kb} b t^2}{\gamma_m 6} = \frac{0.9 \times 1000 \times 100^2}{3.1 \times 6 \times 10^6}$$

$= 0.484$ kN/m per metre.

The applied design moment $m = \alpha W_k \gamma_f L^2$ and since this may be allowed to equal the design moment capacity of the wall then

$$W_k = \frac{M_d}{\alpha \gamma_f L^2} = \frac{0.484}{0.044 \times 1.2 \times 4.5^2} = 0.45 \text{ kN/m}^2$$

The total characteristic wind pressure to be applied is 0.6 kN/m². Therefore, the excess force to be carried by the inner leaf is 0.6 − 0.45 = 0.15 kN/m²

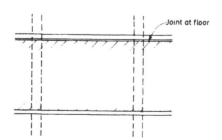

Inner leaf

Try 3.5 N/mm² blocks, 100 mm thick. Aspect ratio as before = 0.75. Orthogonal ratio = 0.25/0.45 = 0.55. Consider self weight as 1000 kg/m³. Then self weight stress at mid-height = 0.025 × 1000/1700 = 0.015 N/mm².

Modified orthogonal ratio

$$\mu = \frac{0.25 + 3.1 \times 0.015}{3.1} = 0.65$$

Now, considering simple support to all edges, from BS 5628, Table 9A (Table 8.7.1) α = 0.0675. Thickness is 100 mm, therefore

$$M_d = \frac{0.45 \times 1000 \times 100^2}{3.1 \times 6 \times 10}$$

$$= 0.242 \text{ kN/metre}$$

Therefore, characteristic wind force that can be carried

$$W_k = \frac{0.242}{0.0675 \times 1.2 \times 4.5^2} = 0.15 \text{ kN/m}^2$$

Total capacity of cavity wall 0.45 + 0.15 = 0.6 kN/m² and is therefore acceptable.

It should be noted that the wall panel could be shown to carry a greater wind force since some degree of fixity could be allowed at the base of both walls. If the mortar jointing were used between the inner block wall and the columns, then some degree of fixity might be considered to be taken at the support due to the anchoring action of the wall. However, the need to allow for movement in the wall would likely negate this.

In the example the total wind load was determined on the assumption that both leaves attain their ultimate stress simultaneously as inferred by the first paragraph to Clause 36.4.5. If, however, the wind load were shared between the two leaves in proportion to their design moments of resistance, then the stress in the leaves would be as follows:

M_d brick = 0.484 M_d block = 0.242, thus W_k to be shared in ratio

$$\frac{0.484}{0.726} \times 0.6 = 0.4 \text{ kN/m}^2 \text{ to brick and}$$

$$\frac{0.242}{0.726} \times 0.6 = 0.2 \text{ kN/m}^2 \text{ to block.}$$

This in turn gives

f_{kp} in bricks

$$= \frac{0.4 \times 1.2 \times 0.044 \times 4.5^2 \times 3.1 \times 6 \times 10^6}{1000 \times 100^2}$$

$$= 0.8 \text{ kN/mm}^2$$

$$f_{kp} \text{ in blocks} = \frac{0.8 \times 0.2 \times 0.0675}{0.4 \times 0.044} = 0.61 \text{ kN/mm}^2$$

This tends to suggest an overstress in the blockwork. However, as we are working in ultimate terms and since the wind pressure will actually be distributed as suction and positive pressure on both leaves, then it would appear reasonable to accept the first method for design purposes. The additional allowance for self weight fixity could be shown to prove the wall even more adequate.

8.4 Design of laterally loaded walls incorporating bed joint reinforcement

BS 5628 Part 2 (Annex A) provides guidance on the design of laterally loaded walls incorporating bed joint reinforcement. Bed joint reinforcement has been traditionally used in walls either to assist in controlling tensile stresses resulting from movement or to reinforce over openings. It has also been used by engineers to enhance the resistance to lateral loading but there was until BS 5628: Part 2 no codified approach.

8.4.1 General design aspects

Since the design of walls containing bed joint reinforcement is still rather in the development stage, there is no single accepted method and as a result the Code adopts a pragmatic approach and gives four alternative methods for design. There are also a number of variations to design as follows.

8.4.1.1 Safety factors

The whole approach to such design is that the reinforcement is being used solely to enhance the capacity of unreinforced walls and therefore the same partial safety factors for the masonry as used in Part 1 are employed (Part 2 contains γ_m values for reinforced masonry but these are not used when designing for bed joint reinforcement.)

8.4.1.2 Mortar

Additionally the design of reinforced sections may employ mortar of designation (iii) unlike that for true reinforced members where mortars of designation (i) or (ii) must be used.

8.4.1.3 Reinforcement

Since mortar may readily carbonate the bed joint reinforcement must be properly protected. This will require the use of galvanised or stainless steel bars, according to the exposure situation E2 – E3 (see Section 6 of BS 5628: Part 2), unless in a dry location (Exposure 1) when ordinary carbon steel can be acceptable. It is important to note that the bond strength of all bed joint reinforcement should be determined from the manufacturer.

8.4.1.4 Panel dimensions

In addition to the limitations imposed on the methods of analysis a limit on slenderness is also imposed. These are slightly greater than those used in BS 5628: Part 1 and are given as follows:

(a) Panel supported on three edges:
   (1) two or more sides continuous:
      height × length equal to 1800 t_{ef}^2 or less;
   (2) all other cases:
      height × length equal to 1600 t_{ef}^2 or less;

(b) Panel supported on four edges:
   (1) three or more sides continuous:
      height × length equal to 2700 t_{ef}^2 or less;
   (2) all other cases:
      height × length equal to 2400 t_{ef}^2 or less.

But no dimension should exceed 60 t_{ef} where t_{ef} is the elective thickness as defined in Part 2: Clause 4.3.2.4.

8.4.2 Methods of design

The four methods of design may be summarised as follows:

(1) Design as a horizontal reinforced member taking all the load.*
(2) Design a horizontal reinforced member taking excess load over plain wall.*
(3) Design by the approach used in Part 1 but modifying the orthogonal ratio to take account of the enhancement provided by the reinforced section.*
(4) Design by determining serviceability cracking load for reinforced member. (Strength and deflection).

8.4.2.1 Method 1

Most walls in which bed joint reinforcement is to be incorporated will, subject to the aspect ratio, tend to span in two directions. This method is taking the reinforced wall as spanning horizontally and is liable to be the most conservative. The unreinforced wall is assumed to do no work and all the load has to be carried by the reinforced section. There is however a limitation in that the capacity must not be more than 50% greater than the same wall unreinforced, unless a serviceability and deflection check is carried out in accordance with A6 of BS 5628: Part 2.

8.4.2.2 Method 2

In this method account is taken of the capacity of the unreinforced wall (including any two way action). Any additional capacity being taken by the reinforcement

*Note: methods 1–3 also require a serviceability and deflection check when the strength exceeds a certain amount. It should also be noted that this section refers to Annex A, BS 5628: Part 2 which is based on a restricted amount of research and should therefore be used with discretion.

subject to a maximum enhancement of 30%, unless a serviceability and deflection check is carried out in accordance with A6 of BS 5628: Part 2.

8.4.2.3 Method 3

The approach used for the third method of design is essentially the same as that used in Part 1, for unreinforced panels, except the strength of bending about the vertical axis and hence the orthogonal ratio is based on the design strength of the reinforced section. With an unreinforced wall the section modulus is normally constant and therefore the orthogonal ratio, in the general case, is equal to the flexural strength along the bed joint divided by the flexural strength perpendicular to the bed joint. Where a wall is reinforced along the bed joints, then the strength of bending in that direction is controlled by the reinforced section and the strength of bending about the bed joint is controlled by the flexural strength of the masonry. In that case the orthogonal ratio is found by determining the moment capacity in the two directions. The maximum enhancement of lateral load resistance above that for the equivalent unreinforced wall should be taken to be 50%, unless a serviceability and deflection check is carried out in accordance with A6 of BS 5628: Part 2.

8.4.2.4 Method 4

In this method the load to cause cracking in the reinforcement wall is taken as being equivalent to load capacity of the unreinforced section. The cracking load is found by dividing the characteristic capacity of the wall (i.e. partial safety factors set to unity) by the partial safety factor for the serviceability limit state ($\gamma_m = 1.5$) Having determined this an ultimate limit state check is made using one of the other previous methods with the cracking load taken as the characteristic wind load. In addition, a check must be made on the deflections. Full analysis by this method is, therefore, enhancement offered by methods 1, 2 and 3.

8.4.3 Design example

The following examples show the enhancement in lateral load obtained by the incorporation of bed joint reinforcement using each of the four methods given in 8.4.2.

The following examples are based on the wall panel as shown below

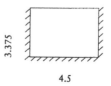

3.375

4.5

Unit strength $= 3.5$ N/mm²
Thickness $= 100$ mm
$\gamma_m = 3.1$
$\gamma_m = 1.2$

8.4.3.1 Unreinforced wall

The capacity of the unreinforced wall is determined as in Part 1 BS 5628.
For 100 mm 3.5 N/mm² block mortar designation (iii)
$f_{kb} = 0.25$, $f_{kp} = 0.45$
Therefore $\mu = 0.55$ and as $h/L = 0.75$, then $\alpha = 0.071$ (Table 8.7.1)

$$M = f_{kp}\, b\, t^2/6\, \gamma_m$$

$$= \frac{(0.45 \times 100^2 \times 10^{-3})}{(6 \times 3.1)} = 0.242 \text{ kN/m}^2$$

$$m = \alpha\, W_k\, \gamma_f\, L^2 \text{ and letting } m = M$$

$$W_k = \frac{M}{\alpha\, \gamma_f\, L^2} = \frac{0.242}{0.071 \times 1.2 \times 4.5^2} = 0.14 \text{ kN/m}^2$$

If self weight (1000 kg/m³) were allowed for, then the modified orthogonal ratio $\mu = 0.65$
Hence $\alpha = 0.0675$ (Table 8.7.1), giving $W_k = 0.15$ kN/m²
Allowing for self weight at the base of the wall would give an effective fixity $\phi_2 = 0.31$
Then $\alpha = 0.063$ giving $W_k = 0.16$ kN/m²
(Note: α obtained by linear interpolation from Tables 8.7.1 and 8.7.2, for $\mu = 0.65$. Analysis from equations in 8.1.1.2 gives same answer).

8.4.3.2 Reinforced Walls

Method 1 Horizontal reinforced member (see Section 9)
Block strength = 3.5 N/mm², $h/t = 2.0$, therefore f_k (Table 2d, BS 5628 Part 1) = 3.5 N/mm².

Using proprietary 50 mm wide bed joint reinforcement (10 mm²) per bar, then $d = 75$ mm

$$\text{Now } z = d \left(1 - \frac{0.5\, A_s\, f_y\, \gamma_m}{d\, f_k\, \gamma_s} \right)$$

Using reinforcement at 450 vertical centres $A_s = 22$ mm²/m.
Then $z = d \left(1 - \left(\frac{0.5 \times 22 \times 485 \times 3.1}{1000 \times 75 \times 3.5 \times 1.15} \right) \right) = 0.95d$

Design moment $M_d = \dfrac{A_s\, f_y\, d}{\gamma s}$

$$= \frac{22 \times 485 \times 75 \times 0.95}{1.15 \times 10^6} = 0.66 \text{ kN m/m}$$

As panel is being designed to span horizontally between simple supports then
$m = W_k\, \gamma_f\, L^2 / 8$ and letting $m = M_d$

$$\text{then } W_k = \frac{0.66 \times 8}{1.2 \times 4.5^2} = 0.22 \text{ kN/m}^2$$

Ignoring self-weight, then capacity increase
$$= \frac{0.22 - 0.14}{0.14} = 57\%$$

Including self-weight, then capacity increase
$$= \frac{0.22 - 0.15}{0.15} = 47\%$$

Allowing for base fixity, then capacity increase
$$= \frac{0.22 - 0.16}{0.16} = 38\%$$

The maximum enhancement must not exceed 50% without a serviceability and deflection check. Since some self-weight will be acting then the enhancement does not exceed 50% and, therefore, a value of $W_k = 0.22$ kN/m² would be acceptable.

This block wall is the same as that used for the inner leaf of example (8.3.6). In that instance the outer leaf had a capacity of 0.45 kN/m² and the inner leaf a capacity of 0.15 kN/m², giving a total characteristic wind load of 0.60 kN/m².

By using reinforcement on the inner leaf (10 mm² at 450 centres) the capacity may be increased to 0.67 kN/m² (0.45 + 0.22). This is approximately 10% more than the capacity of the wall allowing for base fixity.

A load in excess of the maximum 0.21 kN/m² as controlled by the 50% enhancement may be exceeded providing serviceability and deflection check is carried out (see method 3 for such a check).

Increasing the reinforcement to 225 centres would give $M_d = 1.24$ kN m and a value of $W_k = 0.41$ kN/m². This is considerably greater than 50% above the unreinforced capacity and to be justified would require a serviceability check to be carried out.

Finally, a check should be carried out on the wall ties which from example 5 (8.3.6) are seen to be adequate.

Method 2 Excess load taken on reinforcement
The maximum enhancement that can be taken with this method, without a serviceability and deflection check, is 30%. From the example shown before Method 1, it may be seen that the maximum lateral load that may be taken on the unreinforced block wall is 0.16 kN/m².

Now 30% of 0.16 = 0.048 kN/m² which would require a reinforced capacity of

$$M = \frac{W_k\, L^2}{8}$$

$$= \frac{0.048 \times 4.5^2}{8} = 0.15 \text{ kN/m}$$

By comparison with the previous example only about one quarter of the reinforcement would be required. However, both bar and spacing cannot be reduced below 14 mm² at 450 centres minimum requirement (Annex A 2.4, BS 5628: Part 2) therefore in this example the same reinforcement as used for Method 1 would need to be used.

The inner leaf capacity by this method could therefore be increased to about: 0.21 kN/m² (0.16 + 0.048) within the normal enhancement constraint.

A load in excess of 0.21 kN/m² as controlled by the 30% maximum enhancement may be exceeded providing serviceability and deflection check is carried out (see method 3 for such a check).

Method 3 Modifying orthogonal ratio

From the unreinforced example

$$f_{kb} = 0.25, \quad f_{kp} = 0.45 \quad \text{and} \quad \mu = 0.55$$

Now moment capacity horizontal plane

$$= \frac{f_{kb} \, b \, t^2}{6 \, \gamma_m}$$

$$= \frac{0.25 \times 100^2 \times 10^{-3}}{6 \times 3.1} = 0.13 \text{ kN m/m}$$

Again using 10 mm² bars at 450 corners then from the example given under Method 1 the moment capacity about vertical plane = 0.66 kN m/m

Thus the modified orthogonal ratio = 0.13/0.66 = 0.20

This is outside the extent of Table 9 BS 5628: Part 1 and therefore the bending moment coefficient must be determined.

Analysis of this gives = 0.089 (Table 8.7.1)

Now the bending moment coefficient is used in conjunction with the strength about the vertical axis. With an unreinforced panel this is taken to be f_{kx} (f_{kp}) but with a reinforced panel the strength must be based on the reinforced section which has been previously shown = 0.66 kN m.

Now since $m = \alpha W_k \, \gamma_f$ and $m = M$

$$\text{then } W_k = \frac{M}{\alpha \, \gamma_f \, L^2}$$

$$= \frac{0.67}{0.089 \times 1.2 \times 4.5^2} = 0.31 \text{ kN/m}^2$$

Allowing for self-weight $\mu = 0.23$ and $\alpha = 0.087$
then $W_k = 0.32$ kN/m²
Allowing for base fixity as well, then $\alpha = 0.081$ and $W_k = 0.34$ kN/m²
Ignoring self-weight, then capacity increase

$$= \frac{0.31 - 0.14}{0.14} = 121\%$$

Including self-weight, then capacity increase

$$= \frac{0.32 - 0.15}{0.14} = 113\%$$

Allowing for base fixity, then capacity increase

$$= \frac{0.34 - 0.16}{0.16} = 112\%$$

These are all more than 50% and cannot be used unless a serviceability and deflection check is carried out.

The maximum characteristic load for the unreinforced wall from page 112 is 0.16 kN/m².

Therefore, the ultimate load (i.e. failure strength), with $\gamma_f = 1.0$ and $\gamma_m = 1.0$

$$= 0.16 \times \frac{1.2}{1.0} \times \frac{3.1}{1.0} = 0.60 \text{ kN/m}^2$$

As this is at least as great as the cracking load for the equivalent reinforced leaf, the serviceability design load capacity of the reinforced leaf

$$= \frac{0.60}{\gamma_{mm}} = \frac{0.60}{1.5} = 0.40 \text{ kN/m}^2$$

The maximum serviceability design load on the wall = $\gamma_f \times W_k = 1.0 \times 0.34 = 0.34$ kN/m² which is less than the serviceability design load capacity of the reinforced leaf (0.40 kN/m² as above, where γ_{mm} was taken from BS 5628: Part 2) and therefore cracking will not occur.

However, a deflection check must still be carried out (see Annex B of BS 5628: Part 2). This is not simple as it necessitates it being analysed as an elastic plate but assuming that the masonry is unreinforced and has a short term elastic moduli as given by clause 3.4.1.7 of BS 5628: Part 2. This becomes more complex with a two-way spanning element. If for simplicity the wall is taken as spanning horizontally the deflection is determined as:
Deflection = 5/384 ($W_k \, L^4 / E \, 1$)
For a one metre strip: $W_k = 0.35$ kN/m, $L = 4.5$ m, $I = bd^3/12 = 1 \times 100^3/12 = 8.33 \times 10^4$, $E = 900 \, f_k$ (f_k for 3.5 N/mm² with designation (iii) mortar is 0.35 N/mm², table 2d BS 5628: Part 1).
Hence deflection

$$= \frac{5}{384} \times \frac{0.35 \times 4.5^4 \times 10^{12}}{900 \times 3.5 \times 8.33 \times 10^7} = 7.1 \text{ mm}$$

Now span/250 = 4.5/250 = 18 mm (clause 3.1.2.2.1 BS 5628: Part 2).

The estimated deflection is less than the span/250 limit and therefore satisfactory.

Method 4 Cracking load

From the example of the unreinforced panel.
Then W_k allowable without self-weight = 0.14 kN/m²
with self-weight = 0.15 kN/m²
including base fixity = 0.16 kN/m²
Reducing the safety factors to unity, i.e. γ_f, $\gamma_m = 1.0$
Then characteristic cracking load = 0.14 × 1.2 × 3.1
= 0.52 kN/m²
(with self-weight) = 0.56 K/m²
(with base fixity) = 0.60 kN/m²
Allowing for a partial safety factor (1.5) for material strength at the serviceability limit state (Clause 3.5.3.2 BS 5628: Part 2)
gives a cracking load of 0.52/1.5 = 0.35 kN/m²
(with self-weight) (0.37)
(with base fixity) (0.4)
To allow one of these capacities, the wall must be shown to resist the amount by one of the methods 1–3 and the walls deflection must be shown to be less than span/250.

Now as shown under Method 1 this will be achieved by the use of 10 mm² reinforcement at 225 mm centres ($W_k = 0.41$ kN/m²) and by inspection of the example shown under method 3 the deflection will be acceptable.
Deflection = (0.40/0.34)7.1 = 8.4 < 18.

8.4.3.3 Comparison of methods 1 to 4

The characteristic wind load (kN/m²) given by the various methods on the example panel are summarised as follows:

Condition	Unrein- forced	Method 1	Method 2	Method 3	Method 4
No self-weight	0.14	0.21	0.19	0.31	(0.35)
With self-weight	0.15	0.22	0.20	0.32	(0.37)
With base fixity	0.16	0.24	0.21	0.34	(0.40)

(10 mm² at 225d).

This example shows that by incorporating bed joint reinforcement the lateral load capacity of the panel may be increased by up to 150%. Method 4 gave the maximum enhancement but is the most complex method as it requires both the determination of the cracking load and an analysis by one of the methods 1 to 3 as well a deflection check. Methods 1 and 2 are the simplest to use

but only provide an enhancement of 50 or 30% respectively. Method 3 indicated in the region of a 100% increase in lateral capacity, and although it requires a deflection check for enhancements over 50%, has the advantage in that it is basically an extension to the principle method used for unreinforced panels.

8.5 Bending moment coefficient tables

The following tables are an extension to those in BS 5268 and give bending moment coefficients for all combinations of edge fixity. The values for orthogonal ratios between 0.25 to 0.05 are appropriate for wall with bed joint reinforcement as described in 8.4.3.2, method 3. For the purpose of the following tables:

μ is the orthogonal ratio
H is the panel height
L is the panel length
ϕ is the degree of fixity 0.0 pinned/simple support
1.0 fixed support

Table 8.7.1

$\phi1 = 0.0$ $\phi3 = 0.0$ $\phi2 = 0.0$

μ	h/L						
	0.30	0.50	0.75	1.00	1.25	1.50	1.75
1.20	0.029	0.042	0.055	0.067	0.076	0.082	0.087
1.10	0.030	0.044	0.057	0.069	0.077	0.084	0.089
1.00	0.031	0.045	0.059	0.071	0.079	0.085	0.090
0.90	0.032	0.047	0.061	0.073	0.081	0.087	0.092
0.80	0.034	0.049	0.064	0.075	0.083	0.089	0.093
0.70	0.035	0.051	0.066	0.077	0.085	0.091	0.095
0.60	0.038	0.053	0.069	0.080	0.088	0.093	0.097
0.50	0.040	0.056	0.073	0.083	0.090	0.095	0.099
0.40	0.043	0.061	0.077	0.087	0.093	0.098	0.101
0.30	0.048	0.067	0.082	0.091	0.097	0.101	0.104
0.25	0.051	0.071	0.085	0.094	0.099	0.103	0.106
0.20	0.054	0.075	0.089	0.097	0.102	0.105	0.108
0.15	0.060	0.080	0.093	0.100	0.105	0.108	0.110
0.10	0.069	0.087	0.098	0.104	0.108	0.111	0.113
0.05	0.082	0.097	0.105	0.110	0.113	0.115	0.116

Table 8.7.2

$\phi1 = 0.0$ $\phi3 = 0.0$ $\phi2 = 1.0$

μ	h/L						
	0.30	0.50	0.75	1.00	1.25	1.50	1.75
1.20	0.015	0.028	0.041	0.053	0.062	0.069	0.075
1.10	0.016	0.029	0.043	0.054	0.064	0.071	0.077
1.00	0.017	0.031	0.045	0.056	0.066	0.073	0.079
0.90	0.018	0.032	0.047	0.059	0.068	0.075	0.081
0.80	0.020	0.034	0.049	0.061	0.070	0.077	0.083
0.70	0.021	0.037	0.052	0.064	0.073	0.080	0.085
0.60	0.023	0.039	0.055	0.067	0.076	0.082	0.087
0.50	0.026	0.043	0.059	0.071	0.079	0.085	0.090
0.40	0.029	0.047	0.064	0.075	0.083	0.089	0.093
0.30	0.034	0.053	0.069	0.080	0.088	0.093	0.097
0.25	0.037	0.056	0.073	0.083	0.090	0.095	0.099
0.20	0.041	0.061	0.077	0.087	0.093	0.098	0.101
0.15	0.046	0.067	0.082	0.091	0.097	0.101	0.104
0.10	0.054	0.075	0.089	0.097	0.102	0.105	0.108
0.05	0.069	0.087	0.098	0.104	0.108	0.111	0.113

Table 8.7.3

μ		0.30	0.50	0.75	1.00	1.25	1.50	1.75
					h/L			
1.20		0.023	0.033	0.043	0.051	0.057	0.061	0.064
1.10		0.024	0.034	0.044	0.052	0.058	0.061	0.064
1.00		0.024	0.035	0.046	0.053	0.059	0.062	0.065
0.90		0.025	0.036	0.047	0.055	0.060	0.063	0.066
0.80		0.027	0.037	0.049	0.056	0.061	0.065	0.067
0.70		0.028	0.039	0.051	0.058	0.062	0.066	0.068
0.60		0.030	0.042	0.053	0.059	0.064	0.067	0.069
0.50		0.031	0.044	0.055	0.061	0.066	0.069	0.071
0.40		0.034	0.047	0.057	0.063	0.067	0.070	0.072
0.30		0.037	0.051	0.061	0.066	0.070	0.072	0.074
0.25		0.039	0.053	0.062	0.068	0.071	0.073	0.075
0.20		0.043	0.056	0.065	0.069	0.072	0.074	0.076
0.15		0.047	0.059	0.067	0.071	0.074	0.076	0.077
0.10		0.052	0.063	0.070	0.074	0.076	0.078	0.079
0.05		0.060	0.069	0.074	0.077	0.079	0.080	0.081

$\phi 1 = 0.0$ $\phi 3 = 1.0$ $\phi 2 = 0.0$

Table 8.7.4

μ		0.30	0.50	0.75	1.00	1.25	1.50	1.75
					h/L			
1.20		0.013	0.023	0.033	0.042	0.048	0.053	0.056
1.10		0.014	0.024	0.034	0.043	0.049	0.054	0.057
1.00		0.015	0.025	0.036	0.044	0.050	0.055	0.058
0.90		0.016	0.027	0.037	0.046	0.052	0.056	0.060
0.80		0.017	0.028	0.039	0.047	0.053	0.057	0.061
0.70		0.018	0.030	0.041	0.049	0.055	0.059	0.062
0.60		0.020	0.032	0.043	0.051	0.057	0.061	0.064
0.50		0.022	0.034	0.046	0.053	0.059	0.062	0.065
0.40		0.024	0.037	0.049	0.056	0.061	0.065	0.067
0.30		0.028	0.042	0.053	0.059	0.064	0.067	0.069
0.25		0.030	0.044	0.055	0.061	0.066	0.069	0.071
0.20		0.033	0.047	0.057	0.063	0.067	0.070	0.072
0.15		0.037	0.051	0.061	0.066	0.070	0.072	0.074
0.10		0.043	0.056	0.065	0.069	0.072	0.074	0.076
0.05		0.052	0.063	0.070	0.074	0.076	0.078	0.079

$\phi 1 = 0.0$ $\phi 3 = 1.0$ $\phi 2 = 1.0$

Table 8.7.5

μ		0.30	0.50	0.75	1.00	1.25	1.50	1.75
					h/L			
1.20		0.019	0.027	0.035	0.040	0.044	0.046	0.048
1.10		0.019	0.027	0.036	0.041	0.044	0.047	0.049
1.00		0.020	0.028	0.037	0.042	0.045	0.048	0.050
0.90		0.021	0.029	0.038	0.043	0.046	0.048	0.050
0.80		0.022	0.031	0.039	0.043	0.047	0.049	0.051
0.70		0.023	0.032	0.040	0.044	0.048	0.050	0.051
0.60		0.024	0.034	0.041	0.046	0.049	0.051	0.052
0.50		0.025	0.035	0.043	0.047	0.050	0.052	0.053
0.40		0.027	0.038	0.044	0.048	0.051	0.053	0.054
0.30		0.030	0.040	0.046	0.050	0.052	0.054	0.055
0.25		0.032	0.042	0.048	0.051	0.053	0.055	0.056
0.20		0.034	0.043	0.049	0.052	0.054	0.055	0.056
0.15		0.037	0.046	0.051	0.053	0.055	0.056	0.057
0.10		0.041	0.048	0.053	0.055	0.056	0.057	0.058
0.05		0.046	0.052	0.055	0.057	0.058	0.059	0.059

$\phi 1 = 1.0$ $\phi 3 = 1.0$ $\phi 2 = 0.0$

Table 8.7.6

φ1 = 1.0, φ2 = 1.0, φ3 = 1.0

μ	h/L						
	0.30	0.50	0.75	1.00	1.25	1.50	1.75
1.20	0.012	0.020	0.028	0.034	0.038	0.041	0.044
1.10	0.012	0.020	0.028	0.034	0.039	0.042	0.044
1.00	0.013	0.021	0.029	0.035	0.040	0.043	0.045
0.90	0.014	0.022	0.031	0.036	0.040	0.043	0.046
0.80	0.015	0.023	0.032	0.038	0.041	0.044	0.047
0.70	0.016	0.025	0.033	0.039	0.043	0.045	0.047
0.60	0.017	0.026	0.035	0.040	0.044	0.046	0.048
0.50	0.018	0.028	0.037	0.042	0.045	0.048	0.050
0.40	0.020	0.031	0.039	0.043	0.047	0.049	0.051
0.30	0.023	0.034	0.041	0.046	0.049	0.051	0.052
0.25	0.025	0.035	0.043	0.047	0.050	0.052	0.053
0.20	0.027	0.038	0.044	0.048	0.051	0.053	0.054
0.15	0.030	0.040	0.046	0.050	0.052	0.054	0.055
0.10	0.034	0.043	0.049	0.052	0.054	0.055	0.056
0.05	0.041	0.048	0.053	0.055	0.056	0.057	0.058

Table 8.7.7

φ4 = 0.0, φ1 = 0.0, φ3 = 0.0, φ2 = 0.0

μ	h/L						
	0.30	0.50	0.75	1.00	1.25	1.50	1.75
1.20	0.007	0.016	0.027	0.038	0.047	0.055	0.062
1.10	0.007	0.017	0.029	0.040	0.049	0.057	0.064
1.00	0.008	0.018	0.030	0.042	0.051	0.059	0.066
0.90	0.009	0.019	0.032	0.044	0.054	0.062	0.068
0.80	0.010	0.021	0.035	0.046	0.056	0.064	0.071
0.70	0.011	0.023	0.037	0.049	0.059	0.067	0.073
0.60	0.012	0.025	0.040	0.053	0.062	0.070	0.076
0.50	0.014	0.028	0.044	0.057	0.066	0.074	0.080
0.40	0.017	0.032	0.049	0.062	0.071	0.078	0.084
0.30	0.020	0.038	0.055	0.068	0.077	0.083	0.089
0.25	0.023	0.042	0.059	0.071	0.08	0.087	0.091
0.20	0.027	0.046	0.064	0.076	0.084	0.090	0.095
0.15	0.032	0.053	0.070	0.081	0.089	0.094	0.098
0.10	0.039	0.062	0.078	0.088	0.095	0.100	0.103
0.05	0.054	0.076	0.090	0.098	0.103	0.107	0.109

Table 8.7.8

φ4 = 1.0, φ1 = 0.0, φ3 = 0.0, φ2 = 0.0

μ	h/L						
	0.30	0.50	0.75	1.00	1.25	1.50	1.75
1.20	0.005	0.012	0.021	0.031	0.039	0.047	0.054
1.10	0.005	0.013	0.023	0.032	0.041	0.049	0.056
1.00	0.006	0.014	0.024	0.034	0.043	0.051	0.058
0.90	0.006	0.015	0.026	0.036	0.045	0.053	0.060
0.80	0.007	0.016	0.028	0.039	0.048	0.056	0.063
0.70	0.008	0.018	0.030	0.041	0.051	0.059	0.065
0.60	0.009	0.020	0.033	0.044	0.054	0.062	0.069
0.50	0.010	0.022	0.036	0.048	0.058	0.066	0.072
0.40	0.012	0.026	0.041	0.053	0.063	0.071	0.077
0.30	0.016	0.031	0.047	0.060	0.069	0.076	0.082
0.25	0.018	0.034	0.051	0.064	0.073	0.080	0.085
0.20	0.021	0.039	0.056	0.068	0.077	0.084	0.089
0.15	0.025	0.044	0.062	0.074	0.083	0.089	0.093
0.10	0.032	0.053	0.071	0.082	0.089	0.095	0.099
0.05	0.046	0.068	0.084	0.093	0.099	0.103	0.106

Table 8.7.9

φ4 = 1.0
φ1 = 0.0   φ3 = 0.0
φ2 = 1.0

μ	h/L						
	0.30	0.50	0.75	1.00	1.25	1.50	1.75
1.20	0.004	0.009	0.017	0.025	0.033	0.040	0.047
1.10	0.004	0.010	0.018	0.027	0.035	0.042	0.049
1.00	0.004	0.011	0.019	0.028	0.037	0.044	0.051
0.90	0.005	0.011	0.021	0.030	0.039	0.046	0.053
0.80	0.005	0.013	0.023	0.032	0.041	0.049	0.056
0.07	0.006	0.014	0.025	0.035	0.044	0.052	0.059
0.60	0.007	0.016	0.027	0.038	0.047	0.055	0.062
0.50	0.008	0.018	0.030	0.042	0.051	0.059	0.066
0.40	0.010	0.021	0.035	0.046	0.056	0.064	0.071
0.30	0.012	0.025	0.040	0.053	0.062	0.070	0.076
0.25	0.014	0.028	0.044	0.057	0.066	0.074	0.080
0.20	0.017	0.032	0.049	0.062	0.071	0.078	0.084
0.15	0.020	0.038	0.055	0.068	0.077	0.083	0.089
0.10	0.027	0.046	0.064	0.076	0.084	0.090	0.095
0.05	0.039	0.062	0.078	0.088	0.095	0.100	0.103

Table 8.7.10

φ4 = 0.0
φ1 = 0.0   φ3 = 1.0
φ2 = 0.0

μ	h/L						
	0.30	0.50	0.75	1.00	1.25	1.50	1.75
1.20	0.007	0.014	0.023	0.031	0.038	0.044	0.048
1.10	0.007	0.015	0.025	0.033	0.039	0.045	0.049
1.00	0.008	0.016	0.026	0.034	0.041	0.046	0.051
0.90	0.008	0.017	0.027	0.036	0.042	0.048	0.052
0.80	0.009	0.018	0.029	0.037	0.044	0.049	0.054
0.70	0.010	0.020	0.031	0.039	0.046	0.051	0.055
0.60	0.011	0.022	0.033	0.042	0.048	0.053	0.057
0.50	0.013	0.024	0.036	0.044	0.051	0.056	0.059
0.40	0.015	0.027	0.039	0.048	0.054	0.058	0.062
0.30	0.018	0.031	0.044	0.052	0.057	0.062	0.065
0.25	0.020	0.034	0.046	0.054	0.060	0.063	0.066
0.20	0.023	0.037	0.049	0.057	0.062	0.066	0.068
0.15	0.027	0.042	0.053	0.060	0.065	0.068	0.070
0.10	0.033	0.048	0.058	0.064	0.068	0.071	0.073
0.05	0.043	0.057	0.066	0.070	0.073	0.075	0.077

Table 8.7.11

φ4 = 0.0
φ1 = 0.0   φ3 = 1.0
φ2 = 1.0

μ	h/L						
	0.30	0.50	0.75	1.00	1.25	1.50	1.75
1.20	0.005	0.011	0.019	0.026	0.032	0.038	0.043
1.10	0.005	0.011	0.020	0.027	0.034	0.039	0.044
1.00	0.006	0.012	0.021	0.029	0.035	0.041	0.045
0.90	0.006	0.013	0.022	0.030	0.037	0.042	0.047
0.80	0.007	0.014	0.024	0.032	0.039	0.044	0.048
0.70	0.007	0.016	0.026	0.034	0.041	0.046	0.050
0.60	0.008	0.017	0.028	0.036	0.043	0.048	0.052
0.50	0.010	0.019	0.030	0.039	0.045	0.051	0.055
0.40	0.011	0.022	0.034	0.042	0.049	0.054	0.057
0.30	0.014	0.026	0.038	0.046	0.053	0.057	0.061
0.25	0.016	0.029	0.041	0.049	0.055	0.059	0.063
0.20	0.018	0.032	0.044	0.052	0.058	0.062	0.065
0.15	0.022	0.036	0.048	0.056	0.061	0.065	0.068
0.10	0.027	0.042	0.054	0.061	0.065	0.068	0.071
0.50	0.037	0.052	0.062	0.067	0.071	0.073	0.075

Table 8.7.12

φ4 = 1.0
φ1 = 1.0 φ3 = 0.0
φ2 = 1.0

μ	h/L						
	0.30	0.50	0.75	1.00	1.25	1.50	1.75
1.20	0.004	0.008	0.015	0.022	0.028	0.033	0.038
1.10	0.004	0.009	0.016	0.023	0.029	0.034	0.039
1.00	0.004	0.010	0.017	0.024	0.030	0.036	0.041
0.90	0.005	0.011	0.018	0.026	0.032	0.037	0.042
0.80	0.005	0.011	0.020	0.027	0.034	0.039	0.044
0.70	0.006	0.013	0.021	0.029	0.036	0.041	0.046
0.60	0.007	0.014	0.023	0.031	0.038	0.044	0.048
0.50	0.008	0.016	0.026	0.034	0.041	0.046	0.051
0.40	0.009	0.018	0.029	0.037	0.044	0.049	0.054
0.30	0.011	0.022	0.033	0.042	0.048	0.053	0.057
0.25	0.013	0.024	0.036	0.044	0.051	0.056	0.059
0.20	0.015	0.027	0.039	0.048	0.054	0.058	0.062
0.15	0.018	0.031	0.044	0.052	0.057	0.062	0.065
0.10	0.023	0.037	0.049	0.057	0.062	0.066	0.068
0.05	0.033	0.048	0.058	0.064	0.068	0.071	0.073

Table 8.7.13

φ4 = 0.0
φ1 = 1.0 φ3 = 1.0
φ2 = 0.0

μ	h/L						
	0.30	0.50	0.75	1.00	1.25	1.50	1.75
1.20	0.006	0.013	0.020	0.026	0.031	0.035	0.038
1.10	0.007	0.013	0.021	0.027	0.032	0.036	0.039
1.00	0.007	0.014	0.022	0.028	0.033	0.037	0.040
0.90	0.008	0.015	0.023	0.029	0.034	0.038	0.041
0.80	0.008	0.016	0.024	0.031	0.035	0.039	0.042
0.70	0.009	0.017	0.026	0.032	0.037	0.040	0.043
0.60	0.010	0.019	0.028	0.034	0.038	0.042	0.044
0.50	0.011	0.021	0.030	0.036	0.040	0.043	0.046
0.40	0.013	0.023	0.032	0.038	0.042	0.045	0.047
0.30	0.016	0.026	0.035	0.041	0.044	0.047	0.049
0.25	0.018	0.028	0.037	0.042	0.046	0.048	0.050
0.20	0.020	0.031	0.039	0.044	0.047	0.050	0.052
0.15	0.023	0.034	0.042	0.046	0.049	0.051	0.053
0.10	0.027	0.038	0.045	0.049	0.052	0.053	0.055
0.05	0.035	0.044	0.050	0.053	0.055	0.056	0.057

Table 8.7.14

φ4 = 0.0
φ1 = 1.0 φ3 = 1.0
φ2 = 1.0

μ	h/L						
	0.30	0.50	0.75	1.00	1.25	1.50	1.75
1.20	0.005	0.010	0.016	0.022	0.027	0.031	0.034
1.10	0.005	0.010	0.017	0.023	0.028	0.032	0.035
1.00	0.005	0.011	0.018	0.024	0.029	0.033	0.036
0.90	0.006	0.012	0.019	0.025	0.030	0.034	0.037
0.80	0.006	0.013	0.020	0.027	0.032	0.035	0.038
0.70	0.007	0.014	0.022	0.028	0.033	0.037	0.040
0.60	0.008	0.015	0.024	0.030	0.035	0.038	0.041
0.50	0.009	0.017	0.025	0.032	0.036	0.040	0.043
0.40	0.010	0.019	0.028	0.034	0.039	0.042	0.045
0.30	0.013	0.022	0.031	0.037	0.041	0.044	0.047
0.25	0.014	0.024	0.033	0.039	0.043	0.046	0.048
0.20	0.016	0.027	0.035	0.041	0.045	0.047	0.049
0.15	0.019	0.030	0.038	0.043	0.047	0.049	0.051
0.10	0.023	0.034	0.042	0.047	0.050	0.052	0.053
0.05	0.031	0.041	0.047	0.051	0.053	0.055	0.056

Table 8.7.15

μ	h/L						
	0.30	0.50	0.75	1.00	1.25	1.50	1.75
1.20	0.003	0.008	0.014	0.019	0.024	0.028	0.031
1.10	0.004	0.008	0.014	0.020	0.025	0.029	0.032
1.00	0.004	0.009	0.015	0.021	0.026	0.030	0.033
0.90	0.004	0.010	0.016	0.022	0.027	0.031	0.034
0.80	0.005	0.010	0.017	0.023	0.028	0.032	0.035
0.70	0.005	0.011	0.019	0.025	0.030	0.033	0.037
0.60	0.066	0.013	0.020	0.026	0.031	0.035	0.038
0.50	0.007	0.014	0.022	0.028	0.03	0.037	0.040
0.40	0.008	0.016	0.024	0.031	0.035	0.039	0.042
0.30	0.010	0.019	0.028	0.034	0.038	0.042	0.044
0.25	0.011	0.021	0.030	0.036	0.040	0.043	0.046
0.20	0.013	0.023	0.032	0.038	0.042	0.045	0.047
0.15	0.016	0.026	0.035	0.041	0.044	0.047	0.049
0.10	0.020	0.031	0.039	0.044	0.047	0.050	0.052
0.05	0.027	0.038	0.045	0.049	0.052	0.053	0.055

Table 8.7.16

μ	h/L						
	0.30	0.50	0.75	1.00	1.25	1.50	1.75
1.20	0.008	0.020	0.040	0.062	0.085	0.108	0.133
1.10	0.009	0.022	0.042	0.066	0.090	0.114	0.141
1.00	0.009	0.023	0.046	0.071	0.096	0.122	0.151
0.90	0.010	0.026	0.050	0.076	0.103	0.131	0.162
0.80	0.012	0.028	0.054	0.083	0.111	0.142	0.175
0.70	0.013	0.032	0.060	0.091	0.121	0.156	0.191
0.60	0.015	0.036	0.067	0.100	0.135	0.173	0.211
0.50	0.018	0.042	0.077	0.113	0.153	0.195	0.237
0.40	0.021	0.050	0.090	0.131	0.177	0.225	0.272
0.30	0.027	0.062	0.108	0.160	0.214	0.269	0.325
0.25	0.032	0.071	0.122	0.180	0.240	0.301	0.362
0.20	0.038	0.083	0.142	0.208	0.276	0.344	0.413
0.10	0.065	0.131	0.225	0.321	0.418	0.515	0.613
0.05	0.106	0.208	0.344	0.482	0.620	0.759	0.898

Table 8.7.17

μ	h/L						
	0.30	0.50	0.75	1.00	1.25	1.50	1.75
1.20	0.006	0.014	0.029	0.046	0.065	0.084	0.104
1.10	0.006	0.016	0.031	0.050	0.069	0.090	0.110
1.00	0.007	0.017	0.034	0.053	0.074	0.096	0.116
0.90	0.007	0.019	0.037	0.058	0.080	0.102	0.125
0.80	0.008	0.021	0.040	0.063	0.087	0.110	0.136
0.70	0.009	0.023	0.045	0.070	0.095	0.120	0.149
0.60	0.011	0.026	0.051	0.078	0.105	0.134	0.165
0.50	0.013	0.031	0.058	0.088	0.118	0.152	0.186
0.40	0.015	0.037	0.069	0.102	0.138	0.176	0.215
0.30	0.020	0.046	0.084	0.123	0.167	0.212	0.258
0.25	0.023	0.053	0.096	0.140	0.189	0.238	0.289
0.20	0.028	0.063	0.110	0.163	0.218	0.274	0.330
0.15	0.036	0.078	0.134	0.197	0.261	0.326	0.392
0.10	0.049	0.102	0.176	0.255	0.334	0.415	0.496
0.05	0.082	0.163	0.274	0.387	0.502	0.616	0.731

Table 8.7.18

| | φ1 = 1.0, φ2 = 0.0, φ3 = 1.0 | | | | | | |

μ	h/L						
	0.30	0.50	0.75	1.00	1.25	1.50	1.75
1.20	0.004	0.011	0.022	0.036	0.051	0.067	0.084
1.10	0.005	0.012	0.024	0.039	0.055	0.072	0.089
1.00	0.005	0.013	0.026	0.042	0.059	0.077	0.095
0.90	0.005	0.014	0.028	0.045	0.064	0.083	0.102
0.80	0.006	0.016	0.031	0.050	0.070	0.090	0.110
0.70	0.007	0.017	0.035	0.055	0.076	0.098	0.119
0.60	0.008	0.020	0.040	0.062	0.085	0.108	0.133
0.50	0.009	0.023	0.046	0.071	0.096	0.122	0.151
0.40	0.012	0.028	0.054	0.083	0.111	0.142	0.175
0.30	0.015	0.036	0.067	0.100	0.135	0.173	0.211
0.25	0.018	0.042	0.077	0.113	0.153	0.195	0.237
0.20	0.021	0.050	0.090	0.131	0.177	0.225	0.272
0.15	0.027	0.062	0.108	0.160	0.213	0.269	0.325
0.10	0.038	0.083	0.142	0.208	0.276	0.344	0.413
0.05	0.065	0.131	0.225	0.321	0.418	0.515	0.613

Table 8.7.19

| | φ1 = 0.0, φ2 = 1.0, φ3 = 0.0 | | | | | | |

μ	h/L						
	0.30	0.50	0.75	1.00	1.25	1.50	1.75
1.20	0.008	0.018	0.034	0.050	0.066	0.082	0.098
1.10	0.008	0.019	0.036	0.053	0.070	0.086	0.102
1.00	0.009	0.021	0.038	0.056	0.074	0.091	0.108
0.90	0.010	0.023	0.041	0.060	0.079	0.097	0.113
0.80	0.011	0.025	0.045	0.065	0.084	0.103	0.120
0.70	0.012	0.028	0.049	0.070	0.091	0.110	0.128
0.60	0.014	0.031	0.054	0.077	0.099	0.119	0.138
0.50	0.016	0.035	0.061	0.085	0.109	0.130	0.149
0.40	0.019	0.041	0.069	0.097	0.121	0.144	0.164
0.30	0.024	0.050	0.082	0.112	0.139	0.162	0.183
0.25	0.028	0.056	0.091	0.123	0.150	0.175	0.196
0.20	0.033	0.065	0.103	0.136	0.165	0.190	0.211
0.15	0.040	0.077	0.119	0.155	0.185	0.210	0.231
0.10	0.053	0.097	0.144	0.182	0.213	0.238	0.260
0.05	0.080	0.136	0.190	0.230	0.261	0.286	0.306

Table 8.7.20

| | φ1 = 0.0, φ2 = 1.0, φ3 = 1.0 | | | | | | |

μ	h/L						
	0.30	0.50	0.75	1.00	1.25	1.50	1.75
1.20	0.005	0.013	0.025	0.039	0.053	0.066	0.079
1.10	0.006	0.014	0.027	0.041	0.056	0.069	0.083
1.00	0.006	0.015	0.029	0.044	0.059	0.073	0.088
0.90	0.007	0.017	0.032	0.047	0.063	0.078	0.093
0.80	0.008	0.018	0.034	0.051	0.067	0.084	0.099
0.70	0.009	0.021	0.038	0.056	0.073	0.090	0.106
0.60	0.010	0.023	0.042	0.061	0.080	0.098	0.115
0.50	0.012	0.027	0.048	0.068	0.089	0.108	0.126
0.40	0.014	0.032	0.055	0.078	0.100	0.121	0.139
0.30	0.018	0.039	0.066	0.092	0.116	0.138	0.158
0.25	0.021	0.044	0.073	0.102	0.127	0.150	0.170
0.20	0.025	0.051	0.084	0.114	0.141	0.164	0.185
0.15	0.031	0.061	0.098	0.131	0.159	0.184	0.205
0.10	0.041	0.078	0.121	0.157	0.187	0.212	0.233
0.05	0.064	0.114	0.164	0.204	0.235	0.260	0.281

Table 8.7.21

μ	h/L						
	0.30	0.50	0.75	1.00	1.25	1.50	1.75
1.20	0.004	0.010	0.020	0.031	0.043	0.054	0.065
1.10	0.004	0.011	0.021	0.033	0.045	0.057	0.069
1.00	0.005	0.012	0.023	0.035	0.048	0.061	0.073
0.90	0.005	0.013	0.025	0.038	0.052	0.065	0.078
0.80	0.006	0.014	0.027	0.041	0.056	0.069	0.083
0.70	0.007	0.016	0.030	0.045	0.060	0.075	0.090
0.60	0.008	0.018	0.034	0.050	0.066	0.082	0.098
0.50	0.009	0.021	0.038	0.056	0.074	0.091	0.108
0.40	0.011	0.025	0.045	0.065	0.084	0.103	0.120
0.30	0.014	0.031	0.054	0.077	0.099	0.119	0.138
0.25	0.016	0.035	0.061	0.085	0.109	0.130	0.149
0.20	0.019	0.041	0.069	0.097	0.121	0.144	0.164
0.15	0.024	0.050	0.082	0.112	0.139	0.162	0.183
0.10	0.033	0.065	0.103	0.136	0.165	0.190	0.211
0.05	0.053	0.097	0.144	0.182	0.213	0.238	0.260

Diagram at left: $\phi_1 = 1.0$, $\phi_2 = 1.0$, $\phi_3 = 1.0$

8.6 Design to BS 5628: Part 3

The previous part of this Chapter, *Sections 8.1 to 8.4,* dealt with the mathematical solutions to the design of plain masonry walls subject to lateral loading. It is possible to produce lateral design load tables based on the recommendations of BS 5628 but even so it is still necessary to determine a number of basic items such as the design wind pressure, the type of support conditions, and so on, before they can be used. However, by imposing certain conditions it is possible to simplify the sizing of a wall subject to lateral wind pressure even further. Such an approach is given in BS 5628: Part 3[8.3] which provides simple area/thickness rules applicable to buildings up to and including four storeys.

The general approach given in BS 5628: Part 3 is explained in this Section, and covers both free standing walls as well as walls with edge restraint. In the latter case it also includes allowance for openings.

8.6.1 Walls with edge restraint

Clause 18.4.2 of BS 5628: Part 3 starts by stating that non-loadbearing walls may be designed following the recommendations of BS 5628: Part 1. However, providing the building and walls comply with the following conditions, then the walls may be proportioned directly from simple rules.

The restrictions applied are:

(1) the building is not more than four storeys;
(2) the building should be situated in an area of many windbreaks, i.e. in protection category 3 as described in DD 93;
(3) the walls should be free from any doors, windows or other openings, unless either:
  (a) intermediate supports are provided (Figure 8.25(a)) or
  (b) the total area of such openings is not more than 10% of the maximum panel size (Table 8.8) or

2% of the actual panel size, whichever is the less. No opening should be closer to a support (except base) or other opening than half its maximum dimension;

(4) in solid walls the distance between supports should not exceed 40 times the wall thickness;
(5) in double leaf (cavity walls the distance between supports should not exceed 30 times the wall thickness. Also the actual thickness of each leaf should be not less than 100 mm, the cavity width should not exceed 100 mm, and wall ties are used and spaced in accordance with Table 9 of BS 5628: Part 3;
(6) top supported gable ends should be taken as rectangular or measured half way up the triangular portion;
(7) the mortar is not weaker than designation (iii).

To produce simple rules it is necessary to assume certain limited exposure conditions. If the building is located in open farmland, on the top of an escarpment or cliff, or in any other exposed area, the wind pressure should be obtained from CP3: Chapter V: Part 2 and the wall designed following the recommendations of BS 5628: Part 1 or 2. It is also fairly obvious that any support must be sufficient to carry the transmitted load without undue deflection. Having determined the wind pressure it may be possible to select by reference to a particular wind zone. Unfortunately no values are given for the wind pressure in each Zone and therefore those given later may be of use. Characteristic loads of connections are given in *Chapter 3*. The approach used in BS 5628: Part 3 is in three stages as follows:

Stage one

Determine the exposure Zone from Figure 8.26 *appropriate to the position at which the building is to be* located.

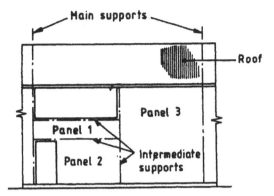

(a) Example of division of a wall into panels with intermediate supports

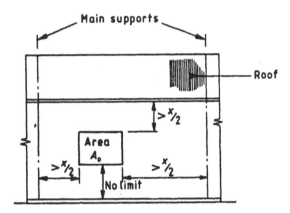

x is the maximum dimension of opening (height or length)
A_0 is the permitted area of opening (see 18.4.2.1(c))

(b) Effect of opening in a wall

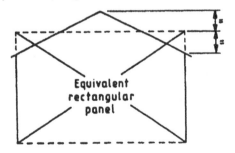

(c) Gable walls

Figure 8.25

Stage two

Determine the number of sides on which the wall is supported and whether the supports are pinned or fixed (see Figures 8.27 to 8.30). Where the conditions for assessing a fixed or pinned support are not met, the edge should be regarded as being free.

Stage three

From the exposure Zone and support conditions determine the permitted area/thickness from Table 8.8 checking that the distance between the supports does not exceed forty times the total thickness of the masonry in the wall.

Figures 8.27 to 8.30 indicating pinned and fixed supports may also be of use in assessing support conditions when designing in accordance with BS 5628 Part 2. In addition, by considering typical low rise buildings and the basic wind speeds in the various exposure Zones, it is possible to show that the dynamic wind pressure for the four Zones applicable to buildings up to four storeys will be in the following general region:

Zone 1 0.55 kN/m²
Zone 2 0.65 kN/m²
Zone 3 0.75 kN/m²
Zone 4 0.90 kN/m²

8.6.2 Free standing walls (without piers)

BS 5628: Part 3 clause 18.4.1 gives guidance on the sizing of free standing walls subjected to only wind loads. These walls may be sized by simple use of a height to thickness ratio as given below.

Height to thickness ratio for free standing single-leaf walls without piers

Wind zone (Fig. 8.26)	Maximum permitted height (a) to thickness ratio R
1	8.5
2	7.5
3	6.5
4	6.0

Thus $h \leq t\,R$ or $t \geq \dfrac{h}{R}$

The use of this approach is however subject to the following conditions:

(a) height – overall height above level of lateral restraint
(b) unit strength min 3.5 N/mm², density min. 1400 kg/m³
(c) protected by wind breaks (protection category 3, DD93)
(d) no dpc.

Where a dpc exists which cannot resist flexure then the thickness should be taken as not less than the greater of:

$$1.33\,\frac{h_d}{R} \quad \text{or} \quad \frac{h_1}{R}$$

where

h_d is the height of the wall above the dpc
R is the ratio in the above table (Table 7 BS 5628) for the appropriate wind zone
h_1 is the height of the wall above the lowest level of lateral restraint below the dpc (e.g. concrete slab)

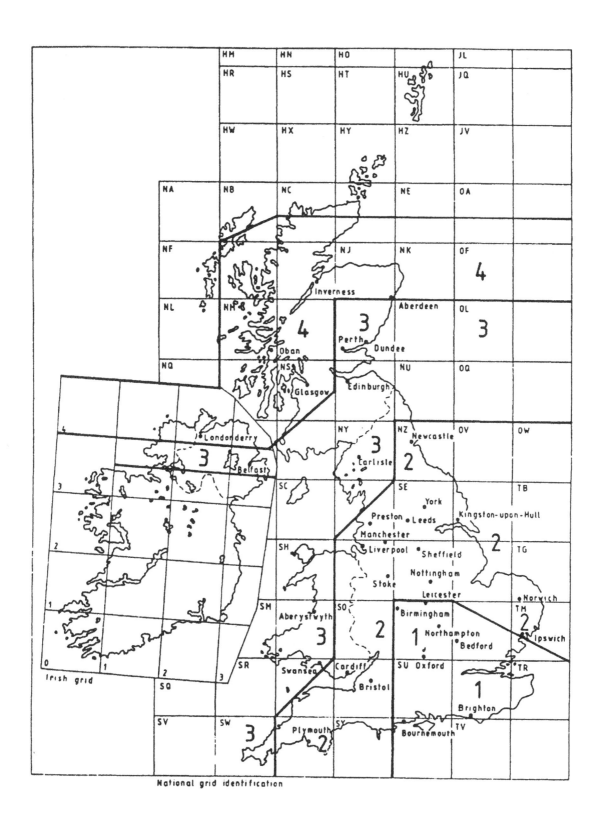

Figure 8.26 Wind zones

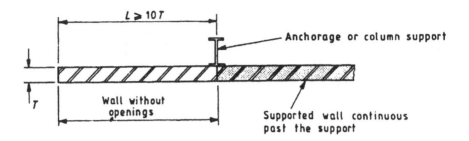

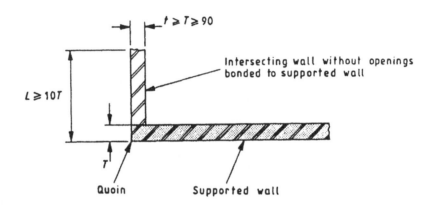

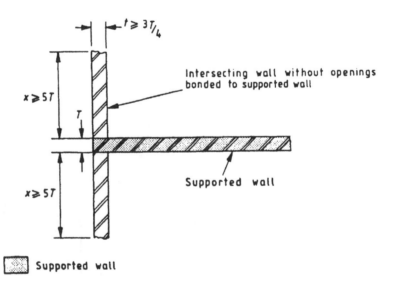

Figure 8.27

124

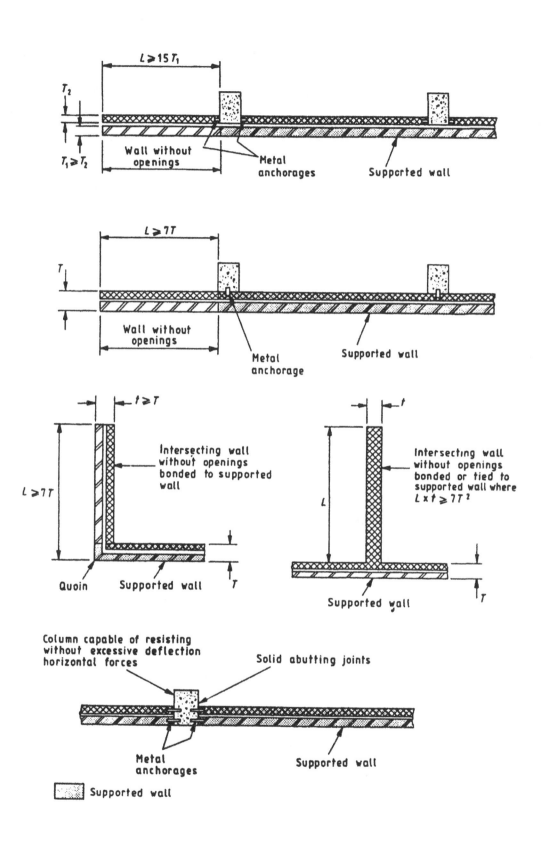

Figure 8.28 Fixed support conditions in cavity walls

Table 8.8

Wind zone	Height (m)	A		C		D		F		G	
		Cavity wall	190 mm solid wall	Cavity wall	190mm solid wall	Cavity wall	190 mm solid wall	Cavity wall	190 mm solid wall	Cavity wall	190 mm solid wall
1	5.4	11.0	13.5	26.5	28.5	20.5	29.0	32.0	41.0	8.5	10.0
	10.8	9.0	11.5	17.5	21.5	15.5	23.5	32.0	41.0	7.0	8.0
2	5.4	8.5	12.0	21.0	24.0	17.5	25.5	32.0	41.0	7.5	8.5
	10.8	8.0	9.5	13.5	17.5	13.0	20.5	28.0	36.5	6.0	7.0
3	5.4	8.5	10.5	15.5	20.0	14.5	22.5	30..5	40.5	6.5	7.5
	10.8	7.0	8.5	11.5	15.5	11.0	17.5	24.5	31.5	5.0	6.0
4	5.4	8.0	9.5	13.0	17.0	12.5	19.5	27.0	35.0	6.0	6.5
	10.8	6.5	7.5	10.5	13.5	9.5	14.5	21.5	27.5	4.0	5.0

Cavity walls   – outer leaf – 100 mm min. thickness 14.0 N/mm² min. strength
                 – inner leaf – 100 mm min. thickness 3.5 N/mm² min. strength
For 140 mm concrete blocks in either leaf, area may be increased by 20%
For 190 mm solid walls min. block strength 3.5 N/mm²
More examples may be found in Table 8 of BS 5628: Part 3.

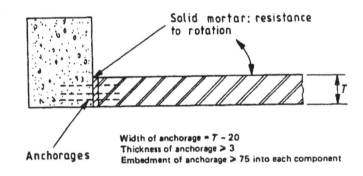

Solid mortar: resistance to rotation

Anchorages

Width of anchorage = T – 20
Thickness of anchorage ≥ 3
Embedment of anchorage ≥ 75 into each component

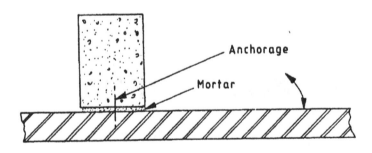

Anchorage

Mortar

All dimensions are in millimetres.

Figure 8.29 Details of fixed supports

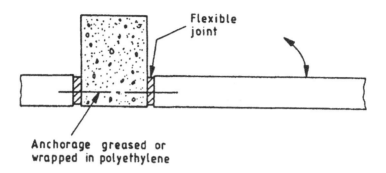

Flexible joint

Anchorage greased or wrapped in polyethylene

Figure 8.30 Details of simple pinned support

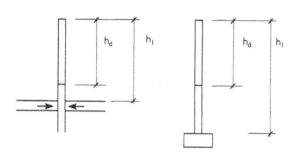

This gives $h_d < 1.33\ t\ R$ or $t > 1.33\dfrac{h_d}{R}$

$$h_1 < t\ R \quad \text{or} \quad t > \frac{h_1}{R}$$

The above guidance relates to walls that are only subjected to wind loads. A particular point is made that other types of walls should be designed to adequately resist any force which may be expected to be exerted on them, and parapets, balustrades, canopy walls or walls to areas where access other than maintenance is envisaged should be designed for loads not less than those given in BS 6180. Stability may be able to be provided by designing adequate lateral support to the top of such walls; this could consist of a capping rail of wood, or metal of adequate transverse strength or of reinforcement in horizontal joints, in either case securely fixed at both ends to an element of structure capable of resisting the forces involved without exceeding the limits of stress allowed for the particular material. When a wall is too long for this method, or other circumstances would render the method inappropriate, vertical supports should be provided at intervals; in some cases it may only be necessary to fix these at the floor if the fixing is sufficiently secure.

8.6.3 Internal walls not designed for wind loading

This is not strictly appropriate to this chapter but is included here for completeness.

The length or height of an internal wall not subject to wind loading may be sized in accordance with the recommendations given in Clause 18.5, BS 5628: Part 3 which is essentially as follows:

(i) wall restrained at both ends but not at the top

   $t \geq L/40$ and $t \geq H/90$ or $t \geq H/15$ with no restriction on the value of L or $t < L/40$ and $t > L/59$ and $t \geq (H + 2L)/133$;

(ii) wall restrained at both ends and at the top

   $t \geq L/50$ and $t \geq H/90$ or $t \geq H/30$ with no restriction on the value of L or $t < L/40$ and $t \geq L/110$ and $t > (3H + L)/200$;

(iii) wall restrained at the top but not at the ends $t \geq H/30$;
   where
   t is the thickness (in mm)
   H is the height (in mm)
   L is the length (in mm)

   Consideration should also be given to:
   (a) accommodation for movement
   (b) openings
   (c) chases
   (d) exceptional lateral loading (dependent on building use)
   (e) wind loads

Plaster may be taken into account (maximum 13 mm) but if included the wall may require temporary bracing prior to plastering.

8.7 References

8.1 The Building Regulations 1972 (and amendments 1976). HMSO, London.
8.2 BRITISH STANDARDS INSTITUTION. CP111: 1970 *Structural recommendations for loadbearing walls*. BSI, London. pp 40.
8.3 BRITISH STANDARDS INSTITUTION. CP 121: Part 1: 1973 *Code of Practice for walling. Part 1: Brick and block masonry*. BSI, London. pp 84.
8.4 BRITISH STANDARDS INSTITUTION. BS 5628: Part 1: 1992 *Code of Practice for use of masonry. Part 1: Unreinforced masonry*. BSI, London. pp 58.

8.5 ISAACS, D V. An interim analysis of the strengths of masonry block walls in respect to lateral loading. *Special Report No 1*, Commonwealth Experimental Building Station. June 1948.

8.6 BRITISH CERAMIC RESEARCH ASSOCIATION. Technical Notes 226, 242 and 248. BCRA, 1974/1975.

8.7 ANDERSON, C. Lateral loading tests on concrete block walls. *The Structural Engineer*, Vol 54, No 7, July 1976.

8.8 FISHBURN, C C. Effects of mortar properties on strength of masonry. *Monograph 36*. United States Department of Commerce. National Bureau of Standards, Washington. 20 November 1961. pp 1–45.

8.9 JOHANSEN, K W. *Yield-line formulae for slabs*. Eyre & Spottiswoode Publications Limited, Leatherhead, 1972. Publication No 12.044. pp 106.

8.10 JONES, L L. *Ultimate load analysis of reinforced and pre-stressed concrete structures*. Chatto and Windus, London, 1962.

8.11 JONES, L L, and WOOD, R H. *Yield-line analysis of slabs*. Thames and Hudson, Chatto and Windus, London, 1967.

8.12 COMITE EUROPEAN DU BETON. *Bulletin D'Information No 35*. CEB, 1962.

8.13 SINHA, B P. A simplified ultimate load analysis of laterally loaded orthotropic model brickwork panels of low tensile strength. *The Structural Engineer*, Vol 56B, No 4, 1978. pp 81–84.

8.14 SINHA, B P. An ultimate load analysis of laterally loaded brickwork panels. *The International Journal of Masonry Construction*, Vol 1, No 2, 1980. pp 57–61.

8.15 HELLERS, B B, and JOHANSON, B. Horizontally loaded masonry. *Report No BER 730637-7*, Swedish State Committee for Building Research, Malmo.

8.16 CAJDERT, A. Laterally loaded masonry walls. Chalmers University of Technology, Division of Concrete Structures, 1980. Publication No 80: 5. Goteburg, Sweden.

8.17 BRITISH STANDARDS INSTITUTION. BS 5628: Part 3: 1985 *Code of Practice for use of masonry*. Part 3: *Materials and components design and workmanship*. BSI, London.

8.18 WEST, H W H., HODGKINSON, H R., HESELTINE, B A and DE VEKAY, R C. Research results on brickwork and aggregate blockwork since 1977. *The Structural Engineer* Vol 64A, No. 11, November 1986.

8.19 ANDERSON, C. Lateral strength from full-sized tests related to the flexural properties of masonry. *Masonry International* Vol 1, No 2, 1987. pp. 51–55.

Chapter 9
Reinforced Masonry

9.1 Introduction

Preparation of the first design guidance for reinforced masonry (more specifically reinforced brickwork) commenced in 1937 but was not issued until 1943 in the form of a British Standard BS 1146.[1] Some guidance on reinforced masonry was provided in CP 111[2] but it was not until the introduction of BS5628: Part2 in 1985[3] that detailed design guidance became available in the UK. Subsequently BS5628 was amended and a new edition published in 1995[4].

This Chapter is presented in a format which is intended to facilitate reference to the Code and the headings in the subsequent sections of this Chapter reflect this. The references have also been grouped together under the sections to which they are relevant. The subject of prestressed masonry has not been covered because it is so rarely used.

9.2 Design to BS 5628 Part 2

Section 1: General

1.1 Scope

BS 5628: Part 2 was prepared to bring together UK design experience and practice of the use of reinforced and prestressed masonry. Where appropriate, overseas experience was introduced to supplement that available in the UK.

The document gives recommendations for the structural design of reinforced and prestressed masonry constructed of brick or block masonry, and masonry of square dressed natural stone. Far more experience was available in the use of reinforced masonry than in prestressed masonry and this is apparent in both the scope and content of these respective parts of the document. Included in the document, in Appendix A, is guidance on design methods for walls containing bed joint reinforcement to enhance their resistance to lateral load and this is covered in Chapter 8.

Since this Code is essentially structural in content, attention is drawn to the need to satisfy other than structural requirements (for example, requirements such as fire resistance, thermal insulation and acoustic performance) in the sizing of members and elements

The Code also assumes that the design of reinforced and prestressed masonry is entrusted to 'appropriately qualified and experienced people' and that 'the execution of the work is carried out under the direction of appropriately qualified supervisors'. This latter requirement is highlighted by the fact that BS 5628: Part 2, unlike Part 1[5], only recognizes the special category of construction control.

1.2 References

This section requires no further detailed comment.

1.3 Definitions

The definition of masonry permits units to either be laid in situ or as prefabricated panels. In both cases the units must be bonded and solidly put together with concrete and/or mortar so as to act compositely. The use of prefabricated panels is not new and has been established on a limited scale for a number of years. Very often, however, such panels are concrete elements to which masonry slips are bonded during the manufacturing process. The design procedures contained in BS 5628: Part 2 should enable efficient reinforced masonry panels to be produced.

There are a number of forms in which units of different types may be bonded together to leave clear channels or cavities which may be reinforced or prestressed. The Code defines the four types of construction most likely to be employed, but the many other possibilities are equally valid. The types defined are:

(a) grouted cavity
(b) pocket type
(c) Quetta bond
(d) reinforced hollow blockwork.

It is interesting to observe that the general definition of reinforced brickwork in BS 1146[1] has now been omitted

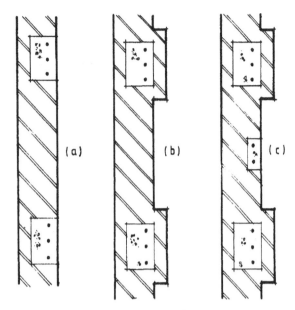

Figure 9.1 Typical grouted cavity construction

in favour of a definition of reinforced masonry which includes all types of masonry unit put together in any form. The first three types of construction previously listed are, however, more commonly constructed of brickwork.

Grouted Cavity Masonry

Grouted cavity construction is probably the construction method with the widest application and may employ virtually any type of masonry unit. Essentially two parallel leaves of units are built with a cavity at least 50 mm wide between them. The two leaves must be fully tied together with wall ties. Reinforcing steel is placed in the cavity which is filled with high slump concrete. The word 'grout' in this context is derived from United States practice. In the UK Code 'infilling concrete' is the term corresponding to the USA term 'grout'. The word grout is reserved for the material used to fill ducts in prestressed concrete and prestressed masonry. A typical grouted cavity construction is illustrated in Figure 9.1.

Earlier guidance on reinforced brickwork[6] did not include the concrete or mortar in the cavity as contributing to the compressive strength of the wall. The reason for this conservative approach was the fear that in the long term, differential movement would lead to a loss of composite action. The Code committee accepted that this approach was unnecessarily cautious but included a restriction on the effective thickness of a grouted cavity wall section. For cavities up to 100 mm the effective thickness may be taken as the total thickness of the two leaves plus the width of the cavity, but for greater cavity widths the effective thickness is the thickness of the two leaves plus 100 mm. Attention should be paid to Clause 32 which specifies the type of steel and cover necessary

130

for a given condition of exposure. In some cases mortar may be used to fill the cavity rather than concrete and, because this reduces the protection offered to the reinforcing steel, steel which has some additional form of resistance to corrosion may need to be specified. Regardless of the type of infill, the minimum permitted cover of concrete or mortar to the steel is 20 mm, except where stainless steel is used.

Pocket type masonry

This type of construction is so named because the main reinforcement is concentrated in vertical pockets formed in the masonry[7]. This type of wall is primarily used to resist lateral forces in retaining or wind loading situations. It is the most efficient of the brickwork solutions if the load is from one side only and the wall section may be increased in thickness towards the base. An example is shown in Figure 9.2.

A particular advantage of the simplest and most common form of the pocket type wall is that the 'pocket' may be closed by a piece of temporary formwork propped or nailed to the masonry. After the infilling concrete has gained sufficient strength, this formwork may be removed and the quality of the concrete and workmanship inspected directly.

Clause 32 specifies the cover, grade of concrete and the minimum cement content to ensure the durability of the steel in a pocket type wall; low carbon steel (mild or high yield) without any surface coating would normally be used.

Quetta bond

The Quetta bond traces its origin to the early use of reinforced brickwork in the civil reconstruction of the town of Quetta in India following earthquake damage[8]. The section produced by this bond is at least one and a half units thick, as shown in Figure 9.3, and the vertical pocket formed may be reinforced with steel and filled with concrete or mortar. The face of the wall has the appearance of Flemish bond. There is also a modified form of Quetta bond in which the face of the wall has the appearance of Flemish garden wall bond. In thicker walls the steel may be placed nearer to the faces to resist lateral loading more efficiently.

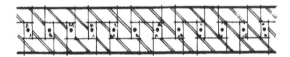

Figure 9.2 Pocket type example

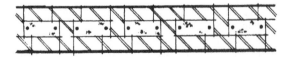

Figure 9.3 Section produced by Quetta Bond

When Quetta bond and grouted cavity construction are employed using similar materials they are treated similarly from the viewpoint of durability and in certain exposure conditions protected reinforcement may be necessary.

Reinforced hollow blockwork

In this form of construction the cores of hollow blocks are reinforced with steel and filled with in situ concrete[9]. The work size of the most common blocks is 440 × 215 × 215 mm, although 390 × 190 × 190 mm blocks are also widely available. Although other sizes of blocks may be available, they are not nearly so common in the UK. In addition to the standard two core hollow blocks, specials such as lintel and bond beam blocks are available and are illustrated in Figure 9.4. For retaining walls up to about 2.5 m high, a single leaf of reinforced hollow blockwork is usually all that is required. It is, therefore, a very cost effective way of building small retaining walls.

1.4 Symbols

The following symbols are used in the Code:

A_m cross sectional area of masonry

A_s cross sectional area of primary reinforcing steel

A_{s1} the area of compression reinforcement in the most compressed face

A_{s2} the area of compression reinforcement in the least compressed face

A_{sv} cross sectional area of reinforcing steel resisting shear force

a shear span

a_d deflection

a_v distance from face of support to the nearest edge of a principal load

b width of section

b_c width of compression face midway between restraints

b_t width of section at level of the tension reinforcement

c ever arm factor

d effective depth [see Clause 2.4]

d_c depth of masonry in compression

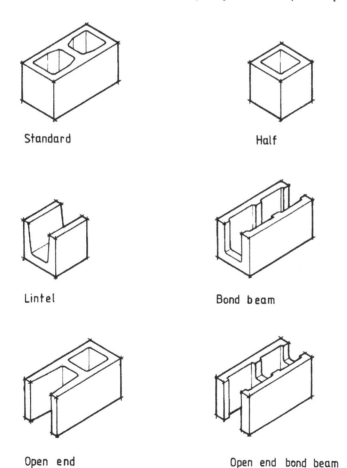

| Standard | Half |

| Lintel | Bond beam |

| Open end | Open end bond beam |

Figure 9.4 Types of hollow concrete block

d_1 the depth from the surface to the reinforcement in the more highly compressed face

d_2 depth of the centroid of the reinforcement from the least highly compressed face

E_c modulus of elasticity of concrete

E_m modulus of elasticity of masonry

E_{mi} initial or short term modulus of elasticity

E_{ml} long term modulus of elasticity taking account of creep and shrinkage

E_n nominal earth or water load

E_s modulus of elasticity of steel

e base of Napierian logarithms [2.718]

e_x resultant eccentricity in plane of bending

F_{bst} tensile bursting force

F_c compressive force

f_b characteristic anchorage bond strength between mortar or concrete infill and steel

f_{ci} strength of concrete at transfer

f_k characteristic compressive strength of masonry

f_{kx} characteristic flexural strength [tension] of masonry

f_{pb} stress in tension at the design moment of resistance of the section

f_{pe} effective prestress in tendon after all losses have occurred

f_{pu} characteristic tensile strength of prestressed tendons

f_s stress in the reinforcement

f_{s1} stress in the reinforcement in the most compressed face

f_{s2} stress in the reinforcement in the least compressed face

f_v characteristic shear strength of masonry

f_y characteristic tensile strength of reinforcing steel

G_k characteristic dead load

g_b design load per unit area due to loads acting at right angles to the bed joints

h clear distance between lateral supports

h_{agg} maximum size of aggregate

h_{ef} effective height of wall or column

I moment of inertia of the section

K_t coefficient to allow for type of prestressing tendon

k constant

L length of the wall

l effective span of the member

l_t transmission length

M bending moment due to design load

M_a increase in moment due to slenderness

M_d design moment of resistance

M_p permanent load moment

M_t total design bending moment at right angles to the bed joints

M_x design moment about the x axis

$M_{x'}$ effective uniaxial design moment about the x axis

M_y design moment about the y axis

$M_{y'}$ effective uniaxial design moment about the y axis

N design axial load

N_d design axial load resistance

N_{dz} design axial load resistance of column, ignoring all bending

p overall section dimension in direction perpendicular to the x axis

Q moment of resistance factor

Q_k characteristic imposed load

q overall section dimension in a direction perpendicular to the y axis

r_{ip} reciprocal of instantaneous curvature due to permanent load

r_{it} reciprocal of instantaneous curvature

r_{lp} reciprocal of long term curvature due to the permanent loads

r_n reciprocal of overall long term curvature

S_V spacing of shear reinforcement along member

t overall thickness of a wall or column

t_{ef} effective thickness of a wall or column

t_f thickness of a flange in a pocket type wall

V shear force due to design loads

v shear stress due to design loads

W_k characteristic wind load

Z section modulus

z lever arm

α coefficient

γ_f partial safety factor for load

γ_m partial safety factor for material

γ_{mb} partial safety factor for bond strength between mortar or concrete infill and steel

γ_{mm} partial safety factor for compressive strength of masonry

γ_{ms} partial safety factor for strength of steel

γ_{mt} partial safety factor for strength of tie connections used around the perimeter of a panel

γ_{mv} partial safety factor for shear strength of masonry

θ rotation

μ coefficient of friction due to curvature in a prestressing duct

ρ $= A_s/bd$

ϕ nominal diameter of tendon

1.5 Alternative materials and methods of design and construction

This section requires no further detailed comment.

References

1. BRITISH STANDARDS INSTITUTION. BS 1146:1943 *Reinforced brickwork*. BSI, London. pp.12.
2. BRITISH STANDARDS INSTITUTION. CP 111:1970 *Structural recommendations for loadbearing walls*. BSI, London. pp.40.
3. BRITISH STANDARDS INSTITUTION. BS 5628: Part 2:1985 *Code of practice for use of masonry*. Part 2 : Structural use of reinforced masonry. BSI. London. pp. 44.
4. BRITISH STANDARDS INSTITUTION. BS 5628: Part 2:1995 *Code of practice for use of masonry*. Part 2 : Structural use of reinforced masonry. BSI. London. pp. 52.
5. BRITISH STANDARDS INSTITUTION. BS 5628: Part 1:1992 *Code of practice for use of masonry*. Part I : Structural use of unreinforced masonry. BSI. London. pp. 40.
6. BRITISH CERAMIC RESEARCH ASSOCIATION. Structural Ceramics Advisory Group. *Design guide for reinforced and prestressed clay brickwork*. Special Publication SP 91, 1977.
7. EDGELL, G J. Design guide for reinforced clay brickwork pocket type retaining walls. British Ceramic Research Association. Special Publication SP 105, 1984.

8. ROBERTSON, R O. Earthquake resistant structures: the seismic factor and the use of reinforced brickwork in Quetta civil reconstruction. *Journal of the Institution of Civil Engineers*. No.3, January 1948. p. 171.

9. TOVEY, A K and ROBERTS, J J. Interim design guide for reinforced concrete blockwork subject to lateral loading only. Cement & Concrete Association, Slough. ITN 6, 1980. pp.44.

Section 2: Materials and components

2.1 General

The materials and components employed to produce reinforced masonry should generally comply with BS 5628: Part 3[1] or BS 5390[2]. If materials not covered by these documents are to be used, they should be carefully specified. Reinforced masonry may require the use of special units, unusual wall ties, and so on, which may not be commonly available and these will need to be carefully described in any specification

2.2 Structural units

General

Units to be used for reinforced and prestressed masonry should comply with the appropriate British Standard. In the case of clay bricks and blocks this is BS 3921[3], whilst concrete masonry units are covered by BS 6073: Part 1[4]. Calcium silicate bricks should comply with BS 187[5]. It is also possible to reinforce cast stone and stonemasonry, and these are covered by BS 1217[6] and BS 5390 respectively. If units have been used previously they should not be re-used in reinforced and prestressed masonry without thorough cleaning and inspection. A check should be made to ensure that re-used materials comply with current recommendations. In addition to complying with the relevant Standards, the units should meet the minimum strength requirements and follow the recommendations of BS 5628: Part 3 or BS 5390 (for stone masonry only) in respect of durability and such like.

Minimum strength requirement for masonry units

This part of the Code of Practice includes values for the characteristic strength of masonry units whose compressive strength is at least 7 N/mm^2. Ideally the elasticity of the masonry and infilling concrete should be matched, but in practice a wide variation in constituent properties does not appear to have caused significant problems. There are a number of reasons why properties are not directly comparable. For example, different characteristic strengths are necessary for bricks and blocks of a given unit strength because smaller and squatter units give a greater apparent strength when tested between the platens of a testing machine – this effect can be clearly demonstrated by comparing the characteristic compressive strength of masonry constructed from 20 N/mm^2 bricks with that constructed from 7 N/mm^2 blocks. Both mortar and infilling concrete are normally tested in the form of cubes, the effect of which is that the apparent mortar or concrete strength may be different to the in situ strength. A further factor which can affect the in situ strength of mortar and infilling concrete is the amount of water absorbed by the units. The unit may absorb a considerable proportion of the water from the mortar or the concrete, thereby reducing the water/cement ratio and increasing the strength. Standard cubes made in metal moulds will have a higher water/cement ratio and indicate a lower strength. In practice the strength of the infill concrete may well be determined by the minimum cement content necessary for adequate protection of the reinforcement against corrosion.

There may be certain circumstances where the specification of a minimum strength for the units is not appropriate, for example in a relatively lightly loaded post-tensioned diaphragm wall. The Code does not preclude the use of lower strength units in these circumstances but the designer should consider this carefully. This relaxation is also particularly appropriate for situations where local reinforcement is provided within a building. It is possible to reinforce locally around openings, to provide an in situ lintel, to provide an alternative path for structural support or to improve lateral load resistance even when low strength units are employed. The use of a low strength unit will, however, mean that only a low characteristic masonry strength may be used even though the infilling concrete is significantly stronger. It may be appropriate, in exceptional circumstances, to consider the brick or block element as permanent non-loadbearing formwork and design the element as a reinforced concrete section based on the area of the infilling concrete[7]. A final point which should be noted is that the block strength is normally measured and quoted on the gross area of the unit. In the case of hollow or cellular blocks it may be necessary to convert the gross strengths to nett strengths (see BS 6073) to check compliance with any minimum strength requirement.

Durability of masonry units

Detailed information on the suitability of different types of unit for various conditions of exposure is provided in BS 5628 Part 3.

2.3 Steel

2.3.1 Reinforcing steel

The steel to be used for the reinforcement of masonry will generally be bar, wire or fabric conforming to the requirements of BS 4449[8] or BS 6744[9], BS 4482[10] or BS 4483[11] respectively. However, in certain circumstances, for example for reasons of corrosion resistance, it will be necessary to use steel other than those covered by the above standards.

Stainless steel

Three types of stainless reinforcing steel are available as a direct substitute for conventional ribbed high yield steel reinforcing bars.

Currently, a solid stainless steel reinforcing bar would cost six to seven times as much as high yield steel, depending upon the type of stainless steel and the bar size. One type consists of solid 18–8 stainless steel and another, Type 316 (18% chromium, 10% nickel, 2½% molybdenum), stainless steel. Stainless steel cold twisted bar is also produced from 18–9 Type 302/304 austenitic stainless steel. Bars are available made from both hot rolled 18–8 and cold twisted 18–9 austenitic stainless steels. Even higher standards of corrosion resistance are achieved when bars made from warm worked Type 316 are used.

A relatively new development is a bar which consists of an outer skin of at least 1 mm thickness of 18–8 (18% chromium, 8% nickel) and a core of high yield steel. The bar has a similar profile to a ribbed high yield bar. The relative cost of this type of bar varies with size but currently 16 mm bars are some 12% cheaper than solid stainless steel bars.

Electrostatically epoxy resin coated reinforcing bar

A number of epoxy coatings have been developed in the USA as a method of affording additional protection to reinforcing steel. An attraction in this approach is the cost, which is only approximately 50% greater than that of conventional uncoated reinforcing steel in the USA.

A number of factors should be considered when evaluating the possible use of coated reinforcing bars for reinforced masonry in the UK. For example, coated bars need to be carefully handled to avoid impact damage and it may not be possible to bend the bars to standard radii.

Types of bed joint reinforcement available in the UK

A number of types of bed joint reinforcement are available in the UK. In Figure 9.5, type I consists of two parallel longitudinal rods welded to a continuous zig-zag cross rod to form a lattice truss. The yield strength of the steel is 500 N/mm². This type of bed joint reinforcement is available galvanized with the addition of an epoxy polyester powder coat (applied after fabrication) or in stainless steel. A much lighter form of bed joint reinforcement is made from 1.25 mm high tensile steel main wires and 0.71 mm mild steel bonding wires, illustrated as type 2 in Figure 9.5. This wire may be obtained galvanized or in stainless steel. The minimum cross sectional area recommended in Appendix A means that this percentage of reinforcement in a Type 2 may be too low to give an enhancement in lateral load performance which can be relied upon for design purposes. A third type of reinforcement consists of parallel drawn steel wires, 3.58mm in diameter with orthogonal cross wires 2.5 mm in diameter as shown in Figure 9.5 (type 3).

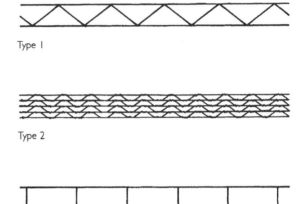

Type I

Type 2

Type 3

Figure 9.5 Type I 'Lattice truss' type bed joint reinforcement
Type 2 'Woven wire' bed joint reinforcement
Type 3 'Ladder type' bed joint reinforcement

2.4 Damp-proof courses

Reference should be made to BS 5628: Part 3 to ensure that the damp-proof course is suitable. In reinforced masonry a damp-proof course may present a particular problem since in some applications it will not be possible to introduce a membrane which will not interfere with the structural behaviour of the wall. Even in more conventional applications, materials which might squeeze out in a highly loaded element should not be used. Care should also be taken to consider the effect of sliding at the damp-proof course as well as adhesion to the mortar when the masonry is acting in flexure[12].

The absence of a damp-proof course in applications such as retaining walls may result in appearance and durability problems with certain facing units, and manufacturer's advice should be sought. Materials such as engineering bricks can be employed as a dpc in some situations. In other applications, such as prestressed diaphragm walls, it will generally be possible to employ one of the more conventional dpcs[13,14]. It may be necessary to provide a vertical membrane between the cross rib and outer face of a diaphragm wall. In this instance it is usual to employ a liquid dpc, either painted directly onto the outer leaf masonry or on the perpend of the cross rib.

2.5 Wall ties

When the low lift grouting technique is employed in conjunction with cavity construction, the vertical twist type of tie complying with BS I243[15] may be used. The requirements regarding length of tie in this Standard are not applicable to reinforced masonry but the designer should ensure that adequate embedment is possible. It is recommended that in situations where the masonry is likely to be wetted for prolonged periods, such as retaining walls, stainless steel ties be employed.

Where the high lift grouting technique is to be used with cavity construction then a more substantial tie should be used to resist the pressure exerted by the infilling concrete during placing. A suitable tie is described in Appendix B to the Code and, again care should be taken to ensure adequate protection against corrosion. Other forms of tie may be used providing they give adequate restraint against the pressure exerted by the concrete.

Whatever type of tie is employed it is clearly necessary to avoid filling the cavity until the leaves have achieved sufficient strength and sufficient bond strength has developed between the mortar and the tie. A minimum of three days is recommended in normal ambient conditions.

Wall ties for prestressed diaphragm wall construction where the cross ribs are not bonded into the outer leaf of the masonry will usually need to be obtained from a specialist supplier. A tie of substantial cross section is required to provide adequate shear resistance.

2.6 Cements

The types of cement which may be used with reinforced masonry are as follows:

1. Ordinary and rapid-hardening Portland cement (BS 12[16])
2. Portland blast-furnace cement (BS 146[17])
3. Sulfate-resisting Portland cement (BS 4027[18])

Neither masonry cement nor high alumina cement are permitted. BS 5628: Part 3 still permits the use of supersulphated cement to BS 4248[19] but this does not seem to have been used in conjunction with reinforced masonry in the UK and has, therefore, been excluded.

Limes which may be non-hydraulic (calcium), semi-hydraulic (calcium) and magnesium, should meet the requirements of BS 890[20].

2.7 .Aggregates

The recommendations of BS 5628 : Part 3 should be followed when considering the suitability of aggregates for mortar. Essentially this means that the fine aggregate should be free from deleterious substances and comply with BS 1200[21]. Marine sands should be washed to remove chlorides. Sands for mortar should be well graded. Single size sands or those with an excess of fines should be avoided if possible, but where their use is unavoidable, trial mixes should be assessed for suitability. Sands to grade M of BS 882 may well be found to be suitable.

Aggregates for infill concrete should comply with BS 882[22], BS 3797[23] or BS 1047[24].

Good mix design practice indicates in general that the largest possible maximum size of aggregate should be used in concrete. In the particular case of reinforced masonry, however, the need to produce a flowing concrete able to fill comparatively small sections without segregation will dictate the maximum size of aggregate which may be employed[25]. In any case the maximum size of aggregate should not be greater than the cover to the steel less 5 mm. The making of trial mixes is recommended to produce the best concrete from the materials available.

2.8 Mortar

2.8.1 General

The recommendations given in BS 5628: Part 3 and BS 5390 should be followed for the mixing and use of mortars. The mix proportions and mean compressive strengths at 28 days are provided in Table 2 of BS 5628 : Part 2. The testing of mortars should be carried out in accordance with Appendix Al of BS 5628: Part 1, which gives information on preliminary tests, the interpretation of test results and site tests. It should be noted that the compressive strength values given in the Table are fairly low and many sands will yield higher strength mortars.

The batching of mortars should be carried out by weight or by the use of gauge boxes. It is not acceptable for reinforced masonry purposes to batch mortar using a shovel since this invariably results in less cement being added than the specification requires.

2.8.2 Readymixed mortars

Readymixed lime: sand for mortars is now widely established and should comply with BS 4721[26]. Care should be taken to ensure that the correct proportion of cement is added on site.

The Code indicates that readymixed retarded mortars should only be used with the written permission of the designer. However, their use is likely to spread because of the convenience factor. Readymixed retarded mortars are delivered to site and placed in small skips which may be mechanically handled near to the point of use. Typically these mortars have a working life of three days, but once the mortar is used in the wall it sets and gains strength in a similar manner to conventional mortars.

2.9 Concrete infill and grout

The minimum grade of concrete infill which may be employed in reinforced masonry is a Grade 30 as described in BS 5328[27]. As an alternative to the Grade 30 mix, a mix of the following proportions by volume of the dry materials may be used or grouted cavity and quetta bond reinforced masonry construction:

1 : 0 – ¼ : 3:2 cement lime : sand : 10 mm maximum size aggregate

It is considered important to use a wet mix to ensure that the units or cavities are completely filled and the concrete properly compacted, but clearly the masonry may absorb a considerable amount of water, thereby effectively reducing the water/cement ratio. One method of keeping the water/cement ratio low whilst still producing a flowing mix is to employ a plasticiser or superplasticiser. The mix has to be produced with a carefully controlled slump, typically of 60 mm, before the admixture is added to give a collapse slump. The concrete then needs to be placed within 20–30 minutes.

To improve the protection offered to the reinforcing steel by the concrete cover, a range of options for a particular exposure condition is given in Table 14 of the Code. In some situations a concrete of a Grade better than 30, up to a Grade 50, may be required.

2.10 Colouring agents for mortar

By choosing the mortar colour with care, a range of effects can be achieved to match or contrast with the units. Using a light sand together with white cement and lime may produce a very light coloured mortar. Even where a coloured mortar is required, white cement will be necessary for some of the lighter mortar colours. White cement is, however, more expensive than ordinary Portland cement.

Pigments can be used to produce a coloured mortar. The final colour will depend not only upon the pigment, but also the cement, lime, sand, and the water/cement ratios. The final colour may also be affected by the water absorbtion of the unit and whether the mortar has been re-tempered.

There is a very wide range of pigments available and these should comply with BS 1014[29] and should be used in accordance with the manufacturer's instructions.

Under no circumstances should the amount of pigment used exceed 10% by weight of the cement in the mortar. In the case of carbon black, the total pigment content should be limited to 3% by weight of the cement.

2.11 Admixtures

2.11.1 General

The term admixtures is taken to include plasticisers for mortar and superplasticisers for infill concrete. The Code indicates that admixtures should only be used with the written permission of the designer. Clearly the manufacturer's requirements should be carefully followed, and if it is intended to use more than one admixture in a mix, then their compatibility should be checked. It is also important to recognize that the effect of an admixture will vary with different types of cement.

Care should be taken to check that any admixture to be used with reinforced masonry does not affect the durability of the units, mortar or concrete, nor should it increase the risk of corrosion of the reinforcement.

To avoid potential corrosion problems the chloride ion content of admixtures should not exceed 2% by mass of the admixture or 0.03% by mass of the cement. In addition the requirements of Table 2 of the Code should be met to limit the total chloride ion content of the mix.

2.11.2 Chlorides

Limits are placed in Table 2 on both the percentage of chloride ion present in sands and in concrete and mortar mixes. The intention is to prevent sufficient chloride ion being present in reinforced masonry to lead to problems caused by the corrosion of the reinforcing steel.

Plasticisers for concrete

There are five types of admixture specified in BS 5075: Part 1:1974, namely:

1. accelerating
2. retarding
3. normal water-reducing
4. accelerating water-reducing
5. retarding water-reducing

Only those of particular relevance to reinforced masonry are considered below in detail:

Normal water-reducing admixtures

Water-reducing admixtures (plasticisers, workability aids) increase the fluidity of the cement paste and, for a given mix, will either increase the workability without increasing the water/cement ratio or will maintain the same workability with reduced water/cement ratio,

Most proprietary admixtures of this type are based on lignosulphonates or solutions of hydroxylated carboxylic acid salts. These work by improving the dispersion of the cement particles. For infilling concrete mixes for reinforced masonry, they offer the following potential benefits:

1. increasing the cohesion and reducing segregation of high workability mixes by lowering the water content whilst maintaining the same workability
2. reducing the water content and hence increasing the strength whilst maintaining the workability

The dosage is usually quite small and trial mixes are recommended. Over-dosage can lead to retardation.

Superplasticisers

A flowing concrete may be produced by using a superplasticiser as a workability agent. Concrete produced in this way can be expected to have a slump of 200 mm or greater and should not exhibit excessive bleeding or segregation. Slumps in excess of 175 mm are generally considered as collapse slump. Superplasticisers may be based on one of the following chemicals:

1. sulphonated melamine formaldehyde condensates
2. sulphonated napthalene formaldehyde condensates
3.. modified lignosulphonates
4. polyhydroxylated polymers
5. mixtures of acid amides and polysaccharides

Mix design of superplasticised concrete

The basic approach to use a superplasticiser is to design

a concrete to have an initial slump of 60–75 mm, which is then dosed with between 1–6 litres per cubic metre (depending on type) of superplasticiser, thereby increasing the slump to collapse. The extent to which a fluid concrete is produced will depend upon the aggregate type, shape and overall grading. The first stage in the design is to use conventional mix design procedures to determine the water/cement ratio and mix proportions needed to give the specified strength with a slump of 75 mm. The proportions of cement, sand and aggregate now need to be checked and adjusted to avoid segregation.

Although most information is available for the use of superplasticisers with OPC cements, rapid-hardening and sulfate-resisting cements may also be used. It would be prudent, however, to check both the time-dependent bulk fluidity and the ultimate strength.

Using a superplasticiser

The superplasticiser needs to be added to the concrete at the point of use and the concrete mixed for a further 2–5 minutes. The concrete should be used immediately since maximum workability is retained for only 30–60 *minutes*. The period during which high workability will be retained is, to some extent, dependent upon the type of mixer and the rate of mixing. The faster the mixing action, the quicker the fall off in high workability.

References

1. BRITISH STANDARDS INSTITUTION. BS 5628: Part 3: 1985 *Code of Practice for use of masonry. Part 3: Materials and components. Design and workmanship.* BSI, London, pp.100.
2. BRITISH STANDARDS INSTITUTION. BS 5390: 1976 *Code of Practice for stone masonry.* BSI, London, pp. 40.
3. BRITISH STANDARDS INSTITUTION. BS 3921: 1985 *Clay bricks and blocks.* BSI, London. pp. 32.
4. BRITISH STANDARDS INSTITUTION. BS 6073: Part 1: 1981 *Specification for precast concrete masonry units.* BSI, London. pp. 12.
5. BRITISH STANDARDS INSTITUTION. BS 187: 1978 *Specification for calcium silicate (sandlime and flintlime) bricks.* BSI, London. pp.12.
6. BRITISH STANDARDS INSTITUTION. BS 1217: 1975 *Cast stone.* BSI, London. pp. 8.
7. BRITISH STANDARDS INSTITUTION. BS 8110: Part 1: 1977 *The structural use of concrete. Code of Practice for design and construction.* BSI, London.
8. BRITISH STANDARDS INSTITUTION. BS 4449: 1997 *Specification for hot rolled steel bars for the reinforcement of concrete.* BSI, London.
9. BRITISH STANDARDS INSTITUTION. BS 6744: 1978 *Specification for austenitic stainless steel bars for the reinforcement of concrete.* BSI, London.
10. BRITISH STANDARDS INSTITUTION. BS 4482: 1985 *Specification for cold reduced steel wire for the reinforcement of concrete.* BSI, London.
11. BRITISH STANDARDS INSTITUTION. BS 4483: 1985 *Specification for Steel fabric for the reinforcement of concrete.* BSI, London. pp.12.
12. BRITISH STANDARDS INSTITUTION. Draft for development DD 86: Part 1:1983 *Damp-proof courses: methods of test for flexural bond strength and short term shear strength.* BSI, London. pp. 8.
13. BRITISH STANDARDS INSTITUTION. BS 8215: 1991 *Code of practice for design and installation of damp proof courses in masonry construction.* BSI, London.
14. BRITISH STANDARDS INSTITUTION. BS 8102: 1990 *Code of practice for protection of structures against water from the ground.* BSI, London.
15. BRITISH STANDARDS INSTITUTION. BS 1243: 1978 *Specification for metal ties for cavity wall construction.* BSI, London. pp. 4.
16. BRITISH STANDARDS INSTITUTION. BS 12: 1996 *Specification for Portland cement.* BSI, London. pp.4.
17. BRITISH STANDARDS INSTITUTION. BS 146: 1996 *Specification for Portland blast-furnace cement. Metric units.* BSI, London.
18. BRITISH STANDARDS INSTITUTION. BS 4027: 1990 *Specification for sulfate resisting Portland cement.* BSI, London.
19. BRITISH STANDARDS INSTITUTION. BS 4248: 1974 *Supersulphated cement.* BSI, London. pp. 24.
20. BRITISH STANDARDS INSTITUTION. BS 890: 1995 *Specification for building limes.* BSI, London.
21. BRITISH STANDARDS INSTITUTION. BS 1199, 1200: 1976 *Specification for Building sands from natural sources.* BSI, London.
22. BRITISH STANDARDS INSTITUTION. BS 882: 1992 *Specification for aggregates from natural sources for concrete.* BSI, London.
23. BRITISH STANDARDS INSTITUTION. BS 3797: 1990 (1996) *Specification for lightweight aggregates for masonry units and structural concrete.* BSI, London.
24. BRITISH STANDARDS INSTITUTION. BS 1047: 1983 *Specification for air-cooled blast-furnace slag aggregate for use in construction.* BSI, London.
25. ROBERTS, J J. The behaviour of vertically reinforced concrete blockwork subject to lateral loading. London. Cement and Concrete Association. February 1975. 42.506.
26. BRITISH STANDARDS INSTITUTION. BS 4721: 1981 *Specification for readymixed building mortars.* BSI, London. pp. 12.
27. BRITISH STANDARDS INSTITUTION. BS 5328: Part 2: 1997 *Methods for specifying concrete mixes.* BSI, London.
28. BRITISH STANDARDS INSTITUTION. BS 1881: 1983 *Methods for testing concrete.* BSI, London.
29. BRITISH STANDARDS INSTITUTION. BS 1014: 1975 *Pigments for Portland cement and Portland cement products.* BSI, London. pp.12.
30. BRITISH STANDARDS INSTITUTION. BS 4486: 1980 *Specification for hot-rolled and processed high tensile alloy steel bars for the prestressing of concrete.* BSI, London. pp. 4.
31. BRITISH STANDARDS INSTITUTION. BS 5896: 1980 *Specification for high tensile steel wire strand for the prestressing of concrete.* BSI, London. pp. 12.

Section 3: Design objectives and general recommendations

3.1 *Basis of design*

3.1.1 *Limit state design*

CP 110:1972[1] stated that the purpose of design is to ensure that all the criteria relevant to safety and serviceability are considered in the design process, these criteria being associated with limit states. This was the first UK Code to adopt limit state design, a philosophy which was applied to BS 5628: Part 1[2] which was published in 1978.

The adoption of limit state design was only possible against a background of a better understanding of performance requirements. Essentially the design process is one of balancing all the factors involved. For example, a wall could be strong enough to withstand a high wind load, but not without deflecting excessively and incurring unacceptable cracking in applied finishes. Conversely, the wall could be designed to minimize deflection but not possess adequate lateral strength to provide an acceptable factor of safety against collapse.

There is insufficient data available to be able to confidently calculate every limit state for every potential reinforced masonry element. From a designer's point of view it is often convenient to be able to use simple sizing rules to ensure that the limit states of deflection and cracking will not be reached and then to carry out a detailed structural analysis of the element for the ultimate limit state. This approach has been adopted in BS 5628: Part 2, although it is possible to exceed the sizing requirements provided that checks are made to ensure that deflection and cracking are not likely to be excessive. In an ideal situation, the probability of reaching a particular limit state should be determined from a full statistical analysis of the behaviour of masonry appropriate to that limit state. In the absence of such comprehensive information, however, BS 5628: Part 2 employs a partial safety factor approach, using characteristic values of strengths.

The characteristic strength of masonry, for example, is defined as the value of the strength of masonry below which the probability of test results falling is not more than 5%. This characteristic strength value is modified by a partial safety factor to give the value (e.g. strength), to be used in design – the design value.

There are two types of partial safety factor employed in BS 5628: γ_f, which is applied to loads, and γ_m, which is applied to materials,

The partial safety factor for loads (γ_f) is intended to take account of:

1. possible unusual increases in load beyond those considered in deriving the characteristic load
2. inaccurate assessment of effects of loading and unforeseen stress redistribution within the structure
3. variations in dimensional accuracy achieved in construction

The partial safety factor for materials (γ_m) takes account of:

1. differences between site and laboratory constructed masonry
2. variations in the quality of materials in the structure

It should be noted that BS 5628 Part 2 allows the designer to design in accordance with BS8110 if the cross section of the infill concrete is substantial. This would involve disregarding the effect of the masonry units, considering them solely as permanent formwork making no contribution to the strength of the element. If the designer chooses to exercise this option, he should ensure that the mix design, method of placing and detailing are also in accordance with BS8110.

3.1.2 Limit states

3.1.2.1 Ultimate limit state

BS 5628: Part 2 indicates that 'The strength of the structure should be sufficient to withstand the design loads taking due account of the possibility of overturning or buckling'. It is thus necessary to show that the strength of the structure is such that there is an acceptable probability that it will not collapse under the load described above. The calculations must take account not only of primary and secondary effects in members, but also in the structure as a whole.

3.1.2.2 Serviceability limit state

3.1.2.2.1 The deflection of a reinforced masonry element may affect not just the element itself in terms of appearance and durability, but also lead to the cracking or loss of bond of any applied finishes. The cracking of a render, for example, might lead to an excess of water entering into a wall and, in the case of some types of clay brickwork, could lead to problems of sulfate attack if the bricks have a high sulfate content.

The Code makes three recommendations to ensure that, within the limitations of the calculation procedures, deflections are not excessive. These may be summarized as:

1. final deflection not to exceed length/125 for cantilevers or span/250 for all other elements
2. limiting deflection span/500 or 20 mm, whichever is the lesser, after partitions and finishes are completed
3. total upward deflection of prestressed elements not to exceed span/300 if finishes are to be applied, unless uniformity of camber between adjacent units can be achieved

3.1.2.2.2 Little guidance is given in the Code on the subject of cracking. Fine cracking is to be expected in reinforced masonry but the crack width should be limited to avoid possible durability problems. The Code also recommends that the effects of temperature, creep, shrinkage and moisture movement be considered and allowed for with appropriate movement joints.

Although the Code does not give any further guidance, the authors' have tried to provide indications of crack widths where this is available. The maximum crack width which the authors consider likely to occur in reinforced masonry designed to the Code is 0.3 mm.

3.2 Stability

3.2.1 General recommendations

The purpose of this note in the Code is to make clear the need for one person to be responsible for the overall design of the structure. This designer has to coordinate the work of other members of the design team to ensure that the stability of the structure is adequate and that the design and detailing of individual elements and components does not impair this stability.

Clearly the layout and interaction of the elements will significantly affect the stability and robustness of the overall design. As a matter of course the inclusion of a significant proportion of reinforced masonry within the structure will tend to improve the overall tieing together of the structure if adequate connections are provided.

It is necessary, as is the case with unreinforced masonry, that the building be designed to resist at any level a uniformly distributed horizontal load equal to 1.5% of the characteristic dead load above that level. In addition, robust connections need to be provided between elements of the construction as detailed in Appendix C of Part I. Finally, of course, compatibility between elements of different materials should be considered when making connections between them.

3.2.2 Earth-retaining and foundation structures

There will, in these situations, be a number of factors to be taken into consideration to ensure the overall stability of structures. For example, although the stem of a retaining wall may be designed according to this Code, there are other considerations to ensure adequate resistance against sliding, overturning and so on. These are essentially geotechnical considerations although, for example, the location, thickness and weight of the wall may be of relevance.

The partial safety factor, γ_f, to be applied to earth and water loads is as for other types of load whether the load is beneficial, e.g. passive pressure on a retaining wall, or not. The designer should only consider revising the value of γ_f if the loads, due to their method of derivation, have already been factored.

3.2.3 Accidental forces

This clause of the Code requires the design to consider the consequences of misuse or accident. It is not expected that the building should be capable of resisting the forces which would result in an extreme case. It is expected, however, that damage resulting from any particular accident should not be disproportionate to the cause of the accident.

The general recommendations of Clause 20.3 of Part 1 are applicable to all building types. In passing it is worth considering the fact that the adoption of a (uniformly distributed) lateral load expressed as a percentage of the total characteristic dead load is a common requirement in seismic regions. In a zone where a significant earthquake risk exists, this percentage should be greater than 1.5%, but the latter should be adequate in the UK where there is only the risk of a relatively minor tremor.

In the case of buildings of five storeys and above (Category 2 in Part 1), it is recommended that either:

1. an assessment is made of the resultant stability and extent of damage following the removal of a loadbearing element or
2. sufficient horizontal and/or vertical tying is provided within the structure

The first approach involves a detailed examination of the structure to calculate the effect of the loss within each 'compartment' of a loadbearing element unless they are designed as protected members. The latter involves (depending on whether option [2] or [3] is taken from Table 12 of Part 1) either analysis for vertical elements only or no further assessment because of the extent of tying.

3.2.4 During construction

This note is intended as a warning to ensure that consideration is given to the need for temporary support during the construction phase. For example, reinforced masonry beams can be readily built in situ off of a horizontal shutter which will need to be propped until the masonry has developed sufficient strength to allow the shutter to be removed.

3.3 Loads

In principle, limit state design requires that the characteristic load on any structure is statistically determined. Regrettably, insufficient data is available as yet to express loads in this way. It is assumed that the characteristic dead, imposed and wind loads may be taken from BS 6399[5]. Nominal earth load (E_n) may be obtained in accordance with current practice, for example, as described in BS 8004[6].

3.4 Structural properties and analysis

3.4.1 Structural properties

3.4.1.1 Characteristic compressive strength of masonry f_k

3.4.1.1.1 General. The purpose of this warning in the Code is to draw attention to the fact that in a reinforced or prestressed element, the units may be loaded in a direction other than that which would normally occur in unreinforced masonry.

The compressive strength of masonry units is determined by applying loads through the platens of a testing machine normal to the bed faces of the unit. The strength so obtained is unique to that direction of loading. Even allowing for the adjustment necessary for the effect of changing the aspect ratio when the unit is tested in a different direction (for example, load normal to the header faces), the strength of the unit is still likely to be different, depending upon the type of unit.

In the case of solid aggregate blocks, variations in strength with unit orientation will be introduced by the method of manufacture, although these will generally be small. In many cases, vertical compaction and vibration during manufacture could lead to a variation in strength over the height of the unit, whereas a few machines mould blocks on end which could lead to variation in properties along the length of the unit. Autoclaved aerated blocks are cut to size from 'cakes' of foamed concrete and here the properties of the units may depend on the orientation in which the units are cut from the 'cake'.

For design purposes solid concrete units and hollow and cellular concrete units filled with concrete are assumed to have the same characteristic strength regardless of the direction of loading, even on end. When unfilled cellular or hollow blocks are employed loaded in directions other than 'normal' the characteristic strength must be determined by test as discussed later.

In the case of some extruded wire cut bricks which have a number of perforations (20~25% of bed area), the strength when loaded through the header faces may be of the order of 10~15% of that obtained when loaded through the bed faces. This is clearly related to the geometrical form of the unit, since when on end the brick is more slender than on bed and platen restraint is reduced. In addition, the perforations act as stress raisers and superimposed on these effects are any directional properties due to the extrusion process. Although this reduction in strength is dramatic, the available test results indicate that when built into an element the strength of the reinforced clay brickwork when loaded parallel to the bed faces is at least 40% of that when loaded normal to the bed faces[7]. Brickwork made from some pressed bricks is stronger when loaded parallel to the bed faces than when normal to them.

The compressive strength of the unit is not, of course, the characteristic strength of the masonry, but the above hopefully illustrates how variations in performance with direction of loading are likely to occur in practice. In the following section the determination of characteristic compressive strength of masonry is discussed.

3.4.1.1.2 Direct determination of the characteristic compressive strength of masonry, f_k. The 'characteristic' masonry strengths presented in Table 3 of the Code are based on those presented in BS 5628 Part 1. Although these are termed characteristic they have not been determined statistically but are in general agreed lower bounds to the masonry strength. The designer may wish to directly determine a value of the characteristic compressive strength of a particular combination of units and mortar. This may be done by deriving a value statistically from test results (see Appendix D).

3.4.1.1.3 Value of f_k where the compressive force is perpendicular to the bed face of the unit.
This section essentially reflects the information provided in Part 1 except that only mortar designations (i) and (ii) are considered. A new table, Table 3(B) and accompanying figure 1(b), have been added which cater for the use of units with a height to thickness ratio of 1.0. This information is useful for reinforced hollow block masonry with filled cores (remembering to use the nett unit strength unless the infill concrete is less strong than the compressive strength of the units, in which case the cube strength of the infill should be used to determine the characteristic compressive strength of the masonry).

3.4.1.1.4 Value of f_k where the compressive force is parallel to the bed face of the unit. This section requires no further detailed comment. Note that filled hollow blocks are treated as solid units and are not covered by this section.

3.4.1.1.5 Value of f_k for units of unusual format or for unusual bonding patterns. This section requires no further detailed comment.

3.4.1.2 Characteristic compressive strength of masonry in bending

This clause indicates that the value derived for the characteristic compressive strength of masonry should be used for both direct and flexural compression. The reason for the statement is that designers familiar with Codes based on permissible stress design, will be used to enhance the maximum permissible compressive stress when this is due to flexural compression. Such enhancements compensate for the inaccurate assumption that the stress distribution is linear across the section and are not necessary for the different assumptions made with limit state design.

3.4.1.3 Characteristic shear strength of masonry

Further information on the provision for shear is given in Clause 22.5.

3.4.1.3.1.1 Shear in bending (reinforced masonry). The value of the characteristic shear strength of masonry, f_v, in which the reinforcement is placed in bed or vertical joints (including Quetta bond) or is surrounded by mortar and not concrete is 0.35 N/mm². No enhancement in shear strength is given for the amount of tensile reinforcement since this type of section has been shown experimentally[10] not to warrant such an enhancement when mortar is the embedment medium. It is not entirely clear why this should be so but is likely to be due to a reduction in the amount of dowel action which can be utilized in such reinforcement. Consequently, there is a reduction in the contribution by dowel action to the average shear strength across the section. It may be noted that 0.35 N/mm² is also the characteristic shear strength assumed for unreinforced masonry.

For simply supported beams or cantilevers an enhancement factor of $2\,d/a_v$ (with a limiting factor of 2) can be applied when a principal load (usually accepted as one contributing to 70% or more of the shear force as a support) is at a distance a_v from the support. This is again demonstrated in the work of Suter and Hendry[11]. The maximum factor of 2 implies a cut off in the shear strength at a ratio $a_v/d = 1.0$.

The Code suggests that in certain walls where substantial precompression can arise, for example, in loadbearing walls reinforced to enhance lateral load resistance, it is often more advisable to treat the wall as plain masonry, i.e. unreinforced, and design to BS 5628: Part 1[2].

For sections in which the main reinforcement is enclosed by concrete infill, an enhancement to f_v is given depending upon the amount of tensile reinforcement, by the formula:

$$f_v = 0.35 + 17.5\rho$$

where $\rho = A_s/bd$ with an upper limit of 0.7 N/mm^2.

For simply supported beams or cantilever retaining walls an enhancement in the shear strength as derived above is given by the formula.

$$[2.5 - 0.25\,(a/d)]$$

Here the shear span is defined as the ratio of the maximum design bending moment to the maximum design shear force, i.e. M/V. This enhancement is similar to that in 19.1.3.1.1, but has been derived on a more rational basis reflecting the greater amount of more specific data on this subject. An upper limit of 1.75N/mm^2 is applied, i.e. a maximum enhancement of 2.5 when a/d = 0; the enhancement factor equals 1.0 when a/d = 6. Much below a/d = 2, the masonry would act as a corbel not a beam, above a/d = 6, the failure mode would be flexural, shear failure being most unlikely. Between these values a 'transition' occurs from shear to flexural failure. This behaviour in shear is analogous to that of reinforced concrete upon which much has been written.

3.4.1.3.2 Racking shear in reinforced masonry shear walls. The first part of this clause deals with walls subjected to racking shear as if they were unreinforced (see BS 5628: Part 1). The increase of 0.6 g_b due to vertical loads both here and in 19.1.3.1.1 is due to an increased 'friction effect' preventing sliding.

A note is given in the Code relating to the effect on shear resistance of damp-proof courses. Some information exists[12], and some general guidance is given in *Sections 2.8* and *6.3*.

3.4.1.4 Characteristic strength of reinforcing steel f_y

The characteristic tensile strength of reinforcing steel is given in the Code as Table 4. The appropriate compressive strength may be obtained by multiplying these values by 0.83.

3.4.1.5 Characteristic breaking load of prestressing steel

This section requires no further detailed comment.

3.4.1.6 Characteristic anchorage bond strength, f_b

Reinforcement exhibits better bond strength in concrete than in mortar and this is reflected in the values given here. The same value is given for bars in compression or tension and any increase due to increase in strength of the concrete is not permitted. This approach is likely to be conservative, but it was felt by the Code Committee that insufficient evidence existed to extend the given values further.

Characteristic anchorage bond strength (N/mm^2) for tension or compression reinforcement embedded in:

	Plain Bars	Deformed Bars
Mortar	1.5	2.0
Concrete	1.8	2.5

The Code contains a note to the effect that these values may not be applicable to reinforcement used solely to enhance lateral load resistance of walls. This is for two reasons:

1. the shape, type and size of certain proprietary reinforcement will differ from the bars normally used as reinforcement
2. normal detailing rules do not generally apply in this situation

The values of f_b apply to austenitic stainless steel for deformed bars only and in other cases values will need to be established by test

3.4.1.7 Elastic moduli

For all types of reinforced masonry the short term elastic modulus, E_m, may be taken as 0.9f_k kN/mm^2. Although the accuracy of this estimate does vary with different types of masonry, it is reasonably well substantiated by experimental work and is consistent with overseas data[13]. It must be noted that this is the 'gross' elastic modulus of reinforced masonry including the concrete infill; an 'effective' modulus should not be calculated based on a transformed section incorporating different values of modulus for the concrete infill and masonry separately. This approach is likely to be somewhat conservative, particularly where relatively high strength concrete is used with relatively low strength units and particularly for blockwork.

The elastic modulus of concrete infill used in prestressed masonry is given in Table 5 of the Code, thus effectively allowing the use of transformed sections. The long term moduli appropriate to various types of reinforced masonry are given in Appendix C.

The elastic modulus of all steel reinforcement is given as 200 kN/mm^2 and that for prestressing steel may be taken from the appropriate British Standard with due allowance made for relaxation under sustained loading conditions.

3.5 Partial safety factors

3.5.1 General

In the comment on *Section 3.16.1*, the role of the partial safety factor is indicated. The partial safety factor for loads, γ_f, is used to take account of possible unusual increases in load beyond those considered in deriving the characteristic load, inaccurate assessment of effects of loading, unforeseen stress redistribution within the structure and the variations in dimensional accuracy achieved in construction. The partial safety factor for materials, γ_m, makes allowance for the variation in the quality of the materials and for the possible difference between the strength of masonry constructed under site conditions and that of specimens built in the laboratory.

3.5.2 Ultimate limit state

3.5.2.1 Loads
The four load cases (a) to (d) in this section indicate the appropriate combinations of design dead load, design imposed load, design wind load, i.e. their corresponding characteristic loads which, together with their attendant values of γ_f need to be considered. These values were selected to produce acceptable global factors of safety.

It will be apparent that load case (a) will be the one which governs the design of many buildings. Case (b) will dominate in the situation where wind load is the primary load. Case (c) considers the combination of all three loads with reduced values of γ_f applied to each due to the fact that it is unlikely that extreme values for all three will occur simultaneously.

There are cases when it may be appropriate to either use different partial safety factors to those recommended or in fact derive design loads in a completely different way. The Code refers to two areas in particular. In the case of farm buildings[14] the design loads are determined on a basis which allows for likely levels of human occupation and is incorporated in a partial safety factor which is described as the classification factor.

3.5.2.2 Materials
When considering the adequacy of a structure or an element to resist design loads, the design strength is considered to be the characteristic strength divided by a partial safety factor. The values to be used have all been recommended on the assumption that the quality of construction on site is what has been described in BS 5628 Part 1 as *special*. Essentially this means that the designer ensures that the construction is in accordance with the Code and any other Specification and that preliminary and site testing of materials is carried out.

The value used for γ_m depends on whether the masonry units are supplied to the normal or special category of construction control. Special category construction control may be claimed by a manufacturer who agrees to provide units which meet or exceed an agreed compressive strength described as the 'acceptance limit' with a specified degree of confidence. To do this the manufacturer must operate a quality control scheme, the results of which may be examined by the purchaser. The scheme must be such that it can be demonstrated to the purchaser that the likelihood of the mean compressive strength of a sample taken from any consignment of units being below the acceptance limit is less than 24%. The procedure for special category control is also described in BS 6073 Part 2[15]. If the manufacturer cannot make the above claim and substantiate it, the designer should choose the slightly larger partial safety factor (γ_{mm}) corresponding to the normal category of manufacturing control.

3.5.3 Serviceability limit state

3.5.3.1 Loads
As when considering the adequacy from a strength point of view, the worst combination of loads should be used when assessing deflections, and the effects of creep, thermal movement, and so on, may need to be considered.

3.5.3.2 Materials
When considering deflections, stresses or cracking, the values of γ_{mm} should be chosen as 1.5 and that of γ_{ms} as 1.0.

3.5.4 Moments and forces in continuous members

In continuous members and their supports it is necessary to consider the effects of pattern loading. It is considered that an adequate assessment will be made of the structure at the ultimate limit state if the two conditions below are considered:

1. alternate spans loaded with maximum combination of dead + imposed load ($1.4\ G_k + 1.6\ Q_k$) and minimum dead load ($0.9\ G_k$)
2. all spans loaded with maximum combination of dead and imposed load.

References

1. BRITISH STANDARDS INSTITUTION. CP 110: Part 1: 1972 *The structural use of concrete*. BSI, London. pp. 156.
2. BRITISH STANDARDS INSTITUTION. BS 5628: Part 1: 1978 *Code of practice for use of masonry*. Part I: Structural use of unreinforced masonry. BSI, London. pp. 40.
3. HASELTINE, B A and MOORE, J F A. *Handbook on BS 5628: Structural use of masonry*. Brick Development Association, Winkfield, Berkshire. May 1981. pp. 118.
4. ROBERTS, J J, TOVEY, A K, CRANSTON, W B and BEEBY, A W. *Concrete masonry designers handbook*. Viewpoint Publications, Palladian Publications Ltd. London, 1983. pp.2
5. BRITISH STANDARDS INSTITUTION. BS 6399: Part 1: 1996 *Code of Practice for dead and imposed loads*. BSI, London.
6. BRITISH STANDARDS INSTITUTION. BS 8004: 1986 *Code of practice for foundations*. BSI, London.
7. HASELTINE, B A. A design guide for reinforced and prestressed brickwork. Proceedings of the North American Masonry Conference, Boulder, Colorado, 1978.
8. BRITISH STANDARDS INSTITUTION. CP 111: 1970 *Structural recommendations for loadbearing walls*. BSI, London. pp.40.
9. BEARD, R. A theoretical analysis of reinforced brickwork in bending. Proceedings of the British Ceramic Society, No.30,1982.
10. SUTER, C and HENDRY, A W. Shear strength of reinforced brickwork beams: influence of shear arm ratio and amount of tensile steel. *The Structural Engineer*, 53(6), 249. 1975.
11. SUTER, C and HENDRY, A W. Limit stale design of reinforced brickwork beams. Proceedings of the British Ceramic Society, No.24, 1975.
12. HODGKINSON, H R and WEST, H W H. The shear resistance of some damp proof course materials. British Ceramic Research Association. Technical Note TN 326, 1981.

13. EDGELL, G J. *Reinforced and prestressed masonry*. Thomas Telford Limited, London, 1982. pp.123.
14. BRITISH STANDARDS INSTITUTION. BS 5502: 1993 *Code of Practice for the design of buildings and structures for agriculture*. BSI, London.
15. BRITISH STANDARDS INSTITUTION. BS 6073: Part 2: 1981 *Method for specifying precast masonry units*. 851. London. pp. 8.

Section 4: Design of reinforced masonry

4.1 General

This section indicates that the subsequent recommendations are based on the assumption that design against reaching the ultimate limit state is critical. As mentioned elsewhere, the serviceability considerations are met using either simple sizing rules or by the detailed calculation of, for example, deflection, at service loads.

4.2 Reinforced masonry subjected to bending

4.2.1 General

This section of the Code deals with the design of elements subjected only to bending. Clearly this applies to a wide range of elements including beams, slabs, retaining walls, buttresses and piers. The design approach may also be applied to panel or cantilever walls reinforced primarily to resist wind forces. Walls containing bed joint reinforcement to enhance lateral load resistance should be designed following the recommendations of Appendix A, which are described in *Chapter 8*. In a few situations it may be appropriate to design a reinforced masonry element as a two-way spanning slab using conventional yield-line analysis. The approach which has been adopted in the Code to the design of members subjected to bending has been developed from the simplified approach developed for concrete[1].

The designer may calculate deflections using the procedure described in Appendix C to check that a member will not deflect excessively under service loads. In many situations, however, it will be sufficient to limit the ratio of the span to the effective depth. The same limiting values should also ensure that cracking in service conditions will not be excessive, although little research evidence is available on this topic. By designing elements within the limiting ratios imposed by the simple sizing rules, it is only necessary to determine that the design resistances exceed the design forces or moments to ensure that there is an adequate factor of safety against reaching the ultimate limit state.

4.2.2 Effective span of elements

The effective span of either simply supported or continuous members may be taken as the lesser of:

1. the distance between the centres of supports
2. the clear distance between the faces of the supports plus the effective depth.

The effective span of a cantilever may be taken as the lesser of:

1. the distance between the end of the cantilever and the centre of its support
2. the distance between the end of the cantilever and the face of the support plus half the effective depth

4.2.3 Limiting dimensions

4.2.3.1 General

Attention is drawn to the fact that the limiting ratios given in Tables 8 and 9 of the Code should not be used when more stringent limitations on deflection and/or cracking are required.

4.2.3.2 Walls subjected to lateral loading

Limiting values of the ratio span to effective depth for walls subjected to lateral loads are given in Table 8 of the Code. In the case of cavity walls, the effective depth of the reinforced leaf should be used. In the case of freestanding walls that do not form part of a building and are subjected primarily to wind loading, the limiting ratios may be enhanced by 30% provided that increased deflections and cracking are not likely to cause damage to applied finishes.

4.2.3.3 Beams

In the case of beams, relatively little data exists to indicate what might be reasonable limiting ratios of span to effective depth. As a result, the same limiting ratios as are used for reinforced concrete have been adopted, although as yet no enhancement based on the level of working stress has been introduced, as it has in the case of reinforced concrete. Further data is required before this can be done, but the evidence available suggests that the recommended values which are given in Table 9 of the Code are fairly conservative.

For simply supported or continuous beams the distance should not exceed the lesser of 60 b_c and 250 b_c^2/d For a cantilever the clear distance from the end to the face of the support should not exceed the lesser of 25 b_c and 100 b_c^2/d. In the case of simply supported or continuous beams, b_c is the breadth of the compression block midway between restraints, in the case of a cantilever it is suggested that bc be taken as the breadth of the compression zone at the support.

4.2.4 Resistance moments of elements

For any singly reinforced masonry section there is a unique amount of reinforcement which would fail in tension at the same bending moment as that at which the masonry would crush. This section is described as balanced and if lower amounts of reinforcement were incorporated the section would be described as under-reinforced. If an under-reinforced section were tested to destruction in flexure the failure would be due solely to that of the steel in tension. In laboratory tests tensile failure often leads to massive deflections and subsequent compressive failure in the masonry. When large amounts of reinforcement are provided, greater than that required for a balanced section, the failures in test beams are due solely to the masonry in the compression zone having inadequate strength. These failures can be sudden, are sometimes explosive and the aim of the Code recommendations is to ensure that all the sections designed using them are under-reinforced.

Some relatively simple assumptions have been made which enable the design moment of resistance of any under-reinforced section to be determined. An upper limit to the design moment of resistance has been set, which is that of the balanced section.

4.2.4.1 Analysis of sections

The mean stress at failure of the masonry in compression is assumed to be f_k/γ_{mm} where f_k is the characteristic compressive strength of masonry and γ_{mm} is the partial safety factor for the compressive strength of masonry. This partial safety factor is intended to allow for the possibility that the masonry in the structural element on site may be weaker than similar masonry constructed in the laboratory. An allowance for other factors which affect the capacity of the section (rather than the masonry in the compression zone) is also included in this partial safety factor and consequently these influences are treated as being equivalent to a reduction in the strength of the masonry. This formulation does not necessarily attribute the various causes of uncertainty in the bending moment capacity to the most appropriate parameters because further evidence of the likely magnitude of the various influences is needed before this can be done. The current recommendations are conservative. The maximum strain in the outermost compression fibre is assumed to be 0.0035 and is reached when the masonry fails in compression. For a balanced section the compression block is considered to have its greatest depth, $d_{c\ max}$ and plane sections are considered to remain plane. This depth is defined by the tensile strain in the steel at failure. This is found from the assumed stress-strain relationship for steel given in the Code.

The short term stress-strain relationship for stocky specimens of brickwork has been established as a curve which may be represented by a parabola with a falling branch[2]. Although less research has been conducted, it is apparent that the stress strain curve for reinforced hollow concrete blockwork is either parabolic or rectangular-parabolic[3]. If the assumption is made that plane sections remain plane, a logical form for the stress block is parabolic. The advantages of the simplicity and familiarity of the rectangular stress block approach are, however, substantial and there is considerable merit for design purposes in replacing the parabola by a statically equivalent rectangle. This is the approach adopted in the Code with the exception that the mean height of the stress block is f_k and not, as may be derived from theory, $0.75\ f_k'$. The accuracy of the simplified approach

adopted in the Code has, however, been demonstrated by experimental work[5].

For those sections which are acting primarily in flexure, but which are also subjected to a small axial thrust, it is considered reasonable to ignore the thrust for design purposes because the flexural stress will dominate. The limiting stress due to the axial thrust which may be ignored in this way is 10% of the characteristic compressive strength of the masonry.

4.2.4.2 Design formulae for singly reinforced rectangular members

This section deals with the design of singly reinforced rectangular members which are sufficiently long (i.e. the ratio of span to effective depth is greater than 1.5) to be acting primarily in flexure. The designer must ensure that the Design Moment of Resistance of the section (which is determined on the basis that it is an under reinforced section) is greater than the bending moment due to the design loads. The design formula is:

$$M_d = \frac{A_s f_y z}{\gamma_{ms}}$$

and this must not exceed

$$\frac{0.4\ f_k\ bd^2}{\gamma_{mm}}$$

where

$$z = d\ \left(1 - 0.5\ \frac{A_s\ f_y\ \gamma_{mm}}{bd\ f_k\ \gamma_{ms}}\right)$$

and:

M_d = design moment of resistance
b = width of the section
d = effective depth
f_y = characteristic tensile strength of reinforcing steel
f_k = characteristic compressive strength of masonry
z = lever arm, which should not exceed 0.95
γ_{mm} = partial safety factor for strength of masonry
γ_{ms} = partial safety factor for strength of steel

For the compression block depths derived for balanced sections on page 35 i.e. $0.53d$ and $0.47d$, the corresponding design moments of resistance are:

$$\frac{0.39\ f_k\ bd^2}{\gamma_{mm}}\ \ \text{and}\ \ \frac{0.36\ f_k\ bd^2}{\gamma_{mm}}$$

In the case of a beam where the width and effective depth have been fixed, possibly by other than structural considerations or by a simple sizing rule, then, if the bending moment due to the design loads is M, the designer must ensure that $M \le M_d$.

4.2.4.3 Design formulae for walls with the reinforcement concentrated locally

4.2.4.3.1 Flanged members. There are a number of situations where reinforced masonry elements may be considered to act as flanged members and the Code includes recommendations for the more usual cases, which are in walls. Naturally, the same principles apply in other cases also.

The width of the masonry which is considered to act as a flange is limited in an arbitrary way so that the design is not extended to cases where the stability of the flanges is critical. Nevertheless, it is important that, when the spacing between concentrations of reinforcement exceeds 1 m, the capacity of the masonry to span between them should be checked.

The thickness of the flange, t_f is taken as the masonry thickness provided that this value does not exceed half the effective depth. The width of the flange is then taken as the least of:

1. for pocket-type walls, the width of the pocket or rib plus 12 × the thickness of the flange
2. the spacing of the pocket or ribs
3. one third of the height of the wall.

In the case of pocket type walls where the pocket is contained wholly within the thickness of the wall, it acts as a homogeneous cantilever. For design purposes, however, it is convenient to group pocket type walls with other walls in which the reinforcement is placed in local concentrations.

The design moment of resistance for under reinforced sections is the same as that for singly reinforced rectangular sections, i.e., given by the design formula. The upper limit for the balanced section is given below:

$$M_d = \frac{f_k}{\gamma_{mm}} b t_f (d - 0.5 t_f)$$

When checking the capacity of the masonry to span between the concentrations of reinforcement, it may be considered to be arching horizontally and justified using Clause 36 of Part 1 to the Code[6]. It is important for the designer to ensure that, at the end of a wall, there is sufficient resistance to the component of the arch thrust that acts in the plane of the wall. The necessary force may be provided by part of an adjacent structure. Alternatively, the end of the wall may be restrained by the provision of additional reinforcement. Similarly the design should not rely on the action of arching forces across movement joints and these are generally located at positions where an additional reinforced rib, pocket or core, have been included in the wall.

4.2.4.3.2 Locally reinforced hollow blockwork. It is possible, particularly in the case of hollow blockwork, that reinforcement is concentrated locally. For example, a hollow blockwork wall may have a few cores reinforced vertically at the centre of a length of walling to divide the horizontal span. In this case the reinforced element is considered to be limited in width to 3 × the thickness of the block.

4.2.5 Shear resistance of elements

This clause deals with the shear requirements of elements in pure bending, although the recommendations are equally applicable to elements subjected to a combination of vertical load and bending where the effect of the moment is much greater than the axial load (i.e. resultant eccentricity, $e_x = \frac{M}{N}$ is greater than $\frac{t}{2}$).

The design for shear in this case would tend to be conservative as there is no method of taking account of the enhanced resistance to shear afforded by the precompression. Shear due to in-plane forces, i.e. racking shear, is dealt with under Clause 25.

4.2.5.1 Shear stresses and reinforcement in members in bending

4.2.5.1.1 Behaviour in shear. The shear stress at any cross section, v, is calculated from the equation

$$v = \frac{V}{bd}$$

where:

b = the width of the section
d = the effective depth (or for a flanged member, the actual thickness of the *masonry* between the ribs if this is less than the effective depth as defined in 2.4)
V = the shear force due to design loads

This equation treats the shear stress as if it were uniformly whole cross section as far as the tensile reinforcement.

This is not strictly true and many researchers have found that, for reinforced concrete without shear reinforcement, the shear resistance is made up of a number of component forces. The situation has been found to be similar for reinforced masonry.

The shear resistance of the section includes contributions from the uncracked part of the section which is primarily in compression, dowel action of the tensile reinforcement and any interlock along the tensile cracks. In reinforced concrete design the shear resistance is increased with an increase in the compressive strength of the concrete and also the amount, but not the grade, of tensile reinforcement. There is no recognized method of allowing for interlock which, in the case of reinforced concrete, is due to aggregates. Also, as dowel action depends for its effectiveness on the tensile strength of the concrete in that the cover must not burst, it should not in general be relied upon. As in practice, however, the figures for shear resistance are derived from tests, there will be a contribution based on both interlock and dowel action included in the design.

Enhancement due to masonry strength. Masonry research in references[7,8] relating to UK work and reference[9] referring to work from Canada, has shown that the shear resistance of masonry depends to some extent on the compressive strength of the masonry and the percentage of reinforcement when the reinforcement is located in bed joints or bond beam units. The former, however, has not been included in *19.1.4.1* since there is insufficient information available. The various types of masonry unit and methods of construction will perform differently in this context and the enhancement is relatively small. The increase in shear strength when the amount of tensile reinforcement is increased is not great and no enhancement is permitted for design purposes. An enhancement due to the percentage of reinforcing steel is included in the formula to be used for reinforced sections in which the main reinforcement is placed within pockets, cores or cavities filled with concrete.

$$f_v = 0.35 + 17.5\,\rho$$

Additional enhancement factors for simply supported beams and cantilever retaining walls include an additional multiplier to allow for the fact that the shear strength of sections increases as the shear span/effective depth ratio decreases, hence:

$$f_v = [0.35 + 17.5\,\rho]\left[2.5 - 0.25\frac{a}{d}\right]$$

where

$$\rho = \frac{A_s}{bd}$$

$\dfrac{a}{d}$ = the shear span/effective depth ratio with a being taken as the ratio of the maximum design bending moment to the maximum design shear force $\dfrac{M}{V}$.

No such enhancement is permitted when the reinforcement is surrounded by mortar instead of concrete due to lack of evidence. The value of f_v can be enhanced in relation to any precompression which exists (see 19.1.3.3).

From the above it is clear that some evaluation has to be made to decide which value of f_v is appropriate. This value is then divided by the partial safety factor for masonry in shear, γ_{mv}, of 2.0 and compared with the value of V obtained from equation (Clause 22.5.1).

Flanged members. The calculation of shear force is simple enough for plane elements but for flanged members b is taken as the width of the flange and as a result a has to be modified. Many other Codes and design guides take (1) as the width of the rib and not the width of the flange. It was felt by the Code Committee that since the shear resistance of the masonry itself (neglecting the contribution from the reinforcement) comes principally from the compression block, then the width should be the full width of the flange. This in turn has an effect on the depth of masonry resisting shear since this depth cannot be the effective depth as is the case for rectangular beams. Tests on pocket type retaining walls have demonstrated that shear failures are extremely difficult to produce[10] and that the Code approach is reasonable, even if not reflecting the true mechanism by which the resistance is mobilised.

Consider a flanged member in which the thickness of the masonry between the ribs is less than d, say 0.5 d: If d was used in the equation (2.5.5.1) for flanged

members, the area of masonry resisting shear would be over estimated. The use of the thickness of masonry between leaves will be slightly conservative since it ignores any contribution of the remaining area of rib.

Provision of shear reinforcement. If $\dfrac{f_v}{\gamma_{mv}}$ is $\geq v$ then for many structures (for example, retaining walls) shear reinforcement is not generally needed. For beams in which $v < \dfrac{1}{2}\dfrac{f_v}{\gamma_{mv}}$, for short span lintels supporting masonry and for shallow depth beams (< 225 mm), shear reinforcement can be safely omitted. Masonry above a lintel will tend to arch over the opening whilst for a shallow beam flexure will generally be the critical design parameter. Shear failure of beams is very rare and even for long spans or deep beams, nominal shear reinforcement may not be required. When the designer has specified the use of nominal links, they should be provided in accordance with Clause 26.5.2.

If the value of v is too large, the designer is faced with a number of alternatives. The mean shear stress could be reduced by increasing the depth of the section and in some cases this is a reasonable solution. For example, in the case of a retaining wall, the thickness can be increased in steps towards the base. In this situation a further advantage is gained since the shear span/effective depth ratio will decrease. In the case of a brickwork beam containing only bed joint reinforcement, increasing the size of the section may well be the only cost-effective solution. A further option for some sections will be to increase the diameter of the main steel since this may enable a higher characteristic shear strength to be used. Where, however, $\dfrac{f_v}{\gamma_{mv}} < v$ and it is not possible to adjust the section as previously described, shear reinforcement should be provided according to the requirement:

$$\frac{A_{sv}}{s_v} \geq b\frac{\left(v - \dfrac{f_v}{\gamma_{mv}}\right)\gamma_{ms}}{f_y}$$

where:

A_{sv} = cross sectional area of reinforcing steel resisting shear forces

b = the width of the section or the rib width in the case of a flanged beam

f_y = the characteristic strength of the reinforcing steel

s_v = spacing of shear reinforcement along the member $\leq 0.75\,d$

v = shear stress due to design loads $\leq \dfrac{2.0}{\gamma_{mv}}$ N/mm^2

This formula has been developed from the truss analogy and has been shown experimentally[11] to be conservative. In the first application of the truss analogy to reinforced concrete it was assumed that the reinforcement and concrete could be considered to behave in a similar way to an N type truss. The tension forces in the truss are carried by the longitudinal and stirrup reinforcement whilst the concrete carries the thrust in the compression zone and the diagonal thrust across the web (when large shear forces are being supported it is possible that the diagonal compressive force could cause failure). Experimental observations of cracking indicated that the inclined compression struts can be taken at 45° to the longitudinal axis of the beam. Thus, to ensure that any crack is crossed by at least one stirrup, their spacing is limited to 0.75 d. Bent-up bars are not included in masonry design since no experimental evidence exists as to their effectiveness and since they are unlikely to be suitable without accompanying stirrups. It may be noted than nominal links of high yield steel or mild steel will provide a contribution to the total shear resistance of not less than 0.43 N/mm². Thus if $v > \dfrac{f_v}{\gamma_{mv}}$ by no more than

0.43 N/mm², then nominal links will suffice. On the other hand, where $v > \dfrac{f_v}{\gamma_{mv}}$ by more than 0.43 N/mm²

links will need to be provided to the formula:

$$\frac{A_{sv}}{s_v} \ge b \frac{\left(v - \dfrac{f_v}{\gamma_{mv}}\right)\gamma_{ms}}{f_y}$$

4.2.5.2 Concentrated loads near supports

When the ratio of shear span to effective depth of a beam is reduced below 2 the shear capacity is considerably increased[11]. When $\dfrac{a_v}{d}$ is much less than 1, say 0.6, then

corbel action takes place and vertical stirrups are not very effective. In this situation horizontal stirrups parallel to the main tension reinforcement become necessary. No information exists on the performance of reinforced masonry corbels.

The clause requires that when a principal load (which is defined as one contributing more than 70% of the total shear force at a support) acts at a distance a_v from the support (where a_v is less than twice the effective depth) then shear reinforcement should be placed over the distance a_v. Generally, however, with simply supported beams and cantilever retaining walls, the designer may wish to take advantage of the enhancement in the shear resistance of the section because of its low shear span/effective depth ratio as described in the previous paragraph, up to a maximum of $2f_v$. The two enhancements must not be applied simultaneously.

4.2.6 Deflection

The calculation of deflections in reinforced masonry is not usual and deflections will generally be acceptable if the sizing rules are obeyed. The values in Table 8 of the Code have been in use for some time and appear to be reasonable, there is little experience of the use of Table 9 but, it is likely to be conservative.

4.2.7 Cracking

Less precise information is available about cracking than deflection and consequently calculations regarding crack

widths are not to be recommended. The designer is advised to use the sizing rules in Tables 8 and 9 of the Code.

4.3 Reinforced masonry subjected to a combination of vertical loading and bending

Research into this aspect of reinforced masonry is somewhat limited. The design methods given in the Code are, therefore, something of a compromise. An eccentricity of 0.05 times the depth of the section in the plane of bending is a common reference point.

4.3.1 General

The members covered by Clause 23 are those which, in accordance with the above, have resultant eccentricities due to simultaneously applied substantial vertical and horizontal or eccentric (eccentricity greater than 0.05 times the depth of the section in the plane of bending) vertical loading.

4.3.2 Slenderness ratios of walls and columns

4.3.2.1 Limiting slenderness ratios

Slenderness ratios for reinforced masonry walls and columns have been limited to the same values as those given for unreinforced masonry in BS 5628: Part 1. These limiting ratios have been adopted because, in the absence of adequate experimental data, they are known to produce satisfactory results. An arbitrary limit, taken from SP 91[13], for cantilever columns of up to 0.5% reinforcement (based upon breadth of section × effective depth) is proposed beyond which special consideration should be given to deflection.

The slenderness ratio is defined, as in BS 5628: Part 1, as the ratio of the effective height to the effective thickness and for rectangular solid sections. A more fundamental approach could be used for other than rectangular sections using a slenderness ratio of effective length to radius to gyration, providing the limits given in Clause 23.2.1 are modified accordingly. The sub-clauses dealing with lateral support, effective height and thickness, again give similar guidance to that given in BS 5628: Part 1.

4.3.2.2 Lateral support

The lateral support requirements are intended to ensure that consideration is given to the overall stability of the structure and to the satisfactory interaction of its elements. Simple and enhanced resistance to lateral movement are described.

4.3.2.3 Effective height

The slenderness of masonry walls and columns is important as it determines their susceptibility to buckling failure based upon the Euler buckling theory. The assessment of effective height by structural analysis referred to in the Code means the analysis of the deflected shape of the member under load and the comparison of it with the idealised deflected profile of a pin ended strut. As an alternative to the more rigorous approach, Table 10 in the Code gives values which may be adopted by the designer. These are the same as given in BS 5628: Part I and reflect semi-empirical but conservative assumptions based, in the case of reinforced masonry, largely on theoretical studies. As with unreinforced masonry, it is assumed that a reinforced masonry column will exhibit a somewhat lower strength for a given height than will a reinforced masonry wall.

4.3.2.4 Effective thickness

The effective thickness of a reinforced masonry wall depends upon its form. For single leaf walls and columns, the actual thickness is used. Where one leaf of a cavity wall is reinforced, the effective thickness may be taken as ⅔ of the sum of the actual thickness of the two leaves, or as the actual thickness of the thicker leaf, whichever is the greater. In the case of the cavity wall, for reasons of practicality, the reinforced leaf will usually be the thicker and its actual thickness will probably be used as the effective thickness, thus avoiding the need to share the load between the leaves and check that the shear between them can be accommodated. For grouted cavity walls the effective thickness is taken as the actual overall thickness with the limitation that the width of the cavity shall be taken as not thicker than 100 mm. This is an arbitrary limitation to prevent the excessive thickening of the concrete infill merely to reduce the slenderness ratio of the wall. The limitation also ensures that the masonry can interact with the concrete infill. If a very wide cavity was desirable it would generally be more economic to design on the concrete section only, regarding the masonry as permanent formwork.

4.3.3 Design

4.3.3.1 Columns subjected to a combination of vertical loading and bending.

This clause deals with the design of short columns (slenderness ratio less than 12) subjected to single axis bending.

4.3.3.1.1 Short columns

It is usually considered sufficient to design short columns for the maximum moment about the critical axis only, even where it is possible for significant moments to occur simultaneously about the axes.

Two methods are given for the design of short columns. The first is based upon first principles in which the cross section of the column is analyzed using strain compatibility to determine the design moment of resistance and the design axial load capacity, and the second is to use the design formulae given. The former method entails the use of assumptions (a) to (e) given in Clause 22.4.1 for the stress and strain distributions in the section being analyzed.

The three sets of formulae given in 23.3.3.1 are also based on the assumptions (a) to (e) referred to previously, including the assumption of a simplified rectangular triangular stress block with an intensity of $\frac{f_k}{\gamma_{mm}}$.

The formula in case (a) is used when the design axial load, N, is less than the capacity, N_d, in the stress diagram (Figure 9.6). The column is then reinforced with a nominal area of reinforcement (see *Section 26.1*). No allowance is made for this reinforcement as the column has been designed as effectively unreinforced (c.f. Appendix B, BS 5628: Part 1). The obvious point that e_x cannot exceed $0.5t$ is made in the Code.

The formulae in case (b) are used when the design axial load, N, exceeds N_d in (a) above. These formulae provide a simple method of design avoiding the complications involved in using the rigorous application of the beam bending assumptions given in 22.4.1. To assist in the use of method (b), guide values for f_{s2} are given in the formulae which vary with the chosen depth of the masonry compression block. These formulae can be used for non-symmetrical arrangements of reinforcement.

In Figure 9.7 the depth of compression block, d_c, is plotted against the stress in the reinforcement in the least compressed face of the column. d_c should not be chosen as less than $2 d'$ to avoid the possibility of the occurrence of a narrow band of masonry forming the compression zone, leading to a local crushing failure.

Case (c) is offered as an alternative to case (b) and permits the design axial load to be ignored when the resultant eccentricity exceeds $\frac{t}{2} - d'$ provided the section

is designed to resist an extra bending moment equal to the design axial load acting at an eccentricity of $\left(\frac{t}{2} - d'\right)$.

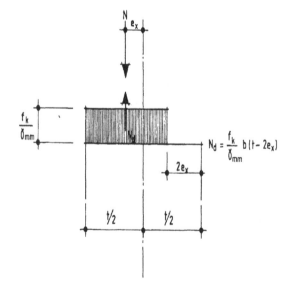

Figure 9.6 Design of short columns ignoring reinforcement

148

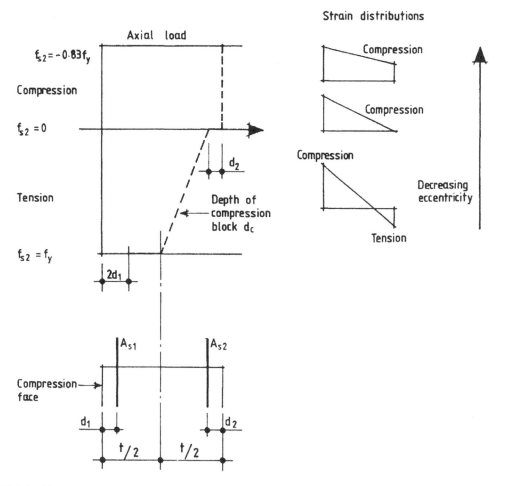

Figure 9.7 Relationship between depth of compression zone and stress in steel in at least compressed face of column

The method permits the area of tension reinforcement resisting this increased moment to be reduced by N. This method may be useful when the direction of bending is irreversible and it is considered necessary to use a symmetrical arrangement of reinforcement.

No design charts have been included in the Code as an alternative to the design formulae and design from first principles. However, charts are a very useful design tool and are, therefore, included in this handbook. The design charts, in the form of interaction curves, given at the end of this Chapter, are generally based upon the same assumption as the beam and column design formulae referred to above. These design charts are based upon the assumption of a rectangular stress block rather than a rectangular parabolic stress block. This has a considerable practical advantage in that each chart can deal with any value of unit characteristic strength, f_k, so that fewer charts are necessary and interpolation between charts is not required. The charts do, however, assume that the reinforced masonry columns being designed have equal amounts of reinforcement positioned at equal depths from the column faces.

In general the assumption of a rectangular stress block will result in a somewhat greater area of reinforcement than would the more complex stress blocks. This is of greatest significance when columns or walls carry predominantly vertical loads, i.e. when the depth of compression zone to depth of section ratio, $\frac{d_c}{t}$ is high.

This is apparent because even when $d_c = t$, the rectangular parabolic and the parabolic stress blocks have some moment of resistance by virtue of their non-symmetrical shape.

A series of tests on eccentrically loaded reinforced brickwork columns by Anderson and Hoffman[14] has shown good agreement with a calculated interaction curve, and as the use of such curves is well established for reinforced concrete design, they should prove to be equally useful for reinforced masonry design.

4.3.3.1.2 Short columns biaxial bending
This sub-clause deals with short columns which are subjected to biaxial bending and is only applicable to symmetrically reinforced rectangular sections, which may

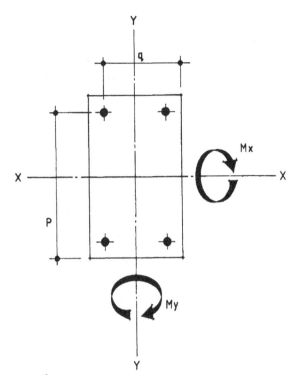

Figure 9.8

be designed to withstand an increased moment about one axis given by the following relationships:

(a) for $\dfrac{M_x}{p} \geq \dfrac{M_y}{q}$ $M_x' = M_x + \alpha \left(\dfrac{p}{q}\right) M_y$

(b) for $\dfrac{M_x}{p} < \dfrac{M_y}{q}$ $M_y' = M_y + \alpha \left(\dfrac{p}{q}\right) M_x$

where:

M_x = the design moment about the x axis
M_y = the design moment about the y axis
M_x' = the effective uniaxial design moment about the x axis
M_y' = the effective uniaxial design moment about the y axis
P = the overall section dimension in a direction perpendicular to the x axis (see Figure 9.8)
q = the overall section dimension in a direction perpendicular to the y axis (see Figure 9.8)
α = a coefficient derived from the following table, Table 4.3.

Table 4.3 Values of the coefficient, α

$\dfrac{N}{N_{dz}}$	α
0	1.00
0.1	0.88
0.2	0.77
0.3	0.65
0.4	0.53
0.5	0.42
>0.6	0.30

where:

N = the design axial load
N_{dz} = the design axial load resistance of the column, ignoring all bending which, for a section of area A_m with symmetrically disposed reinforcement, may be calculated from the expression:
$N_{dz} = f_k A_m$
f_k = characteristic compressive strength of masonry

The initial published version of the Code contains a number of errors in this section. Firstly, the term in the equation for M' has been inverted. The correct equations are given above. Secondly, the expression for N_{dz} contains an expression for the steel, which it should not, and the partial safety factor for the compressive strength of masonry, which should also be omitted: i.e.

$$N_{dz} = f_k A_m$$

This empirical method is based upon the CEB Bulletin D'Information No 141[15] which presents an approximate formulae for symmetrically reinforced sections. Extensive comparisons were made with more rigorous computer based analysis in the drafting of BS 8110[16] which justified the more favourable values of a given in Table 4.3. Table 12 of the Code contains unmodified values comparable to those contained in CEB Bulletin No 141.

4.3.3.1.3 Slender columns
Slender columns are those defined as having a slenderness ratio, $\dfrac{h_{ef}}{t}$, greater than 12. This clause emphasizes that

account of biaxial bending should be taken where appropriate when designing slender columns, and also of the additional moment, M_a, induced by the vertical load and lateral deflection of the column. Columns without simple or enhanced resistance to lateral movement must be considered as cantilevers.

The design of slender columns can be carried out either by analysis of the cross section from first principles to ensure that the design bending moment, including the additional moment, and the design axial load are exceeded by the design moment of resistance and the design axial load capacity respectively, or by use of the Code equations. Alternatively, the design charts may be used, the design moment being modified to include the additional moment. It may be noted that e_{add} for columns at the limit of slenderness ratio for short columns (i.e., 12) correlates with the eccentricity $0.05t$ which may be ignored when designing short columns.

Figure 9.9 indicates the basis of the additional moment concept for slender columns.

4.3.3.2 Walls subjected to a combination of vertical loading and bending
These walls both short and slender as defined for columns, may be designed in the same way as columns subjected to combined loading. Although the sub-clause dealing with short walls refers only to analysis of the section using the assumptions given in 22.4.1, there is no

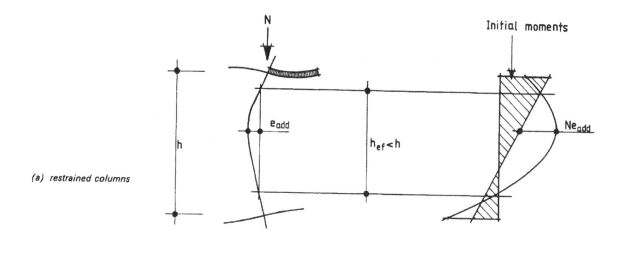

(a) restrained columns

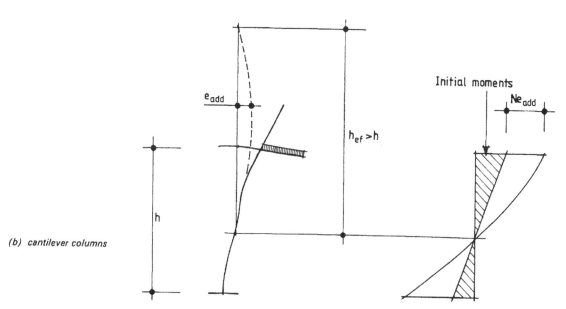

(b) cantilever columns

Figure 9.9 Relationship between effective height of columns and additional moment diagram

reason why the design formulae given in 23.3.1.1 should not be used, or indeed the design charts, taking the width of the section, b, as the unit length of the wall. Often, however, walls in reinforced masonry will be singly reinforced with the reinforcement placed approximately centrally in the section, and it is then necessary to amend the formulae.

4.3.3.2.1 The short wall sub-clause (23.3.2.1) refers specifically to the situation where the resultant eccentricity, e_x, is greater than $0.5\,t$. In this case the axial load may be neglected and the wall designed as a member in bending in accordance with Clause 22.

4.3.3.2.2 Slender walls are treated in the same way as short walls with the exception that the additional moment derived in the same way as for columns is included.

4.3.4 Deflection

This clause refers the designer to the limiting dimensions given in Clause 22.3, but does not make it clear whether Table 8 or Table 9 should be used. Some degree of judgement is required in this matter. Since Table 8 refers to walls, its use seems appropriate, but consideration should be given to the situation where a column of primary structural importance is subjected to a predominantly substantial bending moment, i.e., if the member is designed as a beam in accordance with 23.3.1.1, Table 9 should also be used.

4.3.5 Cracking

If the design vertical load of a wall or column exceeds $\frac{A_m f_k}{2}$ then the eccentricity of the load at a critical cross section is not likely to be great enough to cause cracking due to flexural tension. In more lightly loaded columns reinforcement may be provided to control cracking and this should be provided in the same way as for beams. The recommendations are given in Clause 26.

4.4 Reinforced masonry subjected to axial compressive loading

This clause deals with walls and columns which carry a design vertical load, the resultant eccentricity of which does not exceed 5% of the thickness of the member in the direction of the eccentricity.

In BS 5628 Part 2, the designer is referred either to the equations appropriate for columns subjected to combined loading, or to the design method given in BS 5628: Part I, making no allowance for the reinforcement. Recourse to Part I is also recommended for the design of walls

subjected to concentrated loads, the implication being that the provision of special reinforcement is impractical.

4.5 Reinforced masonry subjected to horizontal forces in the plane of the element

Where walls are used to provide overall stability to a structure, significant horizontal loads can be applied in the plane of the walls. The capability of the element to resist these forces should be checked in respect of both the resistance to racking shear and the resistance to bending.

4.5.1 Racking shear

Walls which are subjected to in-plane horizontal forces and loaded to failure, crack typically in the manner illustrated in Figure 9.10. The cracks are caused by diagonal tension and, although there has been some research into the strength of brickwork when subjected to biaxial loading[19], it is usual to treat the design of walls on the basis of the average stress over the plan area. Thus, if the total design horizontal force is V, the shear stress due to design loads is considered to be V, where:

$$v = \frac{V}{tL}$$

and where t and L are the thickness and length of the wall respectively.

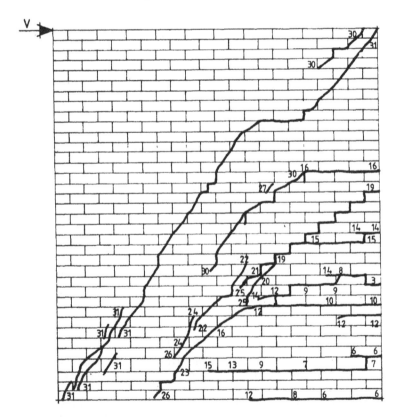

Figure 9.10 Diagonal cracking due to racking load

The Code states that adequate provision against the ultimate limit state being reached must be assumed if the average shear stress is less than the design shear strength, i.e.

$$v \le \frac{f_v}{\gamma_{mv}}$$

f_v is the characteristic racking shear strength taken from 19.1.3.2, i.e. $0.35 + 0.6 \, g_b$ N/mm², where g_b is the design vertical load per unit area of wall cross section due to the vertical dead and imposed loads calculated from the appropriate loading condition. (The maximum value to be taken for f_v is 1.75 N/mm².) Alternatively research[20] has shown that for walls which are reinforced with the main reinforcement in pockets, cores or cavities, a lower bound for the shear resistance is 0.7 N/mm² and this may be used as a characteristic value instead of $0.35 + 0.6 \, g_b$. The value of 0.7 N/mm² was derived from tests on walls with a limited range of shapes and so the use of the value is limited to walls where the height/length ratio is not greater than 1.5.

Where v is greater than $\frac{f_v}{\gamma_{mv}}$, horizontal shear reinforcement should be provided (but in no case should V exceed $\frac{2.0}{\gamma_{mv}}$ N/mm²). This reinforcement should be provided according to Code equation:

$$\frac{A_{sv}}{s_v} \ge \frac{t \left[v - \dfrac{f_v}{\gamma_{mv}} \right]}{\dfrac{f_y}{\gamma_{ms}}}$$

Part of the applied shear force, $V = vtL$, is considered to be resisted by a component of force in the masonry,

$$\frac{f_v}{\gamma_{mv}} tl$$

and the remainder by the total area of horizontal steel acting in tension across any incipient crack. If the crack is assumed to be at 45°, the number of points at which horizontal steel crosses the crack is then $\dfrac{L}{s_v}$

The formula can then be written:

$$V = vtl \le \frac{f_v}{\gamma_{mv}} tL + \frac{A_{sv} f_y L}{\gamma_{ms} s_v}$$

which, rearranged, gives:

$$\frac{\left(v - \dfrac{f_v}{\gamma_{mv}} \right) t}{\dfrac{f_y}{\gamma_{ms}}} \le \frac{A_{sv}}{s_v}$$

Any vertical reinforcement will also help resist shear in racking by dowel action. This is not as effective as the horizontal reinforcement in tension, and so has been ignored. In any event, many shear walls will not require any horizontal steel specifically for shear resistance, particularly where some light horizontal distribution

steel is already provided. In any case of reinforced or unreinforced masonry where the designer is considering the use of shear walls, particular consideration must be given if any type of damp-proof course has been introduced which is likely to produce a plane at the base of the wall along which sliding could occur[21].

4.5.2 Bending

When bending is in the plane of the wall, the analysis and design of the wall should follow the recommendations for flexural members given in Clause 24. The designer should satisfy himself that in designing the wall as an in-plane cantilever, the fixity at the base is adequate. Assumption (f) in Clause 22.4.1 may be ignored.

It is unlikely that bending due to the horizontal forces will be critical, shear is more likely to be so. However, where the slenderness ratio of the wall in either direction exceeds 12, then additional moments may be set up in the wall. It is then necessary to take account of the slenderness at right angles to the plane of the wall by calculating the maximum compressive stress in the wall and checking with Clause 23.3.1.3. This essentially checks bending at right angles to the length of the wall where shear walls support no vertical loads. This approach is likely to be very conservative.

4.6 Detailing reinforced masonry

The previous clauses have covered the basis of design and the analytical procedures to be followed to arrive at the area of reinforcement required to give an adequate margin of safety against failure. As with reinforced concrete, it is the detailing of the reinforcement which is paramount if the calculated design performance is to be achieved in practice. This section explains the requirements and gives guidance on how reinforcement may be incorporated in masonry so that the main steel is effective, any secondary steel economically provided and any cracking controlled.

4.6.1 Area of main reinforcement

The area of main reinforcement that is provided is usually expressed as a proportion of the area defined as the effective depth x the breadth of the section. There are no minimum recommendations in the Code, although many of the early drafts included the following limitation:

$A_s \ge 0.002bd$ for mild steel
$A_s \ge 0.015bd$ for high yield steel

It would be unusual for reinforced sections to include areas of main reinforcement which are much below these values. However, there are a number of situations where the size of the element may be fixed for other than structural reasons and the area of steel supplied does not need to meet such requirements. For example, low grouted cavity retaining walls have an effective depth dictated by the thickness of the units used and the cavity width but may be adequately reinforced using mesh which does not provide an area in excess of the appropriate value above. Another example is where a wall beam is designed according to Clause 22.4.2 where the application of a restriction on the percentage of steel could lead, in the case of hollow blockwork, to extraordinary amounts of steel being required. In this case the reason is the large overall depth of the element.

The omission does lead to certain difficulties and the Code draws the designer's attention to the fact that in some cases a design in accordance with BS 5628 : Part 1, i.e. ignoring the reinforcement, may be appropriate.

It should be noted that when considering the percentage of reinforcement in an element, this may well relate to a locally reinforced section, for example, if some cores of an otherwise unreinforced hollow blockwork wall are reinforced, then the locally reinforced section should be considered for calculating the proportion of reinforcement when designing for flexure or shear.

4.6.2 Maximum size of reinforcement

The limiting sizes given are based on practical considerations. Most mortar joints are designed as 10 mm thick and, therefore, to maintain some cover above and below joint reinforcement, the 6 mm maximum is specified. In most cores and cavities a 25 mm bar is the largest which can be incorporated, particularly if the bars are to be lapped. In pocket type walls, where the pockets can be made large enough, a 32 mm bar can be used. These limitations are based on experience in the UK. In the USA and Canada larger bars are commonly used, but are incorporated in very wide cavities or cores (such as 300 mm wide concrete blocks) and reinforcement is often spliced rather than lapped. Such a wide range of units is not available in the UK.

4.6.3 Minimum area of secondary reinforcement in walls and slabs

Secondary reinforcement is required in walls and slabs to ensure monolithic action. The minimum required is 0.05% of bd and can be provided in any of the following ways:

1. proprietary bed joint reinforcement
2. light reinforcement (6 mm) in bed joints
3. reinforcement in bond beams in reinforced hollow blockwork
4. within the cavity of grouted cavity construction

(Note: in pocket type walls secondary reinforcement is usually omitted.)

Such reinforcement can also perform a secondary function of controlling movements in the masonry (see Clause 34). Particular attention should be paid to the durability requirements of a section especially with respect to steel embedded in mortar.

4.6.4 Spacing of main and secondary reinforcement

The minimum bar spacings are aimed primarily at allowing adequate room for the concrete to flow around

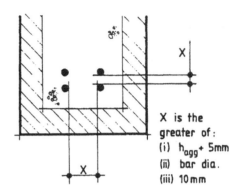

X is the greater of :
(i) h_{agg} + 5mm
(ii) bar dia.
(iii) 10 mm

Figure 9.11

the bars and at obtaining adequate compaction. The particular requirements are illustrated in Figure 9.11. Bars can be grouped in pairs either horizontally or vertically.

Bundling of bars is unlikely to be necessary since the percentage of steel required is comparatively low and this is not generally recommended for reinforced masonry because of the limited size of sections available. Where an internal vibrator is to be used, room should be left between any top bars in beams for its insertion. It is also for this reason that only one bar should be incorporated in pockets or cores whose size is less than 125 × 125 mm. This does not apply at laps of course, but consideration should be given to the use of splices and connectors.

Generally, spacings wider than the minimum should be aimed at, particularly between top bars, to allow the concrete to pass through easily.

The maximum bar spacing of 500 mm is specified for two reasons:

1. to control crack widths
2. to enable walls and slabs to act monolithically

In reinforced hollow blockwork this spacing would typically mean one bar every alternate core. This maximum spacing may be exceeded when the element is designed as a flanged member, but care must be taken to ensure that the masonry between concentrations of reinforcement, where no flange action can occur or where the allowable flange width is exceeded, can span unreinforced between these concentrations. In pocket type retaining walls the spacing between concentrations of reinforcement is likely to be within the range 1.2–1.5 m. The maximum spacing of shear links is 0.75 d (see also Clause 22.5).

4.6.5 Anchorage, minimum area, size and spacing of links

4.6.5.1 Anchorage of links
All links must be anchored to perform their function. Bearing stresses within the bends are not likely to be excessive, but due to the confined nature of certain

reinforced masonry elements, it is suggested that mild steel links be given preference over high tensile steel links.

4.6.5.2 Beam link

The minimum area and spacing of links should be provided such that:

$$\frac{A_{sv}}{s_v} \geq 0.002t \text{ for mild steel}$$

$$\frac{A_{sv}}{s_v} \geq 0.0012t \text{ for high yield steel}$$

The equations differ such that the tensile force provided by the links is approximately equivalent for both mild steel and high yield steel.

Providing the nominal number of links will give a ultimate shear strength over and above that carried by the masonry (i.e. $v - \frac{f_v}{\gamma_{mv}}$) of 0.435 N/mm² and 0.480 N/mm² for mild steel and high yield steel respectively. The maximum permitted value of v is $2.0/\gamma_{mv}$ i.e. 1.0 N/mm² and thus, nominal links will often be adequate even where the masonry shear strength, f_v / γ_{mv}, is substantially exceeded.

4.6.5.3 Column links

Links are required in reinforced concrete columns to prevent buckling failure of the reinforcement and bursting of the concrete cover. In most reinforced masonry columns the loads will be sufficiently low and the confinement provided by the masonry sufficiently high to prevent such a failure. Where the area of steel is greater than 0.25% of the area of masonry, A_m, and more than 25% of the axial load capacity of the column is to be mobilised, then this type of failure is considered to be possible. Links are thus required in this situation and the size and spacing of links are specified as follows:

Minimum size of 6 mm at a maximum c/c distance of:

1. the least lateral dimension of the column
2. 50× link diameter
3. 20× main bar diameter

whichever is the smallest (3 will usually be the limiting dimension). Rules are given to ensure that every bar is adequately restrained.

4.6.6 Anchorage bond

The aim of this clause is to ensure that the forces assumed to be present at the reinforcement level can be safely transmitted to the bars without bond failure occurring. This is achieved by providing a length of bar far enough beyond the point being considered for the calculated stresses in that bar to be developed. The phrase 'design loads' refers to the ultimate limit state and commonly the force required to be 'locked-off' will be $\frac{f_v}{1.15}$ for example, over the supports in continuous beams.

4.6.7 Laps and Joints

4.6.8 Hooks and bends

4.6.9 Curtailment and anchorage

It is necessary for a number of reasons to continue bars beyond the point where they are no longer required to resist bending:

1. to allow for variations in load distribution in which case the shape of the bending moment diagram will be different to that calculated
2. to allow for tolerances in the placement of reinforcement
3. if the presence of stirrups would cause stress in the reinforcement at that point to increase to that corresponding to the moment at a section roughly an effective depth (for 45° cracks) away from that point
4. cracks of above average size may well occur at the points where the bars stop, which may locally reduce the shear strength

The minimum extension beyond the theoretical cut-off point is the greater of the effective depth or 12× the bars size to cater for points 1 to 3, whilst the extra provisions (a) to (c) deal with 4. Provisions (a) and (c) control the size of the crack at the cut-off point and (b) ensures that there is a reserve of shear strength. Provision (c) will be the easiest to apply and is recommended for general use. Provision (b) will often apply where low shear stresses are present and any nominal links provided automatically supply excess shear strength. Extra links can be added to comply with (b) but this is not recommended since extra shear calculation will be necessary and the amount and complexity of the reinforcement involved will generally be more than if the main reinforcement is extended to comply with (a) or (a). It should be noted that Clause 26.6 requires that no bars be cut-off less than an appropriate anchorage length from the last point at which it is assumed to be fully stressed. This will, on occasion, override the requirements discussed above.

No guidance is given in respect of curtailment of compression reinforcement or tension reinforcement which extend into compression zones beyond points of contraflexure. Provisions (a) to (c) can safely be ignored here but bars should extend the greater of 12 bar diameters or an effective depth beyond the point at which they are theoretically no longer required. For anchoring bars at simple supports using hooks or bends, reference should be made to Clause 26.8.

References

1. BRITISH STANDARDS INSTITUTION. CP 110: Part 1:1972 *The structural use of concrete.* BSI, London. pp.156.
2. POWELL, B and HODGKINSON, H R. *The determination of the stress/strain relationship of brickwork.* British Ceramic Research Association, Technical Note TN 249.1976.
3. NEWSON, M J. *A preliminary investigation of the stress/strain relationship for concrete blockwork.* Cement and Concrete Association, Slough, 1983. pp.30.

4. BEARD, R. A theoretical analysis of reinforced brickwork in bending. Proceedings British Ceramic Society, No.30, 192.

5. ROBERTS, J J and EDGELL, G S. *The approach to bending.* Paper presented to Symposium on Reinforced and Prestressed Masonry, Cafe Royale, London, 8 July 1981. Institution of Structural Engineers. pp.12–16.

6. BRITISH STANDARDS INSTITUTION. BS 5628: Part 1:1992 *Unreinforced masonry.* BSI, London. pp.40.

7. RATHBONE, A J. *The behaviour of reinforced concrete block work beams.* Cement and Concrete Association, Slough, 1980. Publication No.42.540.

8. RATHBONE, A J. *The shear behaviour of reinforced concrete block work beams.* Paper presented to Symposium On Reinforced and Prestressed Masonry, Cafe Royale, London, 8 July 1981. Institution of Structural Engineers. pp.17–28.

9. SUTER, G T and KIELER, H. *Shear strength of grouted reinforced masonry beams.* Proceedings 4th International Brick Masonry Conference, 1972.

10. TELLET, J and EDGELL, G J. *The structural behaviour of reinforced brickwork pocket type retaining walls.* British Ceramic Research Association. TN 353, 1983.

11. SINHA, B P and DE VEKEY, R C. *Factors affecting the shear strength of reinforced grouted brickwork beams and slabs.* Proceedings, 6th IBMAC, Rome 1982.

12. BRITISH STANDARDS INSTITUTION. CP 111:1970 *Structural recommendations for loadbearing walls.* BSI, London. pp.40.

13. BRITISH CERAMIC RESEARCH ASSOCIATION/Structural Ceramics Advisory Group. *Design guide for reinforced and prestressed clay brickwork.* Special Publication SP 91, 1977.

14. ANDERSON, D E and HOFFMAN, B S. *Design of brick masonry columns.* Designing, Engineering and Constructing with Masonry Products. Gulf, Houston, Texas, 1969.

15. CEB/FIP. MANUAL ON BENDING AND COMPRESSION. CEB/FIP, Construction Press, Harlow, 1982. pp.111.

16. BRITISH STANDARDS INSTITUTION. BS 8110:1997 *Code of Practice for concrete.* BSI, London.

17. ROWE, R E *et al.* Handbook on the Unified Code for structural concrete (CP 110:1972). Viewpoint Publications, London, pp.153.

18. DAVIES, S R and ELTRAIFY, L A. Uniaxial and biaxial bending of reinforced brickwork columns, Proceeding 6th IBMAC, 1982.

19. HENDRY, A W. Structural brickwork (Chapter 2). Macmillan Press, London, 1981.

20. SCRIVENER, S C. Shear tests on reinforced brick masonry walls. British Ceramic Research Association, Technical Note TN 342, 1982.

21. HODGKINSON, H R and WEST, H W H. The shear strength of some damp-proof course materials. British Ceramic Research Association, Technical Note TN 326, 1981.

Section 5: Design of prestressed masonry

No further comment is provided on this section

Section 6: Other design considerations

6.1 Durability

6.1.1 Masonry units and mortars

The Code refers the designer to BS 5628: Part 3[1] for guidance on durability of masonry units and mortar. To assist the designer, however, some general notes are given here, although where any doubt exists or is indicated, then the fuller guidance in Clause 22 of BS 5628: Part 3 should be consulted. The guidance given relates to the masonry itself and not to the durability of reinforcement, which is covered later.

There are a number of factors which will affect the durability of masonry, including:

1. the degree of saturation
2. the potential for frost attack
3. the susceptibility to sulfate attack
4. the characteristics of the unit

A major factor influencing the durability of masonry is the degree to which it becomes saturated with water. It may become saturated directly by rainfall, indirectly by water moving upwards from the foundations, or laterally from retained material as in a retaining wall. Particular attention should be paid to the choice of masonry units and mortar in the following, and similar, situations, where the masonry is likely to become, and may remain, saturated for long periods:

- in sills, copings and cappings
- in parapets, freestanding and retaining walls
- below dpc, at or near ground level and in foundations.

The durability of masonry depends upon:

1. *Exposure to the weather*: The Local Spell index[2] or the Driving Rain index[3] is a good indication of the general exposure of the site and, therefore, the amount of rain which is available to saturate the masonry. It must be noted, however, that different parts of the same structure may be subjected to different degrees of exposure.

2. *The adequacy of design details and methods taken to prevent the masonry becoming saturated*: External masonry is much less likely to become saturated where projecting features have been provided to shed run-off water clear of the walling, for example, protection to wall heads by roof overhangs or projecting throated copings projecting throated sills bell mouths to rendering, tile hanging and so on.

   On the other hand, where these features are not incorporated, for example, where flush copings are used, increased wetting and potentially longer periods of saturation will occur. External masonry will be maintained in a drier condition by a moderately porous uncracked rendering but dense rendering may lead to entrapment of moisture if imperfections develop or if water is able to get behind the finish via any path. Depending on the masonry substrate, this could lead to frost or sulfate attack.

3. *Exposure to aggressive conditions*:
(a) Frost attack if freezing occurs either during construction or shortly after completion of the work, may cause damage to mortar and even to the masonry units themselves, depending on their type and whether they become saturated. Damage is caused by the volumetric expansion which occurs on the formation of ice within the saturated masonry in

freezing conditions. It is important, therefore, to protect stored masonry units and newly erected masonry adequately, both from saturation and from frost. Frost attack can also occur later in the life of the structure, although the actual process by which this occurs is very complex.

Additional consideration should be given to the choice of any masonry unit and mortar if the walling is liable to be splashed with de-icing salts or if the structure is to be located in conditions of extreme exposure to weather, for example, on the coastline.

(b) *Sulfate attack*: sulfate attack on set mortars is generally due to the expansive reaction of tricalcium aluminate in cement with calcium, sodium or potassium sulfates to form Ettringite. Sulfates may also be derived from the ground, ground-water, hardcore or fill.

The designer should also consider whether advice is required from the manufacturer of the masonry units or from other authoritative bodies, for example, BRE, AACPA, CBA.

4. The characteristic of the masonry units and mortars: Tables 6.1 gives general guidance only and does not necessarily give the minimum requirement (see BS 5628 Part 3 for more detailed information). For example, units other than those indicated may be used, provided that the manufacturer is able to produce authoritative evidence that they are suitable for the intended purpose. It is often desirable to consult the manufacturer in any event. The Tables do not apply where sulfates are present in the ground or groundwater in significant quantities.

6.1.2 Resistance to corrosion of metal components

6.1.2.1 General

As the UK experience of durability of reinforced masonry in terms of resistance to corrosion is limited and the results of the various field and experimental studies are quite variable, the approach which has been adopted in the Code is cautious. Much notice has been taken of the recent experiences with reinforced concrete construction in the UK.

The process by which steel placed in mortar or concrete corrodes is well understood. Steel is thermodynamically unstable and can corrode in both acid and neutral environments when oxygen and moisture are present. The moisture in the pores of the mortar or concrete act as an electrolyte and consequently metal ions are released by steel placed in contact with it and the steel takes up a negative potential relative to the electrolyte. In concrete the electrolyte varies from place to place, as do the properties of the steel, and so it is possible for the steel to adopt different potential differences with respect to the electrolyte along its length. Consequently in the right conditions an electrical current flows and corrosion cells are set up, leading to metal ions being continually released at the anodes and these then combine with hydroxyl ions to form rust.

However, in the alkaline conditions produced during the hydration of the cement in mortar or concrete, the steel is passivated. An oxide film forms on the surface and even in the presence of moisture and oxygen, will not corrode. Problems arise when the passive oxide layer is disrupted and corrosion may then take place. Disruption of the oxide layer occurs due to two main reasons, namely carbonation and chlorides.

Carbonation is the process by which the acid gases in the environment, in particular carbon and sulfur dioxides, neutralise the hydroxides which provide the alkalinity of the concrete or mortar. The rate at which this occurs depends on a number of factors, one of which is the gas permeability of the material. The subsequent loss of alkalinity depends on the initial cement content of the mix. Naturally, factors such as temperature and relative humidity also have an effect. In reinforced concrete it is relatively simple to say that if strong, high cement content concrete with low permeability and hence low water/cement ratio is used, problems are not likely to occur provided enough cover is allowed. It is less simple to do this in reinforced masonry as, for example, in grouted cavity or hollow block construction, the initial water/cement ratio should be high to ensure that the mix is workable enough to fill all the voids adequately. The units do, of course, draw water from the mix, depending upon their own porosity, and the effective water/ cement ratio from the point of view of defining the permeability will be lower than that in the original mix. Once carbonation has taken place, corrosion does not necessarily commence unless both moisture and oxygen

Table 6.1 Recommended qualities of precast concrete masonry units to ensure durability

Location	Strength of unit (N/mm²)		Mortar designation
	Blocks	Bricks	
All internal; external above dpc	Any	15	(iii)
External below dpc; freestanding walls: parapets	3.5 dense† 7.0 lightweight	20	(ii)
Earth retaining walls; sills and copings*	7.0 dense†	30	(ii)

* Where the retaining face is waterproofed and an adequate coping provided, the quality may be that as for freestanding walls. Indeed, most masonry is not waterproof in this situation and the provision of proper waterproofing is always desirable.
† Units other than those indicated may be used provided that the manufacturer is able to produce authoritative evidence that they are suitable for the intended purpose.

are available. Also, if the masonry was saturated its permeability to oxygen should be low and thus corrosion would be limited. Similarly, if the masonry was very dry, the resistance of the concrete or mortar forming the electrolytic cell would be high and the corrosion current restricted.

The other major cause of disruption of the passive oxide coating to steel is the presence of chlorides. It is probably true to say that a major cause of corrosion in reinforced concrete structures has been the excessive use of calcium chloride as an accelerator in the mix, but their use has been excluded in this Code. Chlorides are also available from the environment, such as near the coast or in boundary walls likely to be in contact with de-icing salts.

Although there is some understanding of the mechanisms causing corrosion, there remain a number of situations where there is a fear that unprotected low carbon steel might corrode, even if a high standard of workmanship has been maintained. These situations arise where reinforcement is placed in bed joints or in special units, such as pistol bricks and in grouted cavity or Quetta bond construction. In these situations, the Code recommends the minimum level of protection for reinforcement. The recommendation is linked to the severity of exposure and this is defined in terms of location of the site in relation to the Local Spell index as defined in DD93[2] or the Driving Rain index[3]. An additional exposure situation is also defined which is based on the availability of certain chemicals. Certain other situations, such as reinforced hollow blockwork, can be more easily treated by providing an adequate thickness of concrete cover. In any situation, the designer may choose to do this in any case, for example a grouted cavity wall could be dimensioned such that the cavity would allow the appropriate cover to the steel to be maintained.

6.1.2.2 Classification of exposure situations

Three definitions of site exposure condition (E1, E2, E3) have been defined which relate to wind driven rain, *viz.*

- E1 very sheltered or sheltered
- E2 sheltered/moderate or moderate/severe
- E3 severe or very severe

There are, in addition, certain local conditions to which the masonry may be exposed which can be classified in a similar way to site exposure but are not dependent upon it. Examples are:

- E1 reinforcement in the inner skin of ungrouted external cavity walls and behind surfaces protected by an impervious coating which can readily be inspected
- E2 reinforcement in buried masonry and masonry continually submerged in fresh water
- E3 reinforcement in masonry exposed to freezing while wet or subjected to heavy condensation.

A further set of conditions are so severe that whatever the site classification, the only suitable reinforcement is that which is solid or coated with at least 1 mm of austenitic stainless steel. These conditions are where the masonry is exposed to salt or moor land water, corrosive fumes, abrasion or dc-icing salts. This exposure situation is defined as E4.

6.1.2.3 Exposure situations requiring special attention

Within the broad classification based on Local Spell index, there are buildings with certain features and also certain positions of buildings, which need special consideration. If in any building there are features which are more severely exposed than the general building, for example sills, where run-off is a factor, parapets which are exposed on both faces above the roof line and such like, they should be considered as in site exposure condition E3.

6.1.2.4 Effect of different masonry units

Carbonation and electrolytic corrosion are dependent on the migration of oxygen and moisture through the masonry. Consequently there is a tendency for the protection to the reinforcement offered by the masonry to be greater when low porosity, low permeability materials are used. In this respect the Code is based on an interpolation of a limited amount of evidence, and (if reinforcement is being provided in accordance with Table 13 of the Code) recommends that where bricks of any material have a water absorption of greater than 10% or concrete blocks having a nett density of less than 1500 kg/m^3 are used, the type of reinforcement should be that recommended for the next more severe condition of site exposure.

Selection of type of reinforcement

Once the correct exposure condition is established, the type of reinforcement should be selected from Table 13 of the Code, which is intended to give the minimum acceptable degree of protection for each classification. Alternatively concrete cover may be provided in accordance with Table 14 of the Code when carbon steel is used.

6.1.2.5 Concrete Infill

This has been dealt with in 2.9

6.1.2.6 Cover

The use of protected steels in certain circumstances and the allowance which is made for the degree of protection given to reinforcement by the thickness of masonry above any cover in concrete or mortar, can result in a lower thickness of cover to be specified than in normal concrete structures. Table 14 specifies the minimum cement content and maximum free water/cement ratio for the grade of concrete to be used for a given cover in one of the four exposure categories.

The types of austenitic stainless steels or stainless coated steels which have been recommended do not require any cover to ensure their durability. However, if they are required to transmit force into masonry through their bond with mortar or concrete then there should normally be a cover of at least one bar diameter.

There are certain circumstances which lend themselves readily to a specification for corrosion protection based on concrete cover to low carbon steel reinforcement. For example, in reinforced hollow blockwork construction where the shell is relatively thin, in pocket type retaining walls, possibly at the bottom of grouted cavity beams.

6.1.2.7 Prestressing tendons
Prestressing tendons which are surrounded by concrete or mortar should be treated in the same way as reinforcing steel and all the considerations so far described should be made. In some cases, for example, in diaphragm walls or post-tensioned cavity walls, low carbon steel tendons will be placed in open cavities. In these cases it is recommended that they be protected, for example by galvanising with a minimum zinc coating of 940 g/m². It is also essential that there is some means of draining open cavities.

6.1.2.8 Wall ties
Wall ties should be considered in the same way as reinforcement and protected in the same way as would be necessary for steel in the same location. This can lead to a different type of steel being required for the wall ties to that for the main steel. For example, wall ties in a grouted cavity wall are located in the bed joints and in exposure condition E1 would need to be galvanised with 940g/m² of zinc, although the main reinforcement in the cavity need be carbon steel only. In this situation it is essential that dissimilar metals are not allowed to come into contact. In the example quoted the sacrificial coating of zinc protects the wall ties by being more active electrochemically than the carbon steel. If the wall ties touched the main reinforcement the zinc would act as a sacrificial electrode for the carbon steel of both the wall tie and the main bars and consequently would react more quickly and the protection would not last as long as anticipated. In other situations where dissimilar metals are in contact, bimetallic corrosion cells may be set up, leading to unexpected corrosion.

6.2 Fire resistance
This section requires no further detailed comment.

6.3 Movement
This topic is covered in Chapter 16

6.4 Spacing of wall ties
In ungrouted cavity walls and low-lift grouted cavity walls where bursting forces due to placement of concrete infill are low, the spacing of wall ties should follow the recommendations in BS 5628 Part l4, as follows:

The two leaves of a cavity wall should be tied together securely by metal ties. Where the width of the cavity is more than 75 mm or the ties are for low lift grouted cavity work, only twist type wall ties complying with BS1243[5] should be used. Ties should preferably be embedded simultaneously in both leaves with a slight fall to the outer leaf. They should be placed in the mortar joints as the units are laid – not pushed in after the unit is bedded. They should be embedded at least 50 mm which, for unfilled hollow blockwork, means they must coincide with the web of the block. Cellular blocks are usually laid with their closed end uppermost and so a 50 mm embedment is achievable anywhere along their length. Wall ties of both butterfly and double-triangle type should be laid drip down.

Spacing should be in accordance with Table 6.2, but because of the variety of dimensions of various units, the spacing may be adjusted slightly so as to align the course heights. The total number of wall ties per m² must not be less than the value given in Table 6.2. In addition to their normal spacing, ties should also be provided within 225 mm either side of an opening, movement joint or external corner, at vertical centres not exceeding 300 mm.

In high lift grouted cavity walls, the wall ties should be spaced at not greater than 900 mm centres horizontally and 300 mm centres vertically with each layer staggered by 450 mm. Additional ties should be provided at openings, and so on, at not greater than 300 mm centres vertically.

6.5 Drainage and waterproofing
This section requires no further detailed comment.

6.6 Damp-proof courses and copings
The Code refers the designer to Clause 21 of BS 5628: Part 3 for information on damp-proof courses and copings.

Table 6.2 Spacing of wall ties in ungrouted and low lift grouted cavity walls

Leaf thickness mm	Cavity width mm	Spacing of ties		Number of ties per m²
		Horizontally	Vertically	
Less than 90	50–75	450	450	4.9
90 or more	50–150	900	450	2.5

Damp-proof courses

A warning is given in the Code regarding the possible effects the choice of material will have on the bending and shear strength of the member. BS 5628 Part 3 gives no guidance on the latter and indeed little guidance exists: therein lies the first difficulty in using dpc's in reinforced masonry. The second difficulty is that for horizontal dpc's to function correctly they should form a complete break in the structure. In certain instances, however, this is contrary to what is being achieved structurally, for example, in reinforced hollow block-work cantilever retaining walls where the reinforcement and its surrounding infill must be continuous down to foundation level.

A range of damp-proof course materials is available and these vary in thickness from as little as 0.5 mm up to 2 mm and more. Bitumen polymer dpc's[6] are designed to be used in any position where a flexible dpc is required and are particularly suitable for heavy load situations. Dpc's of this type also tend to retain flexibility at low temperatures. Standard dpc's complying with BS 8215[7] are for general purpose use with moderate loadings, but may need to be handled with care in wintry conditions. Polythene dpc's are not suitable for use in freestanding or the lightly load situations because the strength of the bond with mortar is low. One type of dpc available which will give very good adhesion to mortar is an asbestos based damp course surfaced with a coarse sand finish which gives good resistance to slip along the length of the wall and to tensile stress across the thickness of the wall.

Damp-proof courses can be put into three main groups:

Flexible	Polyethylene
	Bitumen and pitch polymers
	Bitumen – abrasion fibre or asbestos based
	Lead
Semi-rigid	Mastic asphalt
Rigid	Slate
	Dpc brick
	Epoxy resin/sand

This list is in approximate order of adhesion strength and, therefore, shear strength[8], but several may be subject to extrusion under vertical load, namely mastic asphalt and bitumen. It is generally accepted that the former may extrude under pressures above 65 kN/m². Methods of testing dpc's in flexure and in shear are given in DD86[9]. As stated earlier, little quantitative information exists and it is, therefore, always wise to consult the manufacturer or specialist bodies.

The installation of dpc's should comply with BS 8102[10] and BS 743. A dpc is usually placed to extend through the full thickness of the wall. In cold weather bitumen dpc's should be warmed. Bitumen polymer dpc's should be joined with 100 mm overlap. Most BS 743 types of dpc require an 100 mm overlap sealed with a proprietary cement, hot bitumen, or by the careful application of a blow-lamp.

Polythene dpc's should be welted or welded.

Flexible dpc's should be sandwiched between two layers of mortar, the dpc being laid on the first bed whilst still wet and the second preferably being applied immediately. It should extend the full thickness of the masonry and preferably project from it. Flush dpc's may be acceptable in some circumstances if accurate positioning can be relied upon. They should never be recessed behind the face of the mortar. Where a fired-clay dpc brick is used in conjunction with concrete or calcium silicate masonry, the possibility of differential movement should be considered, although in most cases this will not be a problem. BS 5628: Part 3 gives full guidance on positioning of dpc's but is not summarised here since much of it is inapplicable to the type of structure likely to be built in reinforced masonry.

Copings (and cappings)

Freestanding walls, retaining walls, and so on, exposed to the weather should preferably be provided with a coping. The coping may be a preformed unit or it may be built up using creasing tiles. In either case, the drip edge(s) should be positioned a minimum of 40 mm away from the face(s) of the wall. Where, for aesthetic reasons, a capping is used, special care is needed in the choice of materials for capping and for the walling beneath (*Section 32.1*).

A continuous dpc should be used in conjunction with copings or cappings and should be bedded in a designation (i) mortar in the case of fired-clay units, or designation (ii) mortar in the case of concrete and calcium silicate units. In cappings, the dpc may be positioned 150–200 mm down rather than immediately below the capping course to obtain greater weight on the dpc. Dpc's for both cappings and copings should preferably extend 12–15 mm beyond the face(s) of the wall to throw water clear of the wall. Alternatively, a suitable flashing may be used.

Copings may be displaced by lateral loads, vandalism, etc., and consideration should be given in this aspect. L-shaped and clip-over copings may be more satisfactory in these situations, but where necessary any coping should be dowelled or joggle-jointed together and/or suitably fixed down. Provision for movement should be provided in long coping runs; more frequent movement joints may be required owing to increased solar absorption. Any movement joints detailed in the masonry below must be continued through the coping or capping.

Section 7: Work on site

7.1 Materials

The specification of materials to be used in reinforced and prestressed masonry is provided in Section 2.

On site the storage and handling of masonry units and associated materials should follow the recommendations contained in BS 5628: Part 3[1].

Particular attention should be paid to covering masonry units on site, and during construction. Failure to protect many facing units from excessive moisture may well lead to subsequent efflorescence problems and ideally masonry should be covered as building progresses.

7.2 Construction

7.2.1 General

The general requirements for the execution of reinforced and prestressed masonry are similar to unreinforced masonry and are described in detail in BS 5628: Part 3. The following additional requirements should be considered.

The workability of infill concrete should be very high when filling vertical cores or narrow cavities in masonry walls. It is essential that such mixes should be largely self-compacting, although small mechanical vibrators, compacting rods and so on, should also be used to ensure the complete filling of all sections. There are some reinforced masonry elements, such as shallow lintels or beams, in which it is comparatively easy to determine the efficiency of the filling by inspection. Walls filled in fairly low lifts are also reasonably easy to inspect as described below.

The reinforcement should be free from deleterious material as described in the Code. Care should be taken with the fixing and location of reinforcing steel to ensure that the correct cover is maintained and that the steel cannot be displaced during the filling process. This can usually be achieved, in a wall for example, by locating main vertical reinforcement by means of the horizontal distribution steel. Conventional plastic type bar spacers may be used quite readily in beams and other 'open' elements, but should not be allowed to obstruct the core, for example, of hollow blockwork.

7.2.2 Grouted cavity construction

7.2.2.1 General

During the construction of cavity walls, care needs to be taken to keep the cavity clean. For narrow cavities this may be achieved by the use of a timber lathe which may be placed in the cavity and 'drawn up' with the mortar droppings. For wider cavities it will usually be simpler to remove mortar droppings through 'clean out' holes left at the bottom of the wall. All mortar extrusions which infringe into the cavity space should be removed before filling.

7.2.2.2 Low lift

In this method of construction the infill concrete is placed as construction proceeds. usually in lifts of 450 mm, i.e. two courses of blockwork or six courses of brickwork. The 'construction joint' in the core should be at mid-unit height rather than corresponding with the top of the unit. To maintain the appearance of facing masonry, care should be exercised in filling the cores and in preventing grout loss detracting from the appearance. The concrete should be compacted as each layer is placed. It may be necessary to limit the rate of construction and filling to avoid disruption of the masonry due to the pressure exerted by the fresh concrete infill. Any disruption due to the placing process will result in the necessity to rebuild the wall.

7.2.2.3 High lift

The clean out holes at the base of the wall should be at least 150 mm × 200 mm and spaced at intervals of 500 mm. They are used to remove all mortar and other debris prior to placing the concrete. Before the wall is filled, the brickwork must either by replaced in the clean out holes or temporary shuttering fixed to prevent the loss of infill concrete. The latter technique provides a means of checking efficient filling at the base of the wall.

The infilling concrete should not be placed until after three days have elapsed since the brickwork was constructed – longer in adverse weather conditions. The maximum height to be filled by this technique in one pour is 3 m, usually in two lifts. The concrete in each lift should be recompacted after initial settlement due to water absorption by the masonry.

There are examples in the USA where extremely high pours (up to 10 m) have been carried out in a single lift, the mix containing a lot of cement and a great deal of water. However, this is not usual and the practice recommended above is similar to many American recommendations.

7.2.3 Reinforced hollow blockwork

7.2.3.1 General

There are essentially two techniques for filling the cores of hollow concrete blocks. low lift and high lift grouting. In the low lift technique the cores are filled as the work proceeds so that not more than a few courses of blockwork are built up before filling. In the high lift technique the cores are filled in lifts of up to 3 m, care being taken to ensure that the cores are fully filled and that the pressure exerted by the infilling concrete does not disrupt the wall.

7.2.3.2 Low lift

The reinforcing steel within the cores may be located by tying the main steel to the distribution steel. If necessary the face shell of appropriate blocks may be removed to facilitate the tying of vertical steel for laps and so on. The use of plastic spacers which might tend to block up the cores should be avoided. The general aspects applying to low lift grouted cavity construction apply to this technique except that the maximum vertical interval at which concrete is placed may be 900 mm.

7.2.3.3 High lift

In the high lift technique it is particularly important to ensure that all mortar extrusions are removed from the core of the blocks.

This is commonly achieved by leaving clean out holes at the base of the wall. Excess mortar is knocked off the side of the cores and is removed through the holes in the base of the wall. Before filling with concrete these holes need to be securely blocked to prevent the loss of the infilling concrete.

The concrete itself may be placed by hand, skip or pump. Whichever method is used, particular care should be taken with facing work to prevent grout running down the face of the wall. The mixes specified in the Code are such that they are intended to have a high level of workability and should be readily compacted when a 25 mm diameter poker vibrator is used.

Once a wall has been filled using the high lift grouting technique it will be noticed that after a period of some 15 minutes (depending on the mix, absorption of the masonry and weather conditions), the concrete in each core has slumped. At this stage further concrete should be added and some limited recompaction carried out. An alternative approach is to use a proprietary additive in the mix to prevent this slump taking place.

When infilling concrete is placed by a grout pump, the rate of placing should not exceed 0.2 m³ per minute.

Bond beam construction

When using a bond beam within an otherwise unreinforced section of walling, it will be necessary to seal the openings in the bottom of the blocks using an appropriate material. In the USA these are known as 'grout stop' materials. Typical materials used are expanded metal lathe, thick mesh screen and asphalt saturated felt.

Horizontal reinforcing steel will need to be supported to give the appropriate cover by either plastic saddle supports, reinforcing steel or prefabricated brackets. Where it is necessary to splice bars, this should be done vertically (i.e. one bar above and one bar below), rather than side by side, to provide less restriction to the flow of the infilling concrete.

7.2.4 Quetta and similar bond walls

In this method of construction the reinforcement is usually placed progressively, in advance of the masonry. The cavities are filled with mortar or concrete as the work proceeds. In some circumstances, where large voids are produced, either low or high lift techniques may be used.

7.2.5 Pocket type walls

Pocket type walls are usually built to their full height, the starter bars only projecting from the base into the pocket space. The main steel is then fixed and may be held in position using wires fixed into bed joints. Shuttering may be propped against the rear face of the wall, although it has in the past, been successfully fixed to the wall with masonry nails. The concrete is normally placed in lifts with a maximum height of about 1.5 m; this may be vibrated by poker vibrator or compacted using a rod.

7.2.6 Prestressing operations

This section requires no further detailed comment.

7.2.7 Forming chases and holes and provision of fixings

See BS 5628: Part 3:1985, Clause 19.

7.2.8 Jointing and pointing

The Code recommends that joints should only be raked out with the approval of the designer. Deeply raked joints are often considered to provide an attractive finish, but since the mortar is not as well compacted as when finished with a steel, in exposed situations their use could lead to problems of durability. In addition in external work raked joints expose the bed faces of the units which may lead to excessive water being absorbed. The above considerations are equally relevant to unreinforced masonry and reference should be made to BS 5628: Part 3. In sections which are critical in terms of the structural design, it may be necessary for the designer to consider the section as being reduced in size by the dimensional extent of any recess in the mortar joint profile.

References

1. BRITISH STANDARDS INSTITUTION. BS 5628: Part 3:1985 *Code of Practice for use of masonry. Part 3: Materials and components. Design and workmanship.* BSI, London. pp. 100.
2. BRITISH STANDARDS INSTITUTION. BS 8104:1992 *Code of Practice for assessing exposure of walls to wind-driven rain.* BSI. London.
3. LACY, R E. *An index of exposure to driving rain.* Building Research Establishment, Garston, Watford. BRE Digest 127, March 1971. pp.8.
4. BRITISH STANDARDS INSTITUTION. BS 5628: Part 1:1992 *Code of practice for use of masonry. Part 1: Structural use or unreinforced masonry.* BSI. London.
5. BRITISH STANDARDS INSTITUTION. BS 1243:1978 *Specification for metal ties for cavity wall construction.* BSI. London. pp. 4.
6. BRITISH STANDARDS INSTITUTION. BS 6398:1983 *Specification for bitumen damp-proof courses for masonry.* BSI. London. pp.4.
7. BRITISH STANDARDS INSTITUTION. BS 8215:1991 *Code of practice for design and installation of damp proof courses in masonry construction.* BSI, London.
8. HODGKINSON, H R and WEST, H W H. *The shear resistance of some damp-proof course materials.* BCRL Technical Note TN 326.1981.
9. BRITISH STANDARDS INSTITUTION. DD 86: *Draft for Development – Damp-proof courses.* Part 1:1983: *Methods of test for flexural bond strength and short term shear strength.* Part 2:1984: *Method or test for creep deformation.* BSI, London. pp. 8.
10. BRITISH STANDARDS INSTITUTION. BS 8102: 1990 *Code of practice for protection of structures against water from the ground.* BSI, London.

9.3 Design examples

Example I

Problem

Design a 1.8 m high cantilever retaining wall, 215 mm thick, to resist a moment at the base of 9.4 kN m/m run with a shear force of 16.2 kN/m run.

Blocks	440 × 215 × 215, hollow, of unit strength 7 N/mm² with 55% solid
Mortar	Designation (ii)
Reinforcement	f_y = 250 N/mm²

Notes:
1. $\gamma_{mm} = 2.3$
2. exposure condition E3
3. place the steel in centre so that moment may be resisted equally from either side
4. use a concrete grade C40 to BS 5328
5. web thickness of block 40 mm

Solution

Cover required
(Table 14)   = 40 mm
Cover provided = 107 – 40 – 10 – 57 mm ∴OK

The ratio of span to effective depth of this wall should be checked.

Effective span   = length to face of support + ½ effective depth

$$= 1.8 + \frac{0.107}{2} = 1.85 \text{ m}$$

$$\frac{\text{span}}{\text{effective depth}} = \frac{1.85}{0.107} = 17.3 < 18 \therefore \text{OK, and it is not}$$

necessary to check deflection and cracking by calculation.

For 7 N/mm² block with 55% solid, the net strength = 12.7 N/mm²,
thus, f_k = 6.2 N/mm² (Table 3(b))
The maximum design moment, Md, should not exceed that of the balanced section. Hence:

$$M_d = \frac{0.4 f_k \, b \, d^2}{\gamma_{mm}}$$

$$= \left(\frac{0.4 \times 6.2 \times 1000 \times 107^2}{2.3}\right) \times 10^{-6} \text{ kN m}$$

$$= 12.34 \text{ kN m} > 9.4 \therefore \text{OK}$$

Consider now the required area of steel:

$$M_d = \frac{A_s f_y z}{\gamma_{ms}} \quad \text{i.e.} \quad z = \frac{M_d \gamma_{ms}}{A_s f_y}$$

and, $z = d\left(\dfrac{1 - 0.5 \, A_s f_y \, \gamma_{mm}}{bd \, f_k \, \gamma_{ms}}\right)$

Therefore $\dfrac{9.4 \times 10^6 \times 1.15}{A_s \times 250}$

$$= 107\left(1 - \frac{0.5 A_s \times 250 \times 2.3}{1000 \times 107 \times 6.2 \times 1.15}\right)$$

Hence, $A_s = 497$ mm²
Therefore, use R12 every core (225 mm) = 502 mm²/m run

Shear
Design shear force,
V = 16.2 kN/m run

Therefore, $\nu = \dfrac{V}{bd} = \dfrac{16.2 \times 10^3}{1000 \times 107} = 0.15$ N/mm2*

$\rho \qquad = \dfrac{As}{bd} = \dfrac{502}{1000 \times 107} = 0.0047$

Therefore, characteristic shear strength (Clause 19.1.3),

$$f_v = (0.35 + 17.5\rho)\left(2.5 - 0.25\frac{a}{d}\right)$$

$$a = \frac{M}{V} = \frac{9.5}{16.2} = 0.58$$

$$\therefore f_v = (0.35 + 17.5 \times 0.0047) \times$$

$$\left(2.5 - 0.25 \times \frac{0.58}{0.107}\right)$$

$$= 0.49 \text{ N/mm}^2$$

Thus, $\dfrac{f_v}{\gamma_{mv}} = \dfrac{0.49}{2} = 0.245$ N/mm²

Therefore the wall has adequate resistance to shear.

Horizontal steel
The minimum horizontal steel required
$$= 0.0005 \times 107 \times 1000$$
$$= 54 \text{ mm}^2/\text{m run}$$

Therefore, use one 6 mm diameter bar in alternate joints (63 mm²/m) or use proprietary joint reinforcement.

For durability (exposure E3) these must be austenitic stainless steel or carbon steel coated with at least 1 mm of stainless steel.

Detailing
It is possible to calculate the change point for providing, say, 12 mm starter bars to lap with 10 mm bars which run for the full height of the wall, but this may not be economical if the lap length is long. The required anchorage length should also be calculated. The horizontal steel should not touch the vertical steel if they are of dissimilar materials.

Example 2
Problem

Axial load capacity of 2.8 m high wall

Blocks   390 × 190 × 190, hollow, of unit strength N/mm² with 60% solid
Mortar   designation (i)
Reinforcement   one T12 each core, $f_y = 460$ N/mm²

---

*The minimum value of $f_v = 0.35 \therefore \dfrac{f_v}{\gamma_{mv}} = \dfrac{0.35}{3} > \nu = 0.15$ N/mm², which is adequate. The full calculation is shown for information.

Notes:
1. simple lateral support provided top and bottom
2. $\gamma_{mm} = 2.3$
3. exposure condition E2
4. concrete grade C40

Solution

Cover required = 30 mm
Cover provided = 85 − 6 web thickness (35 mm)
= 44 mm ∴ adequate

Simple lateral support provided, therefore $h_{ef} = h$
For single leaf wall $t_{ef} = t$

Therefore, slenderness ratio $= \dfrac{2.8}{0.19} = 14.7$

From Table 7 of Part 1, $\beta = 0.87$
For 21 N/mm² block with 60% solid, the net strength = 35 N/mm²,
thus, f_k = 14.7 N/mm² (Table 3(b))

$$N_d = \frac{\beta b t f_k}{\gamma_{mm}}$$

$$= \frac{0.87 \times 1000 \times 190 \times 14.7}{2.3 \times 10^3}$$

$$= 1056 \text{ kN/m run}$$

Note: This approach makes no allowance for the contribution of the reinforcement. It is possible to use the approach provided for columns in Part 2, but this is unlikely to give a more favourable result.

Example 3

Problem

Design a 4.0 m high column, 390 × 390, with axial load of 500 kN and moment of 60 kN m

Blocks   390 × 190 × 190, hollow, of unit strength 14 N/mm² with 55% solid
Mortar   designation (i)
Reinforcement $f_y = 460$ N/mm²

Notes:
1. lateral restraint in both directions top and bottom
2. $\gamma_{mm} = 2.3$
3. exposure condition E2
4. web thickness of block = 40 mm

Solution (a)

Lateral support is provided, therefore $h_{ef} = h$

∴slenderness ratio $= \dfrac{4.0}{0.39} = 10.3 < 12$

It is therefore a short column
For a 14 N/mm² block with 55% solid, the net strength = 25.5 N/mm², thus, $f_k = 11.4$ N/mm² (Table 3(b))

Assume T20 steel which, for exposure condition E2, requires 30 mm cover with a grade C40 concrete to BS 5328.

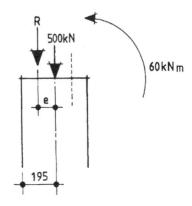

Resultant eccentricity, $e = \dfrac{60}{500} = 0.12$ m $= 120$ mm

$$N_d = \frac{f_k b}{\gamma_{mm}}(t - 2ex)$$

$$= \frac{11.4}{2.3} \times 390 \times (390 - 240) \times 10^{-3}$$

$$= 290 \text{ kN}$$

Design axial load exceeds this, therefore, need to carry out a full analysis.

$$N_d = \frac{f_k}{\gamma_{mm}} b d_c + \frac{f_{s1} A_{s1}}{\gamma_{ms}} = \frac{f_{s2} A_{s2}}{\gamma_{ms}}$$

It is now necessary to choose a value of d_c, which should not be chosen as less than a value of $2d_1$, where d_1 is the depth from the surface to the reinforcement in the more highly compressed face. Assume T20 steel, therefore, with exposure condition E2 (cover = 30 mm) and block web thickness of 40 mm, gives

$$d_1 = 40 + 30 + 10 = d_2$$

Thus, $2d_1 = 160$ mm
Choose $d_c = 250$ mm. This value is between $(t - d_2) = 390 - 80 = 310$, and $\dfrac{t}{2} = 195$ (where d_2 is the depth to the reinforcement from the least compressed face). In this range, f_{s2} is varied linearly between 0 and f_y, i.e. f_y when $d_c = 195$ and 0 when $d_c = 310$.

$$\therefore f_{s2} = \frac{60}{115} \times f_y = 0.52 f_y$$

$$\therefore N_d = \frac{11.4 \times 390 \times 250}{2.3 \times 10^3} +$$

$$\left(\frac{0.83 \times 460}{1.15} - \frac{0.52 \times 460}{1.15} \right)\frac{628}{10^3}$$

$$= \frac{11.4 \times 390 \times 250}{2.3 \times 10^3}$$

$$+ \frac{0.31 \times 460}{1.15} \times \frac{628}{10^3}$$

$$= 483 + 78$$
$$= 561 \text{ kN}$$

$N_d > 500 \therefore$ this is adequate

$$M_d = \frac{0.5 f_k}{\gamma_{mm}} b d_c (t - d_c)$$

$$+ \frac{0.83 f_y}{\gamma_{ms}} A_{s1} (0.5t - d_1)$$

$$+ \frac{f_{s2}}{\gamma_{ms}} A_{s2} (0.5t = d_2)$$

$$= \frac{0.5 \times 11.4}{2.3 \times 10^6} \times 390 \times 250 (390 - 250)$$

$$+ \frac{0.83 \times 460 \times 628}{1.15 \times 10^6} (195 - 80)$$

$$+ \frac{0.52 \times 460 \times 628}{1.15 \times 10^6} (195 - 80)$$

$$= 33.83 + 23.98 + 15.02$$
$$= 72.8 \text{ kN m}$$

$\therefore$ this is adequate. Thus need 4 No T20 bars, one in each core.

Solution (b): alternative solution using interaction curves

Lateral support is provided, therefore $h_{ef} = h$

$\therefore$ slenderness ratio $= \dfrac{40}{0.39} = 10.3 < 12$

It is therefore a short column.
For a 14 N/mm² block with 55% solid, the net strength
$= 25.5$ N/mm²,
thus, $f_k = 11.4$ N/mm²

$$\frac{N}{bt\, f_k} = \frac{500 \times 10^3}{390 \times 390 \times 11.4}$$

$$= 0.29$$

$$\frac{M}{bt^2 f_k} = \frac{60 \times 10^6}{390 \times 390^2 \times 11.4}$$

$$= 0.09$$

$$d = 390 - 40 - 30 - 10$$
$$= 310 \text{ mm}$$

Therefore, $\dfrac{d}{t} = \dfrac{310}{390}$

$$= 0.8$$

From interaction diagram for $f_y = 460$ N/mm²

$\dfrac{\rho}{f_k} = 6 \times 10^{-4}$, where $\rho = \dfrac{A_s}{bt}$

Therefore, $A_s = 6 \times 10^{-4} \times 11.4 \times 390 \times 390$
$= 1040$ mm²
Therefore, use 4 No T20 (= 1260 mm²), one each core.

Example 4

Problem

As Example 3, but design 6.0 m high column.

Solution (a)

Lateral support is provided, therefore $h_{ef} = h$

$\therefore$ slenderness ratio $= \dfrac{6.0}{0.39} = 15.4 < 27$

The slenderness ratio is greater than 12 and it must, therefore, be designed as a slender column with account taken of the additional moment induced by vertical load due to lateral deflection. This may be taken as:

$$\frac{N (h_{ef})^2}{2000\, t} = \frac{500 \times 6.0^2}{2000 \times 0.39}$$

$$= 23.1 \text{ kN m}$$

Assume Y25 steel which, for exposure condition E2, requires 30 mm cover with a concrete grade C40 to BS 5328.

$\therefore d_2 = d_1 = 83$ mm

As before, assume $d_c = 250$ mm. By consideration of previous example, N_d is adequate.

$$M_d = \frac{0.5 \times 11.4}{2.3 \times 10^6} \times 390 \times 250 (390 - 250)$$

$$+ \frac{0.83 \times 460 \times 982}{1.15 \times 10^6} (195 - 83)$$

$$+ \frac{0.52 \times 460 \times 982}{1.15 \times 10^6} (195 - 83)$$

$$= 33.83 + 36.51 + 22.88$$
$$= 93.2 \text{ kN m}$$

This is greater than $60 + 23.1$ kN m, therefore adequate.

Solution (b): alternative solution using interaction curves

Lateral support is provided, therefore $h_{ef} = h$

$\therefore$ slenderness ratio $= \dfrac{6.0}{0.39} = 15.4 < 27$

The slenderness ratio is greater than 12 and it must, therefore, be designed as a slender column with due account taken of the additional moment induced by the vertical load due to lateral deflection. This may be taken as:

$$\frac{N(h_{ef})^2}{2000\,t} = \frac{500 \times 6.0^2}{2000 \times 0.39}$$
$$= 23.1 \text{ kN m}$$

$$\frac{N}{bt\,f_k} = \frac{500 \times 10^3}{390 \times 390 \times 11.4}$$
$$= 0.29$$

$$\frac{M}{bt^2\,f_k} = \frac{(60 + 23.1) \times 10^6}{390 \times 390^2 \times 11.4}$$
$$= 0.12$$

As with Example 3, $\frac{d}{t} = 0.8$, f_y 460 N/mm²

Therefore, from interaction diagram for $f_y = 460$

$$\frac{\rho}{f_k} = 8 \times 10^{-4}, \text{ where } \rho = \frac{A_s}{bt}$$

Therefore, $A_s = 8 \times 10^{-4} \times 11.4 \times 390 \times 390$
$= 1388 \text{ mm}^2$

Therefore, use 4 No T25 (= 1960 mm²) one each core.

Example 5

Problem

Design a beam to span a 3.8 m opening in a blockwork wall. The beam is subjected to a moment of 20 kN m and a shear force of 18 kN.

Blocks   390 × 190 × 190, hollow, of unit strength 7 N/mm² with 55% solid
Mortar   designation (i)
Reinforcement $f_y = 460$ N/mm²

Notes:
1. $\gamma_{mm} = 2.3$
2. exposure condition E1
3. use bond beam blocks with 50 mm of web left intact

Solution

Initial attempt: single course beam
For 7 N/mm² block with 55% solid, the net strength 12.7 N/mm²,
thus, $f_k = 6.8$ N/mm2
Web is 50 mm thick, cover is 20 mm, say 20 mm dia. bar. Therefore, effective depth,

d $= 190 - 50 - 20 - 10$
             $= 110 \text{ mm}$

Using charts:

$$\frac{M}{f_k\,bd^2} = \frac{20 \times 10^6}{6.8 \times 190 \times 110^2}$$
$$= 1.28 > 0.174$$

Try again, two course beam
Effective depth, $d = 390 - 50 - 20 - 10$
             $= 310 \text{ mm}$

Check lateral stability first, 3.8m $\geq 60\,b_c$ or $\frac{250}{d}\,b_c^2$ whichever is the lesser:

$60\,b_c$ $= 60 \times 0.19$
                   $= 11.4 \text{ m}$

$$\frac{250\,b_c^2}{d} = \frac{250 \times 190^2}{310}$$
$$= 29.1 \text{ m}$$

both > 3.8 m ∴OK

Using charts:

$$\frac{M}{f_k\,bd} = \frac{20 \times 10^6}{6.8 \times 190 \times 310^2}$$
$$= 0.16 < 0.174 \therefore OK$$

This gives $\frac{\rho}{f_k} = 5.3 \times 10^{-4}$, where $\rho = \frac{A_s}{bd}$

Therefore, $A_s = 5.3 \times 10^{-4} \times 6.8 \times 190 \times 310$
$= 212 \text{ mm}^2$

Provide 2 No T12 (= 226 mm²)

Shear
Design shear force,

V $= 18 \text{ kN}$

v $= \frac{V}{bd} = \frac{18 \times 10^3}{190 \times 310} = 0.31$ N/mm²

$\frac{A_s}{bd}$ $= \frac{226}{190 \times 310} = 0.0038$

Therefore, characteristic shear strength (Clause 19.1.3.1.2),

a $= \frac{M}{V} = \frac{20}{18} = 1.1$

$\therefore f_v$ $= (0.35 + 17.5 \times 0.0038) \times$
$$\left(2.5 - 0.25\,\frac{1.11}{0.31} \right)$$
$= 0.67$ N/mm²

Thus, $\frac{f_v}{\gamma_{mv}}$ $= \frac{0.67}{2.0} = 0.34 > 0.31$

Consider providing nominal shear reinforcement for 1 m in from each end:

$\frac{A_{sv}}{s_v}$ $= 0.002\,b_t$

A_{sv} $= 200 \times 0.002 \times 190$
             $= 76 \text{ mm}^2$ (two legs)

Therefore, provide nominal throughout, R8 @ 200 (= 100.6 mm² two legs).

Example 6

Problem

Design the stem of a freestanding reinforced concrete masonry perimeter wall to be built from $440 \times 215 \times 215$ mm, two core hollow blocks with a net strength of 10 N/mm^2, made to normal category of manufacturing control. Mortar designation (i) will be employed. The wind load is as follows:

Basic wind speed, v	$= 46$ m/s
Wall height	$= 2.65$ m
Topography factor, s_t	$= 1.0$
Roughness, size and height factor, s_2	$= 0.74$
Statistical factor, s_3	$= 1.0$
s_4	$= 1.0$

Solution

Span $= \text{height} + \dfrac{d}{2}$

Check span/depth ratio $= \dfrac{2.65 + 0.05}{0.107} = 25$

Table 8 permits $18 + 30\%$ for wind load only $= 23.4$
$\therefore$ need to check deflection and cracking:

Design wind speed,

$$V_s = 46 \times 1.0 \times 1.0 \times 0.74 \times 1.0$$
$$= 34.0 \text{ m/s}$$

Dynamic wind pressure $q = \dfrac{0.613 \times 34^2}{10^3}$
$$= 0.71 \text{ kN/m}^2$$

For worst possible case, $C_f = 2$
$\therefore$ total load on wall,

$$F = 2 \times 0.71 \times 2.7$$
$$= 3.83 \text{ kN/m run}$$

$\therefore$ moment to be resisted by wall $= 3.83 \times \dfrac{2.7}{2} \times \gamma_f$

Choose $\gamma_f = 1.2$, not 1.4, because the wall does not affect the stability of a structure.
$\therefore$ moment $= 6.2$ kN m

The wind is incidental from either direction $\therefore$ place steel in the centre of the core

$d = 107$ mm

Consider 1 No 12 mm bar in each alternate core (area of each bar $= 113$ mm^2)
Steel is at 450 centres $= 2.2$ bars/metre

$A_s = 248$ mm^2/m

Now, $z = d\left(1 - \dfrac{0.5\, A_s\, f_y\, \gamma_{mm}}{bd\, f_k\, \gamma_{ms}} \right) < 0.95\, d$

$\therefore z = 107\left(1 - \dfrac{0.5 \times 248 \times 460 \times 2.3}{1 \times 0.107 \times 5.7 \times 1.15 \times 10^6} \right)$

$= 107\,(1 - 0.19)$

$= 87$ mm $< 0.95\, d$

$$M_d = \frac{A_s\, f_y\, z}{\gamma_{ms}}$$
$$= \frac{248 \times 460 \times 87}{1.15 \times 10^6}$$
$$= 8.6 \text{ kN m}$$

which is greater than the moment produced by the wind load.

Check that the masonry strength is adequate:

$$\frac{0.4\, f_k\, bd^2}{\gamma_{mm}} = \frac{0.4 \times 5.7 \times 1 \times 107^2}{2.3 \times 10^3}$$
$$= 11.3 \text{ kN m} \quad \therefore \text{OK}$$

Note: secondary reinforcement of 0.05% bd is required.

Shear

$v = \dfrac{V}{bd} = \dfrac{3.83 \times 10^3}{10^3 \times 107}$

$= 0.036$ N/mm^2

$f_v = 0.35 + 17.5\, \rho \quad \therefore \text{adequate}$

Deflection

Now deflection needs to be checked:

deflection, $a = kl^2\, \dfrac{M}{EI}$

k may be determined using the moment/area theorem – 'the deflection of a point on a member, measured from the tangent at another point on the member, is equal to the moment of the $\dfrac{M}{EI}$ diagram between the two points about the point whose deflection is sought.'

$a = \left(\dfrac{lM}{EI}\dfrac{l}{2} \right) = \left(\left[\dfrac{2}{3}\dfrac{l}{EI}\dfrac{M}{} \right] \dfrac{3l}{8} \right)$

$= \dfrac{1}{2}\, l^2\, \dfrac{M}{EI} - \dfrac{1}{4}\, l^2\dfrac{M}{EI} = \dfrac{1}{4}\, l^2\, \dfrac{M}{EI}$

$\therefore k = 0.25$

The moment applied to the wall $= \dfrac{Wl}{2}$

$\therefore$ deflection, $a = \dfrac{Wl^3}{8EI}$

I is based on the gross cross section ignoring the steel.

$I = \dfrac{bt^3}{12}$

$= \dfrac{1000 \times 215^3}{12}$ mm^4

$= 828 \times 106$ mm^4

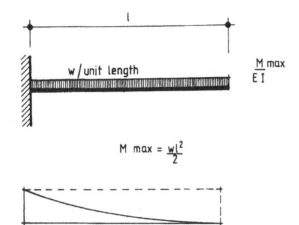

$$M \text{ max} = \frac{wl^2}{2}$$

For short term loading, which is applicable to this example since wind loads are neither continuous nor from one direction only:

E $= 900\, f_k$ N/mm²
$= 900 \times 10^{-3} \times 5.7$ kN/mm²

$\therefore a = \dfrac{3.83 \times 2700^3}{8 \times 900 \times 10^{-3} \times 5.7 \times 828 \times 10^6}$

$= 2.2$ mm

Note that Clause 16.2.2.1(a) is not applicable when short term loading is considered, neither does part (b) of this clause apply if no applied finishes are to be placed on the wall. Assume blockwork is to be rendered (i.e. (b) applies) and limiting deflection is therefore the lesser of $\frac{span}{500} = 5.2$ mm or 20 mm. Thus the deflection in this example is acceptable at $\frac{span}{1080}$. Unacceptable cracking is not likely at this deflection.

Example 7

Problem

Design a two course bond beam to support a uniformly distributed load of 9.75 kN/m run over a span of 2.8 m. The blocks to be used have a block strength of 5.5 N/mm² and are 55% solid. They are to be laid with a mortar of designation (ii). The exposure category is E1. The blocks to be used are of size 440 × 215 × 215.

Solution (a)

Loads
Imposed load = 9.75 kN/m run
Self weight $= 0.215 \times 0.44 \times 2300 \times 10^{-3} \times 9.81$
$= 2.2$ kN m/run

Design load $= 1.6 \times 9.75 + 1.4 \times 2.2$
$= 18.68$ kN/m run

Effective span = the lesser of $2.80 + 0.215 = 3.01$
or $2.81 + 0.354 = 3.16$

168

Therefore take effective span as 3.0 m

$\dfrac{span}{effective\ depth} = \dfrac{3.0}{0.354} = 8.47$

This is less than the value of 20 in Table 9 and no detailed calculation of deflection is required.

The lateral stability requirement is that the clear distance between lateral restraints does not exceed $60\, b_c$ or $250\, \frac{b_c^2}{d}$ whichever is the lesser:

$60\, b_c = 60 \times 0.215 = 12.9$ m

$250\, \dfrac{b_c^2}{d} = \dfrac{250 \times 0.215^2}{0.354} = 32.6$ m

both of which are greater than the span (3 m).

Design bending moment $= \dfrac{18.68 \times 3^2}{8} = 21.02$ kN m

For 5.5 N/mm² block with 55% solid, the net strength = 10 N/mm². For mortar designation (ii), fk = 5.4 N/mm²

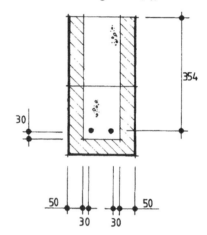

Assume 2 No 12 mm φ bars OK.

$A_s = 226$ mm²

$z = d\left(1 - \dfrac{0.5\, A_s\, f_y\, \gamma_{mm}}{bd\, f_k\, \gamma_{ms}}\right) < 0.95$

$\therefore z = 354 \left(1 - \dfrac{0.5 \times 226 \times 460 \times 2.3}{215 \times 354 \times 5.4 \times 115}\right)$

$= 264$ mm $< 0.95d$

$M_d = \dfrac{A_s\, f_y\, z}{\gamma_{ms}}$

$= \dfrac{226 \times 460 \times 264 \times 10^{-6}}{1.15}$

$= 23.9$ kN m

This is adequate to resist design bending moment but it is necessary to check the capacity of the masonry is not exceeded:

$= \dfrac{0.4\, f_k\, bd^2}{\gamma_{mm}}$

$$= \frac{0.4 \times 5.4 \times 215 \times 354^2}{2.3 \times 10^6}$$

$$= 25.3 \text{ KN m} \quad \therefore \text{OK}$$

Shear

Design shear stress, $v = \dfrac{18.68 \times \frac{3}{2} \times 10^3}{215 \times 354}$

$$= 0.37 \text{ N/mm}^2$$

Shear span $= \dfrac{21.02}{28}$

$$= 0.75 \text{ m}$$

$\therefore f_v \qquad = (0.35 + 17.5\rho) \left(2.5 - 0.25 \dfrac{a}{d} \right)$

$$= 0.35 + \left(\frac{17.5 \times 226}{215 \times 354} \right) \times$$

$$\left(2.5 - \frac{0.25 \times 750}{354} \right)$$

$$= 0.79 \text{ N/m}^2$$

$\therefore \dfrac{f_v}{\gamma_{mm}} \qquad = \dfrac{0.79}{2}$

$$= 0.39 \text{ N/m}^2 \text{ which is greater than the design shear stress}$$

Although shear reinforcement is not necessary to satisfy the calculations, it is suggested that nominal shear reinforcement be provided in a beam of this size. Therefore, provide beam links as indicated in Clause 26.5.2, such that

$$\frac{A_{sv}}{s_v} \qquad = 0.0012 \, b_t$$

for high yield steel, noting that the spacing of the reinforcement, s_v should not exceed $0.75 \, d$.

Solution (b): alternative solution using design charts

B1. Using design chart provided in this chapter

Calculate $\dfrac{M}{bd^2 f_k}$ assuming 12 mm bars as before for the purpose of assessing d

$$= \frac{21.02 \times 10^6}{215 \times 354^2 \times 5.4}$$

$$= 0.144$$

From chart, $\dfrac{\rho}{f_k} = 4.4 \times 10^{-4}$ where $\rho = \dfrac{A_s}{bd}$

$\therefore \dfrac{A_s}{bd} \qquad = 4.4 \, f_k \times 10^{-4}$

$A_s \qquad = 215 \times 354 \times 4.4 \times 5.4 \times 10^{-4}$

$$= 180.8 \text{ mm}^2$$

Therefore, use 2 No 12 mm bars $A_s = 226 \text{ mm}^2$. Otherwise complete example as before.

B2. Using design chart provided in Part 2

$M_d \qquad = Q \, bd^2$

and $Q \qquad = 2c \, (1 - c) \dfrac{f_k}{\gamma_{mm}}$

where $c = \dfrac{z}{d}$

$f_k \qquad = \dfrac{5.4}{2.3}$

$$= 2.35$$

$Q \qquad = \dfrac{M_d}{bd^2}$

$$= \frac{21.02 \times 10^6}{215 \times 354^2}$$

$$= 0.780$$

again assuming d for 12 mm diameter bars.

$\therefore$ from chart, $\dfrac{z}{d} = 0.75$

$A_s \qquad = M_d \dfrac{\gamma_{ms}}{f_y} \dfrac{1}{z}$

$$= \frac{21.02 \times 1.15 \times 10^6}{460 \times 0.75 \times 354}$$

$$= 198 \text{ mm}^2$$

Therefore, use 2 No 12 mm bars $A_s = 226 \text{ mm}^2$. Otherwise complete example as before.

Design charts

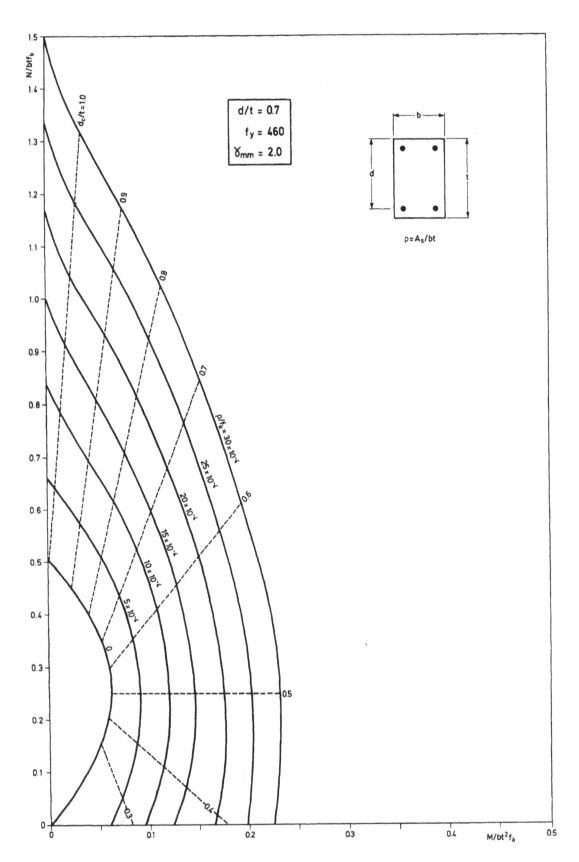

N/btf_k

d_c/t = 1.0

0.9

0.8

0.7

$\rho/f_k = 30 \times 10^{-4}$

25×10^{-4}

20×10^{-4}

15×10^{-4}

10×10^{-4}

5×10^{-4}

0.6

0.5

0

0.3

0.4

d/t = 0.7

f_y = 460

γ_{mm} = 2.0

$\rho = A_s/bt$

M/bt²f_k

172

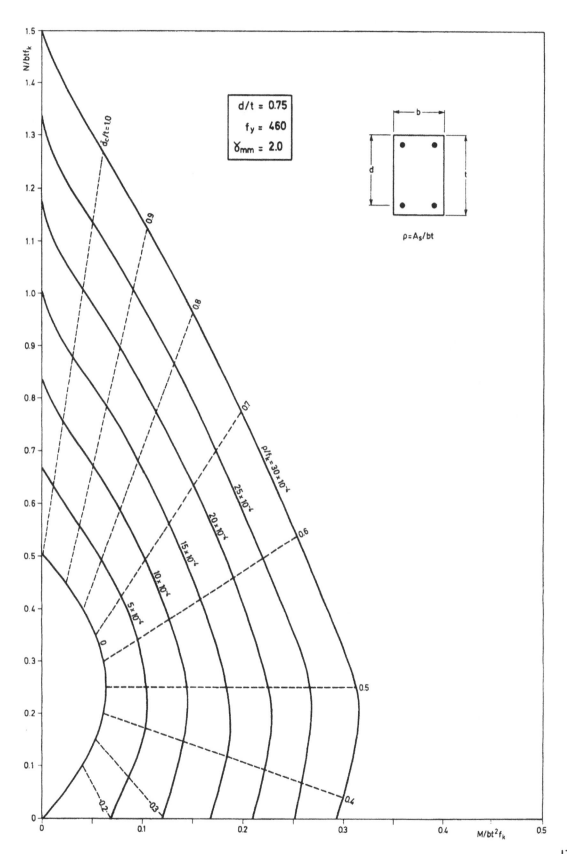

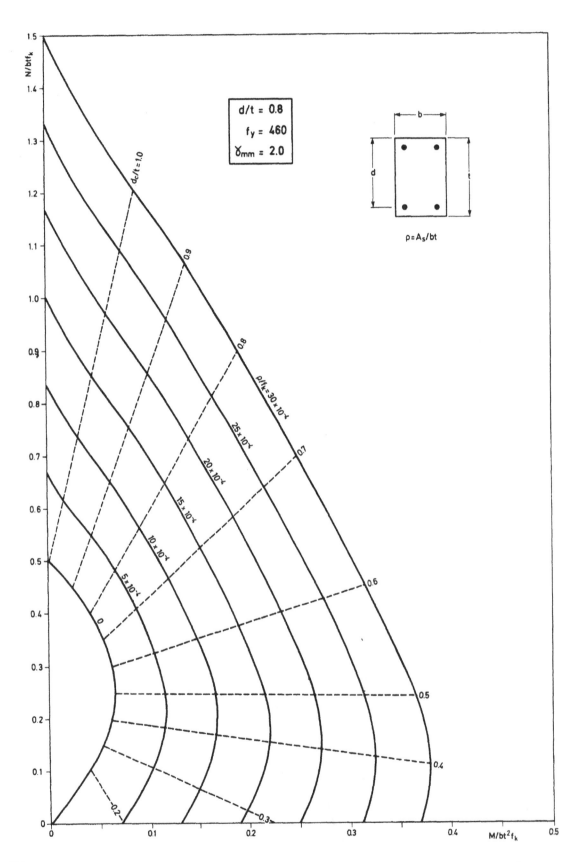

174

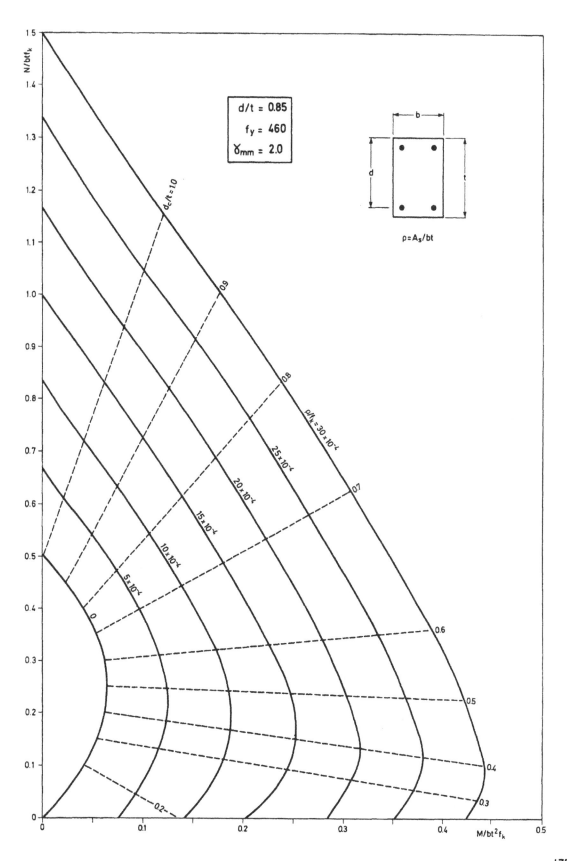

d/t = 0.85
f_y = 460
γ_{mm} = 2.0

$\rho = A_s/bt$

$d_c/t = 1.0$

$\rho f_k = 30 \times 10^{-4}$
25×10^{-4}
20×10^{-4}
15×10^{-4}
10×10^{-4}
5×10^{-4}
0

N/btf_k

$M/bt^2 f_k$

175

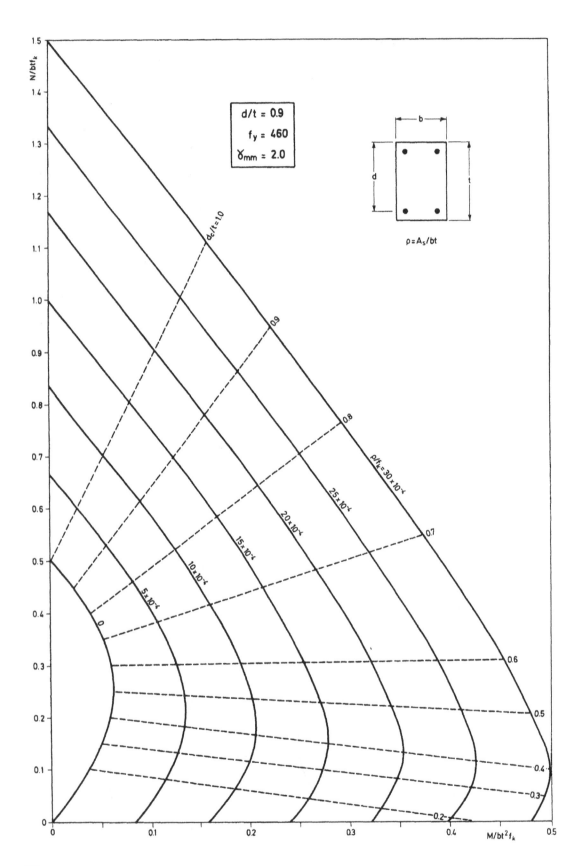

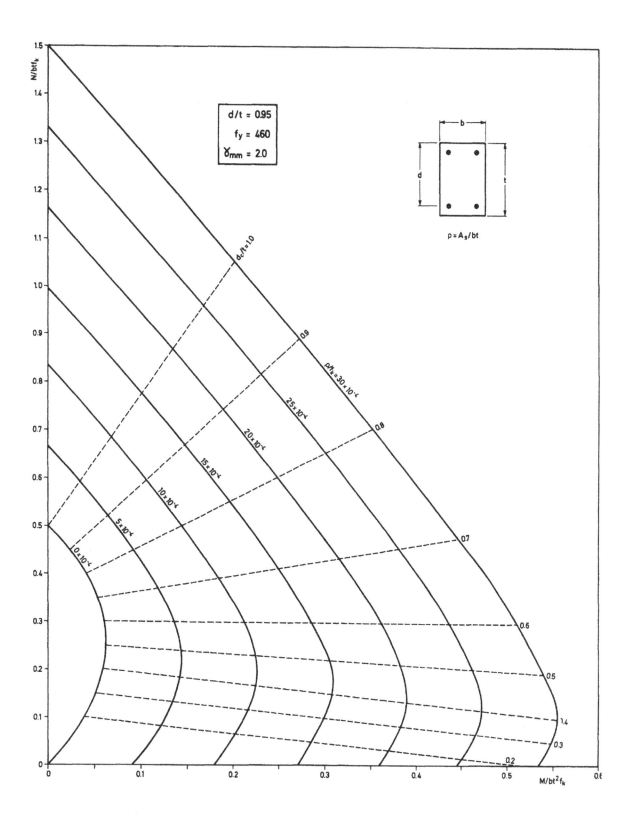

177

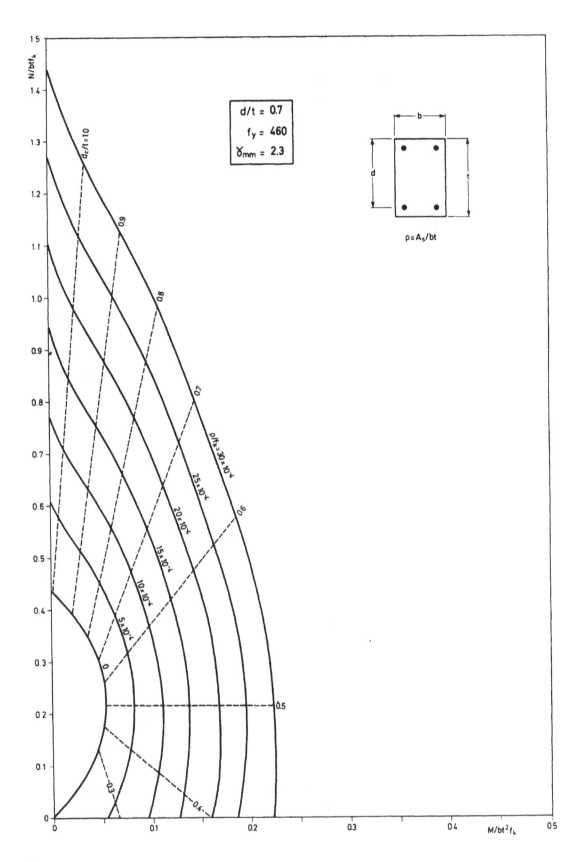

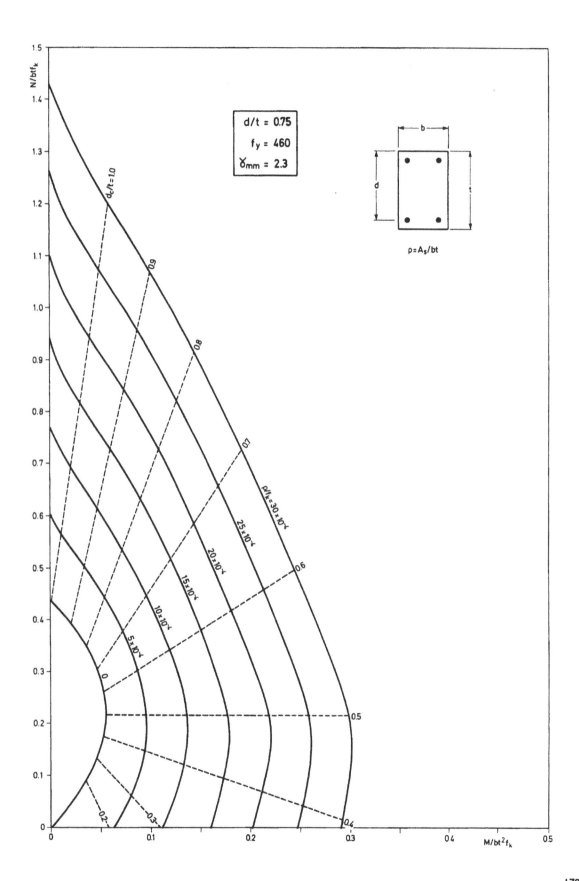

d/t = 0.75
f_y = 460
γ_mm = 2.3

ρ = A_s/bt

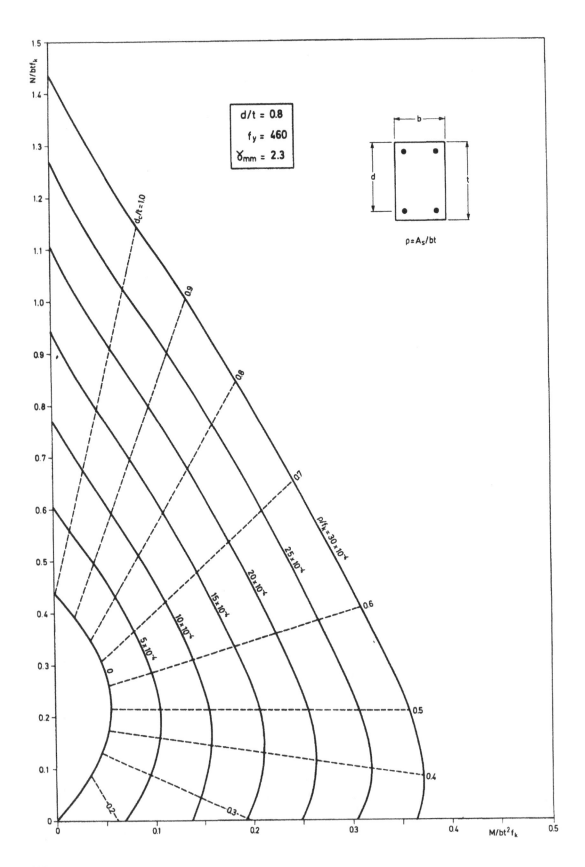

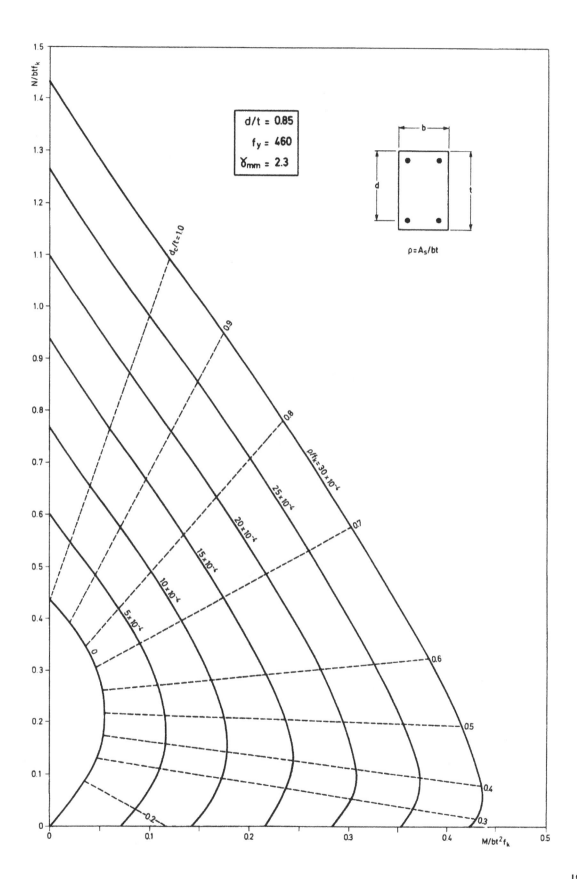

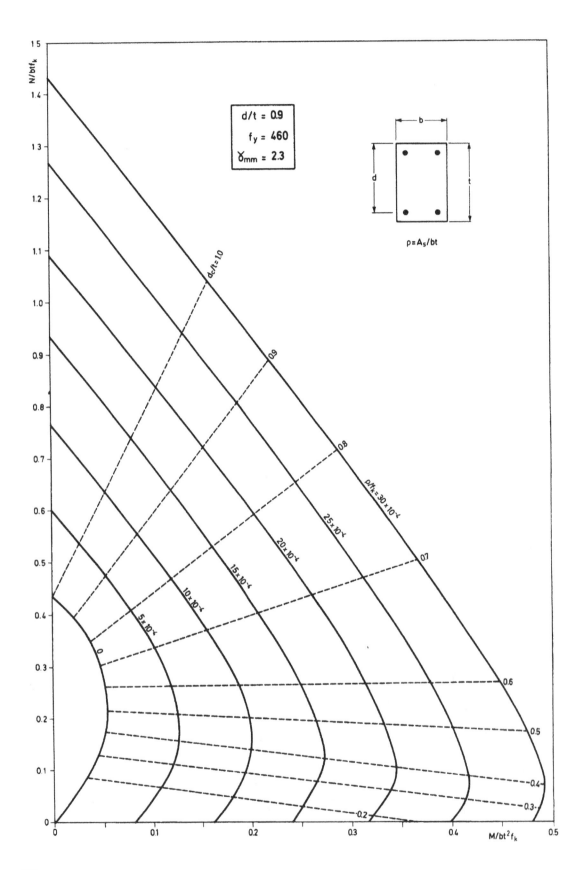

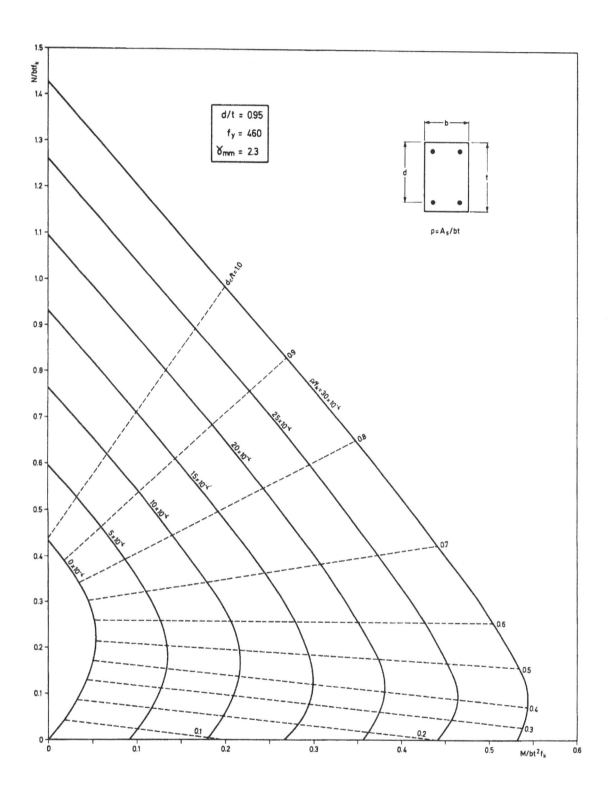

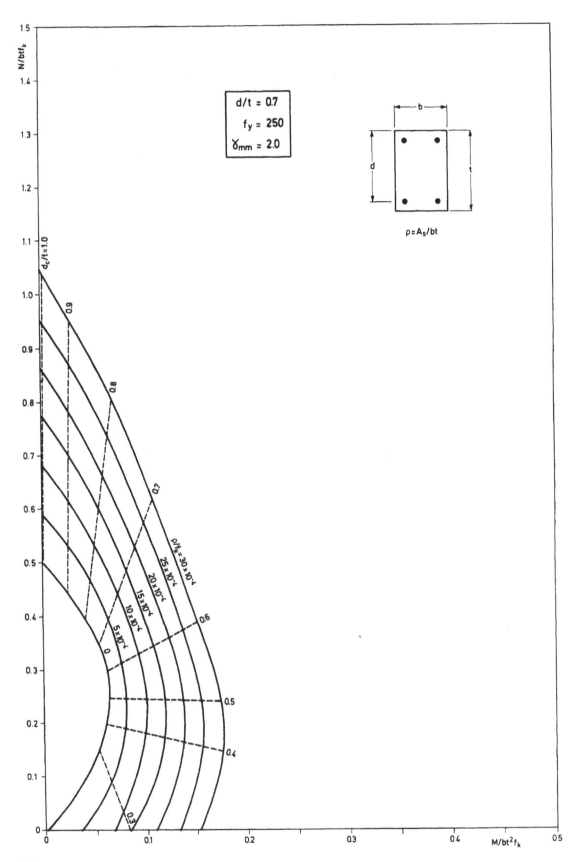

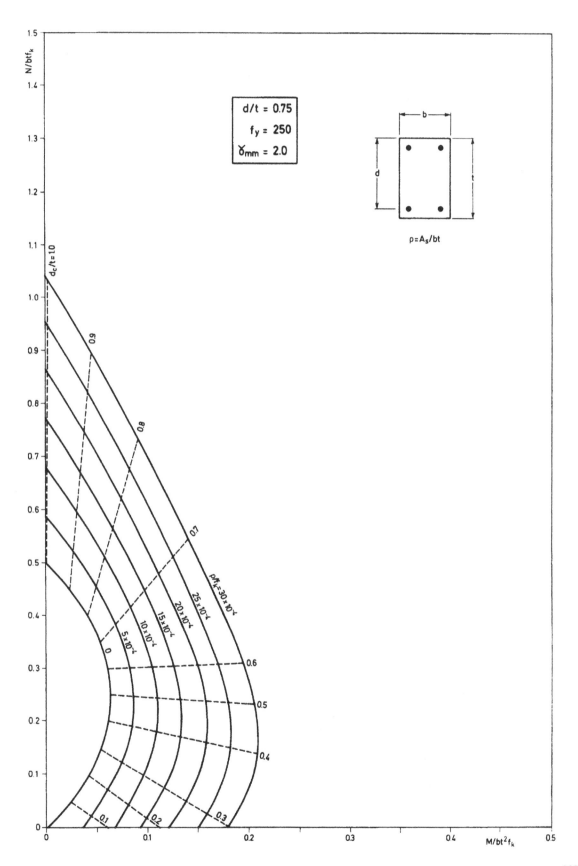

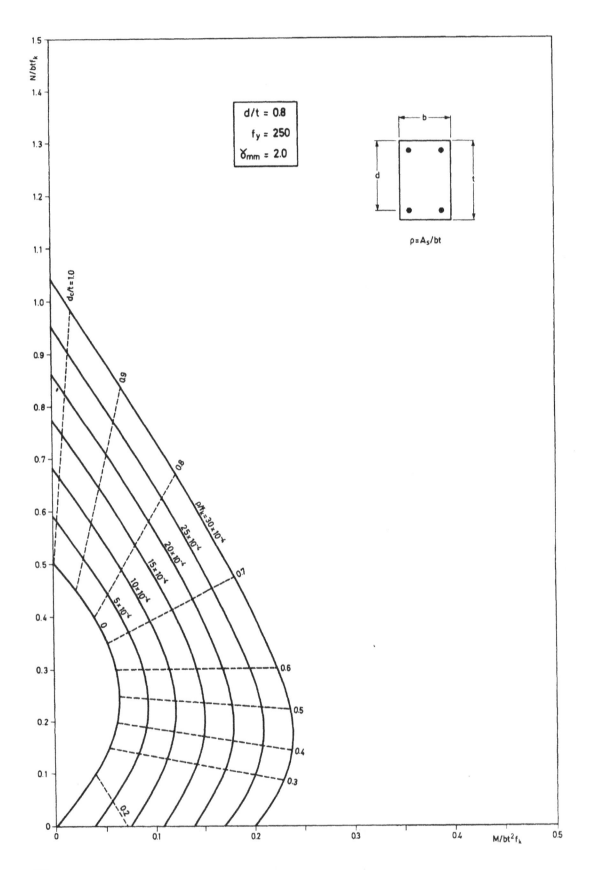

186

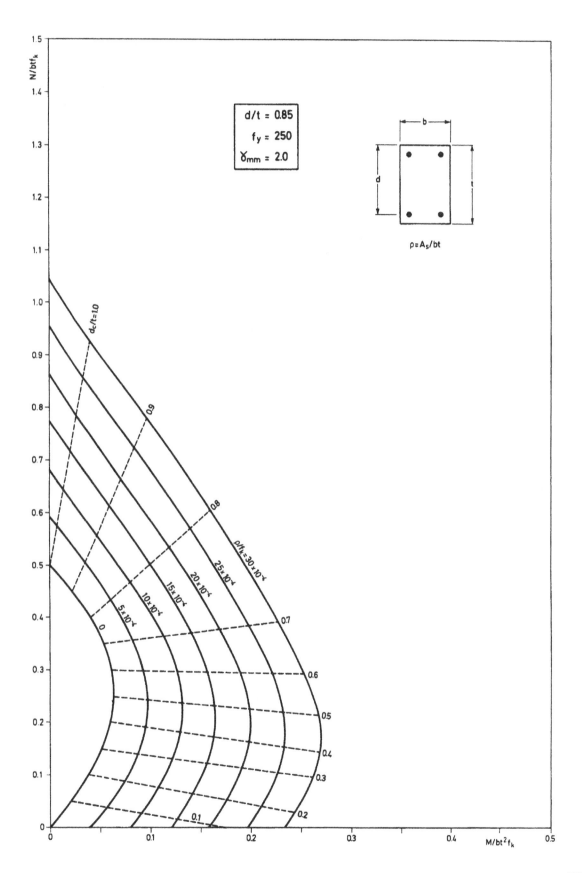

187

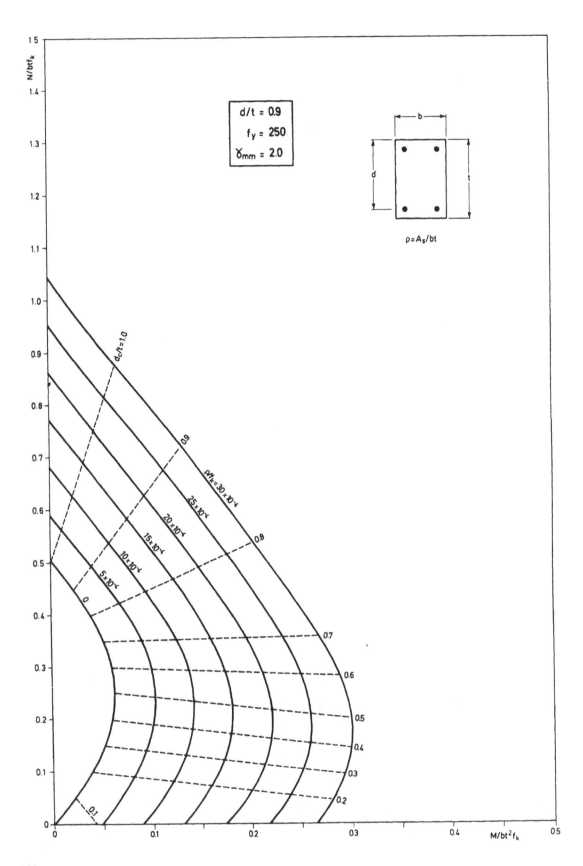

188

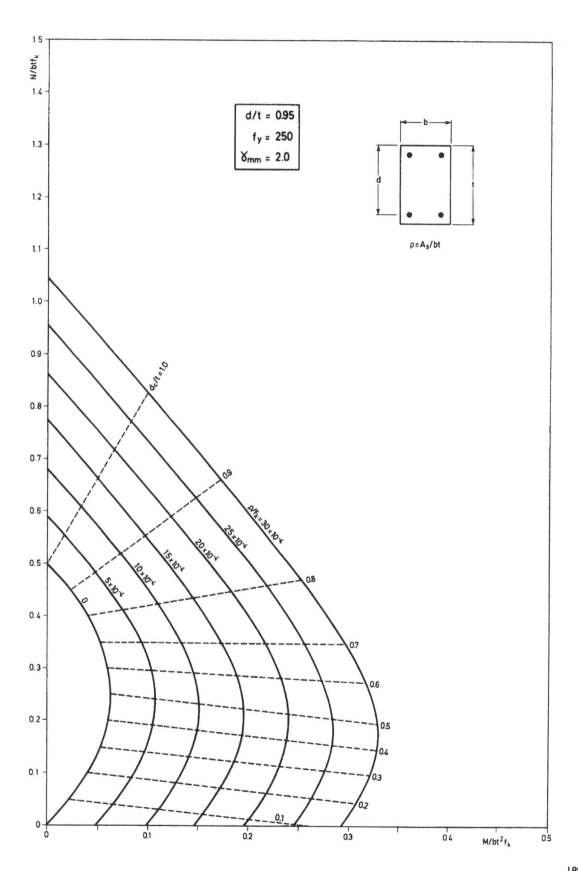

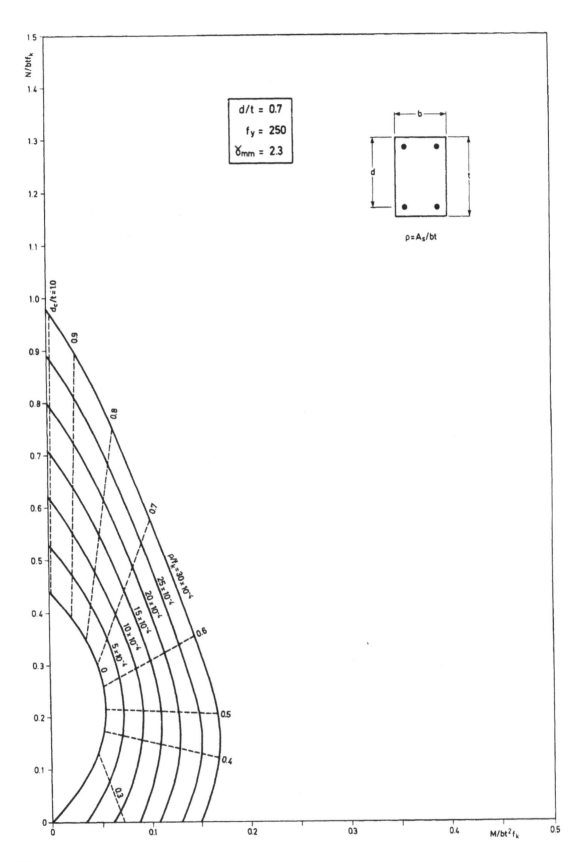

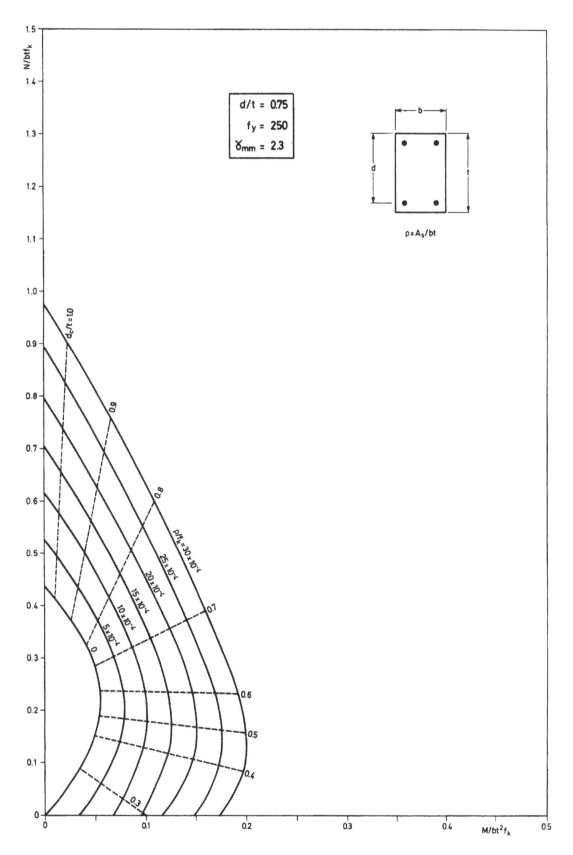

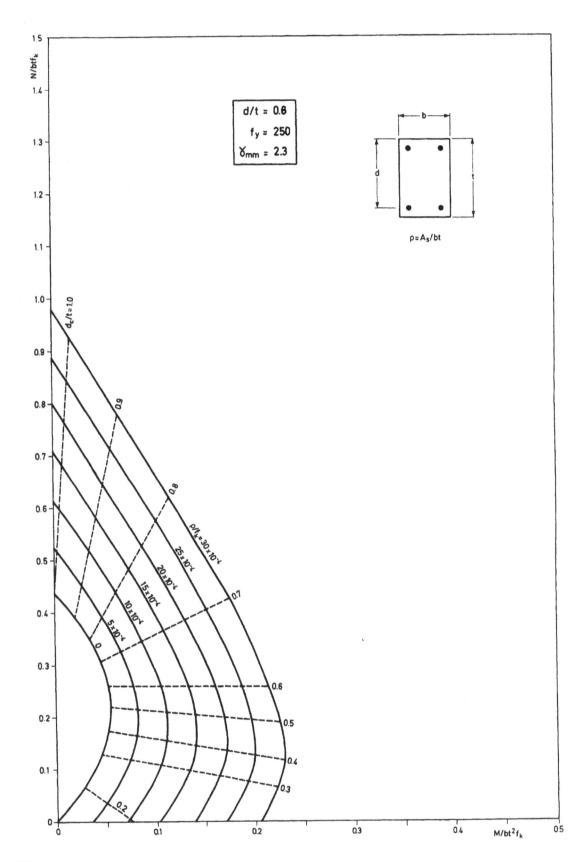

192

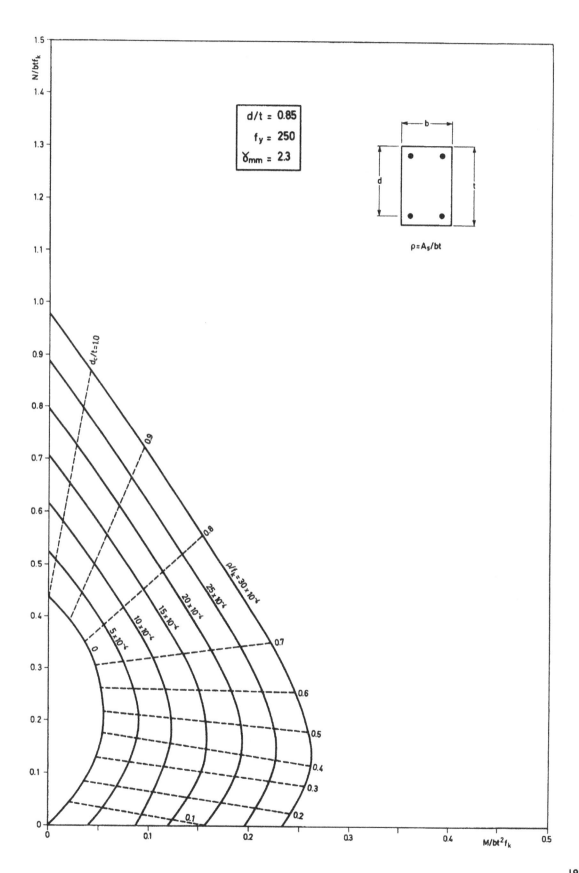

193

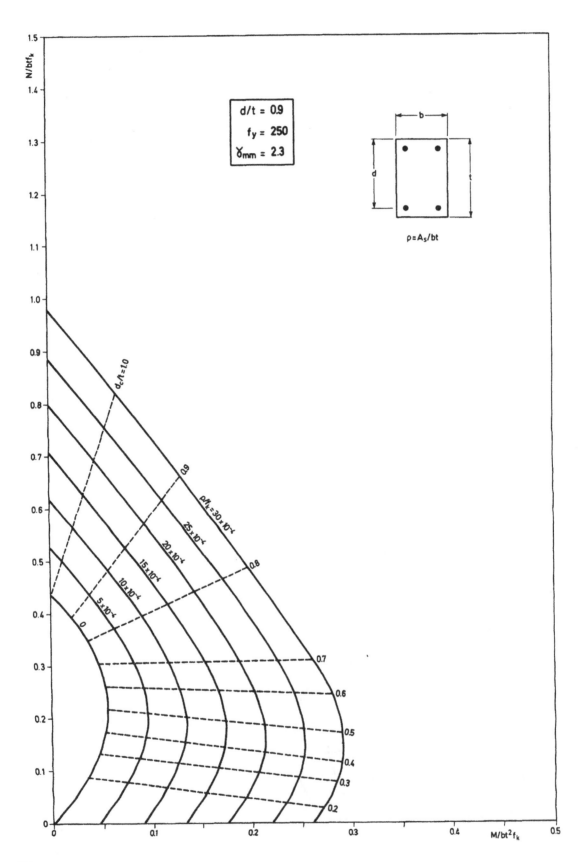

194

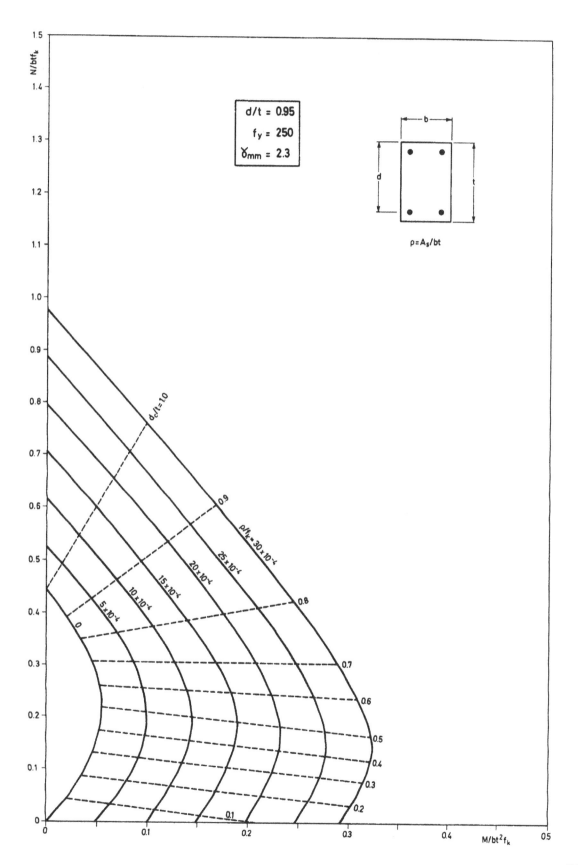

Chapter 10
Concentrated loads

10.1 Background

The concept of allowing increased stresses under localized situations is not new. CP 111:1970 allowed stresses to exceed the basic permissible stresses by up to 50% provided they were of a *purely local nature, as at girder bearings, etc.* There is little published information to support the use of this single figure and, in fact, evidence from unpublished work casts doubt on the validity of the 50% allowance when considering loads applied to the ends of a wall and certain edge conditions.

Experimental evidence shows that compressive failure in concrete and brick masonry is preceded and accompanied by the development of vertical splitting cracks in the units. In concrete the stiff aggregate particles attract load and, because of the random arrangement of the particles, lateral tensile stresses are set up in the cementing matrix. In masonry there is the added complication of the weaker mortar joints which tend *to squeeze out* and thereby add *to the lateral stresses* inflicted on the cementing matrix.

The case for allowing increased stresses under a localized bearing may be illustrated by considering the simple case of a bearing in the centre of a square block of concrete as shown in Figure 10.1. The general form of failure as shown in sketch (b) is to split the block into four parts. By considering section A–A in sketch (c) it is seen that the lateral compressive stresses acting on the cone must be several times the tensile strength of the block since the area to be split is much larger than that of the cone. A significant triaxial stress field develops in the concrete under the localized bearing area enhancing its resistance to vertical stress.

In the case of an edge loading the form of failure will differ depending on whether the element bearing on the edge of the member is able to move laterally as well as downwards or is constructed only to move downwards. The former gives rise to a *sliding wedge* type of failure (Figure 10.2(a)) whereas the latter tends to have a *slip circle* failure (Figure 10.2(b)). Research carried out on the two types of failure, although limited, indicates that the *slip circle* failure will give higher results than the *sliding*

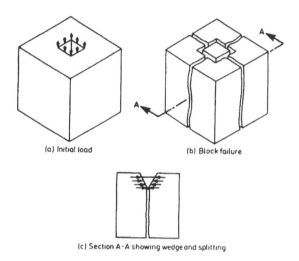

(a) Initial load   (b) Block failure

(c) Section A–A showing wedge and splitting

Figure 10.1 Bearing failure in concrete: (a) initial load (b) block failure (c) section A–A showing wedge and splitting

wedge. More extensive research conducted on concrete shows that at least a 25% enhancement is feasible with the sliding wedge type of failure. *Since it is difficult to predict the exact type of failure it is reasonable to adopt the 25% enhancement for a continuous edge loading.* More recent research on masonry suggest this may not be obtained under certain conditions (see 10.9).

In the case of a part edge loading, as shown in Figure 10.3, an effective wedge forms with rupture planes along the sides of the area. Since the total rupture length is longer than the loaded length, it is reasonable to expect some additional enhancement over the normal edge condition. The additional enhancement will vary according to the ratio of the loaded length to the rupture length. However, by controlling the loaded length a standard enhancement may be specified. This has been done in the Code by allowing for a 50% enhancement for edge loading conditions where the loaded length is

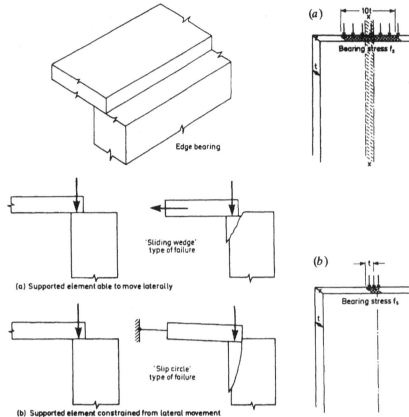

Figure 10.2 Modes of failure in edge bearing: (a) supported element able to move laterally (b) supported element constrained from lateral movement

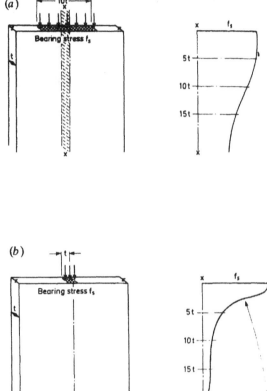

Figure 10.4 (a) Wide bearing width. Intensity of vertical stress hardly changes in this area and therefore if enhanced bearing stresses permitted failure could be expected in main body of wall. (b) Narrow bearing width. Intensity of stress falls rapidly and thus enhanced bearing stress acceptable.

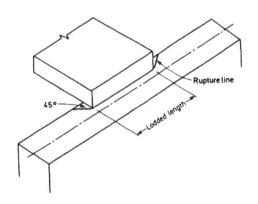

Figure 10.3 Concentrated edge loading

restricted to eight times the width of the bearing. When the load is applied over the full width of the supporting member some enhancement in stress may again be expected but since the load has a smaller area to dissipate into than in the case of the edge loading, some restriction in the extent of loaded area must be applied. If the load

is applied over a length of wall equal to, say, ten times the thickness of the wall then it is probable, most engineers would agree, that an enhancement in stress should not be permitted. This may be seen by considering Figure 10.4(a) which illustrates that the reduction in stress as the load dissipates downwards into the wall is not too rapid, and a normal type of wall failure can initiate in the region of t to $3t$ down from the load. If on the other hand the loaded length of wall is equal to, say, its thickness then the reduction in vertical stress is much more rapid and enhanced bearing stresses are acceptable as wall failure is unlikely to occur except at a very high enhanced stress. Test evidence is limited but an enhancement of 50% would appear to be reasonable in cases where the loaded length is not greater than four times the wall width, since under this condition the rupture length is some 50% longer than the loaded length. By interpolation a 25% enhancement in bearing stress may be expected where the bearing length is about six times the wall width. BS 5628 refers to $6x$ but in a pending amendment this is to be altered to $6t$. Loads on the end of a wall may be dealt with in a similar way but in this

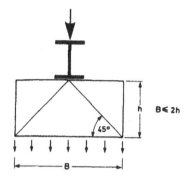

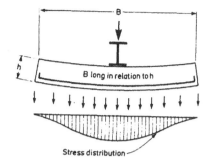

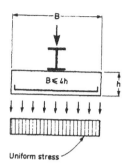

Figure 10.5 Rigid bearing pad

Figure 10.6 Stress distribution under reinforced spreader

instance an enhancement to 1.5 f_k would apply to a bearing length of $2t$ and 1.25 f_k to a length of $3t$.

The enhanced stresses permissible at localized bearing situations allow a heavy beam load to be carried. It is still necessary, however, to ensure that the section of wall is not overstressed at a lower level where the strength is controlled by the capacity reduction factor β. This situation is covered in the Code in Clause 34(b) and is dealt with in detail later.

10.2 Code recommendations

Clause 34 indicates that increased local stresses may be permitted beneath the bearing of a concentrated load of a purely local nature, such as beams, columns, lintels, and so on, provided that the element applying the load is sensibly rigid, or that a suitable spreader is introduced. No guidance is given as to what is considered to be a sensibly rigid beam or suitable spreader and this must therefore be left to the engineer's judgement. Concrete beams or concrete encased steel beams will generally be regarded as rigid so that the stresses beneath the beam can be regarded as being uniformly distributed. In addition, any concrete bearing pad whose length is not greater than, say, twice its height can also be expected to impart a uniform stress to the masonry since the applied load can be distributed at an angle of 45° through it (Figure 10.5)*.

Reinforced spreader beams of a shallower section than shown in Figure 10.5 can be used but some care needs to be taken since the stress distributed beneath the bearing may not be exactly uniform (Figure 10.6), the reason being that the beam theoretically does not act until it has deflected so that a non-uniform stress distribution may occur. It is difficult to be precise on this point but a reinforced spreader, whose length is not greater than, say, four times its height, is likely to impart a reasonably uniform stress to the masonry (Figure 10.6). The reference in the Code that *concentrated loads may be assumed to be uniformly distributed over the area of the bearing* (except bearing type 3) is reasonable if the above guidelines are followed but consideration should be given to the development of non-uniform stresses where the spreader is shallow.

10.3 Permitted bearing stress

The enhanced stress permitted in BS 5628 under beam bearings depends on the position and size of the bearing with respect to the wall and basically follows the technical reasoning explained at the beginning of the chapter. In effect the local load combined with stresses due to other loads should be checked directly beneath the bearing and at a distance below the bearing.

10.4 Directly beneath bearing

The stress directly beneath the bearing should be controlled so that the stresses do not exceed 1.25 f_k/γ_m where the bearing is of a type within those shown in Figure 10.7 (bearing type 1, Figure 4(a) BS 5628) or 1.5 f_k/γ_m where the bearing is of a type within those shown in Figure 10.8 (bearing type 2, Figure 4(b) BS 5628). The conditions for enhanced stresses as shown in Figures 10.7 and 10.8 are also repeated in tabular form in Figure 10.9 which may be easier to follow in certain cases.

10.5 End bearing

In the particular case of beams spanning in the plane of the wall some care needs to be exercised when the bearing is longer than three times the wall thickness. Where the bearing is longer than $3t$ then no enhancement is permitted and the bearing stress must be limited to f_k. However in cases where the bearing is long there will be a tendency for non-uniform stress distribution to occur as indicated in Figure 10.10. This is due to deflection of the lintel and rotation at the bearing. Again it is difficult to be precise but it is reasonable to consider a uniform stress in cases where the bearing length of a reinforced concrete beam or lintel is not greatly in excess of three times its depth (Figure 10.10).

---

*See Solution 3, Chapter 7

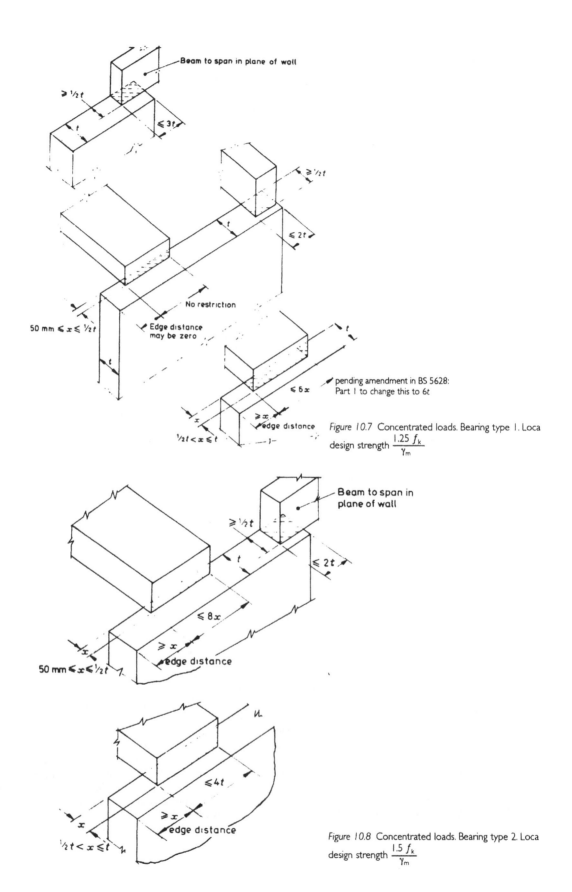

Beam to span in plane of wall

$\geq \frac{1}{2}t$

t

$\leq 3t$

$\geq \frac{1}{2}t$

t

$\leq 2t$

No restriction

50 mm $\leq x \leq \frac{1}{2}t$

Edge distance may be zero

t

t

$\leq 6x$

pending amendment in BS 5628: Part 1 to change this to 6t

x

$\geq x$

edge distance

$\frac{1}{2}t < x \leq t$

Figure 10.7 Concentrated loads. Bearing type 1. Loca design strength $\dfrac{1.25 \, f_k}{\gamma_m}$

Beam to span in plane of wall

$\geq \frac{1}{2}t$

t

$\leq 2t$

$\leq 8x$

$\geq x$

edge distance

x

50 mm $\leq x \leq \frac{1}{2}t$

$\leq 4t$

$\geq x$

edge distance

x

$\frac{1}{2}t < x \leq t$

Figure 10.8 Concentrated loads. Bearing type 2. Loca design strength $\dfrac{1.5 \, f_k}{\gamma_m}$

200

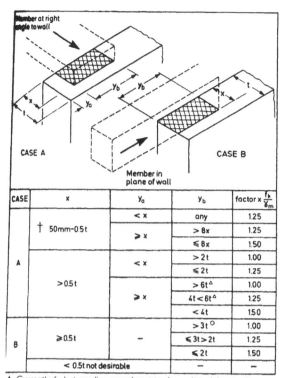

CASE	x	y_a	y_b	factor x $\frac{f_k}{\gamma_m}$
A	† 50mm-0.5t	< x	any	1.25
		⩾ x	> 8x	1.25
			⩽ 8x	1.50
	>0.5t	< x	> 2t	1.00
			⩽ 2t	1.25
		⩾ x	> 6t △	1.00
			4t < 6t △	1.25
			< 4t	1.50
B	⩾0.5t	–	> 3t ○	1.00
			⩽ 3t > 2t	1.25
			⩽ 2t	1.50
	< 0.5t not desirable	–	–	

△ Currently 6x but pending amendment to change it to 6t(see text)
† BS 5628: Part 3 recommends a min bearing of 75 mm (90 mm in the case of concrete floors, but can be reduced at the discretion of the engineer.)

Figure 10.9 Conditions for enhanced stresses

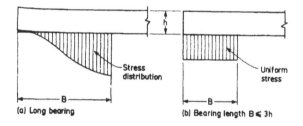

(a) Long bearing

(b) Bearing length B ⩽ 3h

Figure 10.10 Stress distribution at beam ends: (a) long bearing (b) bearing length < 3h

10.6 Stresses at a distance below bearing

Although enhanced stresses are permitted directly beneath a localized bearing, such as beam bearings or spreaders, it is necessary to check the stresses at a lower level to ensure that the lower part of the wall is not overstressed. The stresses need to be checked at a distance 0.4 h below the bearing, which in effect is at the top of the central fifth zone of the wall where the additional eccentricity is assumed to be at a maximum (see Appendix B of BS 5628). Since the capacity reduction factor β is determined with respect to this

position, it follows that the stresses resulting from the local load with any stresses due to other loads need to be controlled so that they do not exceed:

$$\frac{\beta f_k}{\gamma_m}$$

where

f_k is the characteristic strength of the masonry
γ_m is the partial safety factor for the material
β is the capacity reduction factor

For checking the stresses at a lower level the localized bearing stresses may be considered to be distributed at an angle of 45° from the edges of the bearing. Thus, in the situation as shown in Figure 10.11, the reaction of the beam of width B is spread over a length of wall equal to 0.8 h + B. The stresses resulting from this spread of load are then added to any stress also acting along that length of wall, for example, as may be applied from loads above the bearing level. The combined stresses must be checked to ensure that they do not exceed $\beta f_k/\gamma_m$, ie the general stress capacity of the wall.

10.7 End spreader in plane of wall

In the particular case of a spreader beam located at the end of a wall and spanning in its plane (Figure 10.12) the distribution of stress under the spreader may be derived from an acceptable elastic theory. Providing the stress is derived by some elastic analysis the maximum stress which occurs under the loading edge of the spreader as a result of the load applied directly to the spreader combined with stresses due to other loads are allowed to reach a value of 2 f_k/γ_m.

As with the general bearing conditions mentioned previously, the stresses must be checked at a distance of 0.4 h below the bearing, where the stress must not exceed $\beta f_k/\gamma_m$. For this type of spreader, however, the load from the supported beam should be taken as being distributed at an angle of 45° from the edge of the beam (Figure 10.5) rather than from the edge of the spreader as occurs in Figure 10.13.

10.8 Design of spreader

The Code gives no details on design of the spreader and determining the distribution of the bearing stresses. The stress distribution, providing the elasticity of the beam and masonry are known, can be determined by finite element methods or by use of analysis of beams on elastic foundations. The latter method would give rise to the indicative shape of the stress diagram as shown in Figure 10.13 which is explained fully by Timoshenko[10.1].

The design solution for long spreader beams is rather complex and as a degree of uncertainty may occur even after a stress distribution has been derived, it has been decided not to go into the detailed method of design for such spreaders, which in any case may be limited and not very practical. However, the simple case of short spreaders is outlined below since they are often found

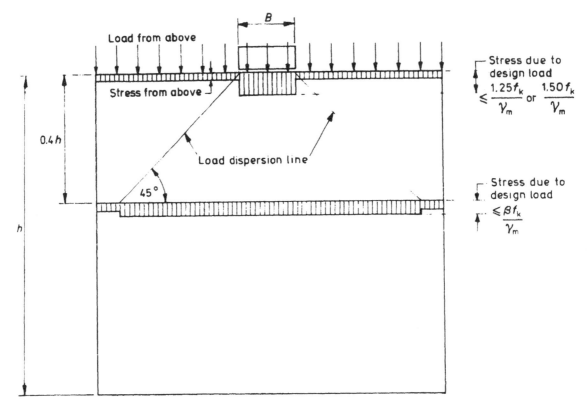

Stress due to design load to be compared with design strength as indicated above.

Figure 10.11 Load distribution for bearing types 1 and 2

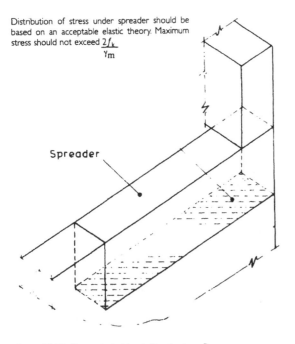

Distribution of stress under spreader should be based on an acceptable elastic theory. Maximum stress should not exceed $\dfrac{2f_k}{\gamma_m}$

Spreader

Figure 10.12 Concentrated loads. Bearing type 3

but, as shown, may not always reduce the bearing condition (Figure 10.15).

With concentrated loads applied to the extreme end of a wall by a stiff beam or perhaps a column (Figure 10.14) the stress may be checked by assuming a rectangular stress block and comparing the values with the enhanced stress permitted for the appropriate bearing condition (type 1 or 2). In certain areas the stress may reach a value whereby it exceeds the enhanced value permitted for localized conditions, in which case a spreader or stronger unit will be required to reduce the stress to an acceptable level. In some instances it may be possible to reduce the stress to an acceptable level by introducing a short spreader so that the load may be distributed linearly. Consider, for example, a situation where a column rests on the end of a wall (Figure 10.16) such that the direct stress assuming a rectangular stress block, produces a stress of $1.5\dfrac{f_k}{\gamma_m}$ from which:

$$P = \frac{1.5 f_k b t}{\gamma_m}$$

If a spreader were introduced of length $2b$ (Figure 10.17) and the stresses determined as in the typical case of a

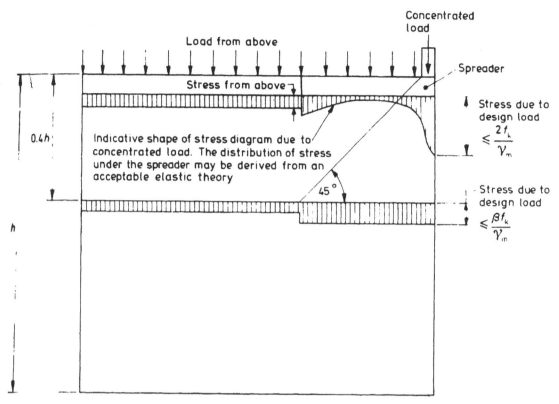

Stress due to design load to be compared with design strength as indicated above.

Figure 10.13 Load distribution for bearing type 3

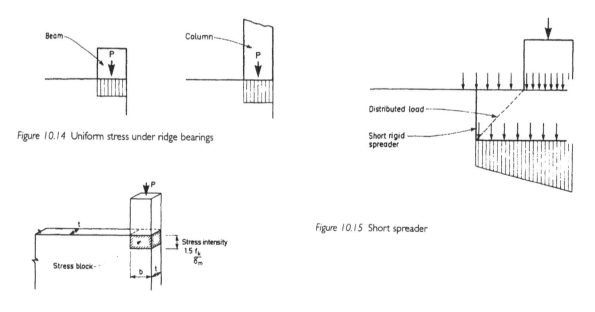

Figure 10.14 Uniform stress under ridge bearings

Figure 10.15 Short spreader

Figure 10.16

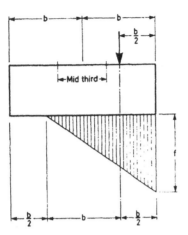

Figure 10.17

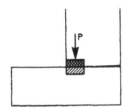

Figure 10.18

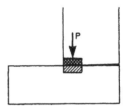

Figure 10.18

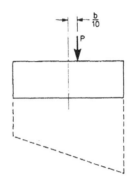

Figure 10.19

foundation design, the eccentricity of load is outside the mid-third of the section, thus giving rise to a triangular stress distribution of which the maximum intensity of stress (ignoring adhesion between spreader and mortar):

$$f = \frac{2P}{3\left(\dfrac{b}{2}\right)t}$$

which gives:

$$P = \frac{3\,b\,t\,f}{4}$$

Now, since the load P must balance:

$$\frac{3\,b\,t\,f}{4} = \frac{1.5\,f_k\,b\,t}{\gamma_m}$$

hence $f = \dfrac{2\,f_k}{\gamma_m}$

which means the spreader designed in this way does not appear to ease the situation. $2\,f_k/\gamma_m$ is the maximum permitted when determined by elastic analysis. This approach uses the typical assumption of the load P being applied axially. In practice, however, and using the limit state concept for the stress block in the column and spreader, it is possible to consider that the load in ultimate terms could be transmitted to the spreader by a

204

stress block concentrated to the rear of the column as shown in Figure 10.18. However, it is important to realize that this condition can only occur where the column is held in position at the top and prevented from rotating, whereas with beams rotation will generally occur, and the load will not be applied in this manner.

The width of the stress block could be determined and the stress under the spreader checked with a revised eccentricity of load. Assuming, for example, that the force P could be carried by the concrete on a stress block of width $b/5$ this would give an eccentricity of $e = b/10$ (Figure 10.19).

Taking

$$P = \frac{1.5\,f_k\,b\,t}{\gamma_m}$$

$$A = 2\,bt$$

$$Z = \frac{(2b)^2\,t}{6}$$

results in a stress diagram with a lower value of $0.525\,f_k/\gamma_m$ and upper value of $0.975\,f_k/\gamma_m$ (Figure 10.20).

Figure 10.20

10.9 Other methods

The current method of dealing with concentrated loads as given BS 5628: Part 1, and the general background to its development is given above in 10.1 to 10.7. Some thoughts on the effect of spreaders is also given in 10.8. The approach currently given in BS 5628 is essentially based on work on concrete and since then a considerable amount of research has been carried out on actual masonry[10.3,10.4,10.5]. The last of these reference 10.5 suggests enhancement values of between 1.0 to 1.60 but which do not fully align with that in BS 5628. In some instances it gives a lower enhancement. The newer research takes into account not only the loaded area but also the effective area of enhancement. Although such additional research has been conducted the only proposed amendment to BS 5628: Part 1 is to amend the value of $6x$ in Figure 5.9(a) to $6t$. However, the reader may wish to review this more recent work, particularly reference 10.5, which was submitted as comments by the United Kingdom on the draft Eurocode EC6 but which is still being discussed.

10.10 References

10.1 TIMESHENKO. *Strength of materials* – Part II – Chapter 1. Van Nostrand Reinhold Co, New York, 1958. pp 1–25.

10.2 WILLIAMS A. *The bearing capacity of concrete loaded over a limited area*. Cement & Concrete Association, 1–25 London, 1979. Publication No 42.526. pp 70.

10.3 ARORA S K. *Review of walls under concentrated load*. Elsevier Applied Science, London, 1988. Proceedings of the eighth International Brick and Block Masonry Conference, Vol 1. Trinity College, Dublin. 19–21 September 1988. pp446–457.

10.4 PAGE A W, HENDRY A W. Design rules for concentrated loads on masonry. ISE, London, 1988. *The Structural Engineer*, Vol 66, No. 17, 6 September 1988, 273–281 pp. and Correspondence: Vol 67, No. 5, 7 March 1989, pp88.

10.5 ARORA S K. Design of masonry walls subjected to concentrated vertical loads. BRE, Garston, 1992. BRE Information Paper 10/92. 4pp.

Chapter 11
Composite action

11.1 Introduction

There are several papers dealing with the subject of composite action of brick masonry walls supported on both reinforced concrete and steel beams and references 11.1 to 11.3 are typical examples. A paper by Levy and Spira[11.4] gives experimental results together with a method of analysis[11.5] for concrete masonry walls, both with and without openings, when strengthened by reinforced concrete elements. A later paper by Davies and Ahmed[11.8] gives a graphical solution of composite wall beams. A detailed review of masonry wall in composite action has been presented by Hendry[11.9], which also makes reference to BS 5628.

Unfortunately, no method of analysis has been fully developed in this country to an extent whereby it can be given total authority and included in a British Standard Code of Practice. Any design for composite action needs not only to assess shear and bending in the supporting beam but account will need to be taken of the serviceability or movement stresses induced in the masonry. Analysis is thus complex and in practice seldom undertaken. This chapter sets out, therefore, only to review the principles of composite construction, to outline the basic recommendations made in the documents referred to and to make a few additional comments for consideration. It is important to emphasize that the published work must be consulted prior to any composite design so that the range of walls that were tested and the limitations imposed on any design recommendations made are clearly understood.

11.2 Treatment of composite action in published work

The traditional method of designing a beam to support a loading from a triangular section of masonry, as shown in Figure 11.1, is a conservative approach, at least when it is only self weight of the wall that is being considered. The assumption that any superimposed load that is applied above the apex of the triangle Figure 11.2) is distributed by some arch action to the supports, thus

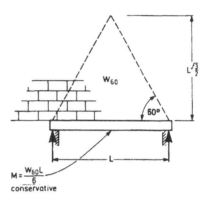

Figure 11.1 Commonly assumed design loading from non-loadbearing walls

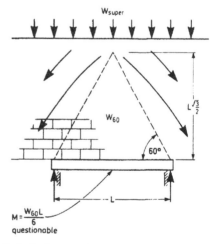

Figure 11.2 Design loading sometimes considered for loadbearing wall

leaving the moment applied to the beam unaltered, is considered by many to be questionable. This method, as indicated by Wood[11.1] when applied to a 225 mm brick

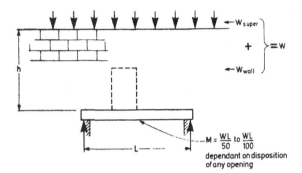

Figure 11.3 Equivalent bending moment method (Woods[11.1])

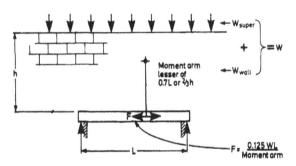

Figure 11.4 Moment arm method (Woods[11.1])

wall of span 3.25 m and height 3.25 m gives bending moments based on the total load as follows:

Superimposed load: 0   10 tonne   20 tonne   30 tonne

Bending moment $\dfrac{WL}{12}$ $\dfrac{WL}{42}$ $\dfrac{WL}{72}$ $\dfrac{WL}{102}$

From test results it was concluded that at low loads this approach over-predicts the bending moment applied to the beam but insufficient data exists to give confidence to the moment predicted under higher loads. Since the simple approach appears unreliable, Woods[11.1] proposes two methods of design. The first (Figure 11.3) is to adopt an equivalent bending moment on the basic span of either $\dfrac{WL}{50}$ where openings occur towards the beam supports, or $\dfrac{WL}{100}$ where there are no openings or where openings occur at mid-span. The second method, Figure 11.4, which is only applicable to walls without openings, is to consider the wall and beam as a composite deep beam which is taken to have an internal moment arm of either $0.7L$ or $\tfrac{2}{3}h$, whichever is the lesser. (L is the effective span, h is the height of the wall.) Sufficient reinforcement needs to be provided to balance the moment produced from considering the total loading applied to a simply supported beam. In both methods the design proposals are only applicable for cases where the height of the wall is not less than $0.6L$, where other limiting conditions are also imposed.

The work by Colbourne[11.2] deals with the problem in a more analytical way by deriving equilibrium equations which were used in a computer programme to predict stresses in the wall and supporting beam. Further work on composite action between brick walls and their supporting beams was carried out by Burhouse[11.3]. This work compares experimental values of the internal moment arm method with the theoretical values calculated in accordance with the method proposed by Colbourne[11.2]. The experimental results for the limited cases investigated appear to give good agreement with moment arm predicted by Colbourne. The results indicate that the method proposed by Wood overestimates the moment arm, although it is fair to say that Wood did propose limited steel stresses. Davies and Ahmed used a modified finite element programme in conjunction with some practical cases to develop an approximate method for analysing composite wall beams[11.7]. Further work by the same authors has simplified the approximate solution to a simple graphical approach[11.8] so that it may be used as the basis for a design procedure.

11.3 The principle of composite action

The experimental data indicates that when a wall (subject to self weight and any applied load at its top) is supported by a beam, some form of composite action takes place. This composite action tends to reduce the distribution of load transmitted to the beam but in so doing increases the stresses in the wall at the support, as shown in Figure 11.5. In many of the experimental cases it was the increased stresses in the masonry near the supports that resulted in failure. The beam itself would take some of the moment but would tend to act more as a tie resisting thrust from the arch.

An unpublished paper on the design of composite action between a wall and its supporting member[11.6] suggests that it could be analogous with that of a beam on an elastic foundation. This report indicated that the relative stiffness parameters of the wall and beam could be an important factor in determining the stress distribution in the wall and the forces applied to the beam.

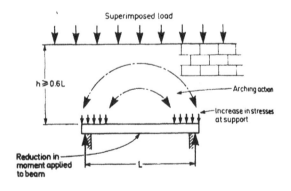

Figure 11.5 Composite action of wall and beam

208

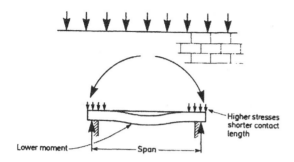

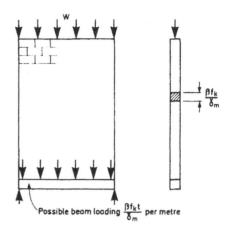

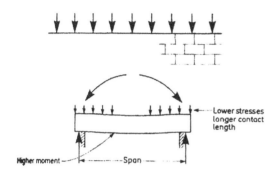

Figure 11.6 Slender beam condition

Figure 11.7 Stiff beam condition

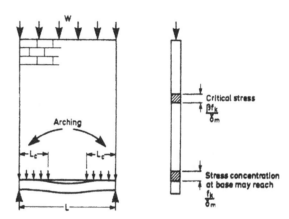

Figure 11.8 Full design stress reached in slender wall

Figure 11.9 Arching in slender wall

The influence of the stiffness of the beam is illustrated in the following Figures. In Figure 11.6 where the beam is slender (ie greater deflection), then the vertical stresses from the wall would be carried over a short length of wall. This would result in higher bearing stresses than would result with a stiff beam, shown in Figure 11.7, since the contact length would be greater. Conversely, since more load is applied towards mid-span with the stiffer beam the bending moment will be higher than for the more slender beam.

The difficulty, particularly when load is applied at beam level, is to ensure that the beam has sufficient capacity to carry the moments but is stiff enough to prevent the masonry from being over-stressed in the region of the beam supports. It is not proposed to go into any further detail in this book with the relative stiffness approach since this is still under consideration and could eventually form the basis of Code recommendations for composite construction of masonry walls and their supporting beams. However, the mathematical solution given by Spira and Levy[11.4,11.5] and Davies and Ahmed[11.7,11.8] and Hendry[11.9] may be of use in estimating the forces and contact stress acting on the wall.

11.4 Limitations of design

There are several limitations imposed upon the published design recommendations and readers should fully understand these before attempting composite design. A

few of the most common limiting parameters determined from experimental results are as follows:

(1) Nearly all the design recommendations indicate that to achieve composite action the height of the wall should be not less than 0.6 of the span ($h > 0.6L$);

(2) Although composite action can be used to reduce the reinforcement in the supporting beams, the loading on the wall in such circumstances should be less than the design load permitted by the structural masonry codes.

In the case of a slender wall (Figure 11.8) the critical stresses within the wall are controlled by the slenderness factor (β in BS 5628). A similar concept of indicating that the beam should carry a loading equal to $\frac{\beta f_k t}{\gamma_m}$ per metre on a simple span could be considered, but since the purpose of the factor is to prevent critical stresses being reached within the mid-height of the wall it is reasonable to suggest that some degree of arch (and hence reduced contact length) is permissible since the stress at the base of the wall could be allowed to reach a maximum of $\frac{f_k}{\gamma_m}$ (Figure 11.9), with no concentrated load enhancement.

From Figure 11.9 it may be seen that the contact length L_c to prevent stresses occurring in excess of those permissible is found from the following:

$$L_c \geq \frac{L \ \beta \ \frac{f_k}{\gamma_m} \ t}{2 \frac{f_k}{\gamma_m} t} \qquad (1)$$

$$L_c \geq \frac{L\beta}{2}$$

Since β will seldom be less than 0.5, then L_c 0.25L. In this case no concentrated enhancement would be permitted (see Chapter 10) and which makes the use of f_k/γ_m correct.

Taking the contact length L_c as so determined suggests a moment applied to the beam:

$$M = \frac{\beta^2 f_k \ t \ L^2}{\gamma_m \ 8} \qquad (2)$$

Now since the design load $W = \frac{\beta f_k \ t \ L}{\gamma_m}$

$$M = \frac{WL}{\left(\frac{8}{\beta}\right)} \qquad (3)$$

This suggests a bending moment of $\frac{WL}{8}$ for a fully loaded short wall and not less than typically $\frac{WL}{16}$ for a fully loaded slender wall (see Figure 11.10).

Should the loading on the wall in either case, or in a short wall, be less than the maximum design capacity then the following moment may be similarly deduced assuming again no concentrated load enhancement:

$$M = \frac{WL}{\frac{8}{\beta} \frac{W}{W_u}} \qquad (4)$$

the term $\frac{W}{W_u}$ simply indicates the ratio between the design load applied to the ultimate capacity.

The predicted bending moments in relation to the ratio $\frac{W}{W_u}$ for various capacity reduction factors (β) are plotted in Figure 11.10. From this Figure it can be seen that the actual design load must be considerably less than the ultimate capacity of the wall when designing the beam to bending moments between $\frac{WL}{50}$ to $\frac{WL}{100}$. Alternatively it may be stated that composite action (reduction in beam moment) can only occur when the wall is lightly loaded.

The bending moment induced while constructing the wall must be considered. There could be a situation where the imposed load is so small that the moment from composite action is less than the moment required to support the masonry during construction.

The work by Wood[11.1] showed that the stresses induced in the beam during construction were equivalent to the beam carrying a bending moment of between $\frac{WL}{20}$ and $\frac{WL}{30}$. Thus after the walling had reached a height equal to about 0.7 of the span, the stresses due to the additional courses were similar to applying a superimposed load to the completed wall. At low loads this gave equivalent moments in the order of $\frac{WL}{300}$ or less. Assuming that the results by Wood[11.1] are typical, then Figure 11.11 could be considered as indicating the moments that might be expected to result from the self weight of the wall. The dotted line represents the equivalent moment considering the normal triangular load as in Figure 11.1.

11.5 References

11.1 WOOD, R H. Studies in composite construction. Part 1: The composite action of brick panel walls supported on reinforced concrete beams. *National Building Studies Research Paper No 13*, HMSO, London, 1952.

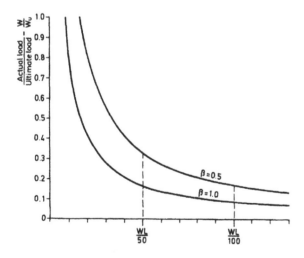

Figure 11.10 Relationship between actual load and ultimate load capacity of wall toi bending moment predicted by equation (4)

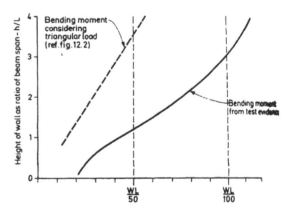

Figure 11.11 Possible bending moment in beam due to self weight of wall (Woods11.1)

11.2 COLBOURNE, J R. Studies in composite construction: an elastic analysis of wall beam structures. *Building Research Station Current Paper 15/69*. BRS, May 1969.

11.3 BURHOUSE, P. Composite action between brick panel walls and their supporting beams. *Building Research Station Current Paper 2170*. BRS, January 1970.

11.4 LEVY, M. AND SPIRA, A. Experimental study of masonry walls strengthened by reinforced concrete elements. International Association for Bridge and Structural Engineering, Publications, 1975. Vol 35–ii. pp 113–132.

11.5 LEVY, M, AND SPIRA, A. Analysis of composite walls with and without openings. International Association for Bridge and Structural Engineering, Zurich, 1973. Mémoires Abhandlugen Publications 33–1.

11.6 BRITISH STANDARDS INSTITUTION. Technical Committee. Unpublished work dealing with composite construction between masonry and supporting beams.

11.7 DAVIES, S R, AND AHMED. A E. An approximate method for analysing composite wall beams. *Proceedings of the British Ceramic Society*. BCS, London 1978. No 27. pp 305–321.

11.8 DAVIES. S R. AND AHMED. A E. A graphical solution of composite wall beams. *The International Journal of Masonry Construction*, No 1, 1980. United Trade Press Ltd. London. pp. 29–33.

11.9 HENDRY, A.W. *Masonry walls in composite action*. Macmillan Education Ltd, Basingstoke, 1990. Structural Masonry pp 218–245.

11.6 Bibliography

WOOD, R H, AND SIMMS, L G. A tentative design method for composite action of heavily loaded brick panel walls supported on reinforced concrete beams. *Building Research Station Current Paper 26/69*. BRS, 1969.

ROSENHAUPT, S. Stress in point supported composite walls. *Proceedings of American Concrete Institute*, Vol 61. 1964.

COULL, A. A composite action of walls supported on beams. *Building Science*, Vol 1, 1966.

YETTRAM, A. AND HIRST, M. An elastic analysis for the composite action of walls supported on simple beams. *Building Science*, Vol 6, 1971.

MALE. S. AND ARBON. P. A finite element study of composite action of walls supported on simple beams. *Building Science*, Vol 6, 1971.

STAFFORD SMITH, B. The composite behaviour of infilled frames, tall buildings. *Proceedings of Symposiuni on Tall Buildings*, 1966. Pergamon Press, Oxford, 1967.

BURHOUSE. P. Composite action between brick panel walls and their supporting beams. *Proceedings of the Institution of Civil Engineers*, Vol 43, 1969.

STAFFORD SMITH, BM KHAN, M A H, AND WICKENS, H G. Test on wall-beam structures. *Proceedings of the British Ceramic Society*, No 27, London, 1978. pp 289–304.

STAFFORD SMITH, B. AND RIDDINGTON, J R. The composite behaviour of masonry walls on steel beam structures. *Proceedings of First Canadian Masonry Symposium*, University of Calgary, Calgary, Alberta, Canada, 1976.

Chapter 12
The thermal performance of masonry walls

12.1 Background to the Regulations

12.1.1 The Building Regulations 1965 and 1972

The thermal insulating requirements of the Building Regulations have become steadily more restrictive. 1965 saw the introduction of the Model Bylaws and National Building Regulations[12.1] in which the requirement under Part F for an external wall was 0.3 Btu/ft^2 h °F (2.08 Wm^2 °C), which was effectively a one thick brick wall (9 inch (225 mm)). Prior to this building control was covered by The Public Health Acts 1936 and 1961. The Regulations were re-cast in 1972[12.2] and the requirement for external walls was improved to 1.7 Wm^2 °C, which was obtained by a simple double-leaf brick wall. In January 1975 Part F of the Building Regulations for England and Wales was amended. The amendment improved the maximum permissible thermal transmittance value (U-value) for external walls of dwellings from 1.7 W/m^2 °C to 1.0 W/m^2 °C. Part F applied only to dwellings and the improvement in U-value was basically intended to reduce the incidence of condensation, but obviously fuel conservation was also achieved. During the following years it became clear, because of the increase in the cost of fuel, that measures would be required to reduce fuel consumption in all heated buildings. One implementation of this policy was the introduction of Part FF to the Building Regulations, which came into operation on 1 June 1979, and which imposed a U-value of 0.6 W/m^2 °C on the external wall to insitutional, other residential, offices, shops and assembly buildings. For factories and storage buildings the required value was 0.7 W/m^2 °C. This extended thermal insulation requirements to all heated buildings with a floor area greater than 30 m^2 (Note: This was better than for dwellings that were still covered by Part F). Previously the Building Regulations contained no thermal requirements for these buildings. Certain multi-use buildings and buildings which, by reason of the proposed use, required only minimal heating, were given special consideration. Purpose Group III (buildings which comprise one or more dwellings) were exempt from Part FF but were still required to satisfy the requirements of Part F. Similar requirements were given in the Building Standards (Scotland Consolidation) Regulations[12.3] and in the Building Regulations (Northern Ireland)[12.4]. There were no thermal requirements in the Building (Constructional) Bylaws that applied in Inner London, although this situation was not expected to continue for much longer. Since the introduction of Part FF designers had become used to the need to achieve good standards of thermal insulation particularly when designing walls to achieve a thermal transmittance value of 0.6 W/m^2 °C. Thus the introduction of more stringent requirements in Part F of the Regulations, on 1 April 1982, did not require new methods of construction, but served to bring the requirements for dwellings into line with those for other buildings. The minimum standard of insulation of roofs in heated buildings was previously controlled by the 'Thermal Insulation (Industrial Buildings) Act, 1957[12.5], but was replaced by the more severe Part FF of the Building Regulations and currently by Part L.

12.1.2 The Building Regulations 1985 and 1991

A major overhaul of the Regulation was made in 1985[12.6] and they were written in mainly functional terms and basically with regard to thermal insulation said that building should simply be designed to have adequate levels of insulation and that condensation should be minimised. Since no specific values were given in the statutory documents in was necessary to issue Approved Documents that contained guidance on ways on meeting the Regulations, which could be updated as necessary to show what was currently considered to provide an adequate level of energy consumption. The requirement with respect to thermal insulation was changed to Part L and when first issued indicated a U-value of 0.6 W/m^2 K. In 1990 Approved Document L increased the elemental requirement for external walls to 0.45 W/m^2 K and limited the area of windows. The Building Regulations[12.7] was re-issued in 1991 (Becoming fully effective from 1 July 1992) and replaced the Building Regulations 1985, consolidating all subsequent revisions to the regulations. They also covered the Inner London area, thus replacing the separate system of building control[12.8], which has operated in Inner London for several hundred years. Some amendments were made to Approved Document L in 1992 and a new version was issued in 1995, which includes three alternative methods for showing compliance with the elemental value remaining at 0.45 W/m² K. The other methods are outlined below which can be used to enable greater U-value to be catered for by calculation.

The main objective of the revision of the 1995 Approved Document[12.9] was to reduce CO_2 emissions but without introducing unacceptable technical risks. The requirements are applicable to dwelling and other buildings whose floor area exceeds 30m². Material changes of use are now included in the Approved Document. For example a conversion of a barn to a house will need to meet the new thermal insulation requirements. Alternatively when an exposed element of an existing dwelling is substantially replaced (i.e. roof, ground floor, exposed wall, windows or semi-exposed wall) it will be necessary to upgrade the insulation to meet the new requirements. When a dwelling is created, or there is a material change of use, a calculated energy rating must be provided using the Government's Standard Assessment Procedure (SAP). Whilst a SAP rating is a necessary requirement, there is no obligation in the 1995 Approved Document to achieve a particular level. Although the resulting value will influence the required U-value for roofs, exposed floors and windows. The requirements for U-values of external walls may under future changes be related to the SAP value or boiler efficiency.

A competent person may certify to a local authority or an Approved Inspector that calculation procedures have been carried out. Any question of competency, it is suggested, should be settled prior to submission of calculations.

Other requirements in the 1995 Approved Document[12.9] are:

(1) Inclusion of the effects of thermal bridging in U-value calculations i.e. the effect of mortar joints.
(2) Double glazing a minimum standard.
(3) Doors are to be included in the glazed area allowance for windows and rooflights.
(4) Improved insulation requirements around openings.
(5) Provisions for reducing air leakage.

The exclusions are small extensions to dwellings not exceeding 10m², which need only be provided with a similar performance to the existing construction. Industrial and storage buildings – where output of space heating system per sq. m is not more than 50 watts/sq m floor area, and any other non-domestic building – where output of space heating system per sq m is not more than 25 watts/sq m floor area, – need achieve no specific insulation standard.

The three compliance methods are:

(a) Elemental Method
(b) Target U-value Method
(c) Energy Rating Method

(a) Elemental Method

This is essentially the same method contained within previous Approved Documents giving prescribed U-values for the main heat loss elements, which must not be exceeded. There is, however, a relationship between the SAP rating of the dwelling and the level of insulation required as shown below:

Table 12.1 Standard U-values for dwellings

Element	For SAP energy ratings of:	
	60 or less	Over 60
Roof	0.2	0.25
Exposed walls	0.45	0.45
Exposed floors and ground floors	0.35	0.45
Semi-exposed walls and floors	0.6	0.6
Windows doors and rooflight	3.0	3.3

* 0.35 where there is no loft

The above values are those in force at the time of writing this publication but are expected to change shortly and readers should consult the latest Approved Document for the current values, which may improve the wall U-value for walls to 0.3 and 0.35 W/m² K depending on the SAP value or boiler efficiency.

All reference to U-values of walls infers that they are calculated by the proportional area method given in Appendix B of Approved Document L, which is covered in more detail later in this Chapter (see 12.6).

This method allows for the effect of mortar joints and other bridging elements. The permitted area of glazing (double not single) is currently 22% of the total floor area, but this includes windows, rooflights and doors. The permitted glazing area may be increased by improving the average performance of the glazed components. For most houses built to the old levels of insulation and a conventional heating system a SAP rating exceeding 60 would be achieved. However, this may not be the case if the system is fuelled by LPG since energy consumption is influenced more by the heating system and type of fuel than by the level of fabric insulation. For the Elemental Method the following design approach is required:

1. Assume SAP is greater than 60.
2. Select appropriate prescribed U-values.
3. Calculate U-values of each fabric element and ensure they do not exceed the prescribed values.
4. Calculate the final SAP rating to check that it corresponds with the assumption made in 1. If the SAP rating is 60 or less either improve the rating or change the fabric and glazing to achieve the higher insulation levels.

(b) Target U Value Method

This method provides flexibility, within limits, of the U-value of particular elements. The objective is to show that the Average U-value of the house is not greater than the Target U-value. The Target U-value is related to the SAP rating as indicated below:

- SAP 60 or less Target U-value = (Total Floor Area × 0.57) + 0.36 (Total Area of Exposed elements)
- SAP over 60 Target U-value = (Total Floor Area × 0.64) + 0.4 (Total Area of Exposed elements)
- The Average U-value is given by:- Total Rate of Heat Loss./Total External Surface Area

Semi-exposed walls and floors are not included in the calculations, but are assumed that they have achieved the required U-value given in the Elemental Method i.e. 0.6W/m²K. The maximum limiting U-values for exposed walls and floors is 0.7 and 0.35 for roofs. For highly efficient heating systems the Target U-value may be increased by up to 10%. The effect of solar gain may also be included by decreasing the glazing used in the calculation. For the Target U-Value Method the following design approach is required.

1. Assume SAP is greater than 60
2. Calculate the Target U Value from the appropriate equation.
3. Calculate the U-values of each element and subsequently the Average U-value.
4. Calculate the final SAP rating. If the rating is 60 or less, take measures to improve the rating or recalculate using the alternative Target U-value.

(c) Energy Rating Method

This method is in effect the Governments Standard Assessment Procedure (SAP Rating) and gives an indication of the annual energy cost for space and water heating of the dwelling. Factors affecting the SAP Rating area:

(i) Fuel – type and cost
(ii) Efficiency of the heating system – type and controls
(iii) Ventilation characteristics
(iv) Thermal efficiency of the building fabric
(vi) Solar orientation

When using this method it is necessary to achieve a SAP of 80 to 85 depending upon the size of the house. The resultant overall specification would be similar to that currently employed by Housing Associations. The maximum limiting U-values for exposed walls and floors is, as with the target U-value method, 0.7W/m² K and 0.35W/m² K for roofs.

12.2 Condensation

Changes to the Approved Document, which have a direct bearing on masonry products include thermal bridging at openings in external walls. There is a specific requirement to limit the amount of thermal bridging and thereby reduce the risk of condensation and mould growth i.e. at reveals. Condensation problems may occur when high insulating materials are used on a building, particularly when the insulation is placed on the inside.

Guidance on the problems of condensation and how to assess whether condensation may occur is given in *BRE Digest 110*[12.10] and in BS 5250[12.11] (although this Code is specific to domestic rooms). A more detailed publication[12.12] is specifically referred to in Approved Document L and which contains comprehensive guidance on methods of avoiding problems with thermal insulation.

12.3 Approved constructions

One way of achieving the requisite U-value for walls is by providing insulation of an appropriate thickness estimated from tables given in Appendix A of Approved Document L[12.9]. Table A5 of Approved Document L reproduced as Table 12.2 below gives the base thickness of insulation required to achieve a given U-value, ignoring any contribution from other materials in the wall construction. However, the thickness of insulation can be reduced by taking allowance for the thickness and type of other materials that are used to make up wall's construction. Tables A6 and A7 in Approved Document L are provided for this purpose, and are reproduced as Table 12.2 and Table 12.3 below. The thicknesses of insulation in these tables is determined by equating it to the equivalent thickness of the indicated material ($l_{insulation}$ = $R_{material}$ × Conductivity of insulant (λ_i) × 10³, Note: R = l/λ for solid materials; for non uniform or bridged element see 12.6 and 12.7). As an example in Table 12.2 the thermal resistance of a cavity (min 25 mm) (Column 1) is 0.18 (See Table 12.13), which is equivalent to a thickness of an insulation material of conductivity 0.02 W/m K of 0.18 × 0.02 × 10³ = 3.6 mm, which rounded to the nearest whole number = 4mm (column 1, row 1). Some of the thermal resistances of the materials given in Table 12.2 are given in Table 12.14, others may be found in reference12.13.

Table 12.2 Base thickness of insulation layer

Design	Thermal conductivity of insulant (W/mK)						
U-value	0.02	0.025	0.03	0.035	0.04	0.045	0.05
W/m²K	Base thickness of insulation (mm)						
	A	B	C	D	E	F	G	H
1	0.30	63	79	95	110	126	142	158
2	0.35	54	67	80	94	107	120	134
3	0.40	46	58	70	81	93	104	116
4	0.45	41	51	61	71	82	92	102
5	0.60	30	37	45	52	59	67	74

Table 12.3 for blocks was derived frrm the proportionate area calculation 1 + (0.934/R_{block} + 0.066/ 0.125) × Conductivity of insulant (λ_i) × 10³, which assumes a block size of 215 by 440 mm (See 12.6). This ignores the effect that any cavity has on the heat flow and provides a simple lower bound solution that might be slightly improved upon by more detailed calculations (See again 12.6). The horizontal row of letters and the left-hand column of numbers are used in examples of the use of the tables (See example following Table 12.3).

Table 12.3 Allowable reductions in base thickness for common components

		Thermal conductivity of insulant (W/mK)						
		0.02	0.025	0.03	0.035	0.04	0.045	0.05
				Reduction in base thickness of Insulating material (mm)				
Component								
	A	B	C	D	E	E	G	H
1	Cavity (25 mm min)	4	5	5	6	7	8	9
2	Outer leaf brick	2	3	4	4	5	6	6
3	13 mm plaster	1	1	1	1	1	1	
4	13 mm lightweight plaster	2	2	2	3	3	4	4
5	10 mm plasterboard	1	2	2	2	3	3	3
6	13 mm plasterboard	2	2	2	3	3	4	4
7	Airspace behind plasterboard dry-lining	2	3	3	4	4	5	5
8	9 mm sheathing ply	1	2	2	2	3	3	3
9	20 mm cement render	1	1	1	1	2	2	2
10	13 mm tile hanging	0	0	0	1	1	1	1

Table 12.4 Allowable reduction in base thickness for concrete components

		Thermal conductivity of insulant (W/mK)						
		0.02	0.025	0.03	0.035	0.04	0.045	0.05
	Density (kg/m³)	Reduction in base thickness of insulation (mm) For each 100 mm of concrete						
	A	B	C	D	E	F	G	H
Concrete inner leaf								
1	600	9	11	13	15	17	20	22
2	800	7	9	11	13	15	17	19
3	1000	6	8	9	11	12	14	15
4	1200	5	6	7	9	10	11	12
5	1400	4	5	6	7	8	9	9
6	1600	3	4	4	5	6	7	7
Concrete outer or single leaf								
7	600	8	10	13	15	17	19	21
8	800	7	8	10	12	14	15	17
9	1000	6	7	8	10	11	12	14
10	1200	4	6	7	8	9	10	11
11	1400	3	4	5	6	7	8	9
12	1600	3	3	4	5	5	6	7
13	1800	2	3	3	4	4	5	5
14	2000	2	2	2	3	3	4	4
15	2400	1	1	2	2	2	2	3

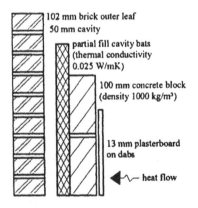

102 mm brick outer leaf
50 mm cavity
partial fill cavity bats (thermal conductivity 0.025 W/mK)
100 mm concrete block (density 1000 kg/m³)
13 mm plasterboard on dabs
heat flow

Figure 12.1 Construction for determination of U = 0.35 W/m² K

Example 1

This example will illustrate the use of the tables in Approved Document L12.9 (Table 12.1and Table 12.2 as given above). The construction illustrated (Figure 12.11) is similar to Example 7 in Approved Document L12.9 but determining for a U-value of 0.35 W/m²K and using insulation with a thermal conductivity of 0.025 W/m K, 100 mm (1000 kg/m³) blocks and plasterboard on dabs. This example may also be compared with Examples 2 and 3 at the end of this Chapter which relate to the same wall but determined by calculation.

Using Table 12.1.

From column E, row 2 of Table 12.2, the base thickness of the insulation layer is 67 mm.

The base thickness may be reduced by taking account of the other materials as follows:

From Table 12.3:

Brick outer lea	column C, row 2	=	3 mm
Cavity	column C, row 1	=	5 mm
13 mm plasterboard	column C, row 6	=	2 mm
Airspace behind			
plasterboard	column C, row 7	=	3 mm

And from Table 12.4

| Concrete block | column C, row 3 | = | 8 mm |
| Total reduction | | | = 21 mm |

The minimum thickness of the insulation layer required to achieve a U-value of 0.35 W/m²K is therefore:

Base thickness less total reduction i.e. 67 − 21 = 46mm (Repeated as calculated examples (12.10, Examples 2 and 3) at end of this Chapter).

12.3.1 Summary

The above tabular method provides an easy way of assessing the requirements of many walls. It is simple to use but it is important to note that although the above Tables can enable the determination of insulation requirements almost in millimetre incruments many of the added insulants will only be available in a limited range of sizes (see Table 12.14) and this needs to be bourne in mind when actually specifying or detailing the wall in question.

12.4 Typical constructions

The purpose of this section is to illustrate typical constructions and provide a general picture of the various ways in which concrete masonry may be used to meet the current and some future requirements. The details provide an indication of the levels of insulation, which may be achieved and the approximate thickness of the various constructions.

General information is given against some of the constructions, including a note of some of the items to be considered by the architect or designer during design and construction. Of necessity this article has treated each element of the building separately, but consideration must be given to ensuring the continuity of insulation between walls, floors and roofs to obviate cold bridging in buildings. Other forms of cold bridge, such as dense concrete and steel lintels, also need to be considered. The values taken for the thermal resistances of various components are presented in a later section (see 12.5.2).

The examples shown in Figure 12.2 indicate in the likely range of wall and insulation thickness for three general double leaf forms of construction: Clear cavity walls with internal insulation; cavity walls with partial fill; and cavity walls with a filled cavity. Figure 12.3 gives example sizes for solid walls. They have been derived for a U-value between 0.3 to 0.6 W/m²K and cover blocks in the range 600–1000 kg/m³ (this should also in essence cover face and void insulated blocks) and insulation materials with a thermal conductivity from 0.025 to 0.035 W/mK.

Each construction, for example, (a) Brick-cavity-block with internal insulation is marked with a string of numbers such as, row three, {0.45}285/30–310/50, which means that for a U-value of 0.45 W/m²K wall widths are likely to be in the range 315–350 mm and contain insulation between 60–90 mm. Since the thermal conductivity of different types and grades of insulation materials varies widely, the designers should carry out their own calculations to verify the thermal transmittance of the construction using given proprietary materials. Thermal resistances of blocks and bricks also need to be checked with manufacturers. It is also important to ensure that other properties such as durability are suitable. The figure indicate what might be achieved for typical constructions but manufacturers will produce individual data illustrating exactly what can be achieved with their particular products. The blocks can include normal density autoclaved aerated concrete blocks and solid lightweight aggregate concrete blocks. Foam or similarly filled aggregate blocks can also provide good thermal resistance (see 12.7 and 12.8). A further development has been the introduction of thermal insulating facing blocks, which provide good insulation without the need for rendering or other coverings. Recent developments in block types include those made with lower density materials than were previously available. Blocks in the form of autoclaved aerated concrete with densities as low as 475 kg/m³ are already available, and further developments can be expected with lightweight aggregate blocks. An alternative approach has been to develop blocks, which are provided with integral bonded insulation on one face. The clauses following Figure 12.2 and Figure 12.3 give further comments on a range of more specific constructions.

12.4.1 Cavity walls retaining a clear cavity

12.4.1.1 Standard thickness masonry

The wall constructions shown indicate the level of insulation that may be achieved with masonry commonly used in the cavity walls. The exact U-value of the various constructions will obviously depend on the products used. The values given against each construction are those which will typically be achieved using standard bricks and 90/100 mm thermal insulating blocks currently available.

To achieve a U-value between 0.3 to 0.6 W/m² K will typically require insulation thicknesses between 10–90 mm as shown. This insulation is usually attached to the inner leaf but this may be dictated by condensation considerations and the properties of the particular material being used. In some circumstances insulation can be added to the inside of the inner leaf, i.e. beneath plasterboard. However, before doing so, the designer should carefully consider the potential problems of interstitial condensation and thermal bridges, which can occur at the junction between the external wall and internal walls and floors.

(1) *Facing brick*: This is perhaps the most common form of construction with the drained cavity maintaining resistance to rain penetration. It is very important to

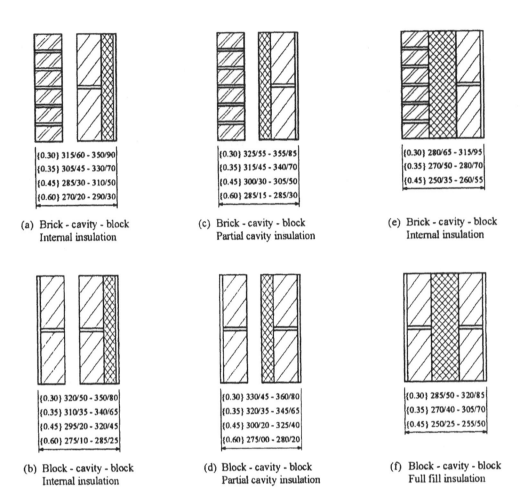

{0.30} 315/60 - 350/90
{0.35} 305/45 - 330/70
{0.45} 285/30 - 310/50
{0.60} 270/20 - 290/30

(a) Brick - cavity - block
Internal insulation

{0.30} 325/55 - 355/85
{0.35} 315/45 - 340/70
{0.45} 300/30 - 305/50
{0.60} 285/15 - 285/30

(c) Brick - cavity - block
Partial cavity insulation

{0.30} 280/65 - 315/95
{0.35} 270/50 - 280/70
{0.45} 250/35 - 260/55

(e) Brick - cavity - block
Internal insulation

{0.30} 320/50 - 350/80
{0.35} 310/35 - 340/65
{0.45} 295/20 - 320/45
{0.60} 275/10 - 285/25

(b) Block - cavity - block
Internal insulation

{0.30} 330/45 - 360/80
{0.35} 320/35 - 345/65
{0.45} 300/20 - 325/40
{0.60} 275/00 - 280/20

(d) Block - cavity - block
Partial cavity insulation

{0.30} 285/50 - 320/85
{0.35} 270/40 - 305/70
{0.45} 250/25 - 255/50

(f) Block - cavity - block
Full fill insulation

Figure 12.2 Wall widths and insulation thicknesses for double leaf walls

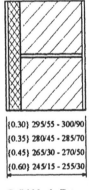

{0.30} 295/55 - 300/90
{0.35} 280/45 - 285/70
{0.45} 265/30 - 270/50
{0.60} 245/15 - 255/30

(a) Soild block, External
or internal insulation

Figure 12.3 Wall widths and insulation thicknesses for insulated solid walls

check for secure fixing of any cavity bats during construction. If allowed to fall forward, rain penetration may occur.

(2) *Fair faced concrete blocks or bricks:* It is possible to include for fair faced concrete bricks or blocks, which may require slightly more insulation than with a clay brick facing. However, some facing blocks may be voided or contain insulation and provide similar or better insulation than a normal brick. All concrete blocks should be protected from rain by covering the stacked blocks on site, but this is particularly important when foam filled blocks are used. Though some potential problems with condensation and pattern staining have been envisaged with this type of block, no actual problems have been recorded. The cavity prevents any direct cold bridging to the outside. In areas of severe exposure the ability of the wall to resist rain penetration should be checked with the manufacturer and good workmanship ensured. There may be cost savings from eliminating finishing trades but, a high standard of blocklaying will be required.

(3) *Rendered double leaf block:* Modifying a cavity wall to include a rendered thermal insulating block on the external leaf gives an improved thermal

218

transmittance value. The precautions mentioned in (1) and (2) need to be applied when bats are positioned, although the render coat will reduce the problems of rain penetration due to misplaced bats. Obviously any render should be applied carefully.

(4) *Tile hung double leaf block*: This construction is similar to (3) above but with the outer render coat replaced by tile hanging. When using this type of construction, consideration should be given to the possible damage or displacement of tiles, particularly those for use at ground and first floor level. Other forms of cladding, such as weatherboarding, may be used as an alternative to tiles.

12.4.1.2 Increasing block thickness

All but one of constructions in Figure 12-2 the were determined to require the use of additional insulation in order to achieve the U-values shown, when employing a 90/100mm block. The wall and insulation thicknesses were determined for precise values but, for convenience, have generally been round up to the nearest 5 mm. However, the required thickness of insulation will not be available in limitless sizes or for that matter in 5 mm increments. It may, therefore, be beneficial to adjust block thicknesses so as to enable the use of economic or available insulation thicknesses. Block thickness are available in a wide range of thicknesses *(see Table 3.3, Chapter 3)*. The typically sizes are 90 mm, 100 mm, 140 mm, 190 mm and 215 mm, although some manufacturers can provide others thickness between 100 mm and 140 mm, which may increase in availability in order to meet the newer insulation standards. When it is necessary to increase the block thickness to achieve the necessary level of insulation it is advantageous to increase only the inner leaf rather than both leaves. The main advantages are that the loadbearing capacity of the wall is increased and better use is made of the insulation since there is a greater amount of material in a protected situation giving better insulation. Using a 140 mm inner leaf in conjunction with a 100 mm outer leaf can produce a U-value in an unfilled double leaf wall down to 0.54 W/m² K using dense plaster; 0.52 using lightweight plaster; 0.49 using plasterboard on dabs and 0.45using foil-backed plasterboard on battens.

12.4.2 Cavity walls with filled cavity

One way of achieving a U-value less than 0.6 W/m² K with standard thickness masonry in a cavity wall is to use a cavity fill material. This is by no means a straightforward solution, however, and many factors have to borne in mind. Although it is always desirable, when using bricks or blocks as the outer leaf of a cavity wall, to ensure good workmanship and supervision, this becomes essential when cavity fill is employed. For an injected fill, the installers are expected to satisfy themselves that the wall is in a suitable condition before using, although this is often very difficult to do. In new construction, on the other hand, if the wall is connected with full width insulation slabs, the construction can be inspected as the slabs are positioned. A number of cavity fill materials are available including urea formaldehyde foam, blown-in mineral fibre, expanded polystyrene in the form of beads or foam, and polyurethane foam. Cavity slab insulation materials include resin-bonded rock wool fibres and glass fibres, and expanded polystyrene. When using a cavity fill material it is important to remember that the outer leaf will be colder and remain wetter than with an unfilled cavity, and thus the outer leaf must be checked for susceptibility to frost damage. Concrete masonry materials are generally not prone to frost damage but manufacturers' experience should be sought, especially in the colder regions of the country; clay facing bricks are more susceptible to frost damage and a risk assessment should be made when employing them on an insulation-filled double leaf wall.

(1) *Facing brick or block*: A similar U-value would be obtained if a facing block replaced the brick outer leaf because the performance is dominated by the insulation. Currently, the incidence of rain penetration problems for filled cavity walls is quoted by the manufacturers as less than 1%, and this is supported by the Building Research Establishment
(2) *Rendered double leaf block*: The illustrations show similar wall thicknesses but less added insulation than for a brick faced wall. Clearly good economic levels of construction can be achieved with two leaves of thermal insulating blocks. The rendering assists resistance to rain penetration. Tile hanging instead of render would enable a reduction in insulation of perhaps 5 mm and improved resistance to rain penetration.
(3) *Double leaf facing block*: Full fill can enable the use of two fair-faced leaves (brick or fair-faced or architectural facing blocks). Good workmanship should be ensured to help resist rain penetration and the type of block used in areas of high exposure to rain given careful consideration (check with manufacturers). Some savings in cost may be achieved by not needing to provide finishes, but a high standard of blocklaying is required. A similar U-value will be achieved with two leaves of concrete brickwork.

12.4.3 Solid walls

The previous constructions were all based on the traditional cavity (double leaf) wall. Though solid walls used in the past were beset by problems of condensation and dampness, there does seem some merit in basically re-examining solid forms of construction, particularly as much higher insulation is now required and there is an increased use of external insulation systems and internal dry linings. Compared with internal insulation and some other methods of improving wall performance, the use of an external insulation system offers the following advantages:

(a) the thermal capacity of the walling material is utilized and the comfort of the occupants thereby improved. Maximum use is made of the structure to reduce solar gain, maintain night-time temperatures and generally reduce temperature variations;

(b) cold bridging problems are avoided;

(c) there is no loss of internal space;

(d) the acoustic performance is not degraded due to longitudinal vibration as might be the case with some lightweight lining systems;

(e) potential interstitial condensation problems are largely avoided;

(f) the fire hazard is potentially much less than in the case of a corresponding internal insulation system in terms of rate of heat build-up;

(g) the exterior protection provided to the insulation should ensure a better resistance to rain penetration than conventional solid wall constructions. Against these advantages must be set the obvious limitations of the type of finish and appearance given to the building, and in some circumstances the vulnerability of particular systems to vandalism. There are basically three types of system:

    (i) lightweight renders;

    (ii) rigid insulation panels;

    (iii) flexible insulation mats.

The lightweight renders are thermally less efficient than the other techniques, but are likely to be cheaper and may be used in marginal cases where the additional insulation provided by the render is sufficient to give an acceptable U-value. Figure 12.3 gives wall and insulation thicknesses for added insulation materials with a thermal conductivity between 0.025 and 0.035 W/mK. Insulating renders are likely to only give U-values around 0.6 W/m² K, which would be adequate for some purposes, although lower values may be possible depending on the thickness of render and type of block. Further details of external insulation systems are given in reference 12.14. While render is often shown as the finishing coat because it is a comparatively cheap solution, claddings could also be used. Additional savings may occur since the overall construction is likely to be reduced in thickness and it is generally cheaper to construct solid walls than cavity (double leaf) walls. Further comments are as follows:

Externally insulated walls – The thickness of insulation may be increased very easily. Satisfactory levels of insulation are given by a construction from as little as 185 mm in overall thickness (using a 140 mm block), a saving in thickness compared with the cavity walls. Careful detailing is necessary around openings and at roof level. The rendered outer coat and the insulation provide an extremely good barrier to rain and thus, in combination with the wall, are likely to provide a weathertight construction. Consideration should be given to thickness of render to give adequate impact resistance if used at ground level. Cladding may also be used to protect the insulation. Figure 12.3 shows between 20 to 90 mm of insulation would be required for block thickness of 140 mm. The insulation would decrease by about 5 mm in the case of a 215 mm block but wall thickness will increase

by 75 mm. Even so the thickness of a single leaf externally insulated walls will be typically less than a double leaf wall, thus offering savings in materials and are easier to construct.

Internally insulated walls – Internally insulated walls would need similar amounts of insulation as externally insulated walls but problems can exist with thermal bridges as well as with rain penetration.

(1) *Lightweight renders*: The exact U-value will depend on the applied density and thickness of render. Some manufacturers suggest thicknesses up to 100 mm. The resistance of proprietary systems to impact damage varies considerably and, in some cases, a strong decorative finish will be needed. The substrate should be prepared as for a traditional render.

(2) *Rigid panels*: There are available rigid panel forms of external insulation that are attached to the substrate either by mechanical fixings or adhesives. The adhesive methods are widely used in other countries, although in the UK preference seems to be given to the mechanical fixings. The panels need to be bonded together with a fibre reinforcing mat or similar and need to be protected with a render or cladding. A grc render can be used to provide the necessary impact protection. Expanded polystyrene boards need to be aged to reduce shrinkage movement and the surface of the board needs to be of a roughened texture so that render may be easily applied.

(3) *Flexible insulants*: With this system the insulation is usually supplied attached to a metal carrier and breather paper. Mechanical fixings are invariably used and again the insulation has to be protected by a render or cladding. When fixing, care needs to be taken not to compress the insulant. A scratch render coat will probably be required to reduce bounce in the metal carrier before the main coats are applied.

(4) *Rendered thermal insulating block*: When external insulation is omitted, a block about 225 mm in thickness is required to achieve a U-value of 0.6 even when foil-backed plasterboard on battens is used. A 190 mm block would give a U-value of less than 0.7 (145 mm is likely to be the minimum to achieve 0.7). The external render will assist in preventing rain penetration. Experience indicates that the construction should be suitable for moderate exposure and, depending on the type of block and thickness, also for severe exposure. Special consideration should be given to detailing at floor level to avoid problems of rain penetration. The use of tile hanging or one of the proprietary cladding panels will considerably improve thermal insulation.

(5) *Posted construction*: A variation on solid wall construction is the posted construction (*Figure 5.2, Chapter 5*). This is an interesting adaptation of the solid wall, which is ideally suited for single storey

structures. In this method the outer solid wall (fair faced or rendered) is used as the loadbearing member and a non-loadbearing isolated studding, supported by a sleeper wall, is provided internally. The cavity is thus completely unbridged and allowed to drain freely at ground level. The plasterboard and studding may be replaced by proprietary partitioning with a consequential increase in insulation. Because of the completely clear cavity (no problems of dirty ties) this construction is suitable even for the most severe conditions of exposure. It is, in fact, a traditional Scottish construction not commonly found in England and Wales. When used on multi-storey structures, attention must be given to the detailing at floor levels to reduce rain penetration problems.

12.5 An explanation of λ, R and U-values

Whilst it is not within the scope of this book to go deeply into the thermal performance of buildings, the designer should at least be aware of the factors which will influence the flow of heat through the fabric. This is of particular importance in meeting either prescriptive requirements of regulation (i.e. the element must have a thermal transmittance value equal to or better than the specified value) or functional requirements (i.e. the resistance to the passage of heat must be adequate). It is apparent that a temperature difference across a building element such as a wall will give rise to a flow of heat through the element, and it is the function of insulation to resist this flow of heat. The process by which heat flows through a material is called conduction and, because air is a poor conductor, insulants tend to comprise of materials in which a lot of air is trapped. Convection will occur in the air within a building due to changes in density because of temperature variations and will lead to heat transfer at various points within the structure. Finally, it should be noted that all bodies emit energy in the form of radiation, the amount of radiation depending upon the temperature of the body and the type of surface.

12.5.1 Thermal conductivity

The ability of a material to conduct heat is known as its thermal conductivity or λ value (Previously referred to as k value). The thermal conductivity of any given material is usually given as the heat flow (W) per unit area (m²) when there is a temperature difference of 1K (°C) across a 1 m thickness of the material (i.e. λ – W/m K). The thermal conductivity of most materials will alter with density, porosity and moisture content. Rain penetration or condensation could affect moisture being present in the fabric of a building and reduce the thermal conductivity of the material. Clearly this would in practice vary form period to period but for normal building design purposes standard moisture contents are taken as shown in Table 12.5 and any test results must be corrected for the moisture content appropriate to its use (see Table 12.6). As an alternative to testing the masonry

material, its thermal conductivity may be obtained from standardised values related to the density of product in protected and exposed situations (Table 12.9). The reciprocal of the thermal conductivity is known as the thermal resistivity and indicates the ability of the material to resist the flow of heat (m K/W), which when multiplied by thickness gives thermal resistance. However, thermal resivitity is seldom used and instead thermal resistance (m² K/W) is obtained directly by dividing by thickness by λ.

Table 12.5 Standard moisture contents for protected and exposed masonry

Material	Moisture content by volume (%)	
	Protected	Exposed
Brickwork	1.0	5.0
Concrete work	3.0	5.0

Protected covers internal partitions, inner leaves separated from outer leaves by a continuous air space, masonry protected by tile hanging, sheet cladding or other such protection, separated by a continuous air space.
Exposed covers masonry directly exposed to rain, unrendered or rendered.
Example: For a typical brick-cavity-block construction the brickwork is considered to have a moisture content of 5% and the blockwork 3%.

Table 12.6 Correction factors to obtain the λ value of a material after a change in its moisture content

Measured moisture content by volume (%)	Factor for material at standard moisture by volume contents		
	1% mc	3% mc	5% mc
0.5	1.12	1.39	1.52
1.0	1.00	1.23	1.35
1.5	0.93	1.14	1.25
2.0	0.88	1.08	1.18
2.5	0.84	1.03	1.13
3.0	0.81	1.00	1.10
4.0	0.78	0.96	1.05
5.0	0.74	0.91	1.00

Example: The λ value at 3% is 1.23 × (λ at 1%)

12.5.2 Thermal resistance

The thermal resistance is a convenient way of measuring the resistance to heat flow of a material or several materials combined. The thermal resistance or R value of a slab of homogeneous material is obtained by dividing its thickness (l) by its thermal conductivity (λ), i.e. $R = l/\lambda$ m² K/W. It is apparent that doubling the thickness of the given material will double the thermal resistance. A list of standard or common thermal resistance for a range of masonry associated products is given in Table 12.14. Other thermal resistances for air voids are given in Table 12.12 and Table 12.13. Additional thermal resistances due to surface effects are also utilised and are shown in Table 12.10 and Table 12.11.

12.5.3 Thermal transmittance

The thermal transmittance (U-value) is the rate of heat flow through unit area of the element when a unit temperature difference exists between the air on each side. The U-value takes into account, as well as the resistance offered by the fabric, the outside and inside surface resistance. In simple terms thermal conductivity is the reciprocal of the sum of the resistance to the passage of heat.

Thus:

$$U = \frac{1}{\Sigma R} = \frac{1}{R_{so} + R_1 + R_2 + \dots R_n + R_{si}} \; \text{W/m}^2\text{K}$$

Where:

ΣR is the sum of the resistance to heat flow
R_{so} is the external surface resistance
R_1 etc are the individual resistances of materials and air-voids that make up the construction
R_{si} is the internal surface resistance

Values for a range of thermal resistances for materials, air voids and surface resistance are given or can be derived from the Tables referred to in 12.5.1 and 12.5.2. Since notionally the U-value provides a direct comparison of the heat flow through a wall or other element, it is the value used to compare the performance of different constructions and to make energy use calculations.

12.6 Calculation procedures

The simple concept of U=1/ΣR applies to single or layered continuous solid homogenous elements. When an element is bridged e.g. for example by studding between insulation or by mortar in the case of masonry the effective thermal resistance and hence U-value of the construction can be adversely affected. The effect of any thermal bridging should be taken into account. The normal procedure is to use a proportional area calculation as indicated in Approved Document L[12.9]. The general procedure is indicated below but can be fairly complex and more detailed guidance may be found in reference 12.13.

12.6.1 Single leaf wall

The U-value of a single leaf wall is calculated as shown in Figure 12.4. For convenience this only shows a block bridged by mortar plus a plaster coat but if other layers exist then the additional resistances are included within U_m and U_b and proportionally summed.

In Figure 12.4:

U is the U-value of the assembly
U_m is the U-value of the mortar fraction
U_p is the U-value of the block fraction
P_m is the proportionate area of the mortar
P_p is the proportionate area of the block
l_1 is the thickness of the block and mortar
λ_m is the conductivity of the mortar
λ_b is the conductivity of the block
R_p is the resistance of the plaster

$$U = P_m U_m + P_b U_b$$

$$U_m = \frac{1}{R_{so} + \dfrac{l_i}{\lambda_m} + R p + R_{si}}$$

$$U_b = \frac{1}{R_{so} + \dfrac{l_i}{\lambda_b} + R p + R_{si}}$$

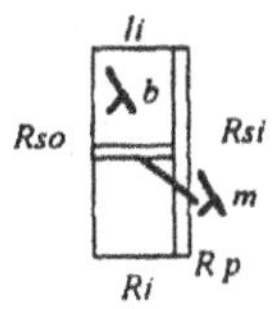

Figure 12.4 Basic equations for single leaf wall

R_{so} is the outside surface resistance
R_{si} is the inside surface resistance

12.6.2 Double leaf (cavity) wall

The U-value of a double leaf wall containing a cavity is determined as indicated in Figure 12.5, where resistances are calculated from each face to the centreline of the cavity. The illustration shows only the inner leaf bridged but allowance for any bridging of the outer leaf would need to be similarly taken into account as shown for the bridged inner leaf.

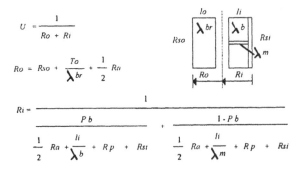

Figure 12.5 Basic equations for a double leaf wall containing a cavity

In the above Figure:

R_i is the resistance of the inner leaf to the centreline of the cavity
R_o is the resistance of the outer leaf to the centreline of the cavity
R_a is the resistance of the cavity
L_o is the thickness of the outer leaf
λ_{br} is the conductivity of the outer leaf

Other symbols are as for the single leaf wall

A complex example, which demonstrates the use of the proportionate area method, may be found in Approved Document L[12.9] but a simpler version is given in Example

3, at the end of this chapter and which is based on based on the same wall as used for the example in 12.3.

12.6.3 Effect of mortar joints

As indicated above the presence of mortar joints must be taken account of but they do not always become critical and technically can give a slight benefit. In 12.5.1 it was indicated that the thermal conductivity varies with density (Table 12.9) and thus a thermal difference or bridge can occur between the mortar and the masonry units. Clearly when the density/thermal conductivity of the mortar and the units is identical no thermal bridge occurs and can be ignored. Now the density of mortar will typically be in the region of 1750 kg/m³ and by reference to Table 12.9 is normally taken to have a thermal conductivity of 0.8 mK/W, when in the inner leaf (protected), which is the reason why Approved Document L states that normal mortar joints need not be taken into account for brickwork that is typically also taken to have a thermal conductivity of 0.80 W/m K (See note to Table 12.9). Technically here it follows that mortar will have a slight beneficial effect when used with masonry units having a density greater than 1750 kg/m³ but the effect is so small that it is generally ignored. The Approved Document L 1995 edition, clause 0.11, also indicates that the effect of thermal bridging can be disregarded where the difference in thermal resistance between the bridging material and the bridged material is less than 0.1 m²K/W. This in effect allows mortar to be ignored when the masonry units have a density of 1280 Kg/m³ in wall with a leaf thickness of 90 mm and 1500 Kg/m³ in a wall with a leaf thickness of 215 mm. An example is shown in Figure 12.6, which is for a 100 mm wall. As this is thickness dependent it complicates the calculation a little and it may be easier to adopt the default value of 1750 kg/m³ which would apply to all situations or 1500 kg/m³ which would apply to leafs up to 215 mm. The effect of mortar joints on lighter density units needs to be allowed for but the effect will be typically insignificant when a continuous cavity bat or full fill is employed as shown by Figure 12.7, which compares U-values for identical constructions with and without the effect of mortar joints.

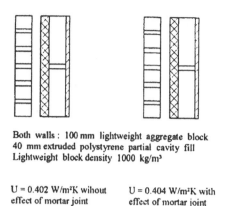

Both walls : 100 mm lightweight aggregate block 40 mm extruded polystyrene partial cavity fill Lightweight block density 1000 kg/m³

U = 0.402 W/m²K wihout effect of mortar joint

U = 0.404 W/m²K with effect of mortar joint

Figure 12.7 Minimal effect of mortar joints in double leaf wall with partially filled cavity

There is a slight complication when mortar joints are included in that the effective thermal resistance of the masonry unit does not become a constant. However, this can be achieved by ignoring the contribution of the cavity and plaster in Figure 12.5. This produces a lower bound value (the inclusion of the cavity and plaster would give a slight benefit) which makes it easier to allow for wall variations. It also enables the effect of mortar joints to be more readily demonstrated as in Table 12.7 in which column 4 shows the thermal resistance of the unit itself and column 9 the thermal resistance of the wall determined by the proportional area method[12.9, 12.13]. This illustrates there is no effect much beyond 1750 kg/m³ unit density. Since added insulation will currently and increasingly in the future dominate thermal design then the effect of mortar joints on the U-value will be small as demonstrated above.

12.7 Calculation methods for multi-slotted blocks

When, in January 1975, Part F of the Building Regulations was amended requiring the external walls of dwellings to meet a U-value of 1.0 W/m² K, not all blocks then in use as the inner leaf to a brick outer leaf were sufficiently good insulators to comply with the mandatory requirements. To improve thermal resistance, some manufacturers incorporated slots into their blocks. Reference to Chapter 4 will indicate that the standard thermal conductivity test in BS 874[12.15] is a test on a solid layer of material and thus there is no way to test the performance of a multi-slotted block without testing walls in the way described in the next section – a non-standard procedure for which few test rigs are yet available. It has, therefore, been necessary to rely on calculation methods to assess the thermal performance of blocks. Several different assumptions may be made with the result that a number of methods of calculation are possible, as illustrated in Figure 12.5.

Thermal bridging can be disregarded when differences is less than 0.1 m²K/W

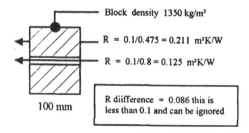

Block density 1350 kg/m³

R = 0.1/0.475 = 0.211 m²K/W

R = 0.1/0.8 = 0.125 m²K/W

100 mm

R diifference = 0.086 this is less than 0.1 and can be ignored

Figure 12.6 Ignoring effect of mortar

Table 12.7 Thermal resistance of blocks with and without bridging effect of mortar

Density Kg/m³	L (mm)	λb W/mK	Rb m²K/W	Fb	λm W/mK	Rm m²K/W	Fm	Rbm m²K/W
1	2	3	4	5	6	7	8	9
600	100	0.19	0.526	0.934	0.8	0.125	0.066	0.435
800	100	0.23	0.435	0.934	0.8	0.125	0.066	0.374
1000	100	0.30	0.333	0.934	0.8	0.125	0.066	0.300
1200	100	0.38	0.263	0.934	0.8	0.125	0.066	0.245
1400	100	0.51	0.196	0.934	0.8	0.125	0.066	0.189
1600	100	0.66	0.152	0.934	0.8	0.125	0.066	0.149
600	100	0.20	0.500	0.934	0.8	0.125	0.066	0.418
800	100	0.26	0.385	0.934	0.8	0.125	0.066	0.338
1000	100	0.33	0.303	0.934	0.8	0.125	0.066	0.277
1200	100	0.42	0.238	0.934	0.8	0.125	0.066	0.225
1400	100	0.57	0.175	0.934	0.8	0.125	0.066	0.171
1600	100	0.73	0.137	0.934	0.8	0.125	0.066	0.136
1800	100	0.96	0.104	0.934	0.8	0.125	0.066	0.105
2000	100	1.24	0.081	0.934	0.8	0.125	0.066	0.083
2400	100	2.00	0.050	0.934	0.8	0.125	0.066	0.052

Where: L is Block thickness, λb is Block conductivity, Rb is Thermal resistance of block Fb is Proportionate area of block = 440 × 215 + 450 × 225, λm is Mortar conductivity, Rm is Thermal resistance of mortar, Fm is Proportionate area of mortar = 1.0 − Fb, and Rbm is Thermal resistance of block mortar combination by proportionate area method.

Method 1: Assessing the average U-value of the wall by determining the U-values of the differing cross sections and proportioning them to their respective areas.

Method 2: Assessing the average U-value of the wall by determining the average conductance of the leaf to the centre line of the cavity with a value for finishes incorporated into the basic formula (the method given in DoE circular BRA/668/68[12.16]).

Method 3: As Method 2 but with the resistance of the finishes added after the conductance of the leaf has been determined.

Method 4: Assessing the average U-value of the wall by determining the average surface to surface conductance of the bridged and unbridged sections (the upper limit of the revised CIBS guide).

Method 5: Assessing the average U-value of the wall by summing the equivalent resistance of the various core and solid strips (the lower limit of the revised CIBS guide and also given by ASHRAE[12.17]) .

12.7.1 *Approved method for multi-slotted blocks*

Because of the variations which occur it is difficult to determine the precise thermal resistance value of a slotted block and, therefore, for the purpose of consistency one method of calculation only should be adopted. The average of Method 4 and Method 5 is that contained in the revised CIBS guide[12.13] and may be adopted by DoE for the purpose of the next Approved Document L[12.1]. The thermal resistance of the air void in a slotted block is given in Table 12.8. Because of the lowering of insulation standards and the tendency for the inclusion of added insulation, which can dominate, the use of multi-slotted may decline in favour of solid lightweight aggregate blocks that are simpler to produce.

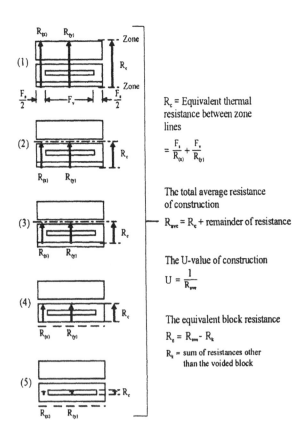

R_e = Equivalent thermal resistance between zone lines

$$= \frac{F_s}{R_{(x)}} + \frac{F_v}{R_{(y)}}$$

The total average resistance of construction

$R_{ave} = R_e +$ remainder of resistance

The U-value of construction

$$U = \frac{1}{R_{ave}}$$

The equivalent block resistance

$R_e = R_{ave} - R_t$

R_t = sum of resistances other than the voided block

Figure 12.8 Different heat flow concepts for voided blocks

Figure 12.9 A hot box rig capable of measuring the thermal transmittance value of 2 m × 2 m wall panels

Table 12.8 Thermal resistances for unventilated divided air spaces for horizontal heat flow, e.g. voids in blocks, R_a

Thickness of air space (mm)	Thermal resistance (m² K/W) for air space of stated width (mm)				
	200 or greater	100	50	20	10 or less
5	0.10	0.10	0.11	0.11	0.11
6	0.11	0.12	0.12	0.12	0.13
7	0.12	0.12	0.13	0.13	0.14
8	0.13	0.13	0.13	0.14	0.15
10	0.14	0.14	0.15	0.16	0.17
12	0.15	0.16	0.16	0.18	0.19
15	0.16	0.17	0.17	0.19	0.21
20	0.17	0.18	0.19	0.22	0.24
25 or more	0.18	0.20	0.21	0.24	0.27

12.8 Foam-filled blocks

In the absence of measured values obtained by testing wall panels in a thermal transmittance rig, the DoE originally accepted calculations based on Method 2. However, the CIBS guide now referred to in Approved Document L uses the mean of Method 4 and Method 5 and gives slightly lower results. This latter method gives better agreement with test results than the resistance obtained from Method 2. In this case the thermal resistance of the air pocket is replaced by the thermal resistance t_b/λ of the insulating material. When considering the use of foam or other insulant filled blocks, a number of factors need to be considered including condensation risk, fire resistance, durability of the wall and properties of the foam, especially in damp conditions.

12.8.1 Condensation risk

Theoretically the presence of high insulation foam materials may lead to condensation within the inner leaf of a cavity wall. This problem will generally only be significant if condensation occurs within the foamed layer, thereby reducing the value of the insulation and possibly causing deterioration of the foam.

12.8.2 Fire resistance

The protection afforded to the insulant by the block will give adequate performance for most low rise applications but each manufacturer will have particular recommendations for the blocks they produce, bearing in mind the insulating material and class of aggregate employed.

12.8.3 Durability of the wall

In an outer leaf situation water will invariably penetrate the block and reduce the insulation value of some materials, possibly affecting their durability. Although most concrete blocks themselves are not likely to suffer from frost attack, a poorly graded sand used to make the mortar could be excessively permeable and suffer frost damage, although this has seldom or if at all, been recorded as a problem in proportioned mortars.

12.8.4 Properties of the foams

Some foam materials are closed cell and do not take up water and therefore change the thermal insulation provided, whereas some materials do take up water and hence reduce the level of insulation. Furthermore, some materials will deteriorate in the presence of moisture. The manufacturer should satisfy themselves as to the adequacy of the materials, which are used. Additional factors to be considered relate to the method of providing the insulation, which may range from spraying in situ to the insertion of rigid panels into the finished blocks. In both cases, the insulation needs to be so well located that it will not be lost or left out during transporting and laying.

Table 12.9 Thermal conductivities of homogeneous masonry

Material and standard moisture content (% by volume)	Bulk dry density (kg/m³)	Thermal conductivity* Protected W/mK	Exposed W/mK
	400	0.15	0.16
	500	0.16	0.18
	600	0.19	0.20
Concrete blockwork and brickwork	700	0.21	0.23
	800	0.23	0.26
	900	0.27	0.30
(Aerated concrete and dense and lightweight aggregate concrete)	1000	0.30	0.33
	1100	0.34	0.38
	1200	0.38	0.42
	1300	0.44	0.49
	1400	0.51	0.57
	1500	0.59	0.65
Note: For foamed slag aggregate concrete multiply by 0.75	1600	0.66	0.73
	1700	0.76	0.84
	1800	0.87	0.96
	1900	0.99	1.09
	2000	1.13	1.24
	2100	1.28	1.40
	2200	1.45	1.60
	2300	1.63	1.80
	2400	1.83	2.00
	1200	0.31	0.42
	1300	0.36	0.49
	1400	0.42	0.57
	1500	0.48	0.65
Fired clay brickwork †	1600	0.54	0.73
	1700	0.62	0.84
	1800	0.71	0.96
	1900	0.81	1.09
	2000	0.92	1.24

* The thermal conductivity of masonry may be taken to be as given for the particular dry density of the material and its position. Alternatively, it may be determined in accordance with BS 874[12.15].
† Brickwork is considered to have a standard a value of 0.84 W/mK and thickness of 102 mm giving a thermal resistance of 0.12 m³K/W.
Protected refers to internal partitions, inner leaves separated from outer leaves by a continuous air space, masonry protected by tile hanging, sheet cladding or other such protection, separated by a continuous air space. *Exposed* refers to masonry directly exposed to rain, unrendered or rendered. The standard thermal conductivities apply to the standard moisture contents in the first column and represent typical values. They are given for a range of densities and it is not implied that the materials are available at all the densities indicated. These values assume mortar joints of similar thermal conductivity and density as those of the bulk material. Typical mortars for walling may be expected to have a density in the range 1600–1800 kg/m³ giving a standard value of 1750 kg/m³

12.9 Test methods for walling

Whilst it is beyond the scope of this manual to go too deeply into methods of test for assessing the U-value or conductance of a wall, the reader should be aware of the test methods available. Essentially there are two techniques which may be employed:

(1) a guarded hot-box rig in which the energy required to maintain a temperature differential across a wall

(all extraneous heat losses having been excluded) is measured;

(2) placing a heat flow meter on the wall to measure the amount of flow of heat per unit area. A typical hot box-rig, as based on ASTM[12.18], is shown in Figure 12.9. It should be noted that this is not a standard method of test in the UK and rig designs and operating parameters differ, which is likely to lead to differences in results obtained. The test result would also include the effect of the mortar joints – this is normally excluded from the calculation method.

As far as the use of heat flow meters is concerned, the principal drawback seems to be that the meters themselves modify the passage of heat through the wall. U-value measurement may be made by modifying the flow of air across a wall to give the appropriate standard resistances and by measuring the air temperature on either side of the wall. Alternatively, the surface temperature may be measured and the conductance of the wall determined.

Table 12.10 Inside surface resistance, R_{si}

Building element direction	Heat flow	Surface resistance (m² K/W) High emissivity	Low emissivity
Walls	Horizontal	0.12	0.30
Ceilings on roofs, flat or pitched and floors	Upward	0.10	0.22
Ceilings and floors	Downward	0.14	0.55

Note: Surface emissivity:
Most materials have a high emissivity. Air spaces lined with low emissivity material such as aluminium foil have a much higher resistance because radiation is largely prevented. However, high emissivity should be assumed unless the air space is known to be lined with such a material giving the normal used value of 0.12 m²K/W

Table 12.11 Outside surface resistance, R_{so}, for stated exposure

Building element	Emissivity of surface	Surface resistance for stated exposure (m² K/W) Sheltered	Normal	Severe
Wall	High	0.08	0.06	0.03
	Low	0.11	0.07	0.03
Roof	High	0.07	0.04	0.02
	Low	0.09	0.05	0.02

The outside surface resistance for standard U-values is based on a wind speed at roof surface of 3 m/s. This corresponds to the values of outside surface resistance to normal exposure given above (0.06).
The effect of differing exposures is usually ignored for opaque structures. However, for glazing, the exposure must be taken into account. The exposure conditions are defined thus:

sheltered – up to third floor of buildings in city centres;

normal – most suburban and rural buildings and fourth to eighth floors of buildings in city centres;

severe – buildings on coastal or hill sites, floors above fifth in suburban or rural districts and floors above the ninth in city centres.

Table 12.12 Standard thermal resistances for unventilated air spaces, R_v

Type of air space		Thermal resistance (m²K/W) for heat flow in stated direction		
Thickness emissivity	Surface	Horizontal	Upward	Downward
5 mm	High	0.10	0.10	0.10
	Low	0.18	0.18	0.18
25 mm	High	0.18	0.17	0.22
or more	Low	0.35	0.35	1.06
High emissivity plane and corrugated sheet in contact		0.09	0.09	0.11
Low emissivity multiple foil insulation with air space on one side		0.62	0.62	1.76
5 mm gap behind normal plaster plasterboard*		0.14*		
20 mm gap behind foil-backed plasterboard		0.35		

*A reduction to 0.08 has sometimes been used to allow for the effect of dabs but the gap will most likely be greater than 5 mm allowing 0.14.
The increase in resistance of the 20 mm air space behind foil-backed plasterboard is due to the reflective nature of the foil.
Linear interpolation is permissible for air spaces intermediate between those given in the Table.

Table 12.13 Standard thermal resistance of ventilated air spaces, R_v

Type of air space (thickness 25 mm minimum)	Thermal resistances m²K/W)
Air space between asbestos cement or black metal cladding with unsealed joints, and high emissivity lining.	0.16
Air space between asbestos cement or black metal cladding with unsealed joints, and low emissivity surface facing air space.	0.30
Loft space between flat ceiling and unsealed asbestos cement sheets or black metal cladding pitched roof	0.14
Loft space between flat ceiling and pitched roof with aluminium cladding instead of black metal or low emissivity upper surface on ceiling	0.25
Loft space between flat ceiling and pitched roof lined with felt or building paper	0.18
Air space between tiles and roofing felt on building paper	0.12
Air space behind tiles on tile hung wall*	0.12
Air space in cavity wall construction	0.18

*For tile hung wall or roof, the value includes the resistance of the tiles.

Table 12.14 Typical thermal conductivity and thermal resistance of common walling materials

Material	Thermal conductivity W/mK	Typical thickness of materials	Thermal resistance m²K/W
Normal mortar	0.8	–	–
Dense facing blocks	1.4	100	0.07
Brickwork outer leaf	0.84	102	0.125
Dense plaster	0.50	13	0.026
Lightweight plaster	0.16	13	0.08
Plasterboard*	0.16	9.5	0.06
	0.16	13.5	0.084
External render	0.70	18	0.026
Tile hanging	0.84	12	0.14
Cavity insulation bats and internal boards	0.025	25	1.0
	0.025	35	1.40
	0.025	45	1.80
	0.,025	46†	1.84
	0.025	50	2.0
	0.035	25	0.714
	0.035	35	1.0
	0.035	45	1.29
	0.035	50	1.43
Cavity insulation fill	0.04	50	1.25
	0.04	75	1.88
	0.04	100	2.50

The above values are often used but should be amended where necessary to suit actual products. This particularly applies to insulations bats and boards, which may vary both in conductivity and thickness from the values given above.
† The inclusion of a 46 mm bat is for the purpose of use with Examples 1 and 2 and does not necessarily indicated an available size.
*When using plasterboard, the resistance of the air pocket between the material and the block may be allowed (see Table 12-12). Thus the resistance of plasterboard on dabs (5 mm gap) = 0.06 for the plasterboard plus 0.14 (but see Table 12.12) for the air pocket = 0.22 m²K/W.

12.10 Calculated worked examples

The following examples show the use of the Table 12-8 to Table 12.14 by adopting a range of materials. Example 1 shows the determination of the U-value of the wall by a simpler method for allowing for mortar bridging, and which may be compared with the fuller method shown in the Example 2. It will become evident, however, that for the wall construction taken there is no significant difference in U-value obtained by the following two methods or the simple tabulated approached, which was demonstrated by way of the example in 12.3, and resulted in a requirement for 46 mm of insulation. Example 1 gives a very slightly worse answer than Example 2 but shows the benefit and simplicity of the method used. Both Examples 1 and 2 use 46 mm of insulation but note this may not be a standard or indeed an available size and is included simply for the purpose of calculation, and for comparison with the tabulated method shown in 12.3 (This observation was also noted in 12.3.1). Analysis using an insulation thickness of 45 mm gives a U-value of 0.356, which is only just outside the requirement for the example and might be considered acceptable as values were generally calculated to two significant figures. Also brickwork is sometimes taken as 105 mm giving a resistance of 0.125, which together with other values determined to three significant places e.g. may achieve a U-value of 0.35 W/m² K. The examples although not necessarily using actual available materials will, nevertheless, help to show more clearly the differences in approach to thermal design and aid understanding for more complex or critical wall configurations.

12.10.1 Example 1 – Determination by basic approach

This example uses the same wall construction as used in 12.3, where the requirements were determined by the tabular approach. It shows a simple method of calculating the U-value of the wall using the basic $U = 1 + \Sigma R$ approach.

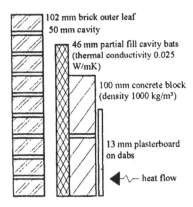

Figure 12.10 Cavity wall, partial fill with mortar bridges between masonry units

Construction: A brick faced double leaf cavity wall containing a 50 mm cavity, 46 mm cavity bat ($\lambda = 0.025$), a 100 mm (440 × 215) lightweight block (density 1000 kg/m³) Rb = 0.300 (see Table 12.7) and plasterboard on dabs on the inner leaf.

Outside surface resistance	Table 12.11	R_{so}	= 0.06
Resistance of brickwork	Table 12.9	R_2	= 0.12
Resistance of cavity	Table 12.13	R_{v1}	= 0.18
Resistance of cavity bat	Table 12.14	R	= 1.84
Resistance of 100 mm block	Table 12.7	R_b	= 0.30
Resistance of air pocket between plasterboard and inner leaf	Table 12.12	R_{v2}	= 0.14
Resistance of plasterboard	Table 12.14	R_3	= 0.08
Inside surface resistance	Table 12.10	R_{si}	= 0.12

The total resistance of the wall construction ΣR = 2.84

Thus $U = \dfrac{1}{\Sigma R} = \dfrac{1}{2.84} = 0.352$ W/m²K

12.10.2 Example 2 – Determination by proportionate area method

This example uses the same wall construction as shown in Figure 12.10 but its U-value is determined by the full

228

proportionate area method. The wall only contains one bridged section unlike the more complex wall shown in Approved Document L[12.9] but is easier to follow and can be compared directly with Example 1.

The wall is to be considered as inner and outer leaves with the boundary taken as half way through the cavity.

Resistance of inner leaf
Resistance through section containing block:

Resistance of half cavity	Table 12.12 0.18 ÷ 2	= 0.09 m²K/W
Resistance of cavity bat	Table 12.14	= 1.84 m²K/W
Resistance of block 1000 kg/m³	Table 12.9 0.10 (0.30	= 0.33 m²K/W
Resistance of air void under pb	Table 12.12	= 0.14 m²K/W
Resistance of plasterboard	Table 12.14	= 0.08 m²K/W
Inside surface resistance	Table 12.10	= 0.12 m²K/W
Total resistance through block section R_b		= 2.60 m²K/W

Resistance through section containing mortar:

Resistance of half cavity	Table 12.13 0.18 ÷ 2	= 0.09 m²K/W
Resistance of insulation	Table 12.14	= 1.84 m²K/W
Resistance of mortar	Table 12.14 0.10 ÷ 0.84	= 0.12 m²K/W
Resistance of air void under pb	Table 12.12	= 0.14 m²K/W
Resistance of plasterboard	Table 12.14	= 0.08 m²K/W
Inside surface resistance	Table 12.10	= 0.12 m²K/W
Total resistance through mortar section R_m		= 2.39 m²K/W

Combination of these resistance:
Fractional area of block section
$F_b = 440 \times 215 \times 450 \times 225 = 0.934$

Fractional area of mortar section
$F_m = 1.0 - 0.934 = 0.066$

The resistance of the inner leaf R_{inner} is then obtained from:
$R_{inner} = 1/((Fb + R_b) + (F_m + R_m)) = 1/(0.934/2.60 + 0.066/2.39) = 2.59$ m²K/W

Resistance of outer leaf:
This is treated as an unbridged element (mortar difference less than 0.1 m²K/W, Approved Document L[12.9] (see also Table 12.7).

Outside surface resistance	Table 12.11	= 0.06 m²K/W
Resistance of brick	Table 12.14 or Table 12.9 0.102 ÷ 0.84	= 0.12 m²K/W
Resistance of half cavity	Table 12.13 0.18 ÷ 2	= 0.09 m²K/W

Total resistance through
brick section R_{outer} = 0.27 m²K/W

Total resistance of wall:
The total resistance of wall is the sum of the resistances of the outer and inner leaves:
$R_{outer} + R_{inner} = 2.59 + 0.27 = 2.86$ m²K/W

U-value of wall:
The U-value of wall is obtained from $U = 1 + \Sigma R = 1 + 2.86 = 0.349$ W/m²K

12.11 Combined method

It should be noted that the combined method of calculation in BS EN ISO 6946[12.] is likely to be adopted for the next revision to Approved Document L. This is essentially the same as given in CIBS Guide A3[12.13]. The basis of this combined method is to calculate an upper and lower limit of the thermal resistance of the bridge part of the structure. The average of these limits is used to determine the U-value of the construction. Thus $U = 0.5(R_U + R_L)$. The equations shown below may give slightly worse answers than that given by the proportional area method as outlined in 12.6.1 and 12.6.2. Allowance for the effect of wall ties may also be included (see 12.12).

$$U = \frac{1}{0.5(R_L + R_U)}$$

$$R_L = R_{se} + \frac{1}{\dfrac{P_m \lambda_m}{l} + \dfrac{P_b \lambda_b}{l}} + R_p + R_{si}$$

$$R_U = \frac{1}{\dfrac{P_b}{R_{se} + \dfrac{l_b}{\lambda_b} + R_p + R_{si}} + \dfrac{P_m}{R_{se} + \dfrac{l_m}{\lambda_m} + R_p + R_{si}}}$$

Single leaf wall – combined method

$$U = \frac{1}{0.5(R_L + R_U)}$$

$$R_L = R_{se} + \frac{l_e}{\lambda_e} + R_a + \frac{1}{\dfrac{P_m \lambda_m}{l_e} + \dfrac{P_b \lambda_b}{l_i}} + R_{si}$$

$$R_U = \frac{1}{\dfrac{P_b}{R_{se} + \dfrac{l_e}{\lambda_e} + R_a + \dfrac{l_i}{\lambda_b} + R_p + R_{si}} + \dfrac{P_m}{R_{se} + \dfrac{l_e}{\lambda_e} + R_a + \dfrac{l_i}{\lambda_m} + R_p + R_{si}}}$$

Double leaf (cavity wall) – combined method

12.12 Effect of wall ties

The current proportional area method used to satisfy the Building Regulations does not require the inclusion of wall ties but this is likely to change when the regulations are next revised with the introduction of the combined method. The correction to be applied in order to allow for the additional heat loss due to wall ties in BS EN ISO 6946[12.19] and CIBS Guide A3[12.13] is $\Delta U_f = \alpha \lambda_f n_f A_f$ but need not be included if the adjustment in the U-value is less than 3%.

Where

α is the scaling factor for mechanical fixings, which is 6 for wall ties.

λ_f is the conductivity of the fixings, which is 17 W/mK for stainless steel or 50 W/mK for galvanised ties.

n_f is the number of fixings per square metre (commonly 2.96/m²)

A_f is the in m² cross sectional area of the fixing.

For typical stainless steel ties $\Delta U_f = 6 \times 7 \times 2.96 \times 0.0000608 = 0.018$ W/m²K.
And for galvanised ties ΔU_f will be in the region of 0.054 W/m²K.
The resulting U-value of the wall is $U = 1/R_T - \Delta U_f$
Where

$$R_T = 0.5 (R_U - R_L)$$

12.13 References

12.1   The Building Regulations 1965. HMSO, London.
12.2   The Building Regulations 1972 HMSO, London.
12.3   The Buildings Standards (Scotland Consolidation) Regulations, 1981. HMSO, London.
12.4   The Building Regulations (Northem Ireland) 1977. HMSO, London.
12.5   HMSO. Thermal Insulation (Industrial Buildings) Act 1057. HMSO, London. 1957.
12.6   The Building Regulations 1985. HMSO, London.
12.7   The Building Regulations 1991. HMSO, London.
12.8   London Building Acts 1930–1939. London Building (Constructional) By-Laws 1972 and 1974 (with subsequent amendments). Greater London Council.
12.9   HMSO Approved Document L, HMSO, London, 1995.
12.10   BUILDING RESEARCH ESTABLISHMENT. *BRE Digest 110: Condensation.* BRE, Garston.
12.11   BRITISH STANDARDS INSTITUTION. BS S2SO: 1975 *Code of basic data for the design of buildings: the control of condensation in dwellings.* BSI, London. 28 pp.
12.12   BUILDING RESEARCH ESTABLISHMENT. Thermal insulation – Avoiding risks. CRC, London, 1994. BR262.
12.13   chartered institute of building services. CIBS Guide, Section A3: Thermal properties of building materials. CIBS, London, 1980.
12.14   ROBERTS, J J. External insulation. Cement & Concrete Association (now British Cement Association), Wexham Springs (now Crowthorne), 1980. Reprint 4/80. 7 pp.
12.15   BRITISH STANDARDS INSTITUTION. BS 874: 1973 *Methods of determining insulating properties, with definitions of thermal insulating terms.* BSI, London. 40 pp.
12.16   DEPARTMENT OF THE ENVIRONMENT. DoE CircularS reference BRA/668/68, 14 September and 30 November, 1976.
12.17   AMERICAN SOCIETY OF HEATING, REFRIGERATING AND AIR-CONDITIONING ENGINEERS. Guide and data book. ASHRAE, New York, 1965.

12.18 AMERICAN SOCIETY OF TESTING MATERIALS. ASTM: C236–66: Standard method of test for thermal conductance and transmittance of built up sections by means of the guarded hot-box. ASTM, 1976.

Chapter 13
The resistance of concrete masonry to rain penetration

13.1 Introduction

BS 5628: Part 3 recognises that rain penetration is one of the commonest building defects. It also rightly indicates that it is essential to consider carefully design, detailing, workmanship and materials in relation to local exposure conditions if the incidence of rain penetration is to be minimized.

When determining the likely exposure of a building, the most exposed part should be given particular attention and this may affect decisions concerning the choice of design and materials for the whole of the building. BS 5628: Part 3 further advises that in cavity walls, some water will inevitably penetrate the outer masonry leaf in prolonged periods of wind-driven rain but proper design and positioning of the damp-proof systems will minimize the risk of penetration further into the building.

13.2 Classification of exposure to local wind-driven rain

The quantity of rain falling on a vertical surface, such as a wall, at any point depends on both the intensity of the rainfall and the wind speed.

Table 13.1, which is a reproduction of part of Table 10 in BS 5628, gives exposure categories defined either in terms of the local spell indices calculated using BS 8104 (see Note to Table 13.1) or in terms of the exposure categories that were given in CP 121: Part 1: 1973. These indices are not precise, since they are derived from inherently variable meteorological data. This variability has been reflected in the definitions of the exposure categories by overlapping the indices at their boundaries. Where exposure categories overlap the designer should decide which is the most appropriate category for the particular case, using local knowledge and experience. (Examples of constructions suitable for particular exposure categories are given in Table 11, BS 5628: Part 3 and which is reproduced in a slightly different form in Table 13.2 later in this chapter.)

During the drafting of BS 5628: Part 3, the technical committee recognized the need to develop a better

Table 13.1 Classification of exposure to local wind-driven rain

1 Exposure category	2 Local spell index calculated as described in BS 8104*	3 Exposure category in CP 121: Part 1: 1973
Very Severe	L/m² per spell 98 and over	Severe
Severe	68 to 123	
Moderate/Severe	46 to 85	Moderate
Sheltered/Moderate	29 to 58	
Sheltered	19 to 37	Sheltered
Very Sheltered	24 or less	

Note: *Reference is made to DD 93 in Table 10 of BS 5628: Part 3 but which has now been replaced by BS 8104.

method of assessing exposure to wind-driven rain than previously used for CP 121, which could be applied to many aspects of the design of masonry.

New data made available by the Meteorological Office made it possible to develop more accurate methods of assessment of local wind-driven rain than were available before.

From research and investigations of rain penetration on masonry walls two distinct aspects of exposure to wind-driven rain emerged:

(a) average moisture content of masonry, for which the local annual index is the most significant factor;
(b) rain penetration through the masonry, for which the local spell index is the most significant factor.

The method of assessment that is now used is given in BS 8104[13.6]. This code of practice represents a standard of good practice and therefore takes the form of recommendations. Although it is seen as being more accurate than the older driving rain index method, used for CP 121, it should be emphasized that neither index is precise enough to enable fine distinctions between degrees of exposure to be made. Hence the user should always take account of local knowledge and experience.

BS 8104 indicates that the quantity of rain failing on a vertical surface, such as a wall, at any point depends on both the intensity of the rainfall and on the wind speed. This Code contains rain maps developed from data amassed over a 33 year period from 1959 to 1991, which is used for the prediction of quantities of rainfall on vertical surfaces.

The features of this code of practice are that it allows calculations of driving rainfall for different orientations, and it allows corrections to be made for ground terrain, topography, local shelter, and the constructional characteristics of the building concerned. The data can be used to determine the annual average values to be calculated as well as quantities for the worst likely spell in any three-year period. Appendix A of the Code gives worked examples using the contained wind-driven rain maps.

Recommendations are given for two methods for assessing exposure of walls in buildings to wind-driven rain: (1) The local wall spell index method (which is a measure of the occurrence of periods of rain, lasting for several hours or even days and used for assessing rain penetration through masonry). (2) The local wall annual index method (which is a measure of the total quantity in a year and used to determine the average moisture content for assessing the risk of mould growth or similar).

The principal items/definitions used in BS 8104 are as follows:

- *Spell*: A period, or sequence of periods, of wind-driven rain on a vertical surface of given orientation. A spell is of variable length and can include several periods of wind-driven rain interspersed with periods of up to 96 h without appreciable wind-driven rain.
- *Airfield index*: The quantity of driving rain that would occur 10 m above ground level in the middle of an airfield, at the geographical location of the proposed wall.
- *Airfield annual index, D_A*: The average annual airfield index for a given direction, in litres per square metre per year.
- *Airfield spell index, D_S*: The airfield index for a given direction during the worst spell likely to occur in any 3 year period, in litres per square metre per spell.
- *Wall annual index, D_{WA}*: The quantity of wind-driven rain in litres per square metre per year at a point on a given wall, based on the airfield annual index and corrections for roughness, topography, obstruction and wall factors.
- *Wall spell index, D_{WS}*: The quantity of wind-driven rain in litres per square meter per spell at a point on a given wall, based on the airfield spell index and corrections for roughness, topography, obstruction and wall factors.
- *Line of sight*: The horizontal view away from the proposed wall, over a sector spanning about 25° either side of the normal to the wall.

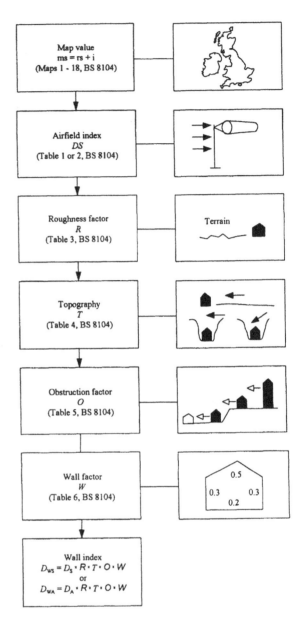

Figure 13.1 Stages and factors for determining the local wall index

- *Terrain roughness category*: A description of the general terrain upwind in terms of the average height and spacing of protective features, e.g. buildings, trees, hedges.
- *Terrain roughness factor, R*: A factor which allows for the general conditioning of the wind speed by the roughness of the terrain upwind of a wall.
- *Topography factor, T*: The effect of local topography on wind speed.
- *Obstruction factor, O*: This relates to shelter from the very local environment, and allows for obstructions, such as buildings, fences or trees, close to and upwind of the wall. It is also concerned with the 'Line of site'.

- *Wall factor, W*: The ratio between the quantity of water falling on the wall and the quantity falling in equivalent unobstructed space.

The procedure for the two methods is essentially the same but involves the selection of slightly different factors. Figure 13.1 shows diagrammatically the stages and factors that need to be determined in order to arrive at a wall's local spell index (two values may be determined – the local wall spell index (litres/m² per spell) and the local wall annual index (litres/m² per year). The important index in terms of rain penetration is the local wall spell index which can be used in conjunction with BS 5628 Part 3 (see Table 13.1 shown earlier) to assess the suitability of a given form of construction. The BS 8104 methods enable the amount of wind driven rain to be determined for each wall elevation according to its actual compass direction and exact geographical location. The procedure is as follows:

- Essentially the process starts by locating the building on one of 18 regional wind-driven rain maps (Figure 13.2), which contain defined subregions and contoured geographical increments (i) – the total range over all the maps is between –3 to + 16 (see example in Figure 13.3). A separate rose (similar to a compass face) is shown for each subregion, each giving 12 values (r_s) – the total range over all the maps is between 13 to 32 corresponding to different orientations. These values are combined to give a map value $m = r_s + i$.
- The next stage is to determine from this 'map value' the airfield (annual or spell) index from tables or from simple equations. The spell index (D_S) will range between 6.3 to 640 litres/m² per spell but more typically will lie between 30 to 300 litres/m² per spell. The table for the annual index (D_A) shows a range of 100 to 10,160 litres/m² per year.
- The next stage in the process is to assess the effect on wind speed of the general terrain upwind of the proposed wall in terms of the terrain roughness factor (R). This will be between 0.75 to 1.15.
- Following this it is necessary to assess the effect of topography (steeper than 1 in 20 upwind of the wall) in terms of the topography factor (T). This will result in another factor, this time, between 0.8 to 1.2.
- The text item is to assess the localized effects of nearby obstructions, giving close shelter to the wall, in terms of the obstruction factor (O). This aspect will give a factor between 0.2 to 1.0.
- The last modifying factor is to assess the effect of the characteristics of the proposed wall (its shape, height, overhangs, etc.) in terms of a wall factor (W). This will be between 0.2 to 0.5.
- Finally the wall spell index D_{WS} is determined from $D_{WS} = D_S \times R \times T \times O \times W$ (litres/m² per spell). The wall annual index D_{WA} is determined from $D_{WA} = D_A \times R \times T \times O \times W$ (litres/m² per year). The product of the above modifying factors indicates that the wall index can vary between 0.024 to 0.69 of the actual airfield index but more likely will be

between 0.2 to 0.4, with the outlying values relating to walls very well protected or very exposed.

13.3 Standard method of test for walling

The rain penetration of testing of masonry walling is covered by BS 4315: Part 2: 1970[13.3], as amended 1983. The Standard gives two methods of assessing the behaviour of a specimen of masonry: A one-minute intermittent spray test and a six-hour continuous spray test.

In the one-minute intermittent spray test water is applied for one minute at half hourly intervals at a rate of 0.5 L/min for each square metre of panel area. The test is carried out with a constant air pressure difference of 250 N/m² (25 mmH₂O) continuously applied across the specimen. This is approximately equivalent to the dynamic pressure head of a 20 m/s wind. The duration of the test is normally be 48 h but may be curtailed or extended according to the extent of penetration.

For the six-hour continuous spray the water is applied for a continuous period of 6 h/day over a number of consecutive days. The spray rate and pressure differential is the same as for the one-minute test. The test is continued until the rates of leakage, measured by the collection of water from the back face of the panel over the final hour of consecutive 6 h spray periods, agree to within 5% or to within 100 mL/(h.m²), whichever is the greater. The specimen may be new or one previously having undergone the one-minute test.

For special types of constructions, different spraying rates and air pressure differences across the specimen may be required.

BS 4315 indicates three means of assessing the performance of a wall:

(1) by using time lapse photography to record areas of dampness as the test progresses;
(2) by measuring the change of weight of the wall and hence the water take up;
(3) by collecting the amount of water passing through the wall.

The less severe test one-minute procedure (Method 1) is used to assess the behaviour of the specimen in terms of the formation of damp areas over the back of the specimen. Either test procedure (Method 1 or 2) can be used when assessing the behaviour of the specimen by change in weight. And the six-minute test procedure (Method 3) is used when assessing the behaviour of the specimen in terms of the rate of leakage measured by collecting the water which runs down the back of a single-leaf panel or down the cavity of a double-leaf panel (Method C). A typical rain penetration rig and panel under test is shown in Figures 13.4 and Figure 13.5.

Unfortunately, BS 4315 does not give any guidance on the interpretation of the results. If a cavity wall is built under laboratory conditions, for example, it is fairly clear that water penetration to the inner leaf is unlikely. In practice, however, a fairly permeable outer leaf may result in high water loads on standard details and may highlight workmanship faults. Some observations on the

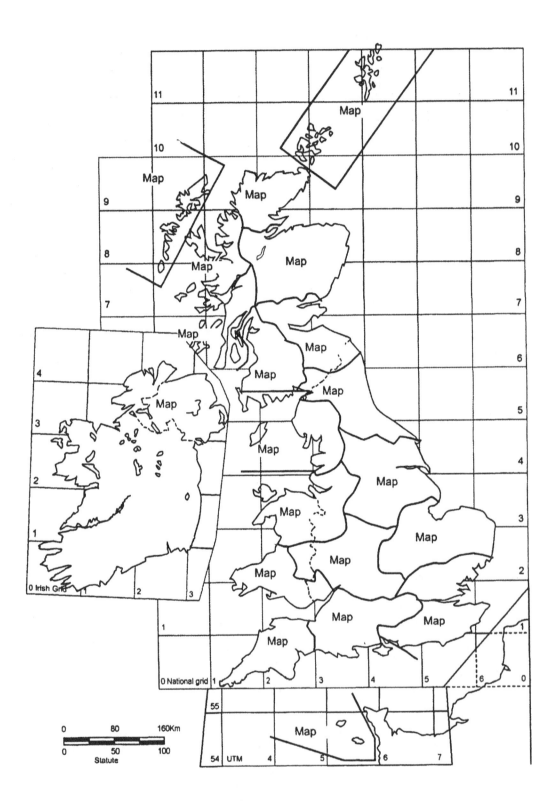

Figure 13.2 Index to wind driven rain maps (ref. BS 8104 – Appendix B)

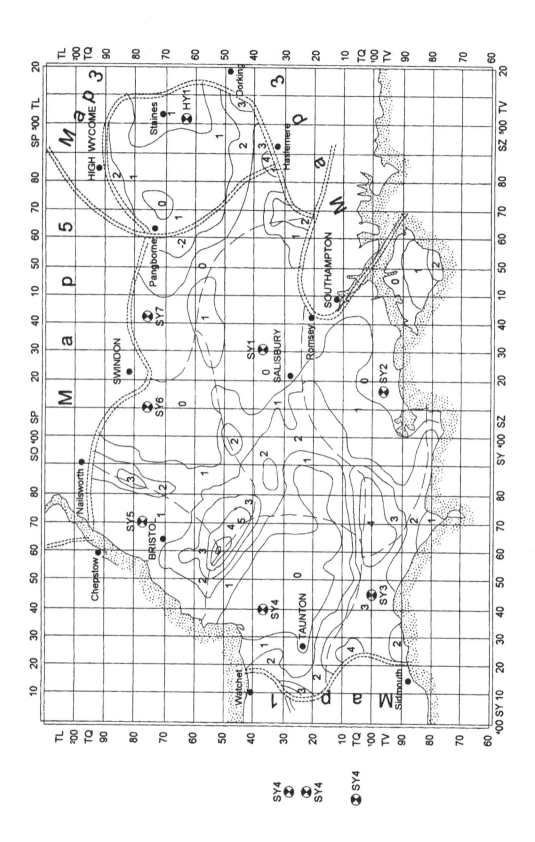

Figure 13.3

Figure 13.4 Control equipment for pressure box

interpretation of results of tests on concrete block masonry are given in the next section.

13.4 Alternative method of tests for walling

BS 4315 is the standard method of assessing the performance of walling against rain penetration. That test is essentially designed for laboratory work. There are problems in extending this to site evaluation, and an alternative air permeance test[13.4] has been proposed, although it has so far not been codified. It was developed to indicate the potential leakage of streams of water through the wall, in conditions of sustained driving rainfall. The method determines the in-situ measurement of the air permeance of walls on site. This makes use of a vacuum box and guard arrangement, which are held on the wall by excess atmospheric pressure. A dry seal is provided by flexible plastic foam. Measurements have been made on the outer leaves of over 80 houses with cavity walls. These had mean air permeances between 100 and 650 l.m^{-2} min^{-1} at 150 mm WG pressure. This implies that in rain storms unrendered outer can be expected to allow considerable volumes of water to enter the cavity. The air permeance equipment has also proved useful, for the preliminary assessment of walls to be used in wetting tests both for full-scale tests on houses and for small-scale tests on wallettes.

13.5 The suitability of various wall types to different exposure conditions

The cavity wall was developed as a simple method to produce rain resistant walling even when permeable masonry was employed. There is no doubt that given perfect workmanship a cavity wall should prevent rain penetration completely. Rain can, of course, penetrate through a single leaf of brick or block masonry by passing through the unit itself, through the mortar, or through any hair cracks which may be present between the unit and mortar, and it is possible for all three of these mechanisms to operate. Providing that the cavity is not bridged, the water may pass into the inner face of the outer leaf and so run down to the cavity tray or below the damp-proof course, as the case may be. This is assuming that the cavity tray is able to cope with the water load placed on it and that provision is made to allow the water to drain from the wall.

In practice, the present levels of workmanship on site mean that particular care needs to be taken in the selection of components. The following factors should be taken into account:

(1) wall ties and cavity trays are frequently found in practice to be extensively bridged with mortar droppings and other debris;
(2) perpend joints are often not properly filled but merely pointed up;
(3) coursing is sometimes lost resulting in wall ties which slope towards the inside of the wall.

It is apparent, therefore, that where local conditions are likely to be severe in terms of rain penetration, careful attention needs to be paid not only to careful site supervision but also to the choice of units so that potential leakage is minimized. Particular care may need to be taken where it is planned to fill or partially fill the cavity with an insulating material. Consideration should, of course, also be given to the use of cavities wider than 50 mm. Rendering of the outer leaf will considerably reduce the potential water load and this form of construction is, therefore, particularly suitable in exposed areas where insulation is to be contained within the wall.

It is possible to build a satisfactory solid wall which will adequately resist severe exposure, but the units themselves need to be resistant to moisture penetration

Figure 13.5 Wall panel in position against the pressure box

and the use of shell bedding may have to be considered. As with cavity walls, a rendering coat will significantly improve the performance of a wall, or a cladding system such as tile hanging will greatly increase the rain resistance of a cavity wall. Some solid wall constructions such as those with a posted lining are also suitable for use in exposed conditions.

Table 13.2, extracted from BS 5628 indicates the suitability of various masonry constructions for a given level of exposure. The table is given for general guidance only; the following factors are assumed to affect rain penetration: thickness, materials, use of rendering, special design and type of bed joint. Thicknesses less than those given may be suitable for some non-domestic construction, such as factory walls. Guidance for construction other than that mentioned below may be derived from experience or tests measuring resistance to rain penetration.

13.6 The suitability of various unit types

Using a BS 4315 test rig a number of rain penetration tests have been carried out on concrete masonry walls13.5. Unfortunately, it is not possible to relate standard block parameters, such as strength or density, to the subsequent behaviour of a wall built from the units. The following are comparisons of the performance of similar open textured facing blacks, considering:

(1) two different types of block from each manufacturer (Figures 13.5, 13.6 and 13.7);
(2) three different manufacturers for each type of block (Figures 13.8 and 13.9);
(3) the repeatability of the construction and test procedure for two of the panels tested.

A summary of the details of the blocks used for each of the above walls is presented in Tables 13.3 and 13.4.

The comparison between the two different types of block for each manufacturer in Figures 13.5, 13.6 and 13.7 shows that manufacturers 1 and 2 produce solid-with-voids blocks which perform better than their respective true solid blocks. The reverse is true for the hollow blocks of manufacturer 3, although there is very little difference between the two and neither performs well. In practice, it has been found that the use of solid-with-voids blocks to form the outer leaf of a cavity wall in areas of high exposure can lead to water build up in the cores and subsequently delay drying out. It is also possible that this water build up can increase the pressure differential across the wall and hence lead to more leakage.

Table 13.2 Assessment of resistance to rain penetration

(A) Thickness of single-leaf walls with or without rendering

Exposure category	Minimum thickness of masonry (excluding rendering and finishes) (see note 1)				
	Clay and calcium silicate masonry		Concrete masonry		
	Rendered (see note 2)	Unrendered (dense concrete)	Rendered	Rendered lightwight aggregate or autoclaved aerated concrete)	Unrendered (see note 2)
Very Severe	Not recommended. Cladding should be used.				
Severe	328	Not recommended	250	215	Not recommended
Moderate/Severe	215	Not recommended	215	190	Not recommended
Sheltered/Moderate	190	440	190	140	440
Sheltered	90	328	90	90	328
Very Sheltered	90	190	90	90	190

Note 1. Thickness of masonry is based on work sizes of masonry units i.e. tolerances are not included.
Note 2. Thicknesses of unrendered walls are based on the use of tooled joints filled completely with cement:lime and mortar.
Note 3. This table is intended to give guidance on the selection of forms of construction from the point of view of resistance to rain penetration only but other factors such as durability should be considered.

(B) Factors affecting rain penetration
BS 5628 provides guidance on a range of factors affecting rain penetration. This includes Applied external finish, Mortar composition, Mortar joint finish and profile and insulation. Each is assessed for increasing probability of rain penetration. (See BS 5628: Part 3 for details).

Table 13.3 Block properties

Block type*	Result of compression tests				
	Block strength (N/mm²)	Standard deviation (N/mm²)	Coefficient of variation (%)	Void content (%)	Material density (kg/m³)
1 (swv)	12.8	0.40	3.15	19.5	2007
1 (ts)	14.8	1.69	11.41	0	2008
2 (swv)	17.7	1.18	6.67	19.8	2124
2 (ts)	22.8	1.90	8.34	0	2095
3 (h)	14.1	0.80	5.63	29.2	2075
3 (ts)	20.8	1.58	7.57	0	2008

Table 13.4 A summary of the time lapse photography results

Wall number	Block type*	Time for 80% wetted area (hours)	% wetted area at			Final % wetted area
			6 hours	12 hours	24 hours	
1	1 (swv)	21.0	17	51	84	93
2	1 (ts)	11.4	37	83	97	100
3	2 (swv)	48.0	16	35	62	70
4	2 (ts)	2.2	97	98	98	100
5	3 (h)	3.2	97	98	100	100
6	3 (ts)	3.2	97	98	100	100
7	2 (ts)	1.4	91	93	97	97
8	1 (swv)	25.5	29	61	79	95

*the number refers to the manufacturer
(swv) = solid-with-voids
(ts) = true solid
(h) = hollow

Of the three different voided blocks, it can be seen from Figure 13.8 that those from manufacturer 3 performed best. In Figure 13.9 it is apparent that the true solid block from manufacturer 1 was the best of the three. Tables 13.2 and 13.3 indicate that in terms of strength and density each group of blocks is very similar and thus these parameters are not good indicators of performance.

The results of two pairs of panels (each pair of identical construction) are presented in Figure 13.10. It is apparent that there is good agreement between the results of the identical panels, one pair of which was built with true solid blocks and one pair with solid-with-voids blocks. Note that it is not possible to carry out repeat tests on the same panels – concrete blockwork walls improve with repeated testing due to internal efflorescence and autogenous healing. This effect has been reported widely in practice, where improvement is also aided by the deposition of dirt in the pores of the blocks.

The results shown in Figure 13.11 indicate that providing the wall does not reach 80% wetted area in the first four or five hours of test, the amount of water passing through a saturated wall is dependent on the area of dampness determined by testing to BS 4315. This must bring into question the usefulness of testing to BS 4315 for concrete block masonry for longer than four to five hours, since it is the quantity of water which is able to pass through the outer leaf of the cavity wall that is of prime importance. It seems fair to speculate that the greater the amount of water passing through the outer leaf, the greater the chance of water crossing to the inner leaf via some defect. An alternative approach to determine acceptable performance would, perhaps, be to record percentage dampness at four hours and set an arbitrary maximum value. Figure 13.12 shows that if the area of dampness at four hours is below 75%, the flow

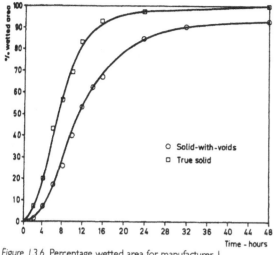

Figure 13.6 Percentage wetted area for manufacturer 1

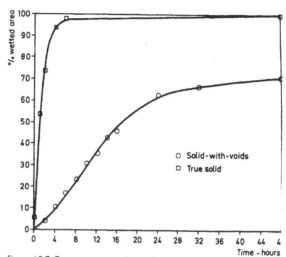

Figure 13.7 Percentage wetted area for manufacturer 2

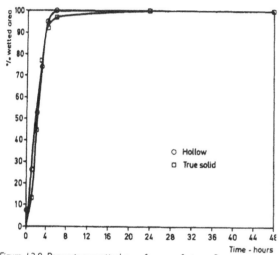

Figure 13.8 Percentage wetted area for manufacturer 3

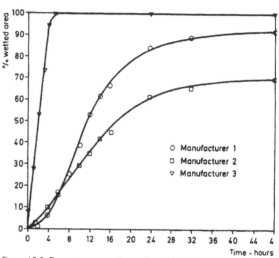

Figure 13.9 Percentage wetted area for voided blocks

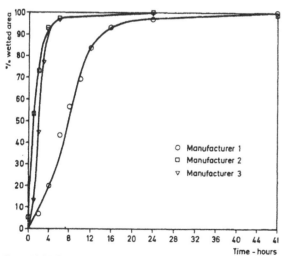

Figure 13.10 Percentage wetted area for true solid blocks

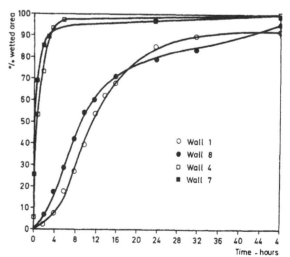

Figure 13.11 Percentage wetted area for repeatability tests

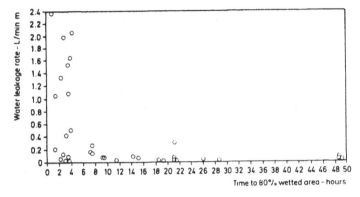

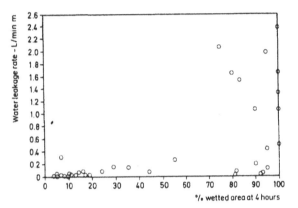

Figure 13.12 Water leakage rate through the fully saturated wall against the time taken for the dry wall to reach 80% wetted area

Figure 13.13 Water leakage rate through the fully saturated wall against the percentage wetted area at four hours

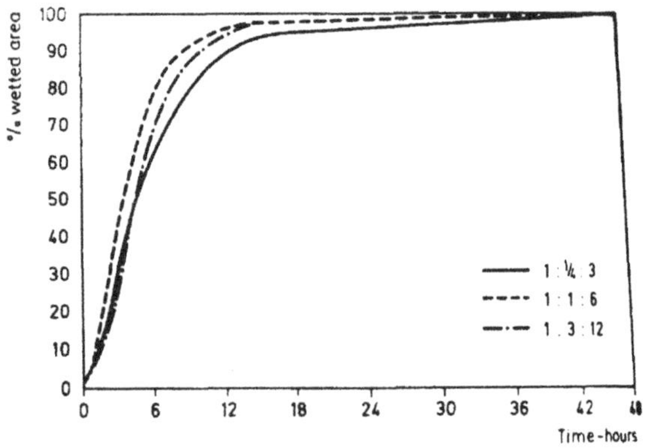

Figure 13.14 Percentage wetted area against time for three walls constructed using different grades of mortar but otherwise identical

rates for water through saturated walls are low and not very dependent upon the area of dampness.

Further limitations of the usefulness of damp area assessment are shown in Figure 13.13, which indicates the effect of mortar designation on otherwise identical walls. It is apparent that the differences between the areas of dampness are not significant, but when fully saturated the flow rates for walls built with 1:1/4:3, 1:1:6 and 1:3:12 (cement:lime:sand) mortars were 0.091, 0.268 and 0.154 litres/minute respectively, i.e. nearly three times as much water passing through the worst in relation to the best, but all giving almost identical time/percentage dampness curves. It must be pointed out, however, that all the suggestions made to date for acceptable performances relate only to the laboratory tests. A paradox exists that even blocks which perform quite badly in the laboratory are known to perform adequately in practice. One reason is doubtless the tendency for the performance of concrete block masonry walls to improve with time.

It has already been indicated that the open textured type of unit is generally the most susceptible to rain penetration. At the other end of the scale, units with very dense faces, as in the case of many bricks, can give rise to a great deal of run-off down the face of the wall due to low absorption of the unit. The results of tests on

concrete brickwork may be found in reference 13.6. The presence of so much water running down the face of the wall highlights any fine cracks which may tend to form between the unit and the mortar making the ingress of water by this path more likely. The use of the weakest grade of mortar practicable must be recommended to limit the size of any potential shrinkage cracks. With all types of masonry construction the use of deeply recessed or raked back joints often gives rise to problems. Mortar joints should be flush and lightly tooled.

13.7 Particular problems due to detailing and specification faults

The following section is not intended to be a fully comprehensive list of problems, rather an indication of some of the pitfalls which can occur in practice:

(1) *Lintel masked with block strips* This design detailing fault has occurred where facing blockwork has been employed and it was required to maintain the blockwork appearance over the face of a concrete lintel, as illustrated in Figure 13.15. If the window is set too far forward, water can penetrate through the porous block and behind the window frame.

(2) *Parapet design* There are a number of parapet details

240

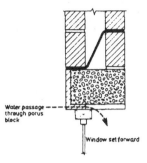

Figure 13.15 Leakage past lintel masked with block slips

in use, but some points to note regarding concrete masonry are as follows:

(a) unless a purpose made, dense impermeable coping is to be employed the coping stone should be laid on a damp-proof membrane extending the full width of the cavity,

(b) where a stepped damp-proof course is positioned, the damp-proof course material should be carried right through the face of the blockwork;

(c) particular care will be needed in the execution of joints in the damp-proof course at corners, and so on;

(d) the junction of the roof membrane with the stepped tray should be carefully considered.

(3) *Stepped trays* Where a wall intersection requires a stepped damp-proof course consideration should be given to providing purpose made preformed stepped trays to avoid the problems of carrying out such work on site.

(4) The damp-proof course material should always pass through the wall, material stuck to the surface of the blockwork is unlikely to be satisfactory.

(5) In areas of high exposure, blocks containing a void may allow water to build up in the cores and hence give rise to differential drying and weathering.

(6) Particular attention should be paid to site supervision so that perpends, and such like, are properly filled.

(7) The grade of sand used should be carefully considered so that potential frost attack is obviated in areas of high exposure where walls may remain wet for long periods.

(8) The cavity must be kept clear of mortar droppings and other debris.

Further information may be found in references 13.8, 13.9 and 13.10. The first of these provides a series of good practice notes. The other two give detailed comment on cavity insulated walls. Guidance on the suitability of wall constructions for use with fully cavity insulation is given in other standards including those of the NHBC[13.11]. Efficiency masonry housebuilding[13.12] provides both general construction details including the provision weatherproofing.

13.8 Remedial measures

Where a problem of rain penetration exists, it is important to establish the factors which are contributing to the ingress of water. A detailed examination of the walling should be carried out with due regard to the factors highlighted in *Section 13.5*. Joints around windows, doors, and so on, provide obvious local defects for the ingress of water, but if rain penetration appears to be through the wall itself, then a more thorough examination will be necessary. External examination will determine whether the mortar joint has been properly pointed and whether cracks have developed between the mortar and the unit. By drilling 10–12 mm diameter holes through the mortar joint it will be possible to insert an optical probe through the wall and examine the cavity for bridged wall ties and other defects which will need to be remedied. If, as a result of investigation, it is concluded that too much water is passing through the masonry, there are a range of options available. A render coating or paint system will be an effective method of reducing rain penetration through the wall, but will change the appearance of the building and will represent a substantial additional cost. A variety of colourless treatments are available but careful consideration will be required to ensure that the particular proprietary system will work with the type of unit used in the wall. Most colourless treatments have a limited life but this can be offset by the ability of concrete masonry to improve in terms of rain resistance with age.

Methods of assessing the effectiveness of masonry waterproofing materials have been described in NBS Technical Note 883[13.7], by the National Bureau of Standards in the USA. Although it is not intended to describe the test procedures contained therein in detail, the following criteria were considered in making the tests on the waterproofing materials:

(1) the material should produce a surface-resistance to water penetration;

(2) the material must be sufficiently permeable to water vapour to prevent accumulation of moisture in the wall;

(3) the material should be resistant to the formation of efflorescence.

Any assessment of a waterproofing product for use as a remedial measure with a given type of walling material should be made with the above points in mind. It should also be noted that the presence of high water content in the masonry can give rise to frost damage of the mortar, particularly in poorly graded or batched mortar, and a check should be made as to the soundness of the mortar, especially in vulnerable joints such as cavity trays in exposed parapets.

13.9 References

13.1 BUILDING RESEARCH ESTABLISHMENT. BRE Digest 127, BRE, Garston.

13.2 BRITISH STANDARDS INSTITUTION. BS 5628: Part 3: 1985. *Use of masonry. Part 3. Materials and components design and workmanship*. BSI, London, 1985, 100 pp.

13.3 BRITISH STANDARDS INSTITUTION. BS 4315: Part 2: 1970 (Amendment No 1, 1983) *Methods of test for resistance to air and water penetration. Part 2: Permeable walling constructions (water penetration)*. BSI, London, pp16.

13.4 NEWMAN, A J AND WHITESIDE, D. Water and air penetration through masonry walls – a device for the measurement of air leakage. *Br. Ceram. Trans. J.* pp.190–195, 1984.

13.5 ROBERTS, J J. Rain penetration problems with concrete blockwork. *Chemistry and Industry*. London, March 1980.

13.6 NEWSON, M J. Rain resistance of concrete brickwork. BCS, Crowthorne, 1987, CECA Reprint 6/87. 3pp.

13.7 NATIONAL BUREAU OF STANDARDS. Technical Note 883. NBS, Washington, USA, 1978.

13.8 BRITISH CEMENT ASSOCIATION. Good practice details. BCA, Crowthorne, 1984, CIW–1, 3pp.

13.9 BRITISH CEMENT ASSOCIATION. Cavity Insulated Walls – Specifiers Guide. BCA, Crowthorne, 1987, CIW–2, 12 pp.

13.10 BUILDING RESEARCH ESTABLISHMENT. Choosing between cavity, internal and external insulation. CRC, London, 1990. GBG5. 6pp.

13.11 NATIONAL HOUSEBUILDING COUNCIL STANDARDS. Section 6. NHBC, Amersham, 1999.

13.12 TOVEY, A K AND ROBERTS, J J. Efficiency masonry housebuilding. Detailing approach. BCA, Crowthorne, 1990. 48.054. 36pp.

Chapter 14
Sound insulation

14.1 Introduction

The general subject of sound insulation is so complex that an in depth study is outside the scope of this handbook. The bibliography at the end of this chapter lists some of the main reference documents related to sound insulation for those needing to make a more detailed study of this topic. This chapter in keeping with the rest of the publication serves to outline the principal objectives of sound insulation and to give general guidance on the sound performance properties of concrete masonry walls.

Sound insulation may be split into two basic parts:

(1) the control of sound between two separate locations, i.e. control of sound between, say, one room and another;
(2) the control of sound within a given location, i.e. control of the sound within a room itself.

More attention will be given to the former since although masonry can be used to aid acoustical control of internal sound, it is better known for its ability to act as a sound separator. The information given, particularly with respect to the sound insulation values of different constructions, is for general guidance and the performance of individual manufacturers' materials may differ from those given.

It is difficult to define sound insulation but it can be considered as the control of noise between the source and the listener. Noise is also difficult to define but in essence may be referred to as *unwanted sound* and it can, therefore, be said that this whole subject simply revolves around the control of unwanted sound. Noise, apart from being most annoying, can under certain conditions, induce stress, damage hearing and will most certainly disturb concentration, thereby affecting not just a person's well being but also their working efficiency. For these and other reasons it is desirable to reduce noise to a minimum wherever possible. Since noise can never be totally eliminated and as people vary widely in their sensitivity to noise, then any limits suggested may have to be adjusted to suit local circumstances ref BS 8233[14.1].

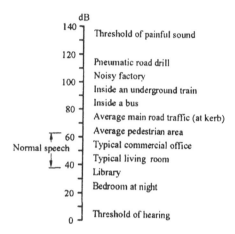

Figure 14.1 Decibel scale of sound intensities

Figure 14.1 indicates the decibel scale of sound intensified from various sources. It is desirable for separating walls between dwellings, and indeed walls between individual rooms of all types of buildings, to possess adequate sound insulation. The sound insulation between dwellings is a mandatory requirement but in other situations, apart from the possible requirements imposed by the Factory Inspector or the local authority, there is seldom any requirement for the control of noise between adjacent rooms and in many cases this is solely the responsibility of the designer, who should note the provision for enhanced sound insulation for dwellings outlined later in this chapter.

14.2 Sound absorption and sound transmission

The control of sound in buildings is dealt with in detail in BS 8233. In simple terms, to prevent an excessive level of noise occurring within a building, it is necessary to consider two properties of the construction material: (a) sound absorption and (b) sound transmission, as shown in Figure 14.2.

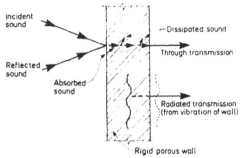

Figure 14.2 Sound transmission

14.2.1 Sound absorption

The sound absorption of a masonry wall is the measure of its ability to reduce the reflection of incidental sound. Masonry is often covered by some decorative material such as plaster and is rarely called upon to act as an absorbent at the surface of the wall. Lightweight aggregate masonry can reduce the reflection of sound by absorbing some of the sound into its porous surface[14.1,14.3,14.4]. The absorption characteristic of lightweight concrete is one of the reasons for the indicated improvement in the sound transmission associated with cavity walls built with this type of unit above that which might be expected from consideration of the mass law. Absorbents may be classified into several types, but with masonry materials these fall into two areas: (a) porous material and (b) cavity resonators. Masonry is not a true surface absorbent in the same way as soft felts but the slightly open texture of the surface of facing units can reduce reflected sound more than a hard, dense, flat surface such as plain concrete or steel sheeting. Hollow concrete blocks with slots or holes in one face can act as cavity resonators and again reduce reflected sound. Most cavity resonators, including those formed of hollow concrete blocks, are only efficient over a narrow band of frequencies and since sound source often covers a wide frequency range this particular form of absorbent is only useful where a noise peak occurs within a close band of frequencies, for example, with certain mechanical plant.

It has been suggested[14.5] that masonry cavity resonators can be designed to act over a broad range of frequencies and considerably reduce the reflected sound and pressure levels in situations where noisy machinery is being used. Such walls have been used adjacent to highways in an attempt to reduce the effect of traffic noise and to reduce the sound from large power transformers[14.9].

14.2.2 Sound transmission

Sound may be transmitted from one location to another in a number of ways as illustrated in Figure 14.3. The two principal sources of sound transmission are: (a) direct transmission and (b) flanking transmission. Both are equally important because adequate sound insulation will be difficult to achieve if one or other of these factors is neglected. It is important to emphasize that the potential sound insulation of, say, a 225 mm separating wall may not be realized if little attention is given to the question of flanking sound. Flanking sound is defined as any sound which travels from the source to the receiving room by paths other than through the separating element. The figures illustrate the general areas where loss of sound insulation may occur but other factors such as cracks or poorly filled perpend joints may explain why one building behaves differently to an otherwise identical building. Unfilled joints have been shown by Levitt[14.6] to reduce the sound insulation of masonry walls by some 3 decibels. In unrendered walls it is important that the units are not of the no-fines type of concrete, since the high permeability of the material will lower its expected insulation[14.6].

14.3 Sound tests

14.3.1 Assessment of sound insulation

Since the sound insulation between one building and another is not only dependent on the separating element but also on its associated structure, it has in the past been necessary to test a structure rather than just the dividing wall in a normal laboratory situation. However, tests[14.13] have shown that it is possible to obtain close agreement between a structure and a specially designed test chamber[14.14]. The sound insulation achieved with walls of lightweight masonry in this test chamber are reported in a CIRIA publication[14.16]. A number of field measurements have also been reported by the Building Research Establishment[14.10,14.11,14.12]. Testing the wall in an appropriate acoustic chamber gives an indication of the potential sound insulation of the wall, but consideration must be given to the possible effects of flanking transmission and workmanship. Section 4 of Approved Document E[14.18] allows test chamber evaluation for new construction in housing as a means of meeting the functional requirements for walls. For details of the test chamber construction consult the Building Research Establishment, Watford WD2 7JR.

14.3.2 Measurement of sound transmission

14.3.2.1 BS 2750 and BS 5821

The measurement of sound transmission is covered by BS 2750: 1980: *Methods of measurement of sound insulation in buildings and of building elements. Parts 1, 3, 4, 6 and 7*[14.7]. The sound insulation of a wall is measured between two opposite facets of the wall at 16 one-third active band frequencies between 100–3150 Hz. The results are then expressed as a single-figure weighted value according to the method given in BS 5821[14.15]. The single-figure value is obtained by moving the reference curve until the sum of unfavourable deviations divided by 16 reaches a value close to 2 without exceeding it, and then reading the reference curve dB value at 500 Hz. A typical result is plotted in Figure 14.4 together with the appropriate reference curve. A number of such tests are used to comply with Building Regulations (see 14.5 statutory regulations).

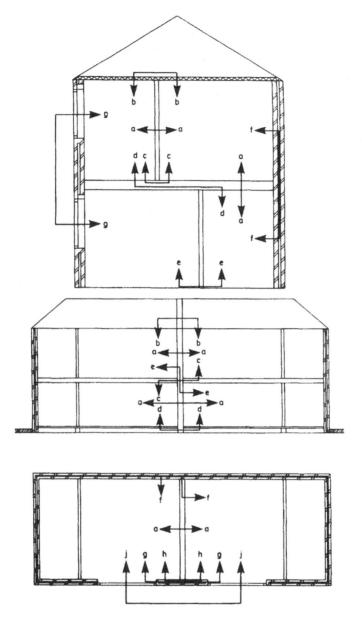

Figure 14.3 Direct and flanking sound within dwellings

Unfortunately the simple method of assessing sound insulation based on the average decibel reduction over the frequency range 100–3150 Hz does not always give a reliable indication of the sound performance of the wall since it is possible for a low value of insulation at one frequency to be offset by a high value at another frequency. In addition this method does not take into account the subjective aspects of insulation.

An attempt to take into account test results to the whole and to some extent the subjective aspects of sound insulation is made in the Building Regulations by requiring the tests to be assessed against mean and individual values. The BS 5821: 1980[14.15] method of rating sound insulation in building elements also

attempts to take into account the subjective aspects of sound insulation. An example of a test result for a 100 mm internal wall and its appropriate single value number from BS 5821 is shown in Figure 14.4. The reference values for airborne sound are shown in column two of Table 14.1.

14.3.2.2 EN ISO 717

The details above relate to determining sound insulation in accordance with the usual British Standard BS 5821. However, other countries use different methods of obtaining a single figure value. These are based on calculating what the sound level difference across a wall

Table 14.1 Showing AIRBORNE calculation of single number quantities and spectrum adaption terms (ISO 717)

Frequency	Reference values for airborne sound	Reference values shifted by -8	$D_{nT,i}$	Unfavourable deviation	Spectrum No. 1 to calculate C	$L_{i1} - D_{n,i}$	$10(L_{i1}-D_{nT,i})/10$	Spectrum No. 2 to calculate C_{tr}	$L_{i1} - D_{nT,i}$	$10(L_{i1}-D_{nT,i})/10$
Hz	dB	dB	dB	dB	dB	dB	dB x 10^{-5}	dB	dB	dB x 10^{-5}
100	33	25	32.8		-29	-61.8	0.07	-20	-52.8	0.52
125	36	28	30.5		-26	-56.5	0.22	-20	-50.5	0.89
160	39	31	32.1		-23	-55.1	0.31	-18	-50.1	0.98
200	42	34	31.9	2.1	-21	-52.9	0.51	-16	-47.9	1.62
250	45	37	32.4	4.6	-19	-51.4	0.72	-15	-47.4	1.82
315	48	40	34.4	5.6	-17	-51.4	0.72	-14	-48.4	1.45
400	51	43	37.5	5.5	-15	-52.5	0.56	-13	-50.5	0.89
500	52	44	40.9	3.1	-13	-53.9	0.41	-12	-52.9	0.51
630	53	45	43.9	1.1	-12	-55.9	0.26	-11	-54.9	0.32
800	53	45	48.5		-11	-59.5	0.11	-9	-57.5	0.18
1000	54	46	48.1		-10	-58.1	0.15	-8	-56.1	0.25
1250	55	47	50.3		-9	-59.3	0.12	-9	-59.3	0.12
1600	56	48	45.6	2.4	-9	-54.6	0.35	-10	-55.6	0.28
2000	56	48	45.8	2.2	-9	-54.8	0.33	-11	-56.8	0.21
2500	56	48	44.1	3.9	-9	-53.1	0.49	-13	-57.1	0.19
3150	56	47	44.1	1.0	-9	-56.0	0.25	-15	-62.0	0.06

Sum of unfavourable deviations = 31.5
Check ^ < 32.0 ? OK

$D_{nT,w}$ = Value of shifted reference curve at 500 Hz = 44

sum = 5.59   -10 lg 5.59 = 42.5   C = 43 - 44 = -1
$D_{nT,w}$ - C = 44 + -1 = 43

sum = 10.29   -10 lg 10.29 = 39.9 C_{tr} = 40 - 44 = -4
$D_{nT,w}$ - C_{tr} = 44 + -4 = 40

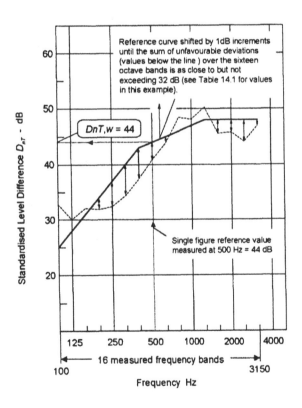

Reference curve shifted by 1dB increments until the sum of unfavourable deviations (values below the line) over the sixteen octave bands is as close to but not exceeding 32 dB (see Table 14.1 for values in this example).

$DnT,w = 44$

Single figure reference value measured at 500 Hz = 44 dB

16 measured frequency bands

Frequency Hz

Figure 14.4 Evaluation of airborne sound tests

or floor would be for a sound source with a particular frequency content. Two common source spectra are used. One is 'pink noise', which gives equal energy at the 16 one-third octave bands, and the other is a typical traffic noise spectrum. The source spectra have a frequency weighting applied to simulate that of the human ear. To standardise the use of both methods of giving a single figure value, the relevant British Standard (BS EN ISO 717-1: 1997[14.18]) has recently been revised. The Standard includes 'spectrum adaptation terms' that can be added to $D_{nT,w}$ to give the other quantities. The spectrum adaptation term for pink noise is known as C, and that for traffic noise as C_{tr}.

The sound level difference for pink noise = $D_{nT,w}$ + C, and the sound level difference for traffic noise = $D_{nT,w}$ + C_{tr}.

The sixteen frequencies used in the definition of $D_{nT,w}$ remain as the fixed values defined some years ago (see Table 14.1). A notable advantage of the methods of giving a single figure value by defining a source spectrum is that the source spectrum can be extended in frequency range if so desired.

The spectrum adaptation terms are expressed in respect to the frequency range used. For example $C_{\sim,100-3150}$ for a frequency range 100–3150 Hz (as used for the Building Regulations) or $C_{\sim,50-3150}$ for a frequency range 50–3150 Hz.

The ISO standard indicates what types of noise the two terms might be used to assess. These are: C – Living activities (talking, music, radio), Children playing, Railway traffic medium and high speed, Highway traffic > 80 km/hour, Jet aircraft (short distance) and Factories emitting mainly medium and high frequency noise, and

C_{tr} – Urban road traffic, Railway traffic at low speed, Aircraft (propeller driven), Jet aircraft (large distance), Disco music and Factories emitting mainly low and medium frequency noise.

Both of the adaptation terms include music: The term C mentions music together with talking and the radio, and the traffic noise adaptation term C_{tr} mentions disco music. The implication being drawn from this is that C is appropriate for music from sources of comparatively low sound power with little bass output such as portable radios, whereas C_{tr} is appropriate for powerful hi-fi systems with a strong bass response. It is the latter type of music that gives rise to most complaints in attached dwellings and the one most likely to be used for Building Regulation purposes in due course.

The sound level spectra to calculate the adaptation terms are given in Table 14.1, column 6 for C, and column 9 for C_{tr}.

The single figure value ($D_{nT,w}$) of the test results of a measurement made in accordance with ISO 140 is again determined by shifting the relevant reference curve in steps of 1 dB towards the measured curve until the sum of unfavourable deviations (values less than the reference value, measured over the 16 one-third-octave bands) is as close as possible but not exceeding 32 dB.

The spectrum adaptation terms, C_j, in decibels, is calculated (using a defined sound spectra – See Table 14.1) from the following equation:

$$C_j = X_{Aj} - X_W$$

where

j is the index for the sound spectra Nos. 1 and 2;

X_W is the single-number quantity calculated according to the procedure;

X_{Aj} is calculated from

$$X_{Aj} = -10 \lg \Sigma 10^{(L_{ij} - x_i)/10} \text{ dB}$$

where

i is the index for the one-third-octave bands 100 Hz to 3150 Hz;

L_i are the levels given in 4.3 at the frequency i for the spectrum j;

X_i is the normalized sound level difference $D_{n,l}$, 4.5 Calculation of spectrum adaptation or standardized sound level difference terms D_{nT}, at the measuring frequency i given to the nearest 0,1 dB.

An example of the determination of $D_{nT,w}$ is given in Table 14.1, which are –1 and –4 below the normal $D_{nT,w}$ value.

14.4 Mass law

An indication of the sound insulation performance of a wall may be estimated by reference to what is known as mass law. In simple terms the mass law states that there is a relationship between the mass of a wall and its sound insulation, the typical relationship commonly taken being shown in Figure 14.5. This law indicates that a wall of a given mass will provide a certain defined average decibel reduction and that walls of similar mass will behave in the same way. In practice, however, this is

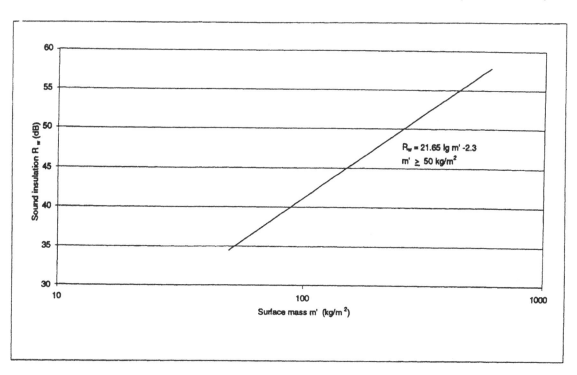

$$R_w = 21.65 \lg m' - 2.3$$
$$m' \geq 50 \text{ kg/m}^2$$

Figure 14.5

247

not absolutely correct, since there are several factors which affect the sound insulation of an element, such as stiffness, permeability, absorption characteristics, and so on, all of which are variables, so that two walls being of the same mass but of different dimensions or elasticity may well perform differently. Despite the variations which can occur between different walls, the mass law is nevertheless a most valuable aid in estimating the likely performance of a wide range of wall constructions.

It is important to note that the mass law applies basically to solid, relatively impervious, well constructed, single leaf walls. The typical variations to these three basic points are dealt with as follows:

Solid/hollow units Where a wall is built from hollow blocks, providing it is still relatively impervious to the passage of air (see Permeability) then it may perform in a similar way to that of an impervious solid wall of the same mass, although some loss may occur depending on the configuration of the block.

Permeability Walls can only be expected to perform as indicated by the mass law if they are relatively impervious to the passage of air. Thus, solid concrete blocks and bricks can be expected to perform as the mass law suggests, but some open textured, permeable, unfinished concrete block walls may fall short of the expected insulation since sound can travel through the fine pores of the material[14.6] (See *Section 14.2.1*). For this reason it is often necessary to seal masonry walls with plaster for them to perform to their maximum efficiency. This will also assist in reducing the effect of unfilled mortar joints. Sealing with two coats of cement based paint has been shown to be effective in an open textured wall (both solid and hollow units) enabling it to perform as expected from the mass law. The effect of using plasterboard on separating walls has been rather varied which may be due to the permeability characteristics of the wall and workmanship factors. Thus plastered walls have tended to be regarded as being more reliable than walls with plasterboard finish, although this was not the case with the tests reported by CIRIA[14.16].

Single leaf/cavity walls The sound insulation of a single leaf masonry wall, as previously explained, is largely related to its surface mass, provided that there are no direct air paths through the wall.

With cavity walls it is indicated that the sound insulation is related not only to the surface mass but also to the width of the cavity and the rigidity and spacing of the wall ties. A cavity wall with a 50 mm nominal cavity with leaves connected by wire ties may be expected to have a resistance to sound transmission similar to that of a solid masonry wall of the same surface mass. An improvement in the resistance to sound transmission can be achieved by lowering the coupling effect between the leaves, i.e. by widening the cavity or by omitting the ties. As an example, using Schedule 2 of the Building Regulations as a basis, a cavity wall having a 75 mm cavity and built with lightweight concrete block leaves connected by wire ties at standard spacings may have an improvement of up to 4 decibels over that expected from a single leaf wall of similar surface mass. Some of this improvement may be due to the lightweight concrete which gives better energy dissipation within the cavity. Some published results[14.6,14.10], however, show that the results may be rather variable. A similar improvement in the resistance to sound transmission can also be expected to occur, whatever the unit, where the leaves are isolated from each other, i.e. no ties[14.6,14.16].

Although no general figures can be quoted, it has been found that increasing the number of ties or using more rigid ties will reduce the resistance to sound transmission through the wall.

Walls with openings Where a gap or hole occurs in masonry walls, such as a window or door opening, there will be sound transmitted through the opening independent of the basic performance of the wall, resulting in a loss of insulation disproportionate to the small area of the opening compared to the area of the wall. An estimate of the value of sound insulation of a wall with an opening may be made by reference to Figure 14.6 which indicates the theoretical loss of sound insulation of a non-uniform partition. Insulation values for doors and windows may be found in CIRIA Report 127[14.21], but typical values are 15 dB for light doors and 20 dB and 40 dB for single glazed and double windows respectively. A typical plastered 225 mm concrete brick wall will have an average sound insulation in the order of 54 dB, but with 50% single glazing (24 dB) the difference in insulation is 54–24 = 30 dB and the ratio of area is 1:1 which indicates that the basic wall insulation will be lowered by some 27 dB. Thus the expected sound insulation of the wall will be 54–27 = 27 dB. If double windows were used (40 dB) the difference in insulation becomes 14 dB for the same area ratio 1:1 which indicates a loss in insulation of some 11 dB and hence the sound insulation of the wall will be 54–11 = 43 dB.

14.5 Statutory regulations

The Building Regulations England and Wales require walls that separate dwellings should have adequate resistance to the transmission of sound. Ways of showing how this requirement can be met may be found in Approved Document E[14.8], issued by the Secretary of State. One approach is to select from a number of construction specifications that are considered to meet the requirement. If these specifications are followed, there is no need to demonstrate the acoustic performance by means of a test. However, industry experience has shown that a number of construction faults occur that affect sound insulation despite the guidance given in Approved Document E. In order to address this matter a BRE report[14.19], has been prepared with the assistance of various industries. This gives:

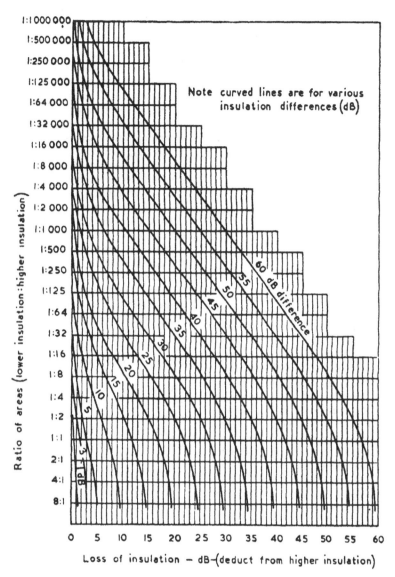

Figure 14.6 Nett sound reduction of non-uniform wall

- More detailed advise based on that given in Approved Document E where experience has shown that confusion arises;
- Advice on design problems that are not covered by Approved Document E;
- Advice for supervisors on the key things that must be done to get good sound insulation; and
- Site check lists to simplify the checking of important sound insulation features.

Approved Document E gives an alternative way of meeting the functional requirement by site testing. This gives designers or builders that wish to provide dwellings that do not follow any of the standard specifications, the option to carry out tests on samples of the construction in order to show that can it meet the required performance targets and hence be satisfactory. However, Approved Document E explicitly states that a failure of a new construction to meet the test requirements is not in itself evidence of failure to meet the requirements of the Building Regulations.

However, if the construction is tested, then evidence of meeting the requirements may be demonstrated if the individual and mean values of standardised level difference $D_{nT,w}$ are not less than those shown in Table 14.2.

The basis of these limits is that all types of construction are expected to have some variability in sound insulation. However, if a large number of examples of any construction type were tested, satisfactory constructions would be expected to have a mean performance of at least the requirements for eight tests. The implication is

Table 14.2 Requirements for airborne sound insulation values of wall (dB)

Mean value		Individual value
Tests on up to 4 pairs of rooms	Tests on up to 8 pairs of rooms	
53	52	49

Note: Floors require 1 dB less

also that if only one test is carried it should achieve a value of at least 53 dB.

Evidence of meeting the functional requirements may be demonstrated by testing the construction in an approved type of test chamber (For details of test chamber construction consult the Building Research Establishment). In this case the prescribed value to be met is 55 dB.

Some of the Approved Document constructions of the Building Regulations are indicated in Figure 14.7. The value of mass required as shown in Figure 14.7 should be taken as the total mass of the wall including the weight of mortar and plaster. It is also generally acceptable to consider the density of material at 3% moisture content by volume, i.e. dry density of material plus 30 kg/m3. For the construction to be acceptable to the Regulations, the walls shown in Figure 14.7 must adjoin or be positioned next to the flanking wall as shown in Figure 14.8(a). For the constructions shown in Figure 14.7 to satisfy the Building Regulations there are certain additional requirements concerning their relationship to the adjacent or structure which are also illustrated.

Additional guidance is given in Approved Document E[14.8] including for step and/or staggered dwellings.

For more information on wall type 3 shown in Figure 14.8 see reference 14.22.

14.6 Enhanced sound insulation

Sound is increasingly becoming an issue and because of this, some organisation have a wish to specify dwellings with higher levels of sound insulation. Information on this topic was not readily available but there is a pending publication[14.20] that will provide guidance on appropriate levels of enhanced insulation and how it can be achieved. The companion volume to reference 14.20 gives detailed advice on ways of securing the expected performance from structures built to comply with the Building Regulations.

The Building Regulation set the minimum requirements for sound insulation of separating walls and floors in dwellings. And warranty organisations such as NHBC and Zurich Municipal have some additional requirements for the airborne insulation of certain internal wall or partitions. There is also some guidance on acceptable levels of internal noise levels due to external sources.

The enhanced insulation levels given in reference 14.20 uses the existing Building Regulation requirements as a baseline, which for airborne sound is 52 dB.

The smallest change in noise level that people can perceive under ideal conditions is 1 dB. However, under normal conditions noise levels have to change by 3 dB to be perceptible. Therefore, the minimum worthwhile increase in sound insulation is 3 dB.

With this in mind then enhanced sound insulation for separating walls in terms of $D_{nT,w}$ would imply a value of 55 dB $D_{nT,w}$ with 95% of results at least 52 dB DnT,w. However, reference 14.20 will go further, and use the level difference given by the wall or floor when the source is A-weighted traffic noise, which is numerically equal to $D_{nT,w} + C_{tr}$ as outlined earlier.

From a study of results for a selection of walls having enhanced sound insulation, the following values would give a sound insulation approximately 3 dB better than the current Building Regulation test requirements and penalise poor performance at low frequencies. Enhanced airborne sound insulation for separating walls is to be defined as: Mean of a large number of tests at least, 50 dB $D_{nT,w} + C_{tr}$, with 95% of results at least,47 dB $D_{nT,w} + C_{tr}$.

The existing requirements for the sound insulation of certain internal partitions (typically limited to walls between a toilet and a habitable room) is that they should obtain a certain level of sound insulation (typically 38 dB) when tested in a laboratory (Note: There are no requirement for insulation tests in-situ). An increase over existing sound insulation of 5 dB is proposed as representing enhanced sound insulation for internal walls/partitions, which would provide results in the range 38–43 dB and more likely 40–43 dB. These values are currently without the additional spectrum adaptation in view of the 5 dB proposed enhanced rather than 3 dB as mentioned earlier for separating walls.

14.7 Estimation of the sound insulation of walls

Adopting the concept of the mass law, together with the information given previously on the effect of cavities and permeability on the sound insulation of a wall, Figure 14.9 may be used to estimate the basic potential sound insulation (sound reduction) of a wall. The mass given relates to the dry density of the wall with allowance being made in the position of the sloping line to cater for the effect of mortar joints and moisture within the wall in service. The basic insulation, as shown in Figure 14.9, may then be modified by the accompanying factors, to give an indication of the potential sound insulation of the wall. Alternatively, an indication of the sound reduction of the wall construction may be determined from Figure 14.5 by including the mass of all wall components, i.e. units, mortar, plaster, and so on. It is important to note the possible significant effect which flanking transmission may have on the actual structure (see *Section 14.2.2*).

When using Figure 14.9 it should be noted that:

(1) The figure applies to walls which are plastered or have two coats of cement based paint. Plasterboard

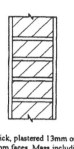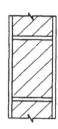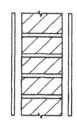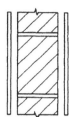

Brick, plastered 13mm on both room faces. Mass including plaster 375 kg/m²

Concrete block, plastered 13mm on both room faces. Mass including plaster 415 kg/m²

Brick, plaster board 12.5 mm on both room faces. Mass including plasterboard 375 kg/m²

Concrete block, plasterboard 12.5 mm on both room faces. Mass including plasterboard 415 kg/m²

(a) Wall type 1 - solid masonry

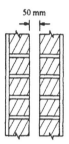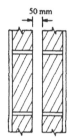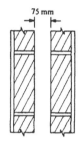

50 mm

50 mm

75 mm

Two leaves of brick with 50 mm cavity, plastered 13 mm on both room faces. Mass including plaster 415 kg/m². May have plasterboard finish where a step and/or stagger of at least 300 mm is used.

Two leaves of concrete block with 50 mm cavity, plastered 13 mm on both room faces. Mass including plaster 415 kg/m²

Two leaves of lightweight aggregate block (max. density 1600 kg/m³) with 75 mm cavity, plastered 13 mm or drylined 12.5 mm on both room faces. Mass including finish 300 kg/m² May be reduced to 250 kg/m² where a step and/or stagger of at least 300 mm is used.

(b) Wall type 2 - cavity masonry

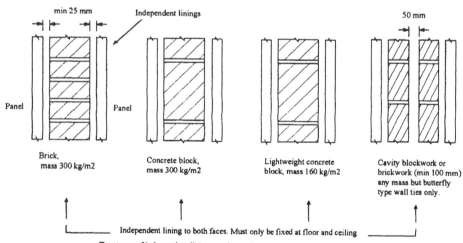

min 25 mm   Independent linings

50 mm

Panel   Panel

Brick, mass 300 kg/m2

Concrete block, mass 300 kg/m2

Lightweight concrete block, mass 160 kg/m2

Cavity blockwork or brickwork (min 100 mm) any mass but butterfly type wall ties only.

Independent lining to both faces. Must only be fixed at floor and ceiling

Two types of indepentdent linings can be used:
(1) Two sheets of plasterboard with a cellular core with a mass of at laest 18 kg/m², or
(2) Two sheets of 12.5 mm plasterboard laminated together, total min mass 20 kg/m², if supported by a framework or if no framework two 25 mm sheets with total mass 25 kg/m²

(c) Wall type 3 - masonry between isolating panels

Figure 14.7 Approved Document E separating wall constructions

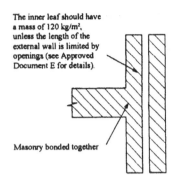
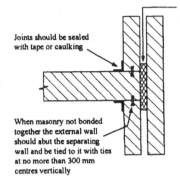

The inner leaf should have a mass of 120 kg/m², unless the length of the external wall is limited by openings (see Approved Document E for details).

Masonry bonded together

Joints should be sealed with tape or caulking

When masonry not bonded together the external wall should abut the separating wall and be tied to it with ties at no more than 300 mm centres vertically

Where the external wall has a cavity, the cavity should be sealed with a flexible closer.

Note: Similar conditions apply when the separating wall is a cavity wall but the cavity in the separating wall should not be stopped by any material which connects the leaves together rigidly. Mineral wool is acceptable.

Figure 14.8 Approved Document E junction requirements

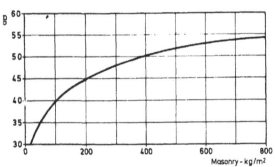

Figure 14.9 Approximate relationship between sound insulation and block weight

may give similar results but some loss in insulation might be expected, particularly with open textured permeable units as compared to a plastered wall.

(2) The figure may be modified typically by:
- (a) −3 dB – for untreated walls
- (b) +4 dB – for cavity walls of lightweight blocks
- (c) + 4 dB – for an isolated cavity (i.e. no ties)

It is important to note that the above is to provide only an indication of performance and would need to be confirmed by testing prior to any adoption for design or consultation.

Where an estimation of the sound insulation is to be made using the basic mass law shown in Figure 14.5, the mortar and render may be taken to have a density of 1800 kg/m³, lightweight plaster 800 kg/m³ and dense plaster 1500 kg/m³.

14.8 References

14.1 BRITISH STANDARDS INSTITUTION BS 8233. Sound insulation and noise reduction for buildings – Code of Practice.

14.2 LENCHUK, P. Noise. How to control it in structures. Florida Concrete and Products Association Inc, Florida, 1970. pp 1–21.

14.3 TOENNIES, H. Sound reduction properties of concrete masonry walls. National Concrete Masonry Association, Arlington, 1955, pp 1–24.

14.4 COPELAND, R E. Controlling sound with concrete masonry. *Concrete Products*, Vol 68, July 1965.

14.5 DIEHL, G M. *Compressed Air Magazine*. USA, 1976.

14.6 LEVITT, M, LEACH, C H C, AND WILLIAMSON J J. Factors affecting the sound insulation of lightweight aggregation concrete block walls. *Precast Concrete*, Vol 2, No 1, March 1971. pp 161–169.

14.7 BRITISH STANDARDS INSTITUTION. BS 2750: 1980 (1993) *Methods of measurement of sound insulation in buildings and of building elements*. Part 1 to 8. BSI, London.

14.8 HMSO. Approved Document E. HMSO, London, 1991. 39 pp.

14.9 Tuning concrete masonry for transformer enclosures. Project Review. Concrete Masonry Association of Australia.

14.10 SEWELL, E C, ALPHEY, R S, SAVAGE, J E AND FLYNN S J. Field measurements of the sound insulation of plastered cavity masonry walls. *BRE Current Paper CP 4/80*.

14.11 SEWELL, E C, ALPHEY, R S, SAVAGE, J E AND FLYNN, S J. Field measurements of the sound insulation of plastered solid blockwork walls. *BRE Current Paper CP 580*.

14.12 SEWELL, E D, ALPHEY, R S, SAVAGE, J E AND FLYNN S J. Field measurements of the sound insulation of dry-lined masonry party walls. *BRE Current Paper CP 6/80*.

14.13 JONES, R D, AND CLOUGH, R H. Sound insulation of house separating and external walls (with lightweight masonry for thermal insulation). *CIRIA Project Record 278*, 1980.

14.14 JONES, R D. A new laboratory facility for the determination of the airborne sound insulation of party walls. *Applied Acoustics*, Vol 9, No 2, 1976. pp 119–130.

14.15 BRITISH STANDARDS INSTITUTION. BS 5821: 1984 (1993) *Method for rating sound insulation in buildings and of building elements.* BSI, London. pp 8.

14.16 JONES, R D. AND CLOUGH, R H. Sound insulation of house separating and external walls (with lightweight masonry for thermal insulation). *CIRIA Report 88*, 1980.

14.17 The Building Regulations 1991 HMSO, London.

14.18 BRITISH STANDARDS INSTITUTION. ISO 717. *Acoustics – Rating of sound insulation in buildings and of building elements. Part 1, Airborne sound insulation. Part 2, Impact sound insulation.* BSI, Milton Keynes, 1996. 16pp/12pp.

14.19 BUILDING RESEARCH ESTABLISHMENT. Quiet homes – a guide to good practice and reducing the risk of poor insulation between dwellings. CRC, London, 1998, pp.71.

14.20 Pending BRE publication, specifying dwellings with enhanced sound insulation.

14.21 BRE/CIRIA Sound control for homes. CTC, London, 1993 or BRE, Milton Keynes, 1993. 129pp.

14.22 WATT, P. Dry-finished concrete masonry party walls. BCA, Crowthorne, 1985. CCA Reprint 2/85. 3pp.

14.9 Bibliography

BUILDING RESEARCH ESTABLISHMENT. BRE Digest 187: *Sound insulation and new forms of construction.* BRE, 1976.

BUILDING RESEARCH ESTABLISHMENT. BRE Digest 252: *Sound insulation of traditional dwellings: 1.* 1981.

BUILDING RESEARCH ESTABLISHMENT. BRE Digest 266: *Sound insulation of traditional dwellings: 2.* BRE, 1982.

BUILDING RESEARCH ESTABLISHMENT. BRE Digest 338: *Insulation against external noise.* BRE, 1988.

CHARTERED INSTITUTE OF BUILDING SERVICES. *Guide Book A3: Sound and vibration.* CIBS. pp A1–11, A1–17.

Sound absorption of concrete block walls. *Concrete building and concrete products*, Vol. 43, No 1, January 1968. pp 13, 15 & 16.

Concrete masonry cuts noise nuisances. *Contract Journal*, Vol 252, No 4882, 19 March 1973. pp 39.

Chapter 15
Fire resistance

15.1 Basic principles

One very important aspect to be considered by the designer is the fire resistance of the structure. This subject is quite complex and forms a major part of the various statutory requirements.

The main objectives of structural fire precautions are that the building should not collapse and that the occupants should be protected from smoke and fire until they can be safely evacuated. The four basic principles to follow are that the building should be designed:

(1) so that the materials used in construction do not assist in rapid development of the fire;
(2) to contain the fire within confined limits both within the building and between buildings;
(3) to provide structural elements with sufficient fire resistance according to the type and size of the building;
(4) to provide a means of escape for the occupants.

Regulations covering all these aspects may be found in Part E of the Building Regulations[15.1], and similarly in the Building Standards (Scotland)[15.2], the London Building Acts[15.3], and the Building Regulations (Northern Ireland)[15.4]. In addition the rules of the Fire Officer's Committee and the Fire Officer's Committee of Ireland are often adopted for insurance purposes.

15.2 Elements of structure

The British Standard covering tests for fire resistance of materials is BS 476[15.6], which has several parts dealing with ignitability, flame spread, etc., but only BS 476: Part 8 prior to 1 January 1988 and BS 476: Parts 20–23 subsequent to this date are generally applicable to masonry walls. Concrete bricks and blocks are essentially non-combustible material and it is generally only necessary to determine the fire resistance of the wall or element. Where blocks contain some form of combustible material in cores or voids to provide extra insulation then additional tests may be necessary.

A significant number of tests to BS 476: Part 1: 1953, BS 476: Part 8: 1972 and BS 476: Parts 21 and 22: 1987

carried out in conjunction with national research projects and sponsored work by manufacturers, have shown that concrete brick and block walling is capable of notional fire resistance periods of between half-an-hour to six hours and of providing several hours protection to other structural elements. The notional fire resistance period of a wall or construction is an indication that it is able to satisfy conditions for loadbearing capacity (stability in Part 8), integrity and insulation for the period specified and is obtained by testing in accordance with a standard time-temperature curve, as shown in Figure 15.1. In this context, *loadbearing capacity* means that the wall does not collapse, *integrity* means that the wall does not develop cracks through which flames can pass, and *insulation* means that the wall does not transmit sufficient heat to ignite flammable materials in contact with the side remote from the fire. The criteria of failure for each condition is as follows:

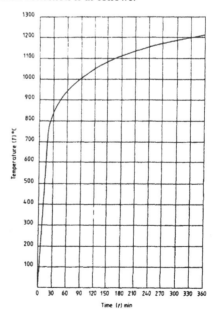

Figure 15.1 Standard time/temperature curve given in BS 476: Part 20: 1972[15.6]

Loadbearing capacity In a non-loadbearing wall which carries only its self weight during the fire test, the loadbearing capacity period is taken as the time taken to collapse. A loadbearing wall is required to support the test load during the prescribed heating period and also 16 hours after the end of the heating period. However, should collapse occur during the post-heating period, or if a residual strength test is not carried out then a reduction in the notional period for loadbearing capacity may be necessary. The amount of reduction may be varied depending on the mode of failure but should not be greater than 20%. The notional maximum period for stability for such a specimen should be construed as 80% of the time to collapse or the duration of heating if failure occurs in the reload test.

Integrity The integrity period is taken to be the time until cracks occur which are large enough to allow flame or hot gases through in sufficient quantity to cause cotton wool pads to ignite.

Insulation The insulation period is taken to be the time period until any point on the surface remote from the fire reaches a temperature greater than 180 °C or the mean surface temperature exceeds 140 °C.

The notional fire resistance of a wall is the period during which the wall fulfils all the relevant requirements. The typical grading periods are ½, 1, 1½, 2, 3, 4 and 6 hours. A test can be terminated at any time before failure occurs if it is expedient to do so. Thus, if a loadbearing wall has stability failure of 140 minutes, an integrity failure of 120 minutes and an insulation failure of 140 minutes, the wall would be given a notional fire resistance period of two hours. In certain cases where one or more of the three conditions can be relaxed, the fire resistance period of the wall may not necessarily be controlled by the lowest of the test conditions. The notional fire resistance periods, however, given in the Regulations and in BS 5628: Part 3[15.7], always cover for all three test conditions.

The insulation criteria of BS 476: Part 8 and Part 20 requires that the surface temperature of the unexposed face of the wall should not exceed a certain maximum and mean value. This temperature is set so that in normal circumstances it would not be high enough to ignite combustible material. It should be noted, however, that temperature beneath highly insulated materials in close contact with the wall may be much higher than the BS limits due to the reduced heat losses from the face. In situations where this is likely to be a problem then advantage may be taken of the better insulating properties of lightweight aggregate and aerated concrete block walls since, although they may be classified as having the same notional fire resistance period as dense walls (the stability and integrity controlling), the improved insulation would result in lower unexposed surface temperatures. For further information on this aspect either the Fire Research Station or individual manufacturers should be consulted.

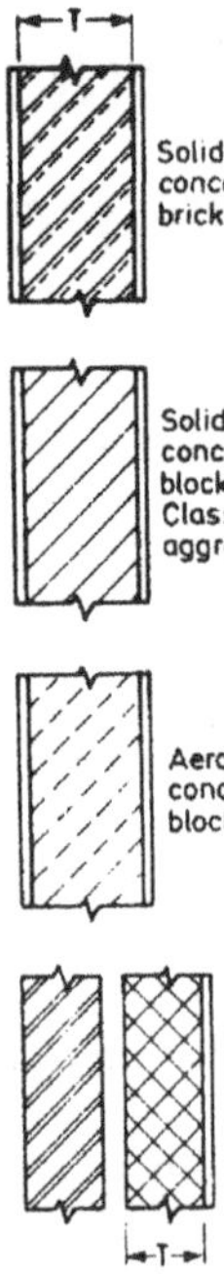

Construction	Performance (hours)	Required thickness (T) - mm	
		Loadbearing	Non-loadbearing
Solid concrete bricks	1	100	75
	2	100	100
	4	200	170
Solid concrete blocks. Class (1) aggregate	1	100	75
	2	100	75
	4	150	100
Aerated concrete blocks	1	100	50
	2	100	62
	4	180	100
Solid concrete bricks or blocks	1	100	75
	2	100	75
	4	100	75

Figure 15.2 Fire resistance of independent walling (walls finished with 12.5 mm cement/sand/gypsum – sand plaster)

Concrete brick and block walling may be used to provide fire resistant structural elements and also the means of providing effective and economical compartmentation to both new and existing buildings. Concrete masonry walling being non-combustible does not produce smoke or toxic gases and, therefore, attention has only to be focused towards decorative coverings such as paints, etc., applied to the surface of the wall. Another application is the upgrading or protection of elements with a comparatively poor fire resistance, such as steel stanchions.

The performance of a given wall is complex since it depends upon a number of factors, the notable items being:

(1) the type of unit, e.g. solid, hollow or aerated;
(2) the type of aggregate*, e.g. Class 1 or Class 2;
(3) the thickness of the wall;
(4) whether the wall is loadbearing or non-loadbearing;
(5) the type of finishes, e.g. none, cement-gypsum plaster, etc.

It is necessary to consider these various factors when determining the type of wall required. It is clearly an advantage to be able to produce a wide range of constructions which can be offered to suit any one particular problem.

*Class 1 aggregate: air-cooled blast furnace slag, foamed or expanded slag, sintered pulverized fuel ash, crushed brick, expanded clay or shale, well burnt clinker, pumice and limestone.

256

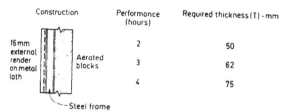

Construction	Performance (hours)	Required thickness (T) - mm
16 mm external render on metal lath — Aerated blocks	2	50
	3	62
	4	75
Steel frame		

Figure 15.3 Fire resistance of non-loadbearing composite walls

The details given in Figures 15.2, 15.3 and 15.4 provide a brief indication of the performance and applicability of walls and constructions using concrete bricks and blocks based on information extracted from the Building Regulations. More extensive information may be found in the deemed-to-satisfy schedules of the individual controlled Standards and BS 5628: Part 3. In particular, information on fire resistance periods up to six hours is given in both BS 5628: Part 3[15.7] and in the Fire Officers' Committee Regulations[15.5].

15.3 Construction

It is important that walls should be correctly constructed so that full benefit is obtained of their strength, sound and thermal insulation, fire resistance and so on. For this purpose the recommendations given in BS 5628: Part 3: *Materials and components, design and workmanship* should be followed and adequate specifications provided. The details given are applicable to walls with mortared joints. Some concrete blocks may contain tongue and grooved ends and to maintain fire performance and sound insulation, it is important that these joints should be filled with mortar and not left butted together.

15.4 References

15.1 The Building Regulations 1976 (amended 1978). HMSO, London.
15.2 The Building Standards (Scotland Consolidation) Regulations 1971 (with subsequent amendments). HMSO, London.
15.3 London Building Acts 1930–1939. London Building (Constructional) By-Laws 1972 and 1974 (with subsequent amendments). GLC, London.
15.4 The Building Regulations (Northern Ireland) 1977. HMSO, London.
15.4 BRITISH STANDARDS INSTITUTION. Rules for the construction of grades 1 and 2. Fire Officers Committee and Fire Officers Committee of Ireland. London, 1978. pp 30.
15.5 BRITISH STANDARDS INSTITUTION. BS 476: Part 1: 1953 *Fire tests on building materials and structures* and BS 476: Part 8: 1972 *Test methods and criteria for the fire resistance of elements of building construction* (pp 16). BSI, London.
15.7 BRITISH STANDARDS INSTITUTION. CP 121: Part 1: 1973 *Code of Practice for walling. Part 1: Brick and block masonry.* BSI, London. pp 84.

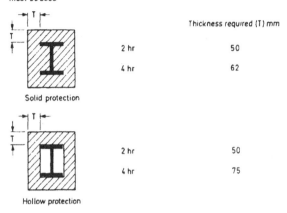

Construction	Performance	Minimum column dimensions	
		Built into wall	Free standing
	1 hr	75*	200
	2 hr	100*	300
600 min	4 hr	180*	450

*In order to effect reduction in column size a wall of appropriate fire resistance must be used

		Thickness required (T) mm
Solid protection	2 hr	50
	4 hr	62
Hollow protection	2 hr	50
	4 hr	75

Note: The above examples are related to steel stanchions weighing less than 45 kg/m and for solid concrete blocks of foamed slag or pumic aggregate. Horizontal joint reinforcement is also required and is given in the Building Regulations

Figure 15.4 Fire resistance of walling used as protection to other structural elements

15.5 Bibliography

Apart from the references (15.1–15.7) given, further information on fire resistance of walling may also be found in the following:

BRITISH STANDARDS INSTITUTION. CP 111: Part 2: 1970 *Structural recommendations for loadbearing walls. Part 2: Metric Units.* BSI, London. pp 40.
MALHOTRA, H L. Fire Note No 6: *Fire resistance of brick and block walls.* HMSO, London, 1966.
DAVEY, N, AND ASHTON, L A. Investigations on building fires. Part V. Fire tests on structural elements. *National Building Studies Research Paper No 12.* HMSO, London, 1953.
FISHER, R W, AND SMART, P M T. Results of fire resistance tests on elements of building construction. Volumes 1 and 2. BRE Reports – HMSO, London, 1975 and 1976.
READ, R E H, ADAMS, F C. AND COOKE, G M E. Guidelines for the construction of fire resisting structural elements. HMSO, London, 1980. pp 37.

257

Chapter 16
Movement in masonry

16.1 Introduction

This chapter, which deals with the general subject of movement in masonry, covers the technical background to the subject and gives general recommendations for the practical accommodation of movement in masonry. Due attention should also be paid to the requirements of *Chapter 6*. Although this handbook is predominantly concerned with concrete masonry, it is recognized that concrete units and fired-clay bricks are often used in the same structure, so that some information has also been included dealing with the use of the latter.

16.2 Technical background

An indication of the various factors which affect movement in masonry, together with typical values for unrestrained free thermal and moisture movement, are presented in this chapter in the manner given in BS 5628: Part 3: 1985[16.1]. It is expected that this method of presentation will provide the designer with a clearer indication of the factors affecting movement, but it should be noted that it is almost impossible to determine mathematically, with any degree of certainty, the extent of movement that will actually occur. The determination of movement is a complex problem and not merely a summation or subtraction of extremes of individual values of thermal and moisture movement, creep, deflection, and so on. For example, as a material expands due to increase in temperature, it will also shrink as moisture is lost. In addition each movement will be controlled to some extent by the degree of restraint to which the masonry is subjected and its orientation and location within the structure. The clauses which follow indicate the various individual movements separately, in such a way as to show not only the general range of values for different materials, but also the implication of other important factors which need to be considered in an attempt to estimate the likely movement within a given construction.

16.3 Thermal movement

From Figure 16.1 the total range of free movement due to thermal effect, which is generally reversible, is equal to the temperature range $(T_{max}-T_{min})$ multiplied by the appropriate coefficient of thermal expansion. However, the movement that actually occurs within a wall after construction depends not only on the range of temperature but also on the initial temperature of the units as laid and on their moisture content. This will vary with the time of year and the weather conditions during the construction period and may, with some materials, be influenced by the age of units. For example, certain steam or similarly cured units have sufficient strength to be delivered relatively fresh from the curing chamber, and may have a higher initial temperature than air cured units. To determine the effective free movement that could occur, therefore, some estimation of the initial temperature and temperature range must be made. The effective free movement, so derived, must still be modified to allow for the effects of restraints.

Table 16.1 indicates typical ranges of coefficients of thermal movement and some estimation must be made of the actual value for the material being used, although most manufacturers should be able to supply more precise values for their own materials. For further information on coefficients of thermal expansion for various materials, together with examples of service temperature ranges of materials, reference should be made to BRE Digest 228[16.2].

Table 16.1 Coefficient of linear thermal movement of units and mortar

Materials	Coefficient of linear thermal movement × 10^{-6} per °C
Fired-clay bricks and blocks (depending on type of clay)	4–8
Concrete bricks and blocks (depending on aggregate and mix)	7–14
Calcium silicate bricks	11–15
Mortars	11–3

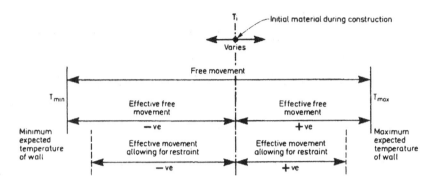

Figure 16.1 Factors affecting thermal movement

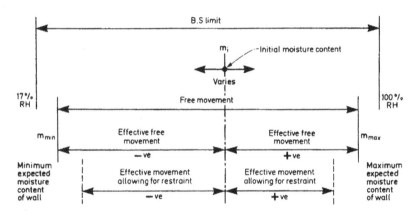

Figure 16.2 Factors affecting moisture movement

The differences between mortars and units can largely be neglected when considering movement along the wall, since the effect of such differences will be controlled by the adhesion of the mortar to the units. Some slight adjustment may be necessary for brick walls due to the greater quantity of mortar present. The units do not restrain the mortar in the vertical direction and therefore the movement in the height of the wall may be determined by multiplying the dimensions of the units and the mortar by the respective coefficients. Also, the move to thin joint construction in Northern Europe, if reflected in the UK would further reduce the impact of mortar on thermal expansion.

16.4 Moisture movement

The movement occurring in a wall as a result of changes in moisture content is basically controlled in the same way as thermal movement outlined above, except that in this instance attention needs to be given to minimum, initial and maximum moisture content rather than temperatures (Figure 16.2).

The values shown in Table 16.2 represent the maximum permitted shrinkage laid down by BS 6073: Part 1: 1981[16.3] for concrete units. These limits are basically for quality control purposes and do not represent practical conditions. The BS test, which is still being reviewed, determines the shrinkage of concrete bricks and blocks between saturation and oven conditions at a relative humidity of approximately 17%, whereas in practice a wall is seldom totally saturated and usually operates in relative humidity conditions between 50 and 85%. Thus the free movement will normally be less than the BS limit (see Table 16.3 and also BRE Digest 228[16.2]).

Table 16.2 Maximum permitted shrinkage of concrete masonry units

Material	Shrinkage %
Concrete bricks and dense and lightweight aggregate blocks	0.06
Autoclaved aerated concrete blocks	0.09

Table 16.3 Moisture movement of concrete and calcium silicate units

Material and type of masonry unit	Shrinkage as a percentage of original (dry) length (%)
Autoclaved aerated concrete masonry units	0.04 to 0.09
Other concrete masonry units	0.02 to 0.06

Note: these figures are obtained from tests carried out as described in BS 1881: Part 5[16.4]

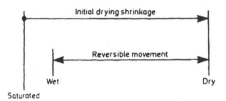

Figure 16.3 Moisture movement of mortars

From Figure 16.2 it can be seen that the effective movement within a wall is related to the moisture content of the units at the time of laying and it is clear that keeping the units as dry as possible before and during construction will reduce subsequent movement. The effective free movement will need to be modified to take restraints into account, but it should be noted that such restraints, particularly at the end of a wall, are likely to increase the tensile stresses in the wall.

The free moisture movement of fired-clay units is generally less than 0.02% and is usually neglected. Attention should be given to the long term movement of fired-clay units due to adsorption of moisture, as the adsorption gives rise to long term expansion, and care should, therefore, be taken when concrete and clay units are used together as, for example, in a cavity wall. Consideration should be given to the effects of differential movement. With this in mind it should also be noted that wire ties have greater flexibility than flat twisted ties. The total long term unrestrained expansion is typically 0.10% but a lower value may be appropriate, depending upon the type of clay.

The free moisture of mortar is similar to that of concrete units, although the effective free movement is likely to be greater, since initial moisture loss will not take place, as shown in Figure 16.3. Typical shrinkage values are given in Table 16.4, although the actual values will depend upon the constituents of the mortar, the mix proportions and the relative humidity. For convenience the lower values in the table may be taken to apply to mortars in external walls and the higher to mortars in internal walls. Reversible movement of internal walls may generally be neglected since they are unlikely to become wet after initial drying out. Very little research on the drying shrinkage of concrete masonry has been carried out. Garvin[16.5] notes in his review on aircrete blockwork that the shrinkage in aircrete masonry was the same with 10 mm or thinner (2–3 mm) joints. However thinner panels shrank more than thicker ones. The overall shrinkage of Aircrete masonry was found to be 0.2 mm/m.

The movement in masonry report[16.6] indicates moisture movements for aircrete range between 0.6–0.7 mm/m. Garvin's work thus underestimates the moisture movement in aircrete when compared with BS 5628: Part 3: 1985[16.1] whilst the movement in masonry report suggests that lower values in the code may be optimistic. The movement in masonry report[16.6] provides significant additional insights into the movement of concrete

masonry. However, in all instances except for aircrete (as noted above), the existing code values are satisfactory or the report recommends further research work.

Table 16.4 Shrinkage of mortars

Initial drying shrinkage (%)	Subsequent reversible movement (%)
0.04–0.10	0.03–0.06

Additional shrinkage of concrete units and mortar can occur as a result of carbonation. The extent of carbonation and the subsequent movement depend on the permeability of the concrete and on the relative humidity. In dense units and in autoclaved units, carbonation shrinkage may be neglected since it is extremely small. In unprotected open textured units and mortar, the shrinkage due to carbonation will still be relatively small and may be neglected for most purposes since such movement is unlikely to exceed 0.2 to 0.3 of the initial free movement.

16.5 Determination of total movement within a wall

To determine the movement likely to take place in a wall it is necessary to combine the individual effective movements due to thermal, moisture and other effects. However, the effective thermal and moisture movements are not directly additive since a wall is unlikely to be at both its maximum temperature and its saturated condition at the same time, so that to estimate the possible maximum movement it is necessary to consider carefully the temperature range over which the moisture movement occurs and make some attempt to combine the thermal and moisture movements on a rational basis rather than just considering the extremes. Since there are so many variables involved, it is extremely difficult to determine with any degree of certainty the actual movement that will occur. This Section basically outlines the factors that affect movement and in general it is simpler to adopt standard rules rather than try to estimate movement. It is hoped that the presentation of the factors in this Section will be of use in instances where some attempt has to be made to mathematically determine the effective movement that may occur.

With fired-clay walls long term expansion usually predominates and a simple solution would be to adopt the effective global movement given in BS 5628: Part 3[16.1]. This indicates that in general, unrestrained or lightly restrained unreinforced walls, e.g. parapets and non loaded spandrals built off membrane-type dpcs can expand up to 1 mm/m during the life of a building. The expansion may be somewhat less in normal storey height walls.

To cater for this movement, expansion joints are usually recommended typically at about 12 m but not exceeding 15 m except where bed joint reinforcement is used and then only when expert advice is obtained. This matter of clay brick expansion is included here so as to highlight the potential problem of differential movement between clay and concrete masonry. With concrete masonry the movement is small and predominantly controlled by the effects of shrinkage and thermal contraction. It is usual to provide 10 mm movement joints at nominal spacings as indicated in Section 16.6 instead of trying to determine the actual movement that may occur and hence determine optimum movement joint spacings. There is no proven mathematical method for determining the optimum spacing of movement joints but readers may find the paper by Copeland[16.7] of some assistance in this respect. This paper is based upon unpublished experimental data by R.W. Carlson and T.J. Reading to determine stress distribution curves that occur in walls with varying length and height. In effect the approach suggested by Copeland is simply to determine the ratio of the effective maximum strain that is likely to occur in the wall, as a result of contraction, to the ultimate strain capacity of the wall. This is then related to a panel size in which the Carlson/Reading curves give a ratio of average stress to maximum stress at the centre line of the panel of the same height to length ratio. This gives rise to the following formula:

$$P_m \leq \frac{e_u}{R\,(e_d + e_{temp})}$$

where

P_m = ratio average stress:maximum stress (Figure 16.4)
e_u = ultimate tensile strain
e_d = drying shrinkage
e_{temp} = temperature contraction
R = factor for degree of restraint

Following the general approach indicated for determination of movement within a wall, this could be rewritten as meaning:

$$P_m = \frac{\text{ultimate strain capacity}}{\text{effective strain}}$$

If this approach is examined with regard to a concrete masonry wall constructed of average units and subject to average weather conditions the following result is found:

Average drying shrinkage = 0.07%
Effective shrinkage, say 50%, giving e_d = 0.0035
Average coefficient of temperature contraction 10×10^{-6} per °C
Effective temperature contraction 20 °C giving e_{temp} = 0.0002
Average ultimate strain capacity e_u = 0.0002

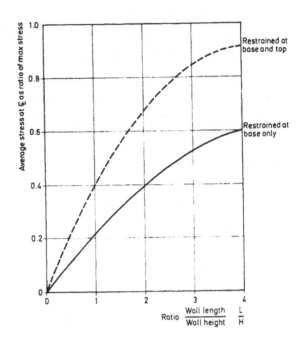

Figure 16.4 Relationship between average stress and maximum stress of walls of various L/H

Taking a value of $R = 0.9$ to allow for the effect of relaxation at damp-proof course (or alternatively this factor could be taken to allow for the fact that the shrinkage and temperature effects are not directly additive), this gives:

$$\frac{e_u}{R\,(e_d + e_{temp})} = \frac{0.0002}{0.9\,(0.0035 + 0.0002)} = 0.4$$

Figure 16.4 indicates that the ratio of length to height for a wall free at the top should not exceed about 2, i.e. $L/h \leq 2$. Taking a typical storey height of 3 m gives the well known recommendation for spacing movement joints in concrete masonry at approximately 6 m centres.

The method outlined by Copeland[16.7] makes a few assumptions which may not be technically correct and suggests that some additional modifications may be necessary to cope with the mass effect of the wall, but there is little published information on the ultimate strain capacity of concrete masonry. However, Copeland's paper is one of the few which tries to deal with the subject in mathematical terms and, together with the suggested approach for determining effective movement, may help to reinforce the engineering judgement upon which this subject must still heavily rely.

16.6 Accommodation of movement

Defects such as cracking are undesirable and difficult to deal with after the event, and it is most important to consider provision for movement at the design stage.

The effects of movement may be reduced by:

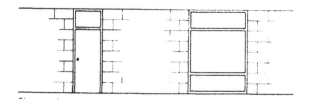

Figure 16.5

(1) designing the building to use discrete panels of masonry;
(2) providing movement joints (control joints);
(3) using the correct mortar;
(4) keeping the units and wall protected during construction;
(5) providing local bed joint reinforcement;
(6) designing to maintain bond pattern.

These factors are considered in more detail, as follows:

Discrete panels: One way of ensuring that the masonry is able to accommodate small seasonal movements is to design the building so that the masonry is separated into discrete panels, for example by use of feature panels at window openings and storey height door openings. This is illustrated in Figure 16.5 and also in the accompanying photographs. In this design the length of masonry panel should be limited to around 6 m and ideally its length should not be greater than about twice its height.

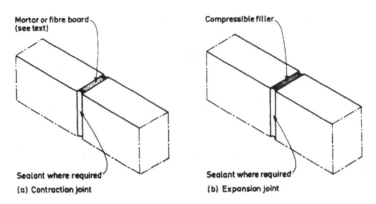

Figure 16.6 Movement joints: (a) contraction joint (b) expansion joint

Movement joints: The provision of movement joints (control joints) has the same effect as the discrete panel method in that joints divide the wall into defined lengths which are able to accommodate the strains arising from temperature and moisture variations. This method is suitable in situations where long walls of masonry occur. The distance between the control joints should not normally exceed about 6 m, but since there are wide variations in the physical properties between different concrete units, some variation of joint spacing is acceptable, although it should be noted that the risk of cracking may be greater where the length of panel exceeds about twice its height. Joint spacing towards 7.5 m may be acceptable for autoclaved dense aggregate units and for certain internal applications. It is, however, always desirable to consult with individual manufacturers before extending joint spacings.

Control joints may be of two types (see Figure 16.6), i.e. contraction joints or expansion joints. Generally the contraction type joints which are able to open to allow the tensile strains within the wall to be accommodated are suitable. Although such joints may have weak mortar within the vertical joint in some low strength units an infill fibre board material would be better and is sometimes required by the manufacturer. In general fibre board is to be preferred as it is suitable for all units and can also accommodate some compression i.e. it can act as a small expansion joint. In some instances, particularly on internal walls, a dry butt joint can be used. Expansion joints will generally only be required where the length of a wall exceeds 30–50 m. These joints must have compressible (not fibre board) material within the joint. The small expansive movement that occurs in the short distance of around 6 m means that fibre material is generally acceptable. All such joints should not be bridged by any service or tie which would prevent free movement occurring. Short flat ties may be used to aid stability of walls at contraction joints providing the wall is still free to move longitudinally, but great care should be taken at expansion joints since any tie must be of a type that would tolerate free compressive movement. This is extremely difficult to achieve in practice and the stability of the walls is best achieved in some other way, for example, by providing top support and sizing the wall accordingly. Alternatively, it may be possible to build the wall into an appropriately sized steel channel incorporating a compressible filter positioned to prevent the wall from making direct contact with the flange. A small factor, but one that is nevertheless important, is that the sealant used must be capable of accommodating the expected movement of the joints. This is not usually a problem in concrete masonry, since the joints are not generally more than about 6–8 m apart and most sealants can accommodate the movement occurring in this length of wall. (Note: with clay masonry the width of the joint should typically be about 30% more than the distance between joints in metres).

The reference to the notional 6 m length applies to walls without end restraint, i.e. the distance between free joints. In the case of a wall which is visibly within this distance, such as a gable wall, it must be noted that the restraint which occurs at the return will limit the ability of the wall to shorten and, thereby, increase the tensile stresses within the wall. In the case of a gable wall or a panel between a control joint and a corner restraint, therefore, the effective length should be considered to be some 25–50% greater than the visual length. When this effective length is much in excess of the notional spacing given previously then joints in the case of the gable, or closer joints in the case of the corner situation, or the provision of· bed joint reinforcement, should be considered. In the particular case of a very rigid vertical support, as in tying to a column, an effective length of twice the visual length should be taken.

In addition to the general rule of spacing joints at certain notional centres, the provision of control joints should also be considered at positions where concentration of changes in stress may occur. Examples of this are given below, some of which are illustrated in Figure 16.7:

(1) change in height;
(2) change in thickness;
(3) at large chases;

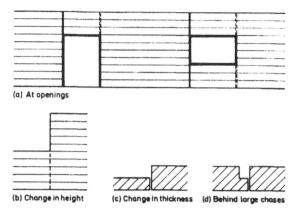

(a) At openings

(b) Change in height **(c) Change in thickness** **(d) Behind large chases**

Figure 16.7 Movement joint locations: (a) at openings (b) change in height (c) change in thickness (d) behind large chases

(4) at expansion or contraction joints in a building or floor slab;

(5) at window and door openings;

(6) at or near to changes in direction.

The use of control joints at changes in direction, unlike most other points, needs a little more explanation. Joints are not generally provided in the extreme corner of external wall returns. It is, in fact, usual to aid stability by placing any control joints required at this location, a short distance from the external corner but within the limit previously given. Similarly, short walls of, say, less than 2–3 m that intersect other walls, will not specifically require a joint at the intersection, although in facing work it will be necessary to provide a joint at that position to maintain the bonding pattern along the main wall (see Figure 16.10). Such joints can also reduce the amount of cutting required and ease construction by avoiding the need to form toothing for partition walls which may be built later. (Note: the butted joint as shown in Figure 16.10(b) may not be acceptable at a separating wall where sound insulation is required – see Chapter 14). Provision of movement joints along and at the ends of walls supported by long span floors or beams will often be beneficial in reducing the effect of any deflection which may occur. Movement joints should be built into the wall during construction and run for the full height of the masonry. The designer is advised against the use of sawn joints since unless very great care is taken they will be ineffective. It is not usual to continue the joints below the ground floor damp-proof course since the amount of movement below this level is minimal. However, in the case of a sloping site where the damp-proof course is well above ground level, the control joint should be continued down to ground level.

For aesthetic reasons, the designer may wish to conceal joints as much as possible and this can be achieved by placing the joints at the position of down pipes, by matching the sealant with the mortar and units and by using stack bonded walls.

Specification of mortar: It is important to note that mortar influences the way in which a wall accommodates movements. A masonry wall built with a relatively low strength mortar is better able to accommodate the stresses developing in a wall as a result of movement than a wall with a very strong mortar, as the weaker mortar relieves the stresses that may otherwise cause problems. As a result it may be taken that the general recommendations on the accommodation of movement, and hence control joint spacings, apply to situations where the most appropriate mortar is employed – namely a mortar of designation (iii). Where stronger mortars are required for structural or durability reasons, then some modification to the recommendations may be necessary. Higher strength mortars can normally be tolerated in vertically reinforced walls due to the presence of the reinforcement. For Aircrete blocks Garvin[16.5] recommends using designation (iv) mortar except where there is a high likelihood that the blockwork will remain wetted for long periods such as below a dpc, or in cavity wall construction where a designation (iii) mortar represents the best compromise between the requirements of the aircrete block inner leaf and the brick outer leaf.

Storage and protection of masonry: For cracking to occur within a masonry wall tensile stresses must be present. These stresses chiefly occur as a result of moisture loss, differential movement and thermal contraction. Since excessive shrinkage could obviously cause problems the British Standard for concrete blocks and bricks, BS 6073: Part 1: 1981[16.3] lays down certain maximum drying shrinkage values for the units. This reduces the risk of excessive shrinkage occurring but it will be quite apparent that the higher the moisture content at the time of laying, the greater the shrinkage will be, and to reduce subsequent shrinkage it is desirable to protect units from the rain while they are stored and during construction. As thermal contraction can also set up tensile stresses it is important that units are not used hot from an autoclave or curing chamber.

Reinforcement: The use of bed joint reinforcement was mentioned in the section dealing with the provision of control joints. Bed joint reinforcement can be used to modify joint spacings and to assist in situations where high stress concentrations occur as, for example, at window openings (Figure 16.8). The reinforcement should be of the tram-line type, preferably with an effective diameter of between 3–5 mm, with a cover of at least 20 mm from the face of the mortar and should be of austenitic stainless steel where it is to be used in a wall which is exposed to the weather. Galvanised mild steel may be appropriate for use on internal walls. Such reinforcement should extend at least 600 mm past the opening. Reinforcement may also be used to increase the distance between movement joints, and care should again be taken whenever spacings in excess of about 6 m are used. The provision of reinforcement will not ensure that cracking does not occur but it will considerably lessen the risk.

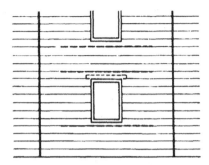

Figure 16.8 Bed joints at openings

Little research evidence is available to support the use of reinforcement to control cracking and the only information that can be offered is that issued by the National Concrete Masonry Association of America which is given in Table 16.5. The advice is apparently based on more practical experience than mathematics and hence needs to be used with some degree of caution, particularly with regard to the suggestion that no reinforcement is required for cases where $L/h \leq 2^*$.

Table 16.5 Movement joint spacing for reinforced walls

Ratio of panel length to height $\dfrac{L}{h}$	Vertical spacing of joint reinforcement mm	Maximum length regardless of height m
2	None*	12
2½	600	14
3	400	16
4	200	18

The reinforcement should consist of two parallel mild steel bars of nominal size 3–5 mm diameter, or equivalent strength in high yield bars

Bonding pattern: Although maintaining the bond pattern is an automatic criterion for facing work, it is often found that little attention is paid to this aspect in work that is to be rendered or plastered. This may seem to be of little consequence but in practice the lack of initial design, together with poor workmanship, can result in a series of virtually aligning perpend joints as depicted by Figure 16.9, with the result that the wall is considerably weakened and cracking is more likely to occur. Similarly it is of benefit to design the building so that wherever possible lintels can be supported by a full length block.

Wall bonding: With concrete masonry it is common practice not to bond intersecting walls together by toothing one wall into the other. The intersection of such

(a) Bond broken

(b) Bond maintained

Figure 16.9 Maintaining bond patterns: (a) bond broken (b) bond maintained

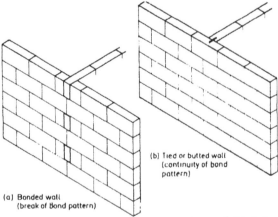

(b) Tied or butted wall (continuity of bond pattern)

(a) Bonded wall (break of Bond pattern)

Figure 16.10 (a) bonded wall (break of bond pattern) (b) tied or butted wall (continuity of bond pattern

walls, as explained in the section dealing with the accommodation of movement, is often a desirable location at which to incorporate either a contraction or expansion joint. Even if a true movement joint is not required it will often be beneficial to use a tied butt joint at such locations since this considerably reduces the amount of cutting and enables the bond pattern of the passing wall to be maintained (Figure 16.10). (However,

*It is suggested for conditions where $L/h \leq 2$, that, due to mass effects, some reinforcement may be required when movement joints are spaced at centres exceeding 6 m.

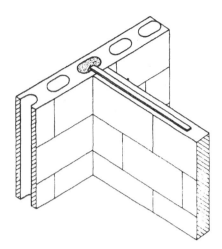

Figure 16.11 Fully tied joint

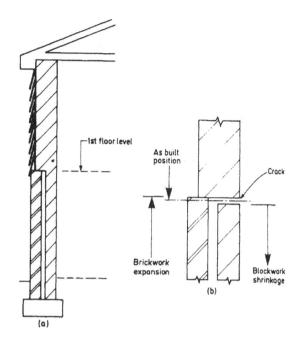

Figure 16.12 Differential movement problem

note again separating walls and sound insulation, Chapter 14). If the intersection is not required to act as a full movement joint it may be tied with either lengths of expanded metal or strips of steel in each course. It is important, however, to ensure that such joints are not confused with movement joints. Tied joints of this form will not generally allow movement to take place, particularly in the example shown in Figure 16.11 and, therefore, care should be taken in the location of such joints. Joints tied with expanded metal are not as rigid as steel strips. A simple approach would be to not use fully tied joints unless the length of the wall from the intersection does not exceed 2m. In many instances in true facing work it will be safer to provide actual movement joints at all locations and design the panels to have adequate stability in some other way, for example, by providing support to the top of the intersecting wall.

16.7 Differential movement

In addition to the general points made previously, to assist in reducing the occurrence of cracking particular attention should be given to the effects of differential movements between various materials. The differential movement between the various types of concrete masonry, i.e. dense aggregate, lightweight aggregate and autoclaved aerated concrete blocks and concrete bricks, will generally be fairly small, but consideration should nevertheless be given to the possible effects of this differential movement. In most instances with cavity walls this movement between inner and outer leaves can be accommodated by the flexibility of the wall ties.

Butterfly and double triangular wire ties are more flexible than flat twisted ties and are thus better at accommodating any differential movement which may occur. This aspect is particularly important where more substantial differential movement may occur, as for example when one leaf is of fired-clay masonry and the

other is of concrete masonry. It is essential that the effects of differential movement are considered since serious problems may otherwise arise. In principle the wall details should be checked to ensure that any differential movement is free to take place. An example of a detail which has caused problems in the past is illustrated in Figure 16.12, where (a) shows the basic wall detail and (b) indicates the effect that the differential movement could have on the wall. When designing with both fired-clay and concrete masonry the essential point to remember is that fired-clay walling has a general tendency to an expansive movement while concrete masonry has a tendency to shrink.

16.8 References

16.1 BRITISH STANDARDS INSTITUTION. BS 5628: Part 3. Code of Practice. *Use of masonry. Materials and components, design and workmanship.* BSC, London, 1986. pp.110.
16.2 BRITISH RESEARCH ESTABLISHMENT. BRE Digest 228: *Estimation of thermal and moisture movements and stresses.* Part 2. BRE, 1979. pp 8.
16.3 BRITISH STANDARDS INSTITUTION. BS 6073: Part 1: 1981 *Precast concrete masonry units.* Part 1. *Specification for precast concrete units.* BSI, London, p.3.
16.4 BRITISH STANDARDS INSTITUTION. BS 1881: Part 5: 1970. *Methods of testing hardened concrete for other than strength.*
16.5 GARVIN, S L. Mortar and movement in aircrete brickwork – a review. *Building Research Establishment Occasional Paper,* Nov 1994.
16.6 CERAM BUILDING TECHNOLOGY, *Movement in Masonry,* 1997.
16.7 COPELAND, R E. Shrinkage and temperature stresses in masonry. *Journal of the American Concrete Institute,* 1957.

16.9 Bibliography

BESSEY, G E. Shrinkage and expansion in brickwork and blockwork, causes and effects. *Journal of the British Ceramic Society*, Vol 6, No 2, 1969.

BRITISH STANDARDS INSTITUTION. BRE Digest 227 and 229: *Estimation of thermal and moisture movements and stresses* (Parts 1 and 3). BRE, 1979. pp 8.

ALEXANDER, S J AND LAWSON, R M. Design for movement in buildings. Technical Note 107, CIRIA, London, 1981. pp 54.

Chapter 17
Specification and workmanship

17.1 Object of specifications

This section deals with specifications for concrete masonry and indicates the main design and workmanship aspects that need to be covered. It is important that masonry, particularly where it has been designed structurally in accordance with BS 5628: Parts 1[17.1] and 2[17.2], should be built correctly. Indeed, these standards and the Statutory Instruments indicate that materials, components and workmanship should comply with the appropriate recommendations of BS 5628: Part 3[17.3]. BS 5628: Part 3[17.3] provides the designer and contractor with a lot of practical information on the subject of construction with masonry. Although it is generally indicated that work should comply with the recommendations of BS 5628: Part 3[17.3], it is not practical to refer to this as the sole contract document for masonry, since the clauses of the Code are not always explicit, and there are often several recommended alternatives to a particular item. The Code does not necessarily cover all possible acceptable ways of dealing with any particular problem. In addition the Code explains general recommendations and not necessarily mandatory items; use is therefore made of the word *should* and not *shall*.

The object of the specification is to ensure that the most important aspects of design and workmanship are simply, clearly and explicitly stated. Thus Part 3 of BS 5628[17.3] provides the designer with extremely useful information on materials, design and workmanship for masonry. The recommendations given may be used or developed for design purposes and can also form the basis for production of the contract specifications.

There are several ways of tackling the question of specifications – they can be individually written for each contract or a general office specification can repeatedly be used. The latter method, which unfortunately is still common, can lead to serious problems since such specifications are often noticeably out of date and can contain errors which are repeated in every contract, but due to the pressure of work in the design office, often insufficient time is available to take the former approach.

A good method of writing specifications is to use a standard specification document such as the National Building Specification (NBS)[17.4] which is in essence a very comprehensive series of standard specification clauses to cover most building trades and operations. The NBS, in particular, was written in very clear, simple and explicit terms and has the major advantage that the text is continually being revised and updated as codes change. The current version of the NBS is in CD Rom format and has a good index listing. When using this form of specification the designer can quickly select the clauses most appropriate, so that the bulk of specifications can be compiled with comparative ease, allowing more time to deal with clauses of particular importance. The outcome of using such standard specifications is that time is available to compile individual specifications for each contract.

Another important factor is that the specification should be correct, which is sometimes difficult to ensure with the normal office specification and again, an advantage of texts such as NBS is that the designer can confidently rely on the accuracy of the wording in the general specification clause and has only to check that any insertions, such as dimensions, reference numbers, etc., have been written correctly.

It was suggested at the beginning of the chapter that specifications should ideally be simple, clear and explicit. Often it is found that they are far too long and complex and that the principal point to be made is lost in a long meaningless passage. It may be more difficult to write short specifications than long ones that effectively say no more, but the shorter the specification is the more likely it is to be clear, understandable, and therefore, the simpler it will be to read and to be followed. The small additional effort is worthwhile from every point of view.

One way in which a specification may be simplified is considered in the following text, which is not uncommonly found and which will have a degree of familiarity:

The coarse aggregate shall be clean, natural or crushed stone complying with BS 882[17.5] and shall be free from chalk, clay, organic or other deleterious material. The particles shall be well and evenly graded to conform to the limits given in Table 1 of BS 882 for the maximum size aggregate appropriate to the concrete mix.

A great deal of this text is simply repeating that which is already a stated requirement within the British Standard, in that the comments regarding *free from chalk*, and such like, are part of the British Standard, so why repeat it? With this in mind, it can be seen that the wording could adequately be reduced to:

The coarse aggregate shall comply with the requirements of BS 882 and be graded in accordance with Table 1.

Examining the British Standard further, it will be found that only graded aggregate is covered by Table 1. Additionally reference to BS 882 can only mean that the aggregate must comply with its requirements. Adopting the technique used for the National Building Specification[17.4], the specification could be reduced to a very clear, simple and explicit form such as:

Coarse aggregate: to BS 882, graded.

Accuracy of specifications is something else to be considered, as errors can easily occur, but they are far easier to find when the specification is uncluttered. The two parts of BS 6073[17.6,17.7] provide very clear guidance on writing specifications for concrete masonry units. To facilitate this, appendix A of BS 6073: Part 2 provides purchasers with the essential details to be given to the manufacturer for an enquiry or order to be fully understood. The information also covers the essential topics when writing a specification. Because the Appendix provides a check list of items, specifiers will be unlikely to omit requirements and errors will be more easily spotted.

The following example illustrates how a simple specification can be formed using appendix A of BS 6073: Part 2[17.7], by following the check list in the code.

Requirements: Fair faced blocks for painting, strength 7 N/mm^2, size 390 × 190 × 100 mm, completely solid.

Specification: Concrete blocks to BS 6073: Part 1
Size: 390 × 190 × 100 mm
Type: Solid
Properties: Super face
Manufacturer: Quality Block Company Limited

To summarize, when specifications are written it is desirable that they are not over-wordy, thereby losing the impact of the clause, and it is essential that they are correct.

17.2 Cross referencing specification/drawings

Having developed simple and correct specifications, a very useful exercise is to cross reference them with the drawings and to schedule the various types of wall specifications at

270

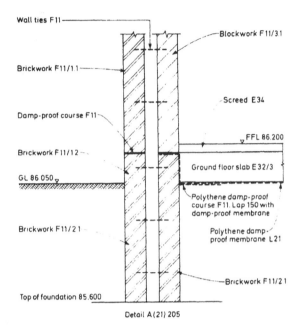

Figure 17.1 Use of type numbers on drawings

the beginning of the masonry section. An illustration of a completed schedule page is given in Table 17.1. From a quick reference to this schedule the contractor is able to identify the position and various types of wall on the project. It is also simple to determine the various types of mortar, and such like, required. The type numbers can be used on the drawings as shown in Figure 17.1 to directly correspond to the specification for the wall.

17.3 Items for specification and general specification clauses

The essential information to be given when ordering concrete masonry units, as recommended by BS 6073: Part 2, is as follows:

Item	Information
Quantity	Number of units or area of walling. Note that allowances for breakages or other wastage on site is the responsibility of the purchaser.
Size	*Work* size required, given in terms of length, height and thickness
Strength	(a) for blocks less than 75 mm, no compressive strength need be specified
(b) for blocks equal to or greater than 75 mm, the minimum compressive strength required in N/mm^2
(c) for bricks, the minimum compressive strength required in N/mm^2 or special purpose bricks where a minimum strength of 40 N/mm^2 and 350 kg/m^2 cement is required

Table 17.1 Completed schedule – brick/block walling

	Use or location	Brick/block	Mortar type[1] and designation	Bond F11:3	Jointing/pointing F11:4
Type F11/1.1	External leaf above damp-proof course	Facing bricks F710 F725	Designation (iii) coloured for pointing	Stretching ⅓ lap	Flush pointing
Type F11/1.2	External leaf below damp-proof course	ditto	Designation (iii) sulphate-resisting cement coloured for pointing	ditto	ditto
Type F11/1.3	Screen walls above damp-proof course	ditto	Designation (iii) coloured for pointing	Flemish garden wall	ditto
Type F11/1.4	Screen walls below damp-proof course	ditto	Designation (iii) sulphate-resisting cement coloured for pointing	ditto	ditto
Type F11/2.1	External walls below ground	Common bricks F605	Designation (iii) sulphate-resisting cement	Stretching	–
Type F11/2.2	Screen walls below ground	ditto	ditto	English	–
Type F11/2.3	Manholes	Engineering bricks F810	ditto	English garden wall	Flush jointing
Type F11/3.1	Inner leaf and partitions, factory	Fair faced blocks F150	Designation (iii)	Stretching ½ lap	Pail handle jointing
Type F11/3.2	Inner leaf, offices	Thermal blocks F160	ditto	ditto	For plaster
Type F11/3.3	Partitions, offices	Common blocks F125	ditto	ditto	ditto

[1] All mortars to be cement : sand mixes with plasticizers.

Type	(a) for blocks, the type of units required, i.e. solid, cellular or hollow (b) for bricks, the type of units required, i.e. solid, perforated, hollow or cellular, or for special use 'fixing bricks'
Materials	To specify to the manufacturer whether any particular material is to be included or restricted
Specials	(a) whether special shape units, such as quoin or cavity closure blocks, are required (b) can also be used to indicate whether any special requirements with regard to tolerances are required
Properties	The manufacturer shall be informed of any requirements with regard to colour, texture, thermal properties, etc.
Quality control	If special category quality control in accordance with the recommendations of BS 5628: Part 1 1992[17.1] is required, this must be conveyed to the manufacturer
Identification	Where additional means of identifying the masonry units are required
Handling	To convey to the manufacturer any special handling requirements, such as paletization, strapping, mechanical off-loading, and so on.

It is unlikely that all these requirements will need to be covered in each specification clause and it is, therefore, necessary to be selective. It may also be necessary to include reference to the *Manufacturer* in the specification and perhaps the location of particular units within the structure if not covered by use of a schedule as adopted by NBS[17.4]. Thus, a general specification clause for normal blocks could read as follows:

Concrete blocks to BS 6073
Size: 390 × 190 × 100 mm
Strength: 3.5 N/mm^2
Type: Solid

A specification for facing blocks could be:

Concrete blocks to BS 6073
Size: 390 × 190 × 100 mm
Strength: 7 N/mm^2
Type: Solid
Materials: No calcium chloride
Properties: Buff fair faced
Manufacturer: Modern Block Company Limited

Most fair faced blocks are produced to tolerances well within the British Standard values. If special or specific requirements on tolerances are needed they must be discussed and agreed with the manufacturer.

The previous information in this chapter has hopefully given some indication as to how to tackle the general topic of specifications. It is not appropriate in this handbook to attempt to cover all the possible specifications which could be required for a masonry contract. Instead a series of basic points and some main specifications have been included as being the topics most likely to be required. The specifications themselves are not intended to be for simple extraction and are at times more wordy than is necessary. The clauses are provided as a check list of the most important features that need to be considered for specification, and guidance notes are provided for particular points.

17.3.1 Items for specification

To write a specification the designer needs to understand the material, whether it is clay brick, concrete brick or block, plastic, timber, and so on. For concrete masonry some of the most important items to be considered are given below. Specification of units may also need to take account of the following:

Materials	Autoclaved aerated concrete
	Lightweight aggregate concrete
	Dense aggregate concrete
Blocks	Standard blocks
	Facing blocks
	Thermal blocks
	Special blocks – quoin blocks, etc.
Bricks	Standard bricks
	Facing bricks
	Fixing bricks
	Special purpose bricks
Block materials	Use and durability
Block strength	2.8 to 35 N/mm² strength
Brick strength	7 to 40 N/mm² strength, use and durability
Block forms	Solid
	Hollow
	Cellular
Brick forms	Frogged
	Unfrogged
Performance requirements	Thermal insulation
	Fire resistance
	Sound insulation
	Rain resistance
Other aspects	Weight
	Speed of construction
	Handling
	Identification

Specifications for design and workmanship may need to take account of:

Standard work	Mortar mixes
Facing work	Joint profile
Control joints	Handling of blocks
Stability	Storage of materials
Bonding	Rendering
Wall ties	Testing
Cutting and chasing	

17.3.2 Specification clauses

17.3.2.1 Materials and properties

(1) Concrete blocks

All blocks to comply with BS 6073: Part 1[17.6] and any additional requirements as specified

(a) Standard blocks
The concrete blocks .... (location) shall be .... (description or manufacturer's designation), to be obtained from .... (manufacturer), or other equal approved, and be .... (solid, cellular, hollow) having a minimum compressive strength of .... N/mm² in the following size(s) ....

A standardized method of specifying precast concrete masonry units is given in BS 6073: Part 2[17.7]. The basic items which should be given in any specification for concrete masonry units are the size and compressive strength, where this is required to be greater than the minimum average permitted by Part 1 (2.8 N/mm²). This should be followed where necessary by details of any additional or specific requirements relating to such things as type, materials, shape, density, colour, and so on, and whether the units are to be of special category (BS 6073: Part 2). Table 13 in Part 3 of BS 5628[17.3] gives recommendations on minimum quality of units for durability.

The purpose of including a reference to the units location indicates that it is of benefit to make it quite clear where each type of unit is located. It would not strictly form part of the specification clause. Describing the unit or including the manufacturers designation will be necessary particularly where special requirements such as thermal and/or sound insulation are required. In some instances the manufacturer's designation may automatically cover the question of block type (solid, cellular, hollow), but in situations such as for reinforced work it is essential to specify hollow. When non-loadbearing partition blocks are required (less than 75 mm thick) the information could be reduced to: *The concrete blocks .... (location) shall be obtained from .... (manufacturer), or other equal approved, and be .... (solid, cellular) in the following size(s) ....*

(b) Facing blocks
The facing blocks .... (location) shall be .... (description or manufacturer's designation), obtained from .... (manufacturer), or other equal approved, and be .... (solid, cellular, hollow) having an average compressive strength of .... N/mm² in the following size(s) .... and comply with the approved sample panels located at ....
The same guidance notes as previously used similarly apply to this clause, except that emphasis in this instance is also on facing units. The inclusion of reference to the sample panel is simply to indicate the importance of this and it would generally form a

separate clause. The purpose of the sample panel(s) is to establish such physical factors as range of sizes and variations in colour, texture, pattern and so on. The panel must be of sufficient size and contain normal and corner blocks, etc., for these factors to be established. The panel(s) are also used to indicate the acceptable workmanship and therefore should also be of such configuration as to fairly cover intricacies of the work.

(c) Special blocks
Special blocks shall be to drawing(s) .... and be within the following tolerances ....
The inclusion of tolerances in this clause is to bring to the readers attention the fact that BS 6073[17.6,17.7] basically applies to normal rectangular units. Generally it will not be necessary to bother with tolerances since the manufacturer will adhere to the Standard as far as is practicable. If for very high quality work it is felt necessary to specify tolerances for special blocks, such as corner blocks, then it is advisable to discuss the matter first with the manufacturer, since such blocks may be manufactured on different machinery to normal blocks. In addition a method of measurement may also be necessary, particularly for blocks with exposed aggregate or split faces.

(2) Concrete bricks
All bricks to comply with BS 6073: Part 1[17.6]
The general clauses for concrete bricks are liable to be similar although not containing as much data, since units will generally be solid (with or without frogs) and the strength designation is also used to define durability – see BS 5628: Part 3: 1985[17.3].

(3) Mortars
(a) Materials for mortar
(i) Cement
The cement for mortar shall be to BS ....
For guidance on cements for mortars see Chapter 3 and BS 5628: Part 3: 1985[17.3]. Ordinary Portland cement to BS 12: 1996[17.8] will be the most commonly used, although Portland blast furnace cement (BS 146: 1996[17.9]) and masonry cement (BS 5224: 1995[17.10]) are also acceptable. Sulphate-resisting Portland cement to BS 4027: 1991[17.10a] may also be used but the subject of sulphate attack should be carefully considered before specifying this cement (see Section 3.4.2 and BS 5628: Part 3: 1985[17.3]).

(ii) Lime
The lime for mortar shall be to BS 890: 1995[17.11]
Building limes normally used for mortar are non-hydraulic (calcium) limes or semi-hydraulic (calcium) and magnesium limes.

(iii) Sand
The sand for mortar shall be to BS 1199 and BS 1200: 1976(1996)[17.12]

Other sands, apart from sands to BS 1199 and BS 1200, can produce acceptable mortars but this needs to be checked with local experience. It should be noted that very finely graded sands can result in mortars more friable and permeable than would normally be accepted.

(iv) Water
The water shall be from normal mains supplies or approval obtained before use
Where the quality of supply is doubtful the water should be tested in accordance with BS 3148: 1980[17.13].

(v) Admixtures
Admixtures may be used subject to approval in writing. Calcium chloride shall not be permitted
Before approving the use of an admixture the architect should first check the appropriate manufacturer's recommendations and then ask the contractor to submit evidence of satisfactory performance of the admixture when used correctly, and details of the arrangements for the use of admixtures. The architect should also confirm in his written approval the agreed procedure for use.

Where admixtures are required they should be specified in accordance with BS 4887: Part 1: 1986[17.14] for air entraining (plasticising) admixtures and with BS 4887: Part 2: 1987[17.15] for set retarding admixtures. There should be no requirement for superplasticizers.

(b) Preparation of mortars

(i) Recommended mortar
Mortar .... (location) *shall be* .... (mix proportions) *by* .... (weight/volume) *of dry materials*
In specifying the mortar mix consideration should be given to its strength, durability and ability to accommodate movements. Attention is drawn to BS 5628: Part 3: 1985[17.3]. The most common mortar for both concrete blockwork and brickwork is of designation (iii), i.e. 1 : 1 : 5–6 or equivalent. Mortar proportions should be specified in one of three ways. (a) The designation and type of mortar should be specified. (b) the actual mix proportions for a particular sand should be specified or (c) the lowest mix proportions for each type and designation should be specified. Where special category construction is to be used in accordance with BS 5628: Part 1: 1992[17.1] then appropriate clauses for strength and sampling will need to be included. Volume proportions will be the norm but increased accuracy will result when batching is by weight.

(ii) Equivalent mortar mixes
Alternative mortar mixes may be used subject to written approval
The contractor's proposed alternative mix should be assessed by the designer for compliance with strength, durability and so on, prior to written approval being

given to the equivalent mortar mix. Some guidance on the strength equivalence of mortars is given in Appendix A of BS 5628: Part 1: 1992[17.1].

(iii) Batching of mortars
Measure materials accurately to specified mix proportions either by weight or gauge boxes

(iv) Mixing of mortars
The mortar shall be mixed by machine and be used within two hours of mixing. Mortars, except coloured mortars, may be retempered during the two hour period
Retempering of coloured mortars may alter colour.

(4) Concrete for core filling
(a) Materials
(i) Cement
The cement for concrete shall be to BS ....
In normal circumstances use Ordinary Portland cement complying with BS 12: 1996[17.8].

(ii) Fine and coarse aggregates
Fine aggregate shall comply with BS .... Coarse aggregate shall comply with BS ....
The type of aggregate should be specified according to the performance requirements. Generally this will mean using an aggregate smaller than that used for the concrete block. Relevant British Standards include BS 882: 1992[17.5] and BS 3797: 1990 (1996)[17.16].

(iii) Water
The water shall be from normal mains supply or approval obtained before use
Where the quality of supply is doubtful water should be tested in accordance with BS 3148: 1980[17.13].

(iv) Admixtures
Admixtures may be used subject to approval in writing
Before approving the use of an admixture the architect should first check the appropriate manufacturer's recommendations and then ask the contractor to submit evidence of satisfactory performance of the admixture when used correctly, and details of the arrangements for the use of the admixtures. The architect should also confirm in his written approval the agreed procedure for use. Calcium chloride must never be used in reinforced masonry. Relevant British Standards include BS 5075: Part 1: 1982[17.17] and BS 5075: Part 2: 1982[17.18].

(b) Preparation of concrete

(i) Recommended mix
Concrete for core filing shall be .... (strength and/or proportions) with a slump of .... mm
The concrete should be specified by the designer to satisfy the performance requirements of the wall, e.g. strength, sound insulation, thermal insulation and so on. A typical general mix for such purposes is 1 : 2½–3 : 2 with a slump of 125 mm.

(ii) Alternative mix
Alternative concrete mixes may be used subject to written approval

(5) Reinforcement, wall and bonding ties
(a) Reinforced concrete reinforcement
The reinforcement for .... (location) shall be .... (type/size) to BS ....
For type indicate type of reinforcement (mild steel, high yield steel, etc.). For size indicate length and diameter (reference to drawing/schedule often required). Reinforcement for cores should generally not exceed 16 mm. Relevant British Standards include BS 4449: 1988[17.19] and BS 4482: 1985[17.20].

(b) Bed joint reinforcement
The bed joint reinforcement for .... (location) shall be .... (type/size/reference number) and comply to .... (BS or other).
For type indicate mild steel, high yield steel, etc. For size indicate length, diameter (reference to schedule) and for proprietary bed joint reinforcement indicate width and/or manufacturer's reference number and manufacturer. Relevant British Standards include BS 4449: 1988[17.19] and BS 4482: 1985[17.20].

(c) Bonding ties
Metal strips or mesh for bonding .... (location) shall be .... (material and BS) of dimension ....
The particular bonding tie to be specified must take account of the conditions of exposure to which it will be subjected. Such ties, when used at control joints, should not affect the performance of joints. Quote dimensions or refer to drawing.

(d) Wall ties
Wall ties shall be .... (description) to .... (BS or other)
The particular wall tie to be specified must take account of the conditions of exposure, sound insulation and structural requirements to which it will be subjected. Butterfly ties to BS 1243: 1978[17.21] will be the most common tie for concrete masonry. Where stiff ties are needed for special structural requirements, additional consideration should be given to the question of accommodation of movements (*Chapter 16*) and its effect on sound insulation (*Chapter 14*).

(6) Handling
(a) Cement
Cement should be stored in dry conditions and be used in order of delivery

(b) Sand
Store sands separately according to type, prevent contamination

(c) Metals
Ties should be stored to prevent metal becoming rusty and contaminated

Reinforcement to be free from loose mill scale and rust

(d) Blocks
Facing blocks shall be carefully unloaded to avoid damage. All blocks to be stacked on prepared level areas to avoid ground contamination. Stacks to be covered to prevent saturation and facing blocks to be protected to avoid becoming stained or marked.
Covering and protection of blocks is desirable to minimize effects of subsequent shrinkage, loss of bond and to maintain appearance.

(7) Testing
(a) General
Independent testing of blocks shall be carried out strictly in accordance with Clause 13 of BS 6073: Part 1

(b) Special category manufacturing control
Quality control and compliance procedure shall be in accordance with Clause 6.5 of BS 6073: Part 1

17.3.2.2 Workmanship

(1) General
(a) Dimensions
All blockwork shall be set out and built to the respective dimensions, thickness and height indicated
When detailing, consideration should be given to the size and position of openings to allow for a full unit to be positioned correctly directly beneath a lintel bearing.

(b) Uniformity
All work to be plumbed and levelled as the work proceeds

(c) Bond
The work shall be built to the bond indicated on the drawings or schedule. Where no bond is indicated, the units shall be laid in stretcher bond, half lap. Where possible the coursing to be arranged to allow a full block to be positioned directly beneath a lintel bearing
The purpose of arranging a full block beneath a lintel is to reduce vertical alignments of perpend joints which may induce cracking.

(d) Cutting
Blocks used for facing shall be cut with a masonry saw. Where cut wet they shall be allowed to dry before use

(e) Chases
Chasing shall be as indicated on the drawings and carried out with a chasing tool unless otherwise approved
Chasing should be specified in accordance with BS 5628: Part 3: 1985[17.3] in order not to affect the serviceability of the wall.

(f) Weather
Protect units separated from stacks before use. Take precautions during cold weather to prevent frost damage to fresh mortar. Specify standards of workmanship (and design, detailing and materials) to ensure rain does not penetrate the wall. Brace constructions to prevent damage by winds or other causes
The general precaution during cold weather is to either stop laying when the air temperature is close to freezing, at say 3°C or less, or ensure a minimum temperature of 4°C in the work when laid and thereafter prevent the mortar from freezing until hardened. Problems can still exist even when initial air temperatures are above 3°C if overnight temperatures are expected to be low, or the units are very cold (BS 5628: Part 3: 1985[17.3]). Rain penetration is one of the commonest building defects and is affected by the quality of workmanship achieved on site (BS 5628: Part 3: 1985[17.3]).

(g) Laying
All units to be laid and adjusted to final position while mortar is still plastic

(2) Mortar joints
(a) Bedding
Units shall be laid on a .... (full, shell mortar bed). Vertical joints shall be filled. Joints to be nominally .... mm thick
Most masonry will normally be laid on a full bed of mortar of normal thickness 10 mm. Shell bedding is used occasionally in an attempt to improve the rain resistance of single leaf walls but it will reduce the loadbearing capacity of the wall. In Germany and other European countries the use of thin mortar joints is well established. Specialist contractors and a separate detailed specification would be required if used in the UK.

(b) Joint types
(i) Facing work
Joints for .... (location) shall be .... (specified profile).
The tooling of joints shall be carried out to the specified profiles while the mortar is thumb-print hard

(ii) Standard work
Joints shall be .... (struck or raked) for .... (plastering, rendering, etc.) at .... (location)
Advice as to whether the wall needs to be struck or raked can be obtained from the block manufacturer.

(c) Excess mortar
Any mortar which extrudes from the joints of fair faced units shall be cut away and not smeared onto the face of the block

(d) Reinforced walls
The cores shall be kept clear and clean of mortar droppings and any extruding mortar shall be

removed while soft. Clean out holes to be provided at the bottom of core

A simple method for providing for clean out is to sit the ends of the first row of blocks on bricks.

(3) Control joints

Control joints shall be constructed as indicated on the drawings. Expansion joints shall be cleaned out to ensure the mortar does not bridge the joint

Attention is drawn to Chapter 16 and BS 5628: Part 3: 1985[17.3] with reference to thermal and other movements. It is important that control joints should be clearly drawn to show positions and details of any ties.

(4) Double leaf (cavity) walls

(a) Wall ties

The walls shall be built with cavities of the width shown on the drawings and tied together with ties embedded in the mortar at least 50 mm spaced in accordance with the following table:

Least leaf thickness (mm)	Cavity width (mm)	Spacing of ties	
		Horizontally (mm)	Vertically (mm)
Less than 90	50–75	450	450
90 or more	50–150	900	450

The spacing may be varied providing that the number of ties per unit area is maintained. Additional ties shall be provided in every course within 225 mm of openings and on each side of control joints. Ties shall be laid falling to the external leaf

(b) Cavities

The cavity and ties shall be kept clean and any extruding mortar shall be struck off flush. No cavity shall be sealed off until instructed

(c) Weepholes (cavity walls)

Weepholes 10 mm wide by 75 mm high shall be provided at not more than 1000 mm centres through the vertical mortar joints of the outer leaf, at ground level and at positions where the cavity is bridged

Weepholes enable the water which penetrates the outer leaf to escape.

(5) Partitions

Partitions shall not be built on suspended slabs until after the props have been removed

Allowance may need to be made for the deflection of the slab or beam.

(6) Reinforcement

(a) Reinforcement

The reinforcement shall be of the size and number as shown on the drawings and be positioned

accurately, to maintain the specified cover

Reference should be made to *Chapter 9* for the recommended cover.

(b) Bed joint reinforcement

Bed joint reinforcement shall have a side cover of not less than ...., be continuous except at control joints or where otherwise indicated, and be located as shown on the drawings

(7) Core filling

Cores shall be filled in lifts not exceeding .... (height). The concrete to be fully compacted by tamping.

To ensure that the cores can be cleaned and the concrete fully compacted the designer should, before allowing or specifying lifts in excess of 450 mm, insert additional clauses or confirm in writing to the contractor the agreed procedure for core filling. Full height filling, maximum 3 m, will ease construction and reduce costs.

(8) Lintels

Concrete block lintels shall be positioned and reinforced as shown on the drawings and filled with concrete as specified. Block lintels are to be propped during construction. All lintels shall have a minimum bearing length of .... mm

Where lintels other than block lintels are to be used, they should be specified in this section. The bearing length will depend on structural requirements. A nominal minimum length of 150 mm is reasonable.

(9) Protection

(a) Stability

Precautions shall be taken to ensure stability of walls during backfilling and concreting operations

(b) Finished work

The tops of constructed walls shall be protected from rain and in addition fair faced work shall be protected against staining from construction activities

(10) Making good

At the completion of the work all temporary holes in mortar joints of fair faced work shall be filled with mortar and suitably tooled. Any damaged blockwork shall be repaired with approved materials or replaced

17.3.2.3 Related work

The following items, although not part of the masonry, are connected with masonry construction and should, unless included in the masonry section, be adequately covered elsewhere in the main specification.

(1) Sealing

Joints around door and window frames, control joints, abutting joints at external columns and other joints where sealing is indicated shall be brush painted with

.... (type or name) *primer and filled with* .... *(type, colour or name)* *sealant to manufacturer's recommendations*

It is important to ensure that the specified sealant is suitable for the particular width and tolerance of the joint to be sealed.

(2) Flashing

Wall flashings shall be built into or secured to the masonry as shown on the drawings and be provided with .... mm laps

It is important that to ensure the flashings are effective they should be clearly detailed. The use of certain dissimilar materials can cause problems (attention is drawn to BS 5628: Part 3: 1985[17.3]).

(3) Damp-proofing

(a) Damp-proof courses

Horizontal damp-proof courses shall be positioned so as to fully cover the thickness, be laid on an even bed of fresh mortar and covered by mortar so as to maintain regular joint thickness. While exposed, they shall be protected from damage. Stepped damp-proof courses at openings shall extend beyond the end of the lintel by at least 100 mm. Vertical damp-proof courses shall be of .... (width) and be fixed so as to separate the inner and outer leaves of the wall. The material for damp-proof courses shall comply with BS 743: 1970[17.22]

Particular care is to be taken in detailing and positioning of damp-proof courses and attention is drawn to BS 5628: Part 3: 1985[17.3] and BRE Digest 77[17.23].

(b) Tanking

Tanking and water-proofing of basement walls or retaining walls shall be carried out to the details as shown on the drawings and all materials are to be used in accordance with the manufacturer's recommendations

(4) Backfilling

Backfilling shall not be placed against concrete masonry walls within .... days of completion of the construction. Vehicles shall not be closer to the wall than a distance equal to the height of the wall

(5) Painting

Painting shall not commence until the surface of the walls has been allowed to dry out and has been cleaned down to remove all dust, dirt and mortar dabs. Where efflorescence occurs, it shall be removed with a cloth or stiff brush prior to painting

For guidance on painting of walls see BRE Digests 163[17.24], 197[17.25] and 198[17.26]. Recommendations on the painting of buildings are given in BS 6150: 1991[17.27]. Paint for use directly in contact with masonry needs to be alkali resistant.

(6) Rendering

Newly applied rendering, including stipple and spatter-dash coats, shall be kept damp for the first three days. A second coat shall be not applied until the previous layer has hardened for seven days. The surface of the rendering shall be finished as specified. The block surface and subsequent rendering coats may be damped sufficiently to control suction but not saturated. Rendering shall not be applied to frost bound walls or during frost conditions. Any rendering shall be discontinuous at control joints

The type of rendering and preparation required should be related to the background material and exposure conditions. Clauses given here are of a general nature. As many materials exist it is suggested that the manufacturer of the blocks, BS 5628: Part 3: 1985[17.3] or BS 5262: 1991[17.28] should be referred to and suitable clauses specified. General advice is that each subsequent coat should not be thicker than that which precedes it. Designation (iii) renders are usually adequate for most purposes. Strong renders should be avoided.

(7) Plastering

Before plastering, all dirt, dust and efflorescence shall be removed. The walls shall be treated and plastered in accordance with the manufacturer's recommendations. Any plastering shall be discontinuous at control joints

As many materials exist it is suggested that the manufacturer of the blocks, or BS 5492: 1990[17.29] should be referred to and suitable clauses specified.

(8) Wall tiling

Before tiling all walls shall be allowed to dry to the level recommended by the tiling manufacturer. Movement joints shall be provided at control joints and other locations recommended by the tiling manufacturer

As many materials exist it is recommended that the manufacturer of the blocks or BS 5385: Part 1: 1995[17.30] or Part 2[17.31] should be referred to.

17.4 Enforcement of specifications

In simple terms, specifications, apart from their legal aspect, are an attempt to ensure that certain important aspects are conveyed to the site. Although it could be argued that having written the specification the designer need pay them no more attention until a problem occurs, it would be wise to watch for non-compliance as work proceeds. For example, in a case where problems eventually arise with mortar, where, perhaps in the first instance it appears too soft. A long and often complicated period then follows for all parties and, in many instances, although not always fully justified, results in hardened analysis of the mortar which for arguments sake let's say indicates incorrect mix proportions. Assuming also that this was one of those jobs where the specification called for the mortar to be batched by use of gauge boxes. Eventually there will no doubt be the familiar, 'Well, of course, I've never seen a

277

gauge box on this site anyway'. 'Oh yes you have, they make the tea on it'.

Surely if the designer considers that the specifications are important it would be far better to bring non-compliance to the attention of the contractor during and not after the event. Equally well, the contractor should accept that the designer has written the specifications for a good reason. This again returns to one of the original points made, that if the specifications are clear, simple and explicit they are more likely to be used.

Since the designer will be unable to pay attention to all specification points during the contract, the following is a short list of faults often encountered and which cause problems:

(1) unfilled perpends – particularly important in separating walls;
(2) mortar incorrectly batched;
(3) render incorrectly batched or wrong thicknesses;
(4) control joints (movement joints) incorrectly formed;
(5) units not covered or stored correctly;
(6) units used too soon after manufacture, or whilst too warm;
(7) correct bonding pattern not maintained in walls to be rendered;
(8) portions of blocks used under lintel bearings;
(9) wrong type of bed joint reinforcement or inadequate length.

17.5 References

17.1 BRITISH STANDARDS INSTITUTION. BS 5628: Part 1: 1992. *Structural use of unreinforced masonry.*

17.2 BRITISH STANDARDS INSTITUTION. BS 5628: Part 2: 1995. *Structural use of reinforced and prestressed masonry.*

17.3 BRITISH STANDARDS INSTITUTION. BS 5628: Part 3: 1985. *Materials and components, design and workmanship.*

17.4 NATIONAL BUILDING SPECIFICATIONS. Specification Writer. NBS Services, Mansion House Chambers, The Close, Newcastle upon Tyne, NE1 3RE.

17.5 BRITISH STANDARDS INSTITUTION. BS 882: 1992. *Specifications for aggregates from natural sources for concrete.*

17.6 BRITISH STANDARDS INSTITUTION. BS 6073: Part 1: 1981. *Specification for precast concrete masonry units.*

17.7 BRITISH STANDARDS INSTITUTION. BS 6073: Part 2: 1981. *Method for specifying precast concrete masonry units.*

17.8 BRITISH STANDARDS INSTITUTION. BS 12: 1996. *Specification for Portland cement.*

17.9 BRITISH STANDARDS INSTITUTION. BS 146: 1996. *Specification for Portland blast furnace cement.*

17.10 BRITISH STANDARDS INSTITUTION. BS 5224: 1995. *Specification for masonry cement.*

17.10a BRITISH STANDARDS INSTITUTION. BS 4027: 1991 *Specification for Portland sulphate-resisting cement.*

17.11 BRITISH STANDARDS INSTITUTION. BS 890: 1995. *Specification for building limes.*

17.12 BRITISH STANDARDS INSTITUTION. BS 1199 and BS 1200: 1976 (1996). *Specification for building sands from natural sources.*

17.13 BRITISH STANDARDS INSTITUTION. BS 3148: 1980. *Methods of test for water for making concrete (including notes on the suitability of water).*

17.14 BRITISH STANDARDS INSTITUTION. BS 4887: Part 1: 1986. *Specification for air-entraining (plasticising) admixtures.*

17.15 BRITISH STANDARDS INSTITUTION. BS 4887: Part 2: 1987. *Specification for set retarding admixtures.*

17.16 BRITISH STANDARDS INSTITUTION. BS 3797: 1990 (1996). *Specification for lightweight aggregates for masonry units and structural concrete.*

17.17 BRITISH STANDARDS INSTITUTION. BS 5075: Part 1: 1982. *Concrete Admixtures. Specification for accelerating admixtures, retarding admixtures and water reducing admixtures.*

17.18 BRITISH STANDARDS INSTITUTION. BS 5075: Part 2: 1982. *Concrete Admixtures. Specification for air entraining admixtures.*

17.19 BRITISH STANDARDS INSTITUTION. BS 4449: 1988. *Specification for carbon steel bars for the reinforcement of concrete.*

17.20 BRITISH STANDARDS INSTITUTION. BS 4482: 1985. *Specification for cold reduced steel wire for the reinforcement of concrete.*

17.21 BRITISH STANDARDS INSTITUTION. BS 1243: 1978. *Specification for metal ties for cavity wall construction.*

17.22 BRITISH STANDARDS INSTITUTION. BS 743: 1970. *Specification for materials for damp-proof courses.*

17.23 BRE DIGEST 77: Damp proof courses.

17.24 BRE DIGEST 163: Drying out buildings.

17.25 BRE DIGEST 197: Painting walls.

17.26 BRE DIGEST 198: painting walls.

17.27 BRITISH STANDARDS INSTITUTION. BS 6150: 1991 *Code of practice for painting of buildings.*

17.28 BRITISH STANDARDS INSTITUTION. BS 5262: 1991. *Code of practice for external rendering.*

17.29 BRITISH STANDARDS INSTITUTION. BS 5492: 1990 *Code of practice for internal plastering.*

17.30 BRITISH STANDARDS INSTITUTION. BS 5385: Part 1: 1995. *Wall and floor tiling. Code of practice for the design and installation of internal ceramic and natural stone wall tiling and mosaics in normal conditions.*

17.31 BRITISH STANDARDS INSTITUTION. BS 5385: Part 2: 1991. *Wall and floor tiling. Code of practice for the design and installation of external ceramic wall tiling and mosaics (including terra cotta and faience tiles).*

Chapter 18
Blemishes, faults and problems

18.1 Introduction

The purpose of this chapter is to provide general information on some of the problems which can occur in masonry construction from time to time. Detailed information on common defects in buildings and remedial measures can be found in another publication[18.1]. This chapter deals with cracks in masonry, colour variation in walling, stains, efflorescence and lime bloom, chemical attack and rendering.

18.2 Cracks in masonry

If a building is adequately designed and there is no excessive differential foundation movement, significant cracking should not develop. There are, however, occasions when cracks do develop and this section is intended to give guidance as to the cause of particular problems. Information on the provision of movement joints and other measures to cater for the effects of movement are presented in *Chapter 16*, which also contains some guidance on the differential movement of the various masonry materials (concrete, clay and calcium silicate) and some design and construction faults which can give rise to cracking. This section is mainly concerned with foundation and other structural movements.

To identify the cause of cracking one of the first steps should be to try to discover the mechanism of the movement. In many instances cracks will be found to taper over the wall height. A slight taper can often be expected as the top of a wall will generally be less restrained than its base. However, when the crack size is greater than can be predicted by consideration of normal linear movements, it may be used to indicate a failure mechanism. The narrowest width of the crack is taken as a fulcrum point from which the likely direction of movement can be postulated. A number of examples are given in Figure 18.1, which it is hoped can be developed to deal with the particular building under investigation.

Depending upon the relative sizes and strengths of units and mortar, the crack may appear either as a stepped crack running from bed joint to perpend alternately up the wall or it may run vertically or diagonally through the units. Before repair to any cracking can be attempted, the cause must first be determined and then a decision made as to whether or not the movement has stopped. In the case of foundation movement, remedial measures, such as underpinning, may be needed to prevent further problems. Assessment of damage in low rise buildings is given in BRE Digest 251[18.2].

Cracks due to shrinkage and thermal contraction in concrete masonry tend to be fine and are generally best left alone, particularly where the surface is rendered. Attempts at repair often make the cracks more conspicuous, unless of course the rendering is to be repainted. With rough cast type renders any cracking will normally go unnoticed. In the particular case of cracking of rendered fired-clay brickwork some consideration should be given to remedial measures since the sulphates which are present in many bricks can, as a result of moisture entering through the crack, attack the rendering. All building materials, such as bricks, blocks, timber, and so on, are continually undergoing some small movement due to changes in moisture content or temperature and complete cure of cracking is, therefore, very difficult. However, subsequent cracking will usually be far less of a problem.

Where cracks along the mortar are to be made good, the mortar should be raked back at least 25 mm, and preferably 50 mm, and slightly undercut if possible. The gap should be brushed to remove loose material and lightly wetted. A mortar of the same designation (assuming it to be correct in the first place) and materials should be pushed firmly into place and when thumb print hard should be tooled to the required profile. The repaired area should then ideally be kept moist or covered with polythene to prevent rapid drying. Another method which can be tried, particularly where further small movement is still expected, is to rake back 25–30 mm and fill with a mastic of similar colour to the existing mortar. Some of the mortar removed as a result of the cutting back can then be tooled into the mastic, and with care a good match can be achieved.

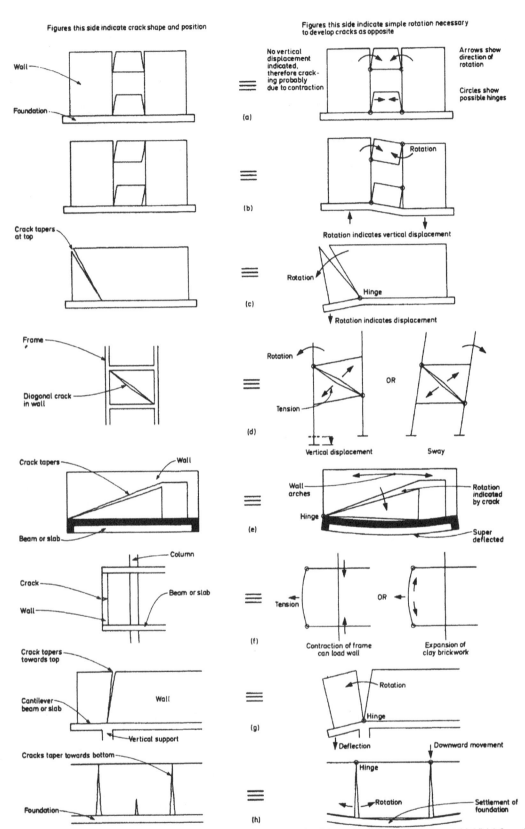

Figures this side indicate crack shape and position

Figures this side indicate simple rotation necessary to develop cracks as opposite

Figure 18.1 The rotation concept: (a) no displacement indicated possible contraction (b) rotation indicates displacement (c) (d) le) floor deflection (f) (g) cantilever deflection (h) floor/foundation settlement

18.3 Colour variation in walling

The use of units from different manufacturing batches can lead to distinct 'banding' or 'patches' in the finished wall and is the most common cause of colour variation in masonry. The problem is, of course, most apparent with fairly smooth faced, evenly textured units, and ideally pallets should be opened and the blocks mixed to ensure an even dispersion of colour variation. With many units natural variations occur or variations are introduced during manufacture to produce a wall with a less uniform appearance.

A second source of problems stems from the incorporation of water proofers during manufacture. If for any reason individual blocks or parts of batches have received a different dosage or no dosage of water proofer leading to more absorbent units being built into the wall, unsightly differential water retention at points where these blocks have been used will appear. The same phenomenon can also occur when blocks are made with markedly different absorption characteristics, but modern quality control procedures have made this comparatively rare.

A third source of colour variation in masonry walling occurs in the mortar joint. With some sands differences in water content of the mix can lead to very pronounced colour variations in the hardened mortar, and strict control of batching on site is needed, not allowing retempering and prolonged wet mortar life. Where coloured mortars are to be produced by means of additives the problem is even more acute and consideration should be given to the use of pre-bagged mortars to achieve greater consistency during mixing.

18.4 Stains on masonry

As a result of site activities or vandalism stains can occasionally occur on masonry and this section has been included to give some guidance on the methods available for attempting to remove them. The removal of a stain from masonry is often a difficult process and sometimes produces disappointing results. Care has to be taken to avoid damage or spoiling of the surface of the masonry. The difficulties which may arise can be categorized by three factors:

(1) masonry surfaces are generally porous so that both stains and applied cleaning solutions tend to become absorbed in the surface texture;
(2) concrete units and mortars are chemically reactive and can dissolve in some cleaning solutions more readily than the materials causing the stain;
(3) local treatment, although removing the stain, may alter the texture or colour sufficiently to cause a different blemish.

For these reasons it is wise to remove the stain as soon as possible. The two basic methods of removing stains are chemically and mechanically.

18.4.1 Chemical cleaning

Materials which may be used for removing stains from masonry include acids, organic solvents and emulsifying agents. To combat the tendency of the masonry to absorb the cleaning liquids the materials are sometimes mixed into a paste with talc, whiting or other powder. Applied to the masonry, the mixture draws the stain out to the surface. Such pastes often need to be covered with polythene to prevent the solvents from drying out too quickly.

Corrosive and poisonous chemicals *must* be handled with extreme care. It is advisable to wear rubber gloves, goggles and so on, when applying them and indeed to seek expert advice before using the materials; particularly if large quantities are to be used. A few chemical methods for removal of the more common stains encountered are given below. It is obviously desirable to experiment on small discrete areas initially to assess the effect upon the stain and any other possible side effects. Where chemical methods are employed it is advisable to wash off residues with water after treatment.

18.4.1.1 Rust

Rust is difficult to remove as it is more chemically inert than the concrete surface but light staining can sometimes be removed by lightly etching the surface with 5–10% solution of hydrochloric acid for a few minutes and then washing down with water. Alternatively, wetting the surface thoroughly with several applications of 15% sodium citrate solution and applying a layer of sodium dithionite crystals held in place by a paste of whiting and water may work.

18.4.1.2 Oil

Particularly if recently spilt, oil can be removed by brushing the surface with a proprietary engine cleaning fluid and then flooding with water to flush away the emulsified oil. Alternatively, scrubbing the surface with a strong detergent solution and then washing with water can be effective. Another method which has been suggested is to make a smooth stiff paste from petrol and talc, apply to the stain and cover with polythene or similar to prevent rapid evaporation. Several applications may be needed to completely remove the stain.

18.4.1.3 Creosote

Creosote can be removed by applying a paste of talc and petrol and then scouring the surface to complete the removal.

18.4.1.4 Bitumen, asphalt and coal-tar products

Again these can be removed by application of a paste of talc and petrol, but as much of the material as possible should be scraped off prior to application of the paste.

A certain amount of consideration is needed in the removal of paints since there are a variety of types. Oil-based paint can often be removed with a proprietary solvent or alkaline paint remover. Chlorinated rubber paints may need special strippers such as naphtha. Cement paints are obviously difficult to remove without affecting the concrete surface. The information given in BRE Digest 197[18.3] may be of use in some cases. Special graffiti removing products are now also available.

Other stains can be treated with organic solvents, bleach solutions and acids to ascertain which is the best chemical to remove them. Chemical cleaning methods are basically applicable to sudden stains caused by site activity and such like. Some long term stains resulting from dirt or algal or fungal growths are often removed by mechanical methods, although algal and fungal growths can be removed chemically by applying a solution of domestic bleach and, after a few days, scrubbing off the dead material.

18.4.2 Mechanical cleaning

Mechanical methods, including high pressure water cleaning and steam cleaning, are often used to remove dirt, grime and fungal growths. Gritblasting can be used very effectively in the removal of dirt and grime, but this method also removes some of the surface, thus altering the surface texture.

18.5 Efflorescence and lime bloom

Efflorescence is the term commonly used to describe the white discolouration which forms on masonry and concrete surfaces. The formation of the discolourations vary depending upon the soluble salt or salts from which it originates. With salts such as calcium sulphate, sodium sulphate, potassium sulphate and magnesium sulphate (particularly when in combination), the salt goes into solution in water and is carried to the surface where it crystallizes as the water evaporates. Some of these salts are often present in fired-clay bricks and can be the cause of the white deposits on brickwork. This crystal formation is the medium normally responsible for the efflorescence on fired-clay brickwork.

With concrete units and mortar the predominant salt is calcium hydroxide which is formed as a product of the chemical reaction between the cement and water. Calcium hydroxide is slightly soluble in water and under certain conditions it can be carried through to the surface and there reacts with carbon dioxide from the atmosphere to produce a surface deposit of calcium carbonate. This can, and normally does, take place within the wet surface. The surface deposit is normally extremely thin which is demonstrated by the fact that when the surface is wetted, the film of water on the surface commonly renders the deposit translucent and it seemingly disappears. This deposit is thus better and more commonly referred to as 'lime bloom' or 'lime running'.

Efflorescence and indeed lime bloom or running is a function of many interrelated factors and although steps can be taken to reduce the likelihood of its occurrence, total prevention is very difficult to achieve. Where the masonry is exposed to the weather, the rainwater, which may once have helped in the formation of the deposit, being also slightly acidic, slowly dissolves the deposit which is thus reabsorbed into or washed from the wall. The continuing rainwater carrying the dissolved material will penetrate into and down the wall. The process continues until the surface deposit is removed and equilibrium conditions exist at which time any remaining salts having been washed in by the rain are insufficient to return to the surface. This condition is normally reached within a year. Occasionally, disturbance of the equilibrium conditions can cause a recurrence of the deposits.

The cause of efflorescence or lime bloom on some new structures can often be traced to poor storage on site and construction/design faults, which permit vast quantities of water to run down and through the stored materials or constructed walls. It is impossible to prevent the normal amount of rain falling on the newly constructed walls but particular attention should be given to the possible discharge from the roof prior to the structure being waterproofed. In some designs, delays in waterproofing the roof can result in large quantities of water pouring down through the walls and causing even worse efflorescence problems. Although the deposits on the external walls may eventually be removed by rainwater, the internal walls, particularly those which are left fair faced, will have the white deposits, some of which can be brushed or washed off. Lime bloom or lime runs (calcium carbonate) can be removed by washing with 5–10% solution of hydrochloric acid but this can cause surface etching. It will also be necessary to pre-wet the wall to prevent the acid being drawn into the body of the wall.

In some instances it may be necessary and even preferable to paint the surface to achieve an acceptable appearance, but it is important to determine and rectify the cause of efflorescence before painting, otherwise further development of the salts can force the paint from the surface. The most practical solution is to apply a white latex or emulsion paint, since they can be applied to damp surfaces and, more importantly, will still allow the wall to breathe. The efflorescence can then form on the surface without damaging the paint, and since most efflorescence is white, it should not be noticeable on the surface. Occasional brushing may also be beneficial.

18.6 Chemical attack

Chapter 3 which dealt with materials provided general guidance on the quality of materials recommended for normal durability situations. One item in particular which should be emphasized is that Portland cement which is used for both units and mortar is not resistant to acid attack. In such situations, for instance in a silage retaining wall, where mild acid attack may occur, the wall should be covered by two coats of chlorinated rubber or similar protective paint system.

Oil paints should not be applied to concrete surfaces without first applying an efficient alkali-resisting primer. Such paints, however, are not normally resistant to acid attack.

18.7 Rendering

If cracking occurs in the rendering but does not continue into the supporting masonry, the problem is likely to lie in the preparation and materials used for the render. This section is intended to give basic guidance on the principal items necessary for good rendering. Consideration of these items may be used for initial assessment of the problem. More comprehensive information can be found in the publication *External Rendering*[18.4], published by the British Cement Association.

18.7.1 Materials

In general the materials which should be used for rendering are similar to those required for mortar except that the sand will normally need to be coarser and comply with BS 1199[18.5].

18.7.2 Preparation of background

Since there is a wide range of masonry materials, each with its own texture, size and absorption or suction characteristics, it is necessary to carefully consider each material to assess the amount and type of surface preparation required. There are so many materials that it is only possible to write in very broad terms. In general, however, the masonry can be classified into the following categories: dense strong and smooth materials; moderately strong, porous materials, and moderately weak, porous materials. Many of the materials will benefit from having the mortar joints raked out.

18.7.2.1 Dense, strong and smooth materials

Dense concrete blocks, concrete bricks, and dense clay blocks not provided with a key to take the rendering may fall into this category. Where the surface is smooth, a stipple coat may be required, particularly with larger units and where the joints are not raked. A stipple coat is a thick slurry (1 part cement to 1½ parts clean sharp sand gauged with water to which an equal quantity of 'bonding' agent has been added), vigorously brushed into the surface and immediately stippled with a bannister brush to form a close textured key.

18.7.2.2 Moderately strong, porous materials

Most lightweight aggregate blocks fall into this category, as do clay and sand lime bricks. These types of masonry material generally have fair suction and good mechanical key. Occasionally, where suction is too great or uneven suction exists, a stipple coat or spatterdash may again be required.

18.7.2.3 Moderately weak, porous materials

Autoclaved aerated concrete blocks and some lightweight aggregate blocks may fall into this category, as do some of the softer types of fired-clay brick. These materials are rather varied and a satisfactory key may be obtainable without additional preparation. Where excessive suction exists the wall may require damping down' (not soaking) prior to rendering, and again a stipple coat or spatterdash can be used to control suction. With some materials a water retentive admixture, such as methyl-cellulose can be used in the rendering to prevent too rapid drying out and to enable it to be worked satisfactorily.

18.7.3 Mix proportions and thickness

The type of render will primarily depend upon the background material, degree of exposure and a few other factors which may also apply. In general a type (iii) cement:lime:sand (1:1:5–6) is the type of render used on masonry walls. Richer renders than this can cause problems due to higher drying shrinkage. Obviously there are also a number of factors governing thickness of render, but the general rule is that no coat should be stronger than the coat which precedes it. In fact, ideally, each successive coat should be made thinner than the immediately preceding coat. The undercoat should be of maximum thickness 15 mm and should be combed to prevent undue stresses forming as the rendering hardens and shrinks and to provide a key for subsequent coats. With some dense, strong materials a maximum overall thickness (undercoat plus final coat) of 15 mm may be called for.

18.7.4 Curing

It is obviously very important to cure the rendering adequately at all its various stages. Stipple and spatterdash coats, undercoats and final coats should all be prevented from rapid drying out, particularly during hot weather. Ideally they should be kept damp for three to four days and then be allowed to dry out for a few days before application of a further coat. Protection from rain and frost may also be necessary.

18.8 References

18.1 ELDRIDGE, H J. Common defects in buildings. HMSO, London, 1976. pp 486.
18.2 BUILDING RESEARCH ESTABLISHMENT. BRE Digest 251: *Assessment of damage in low rise buildings.* BRE, Garston, 1990, pp 8.
18.3 BUILDING RESEARCH ESTABLISHMENT. BRE Digest 197: *Painting walls.* Part 1: *Choice of paint.* BRE, Garston, 1982, pp 8.
18.4 MONKS, W, AND WARD, F. Appearance matters – 2: External rendering. British Cement Association, Slough. Publication No 47.012. pp 32.
18.5 BRITISH STANDARDS INSTITUTION. BS 1198 and 1200: 1976 (1996) *Building sands from natural sources.* BSI, London. pp 8.

Chapter 19
European Standards for Masonry Products

19.1 Introduction

A key aim of the Treaty of Rome is the free movement of goods and services across the borders of the member countries of the European Union. The foremost aim of European Standardisation is to facilitate the exchange of goods and services through the elimination of technical barriers to trade. The European Standards for masonry are related to European legislation in the form of the Construction Products Directory (CPD). Standards which are produced to meet the requirements of the CPD are referred to as Harmonised Standards and these Standards must be met by manufacturers before products can be traded legally within the Single Market.

Responsibility for producing the European Standards for masonry rests with CEN (European Committee for Standardisation). CEN is a legal association, the members of which are the National Standards Bodies of 19 European countries and six Associates. The European Standards for masonry materials are the responsibility of CEN Technical Committee 125 which commenced its work in 1988. The principal outcomes of the work of the committee are European Standards (EN) which must be published by each of the National Standards Bodies as an identical National Standard. Any pre-existing National Standards in conflict have to be withdrawn.

The scope of CEN TC 125 covers standardisation in the field of masonry units of clay, calcium silicate, dense aggregate concrete, lightweight aggregate concrete, auto-claved aerated concrete, natural stone, manufactured stone, mortar for masonry, ancillary components for masonry, and associated test methods.

CEN TC 125 has developed a range of product specifications and associated, but separate, test methods which, as far as practicable, are common to more than one product and have taken into account other test methods and apparatus relating to similar products. It should be noted that the initial product specifications will comprise 'ordinary CEB Standards' which will not enable conforming products to achieve CE marking. A range of amendments to each Standard will address the EC mandate requirements and result in 'Candidate Harmonised Standards'. Such Standards, once approved by the EC and listed in its Official Journal, will enable the complying products to be CE marked.

Many of the proposed CEN Standards for masonry are still in draft form. The purpose of this chapter is to indicate the current state on progress and reflect the changes in requirements in the European Standards compared with current practice in the UK.

19.2 Standards for Masonry Units

Masonry Unit Standards

The six Masonry Unit Standards have been drafted on the basis of designation and it is expected that the physical properties of the unit are declared by the manufacturer. The specifier is expected to indicate the appropriate performance requirements that are needed for the intended use of the units. It is clear that the user should not specify more performance requirements than are necessary for the proposed project.

The format of the Standards is consistent and much of the terminology has now been standardised. The conflict in current British Standards between width and thickness, for example, has been harmonised and all units are specified by length, width and height in mm, and in that order.

The descriptors brick and block are not used in the CEN Standards. Instead, both are known as masonry units. Six Standards for masonry units are planned and these are:

EN 771–1 Specification for masonry units – Part 1: Clay masonry units.

EN 771–2 Specification for masonry units – Part 2: Calcium silicate masonry units

EN 771–3 Specification for masonry units – Part 3: Aggregate concrete masonry units (Dense and light-weight aggregates)

EN 771–4 Specification for masonry units – Part 4: Autoclaved aerated concrete masonry units.

EN 771–5 Specification for masonry units – Part 5: *Manufactured stone masonry units.*

EN 771–6 Specification for masonry units – Part 6: Natural stone masonry units

19.2.1 Clay masonry units

EN 771–1 Specification for masonry units – Part 1: Clay masonry units

When published, this standard will specify the characteristics and performance requirements for masonry units manufactured from clay for use in masonry construction. It is intended to cover all fired-clay masonry units including those of an overall non-rectangular parallel piped shape.

The following performance requirements are defined:

* dimensional tolerances
* geometry
* strength
* density
* freeze/thaw resistance
* content of active soluble salts
* water absorption
* initial rate of water absorption
* thermal properties
* conventional moisture expansion

The standard will also include provision for the evaluation of conformity of the product and marking.

The standard will not specify standard sizes for clay masonry units nor information regarding specially shaped units.

19.2.2 Calcium silicate masonry units

EN 771–2 Specification for masonry units – Part 2: Calcium silicate masonry units

When published, this standard will specify the characteristics and performance requirements for calcium silicate masonry units.

The following performance requirements are defined:

* dimensions and tolerances
* geometry, shape and features
* dry density and tolerances
* compressive strength
* thermal properties
* freeze/thaw resistance

The harmonised standard will also include provision for the evaluation of conformity of the product and marking.

The standard will not cover the following:

* units with more than 60% volume of voids
* products made from shale units intended for use as a damp proof course

19.2.3 Aggregate concrete masonry units (Dense and light-weight aggregates)

EN 771–3 Specification for masonry units – Part 3: Aggregate concrete masonry units (Dense and light-weight aggregates)

When published, this standard will specify the characteristics and performance requirements for the physical and mechanical properties of aggregate, lightweight aggregate or a combination of both.

The following performance requirements are defined:

* strength
* density
* dimensional accuracy
* geometry
* thermal properties
* water absorption
* moisture movement

The standard will also include reference to fire, acoustic, frost resistance, and sulfate resistance. The standard will not include standard sizes or units intended for use as a dpc.

The standard will also include provision for the evaluation of conformity of the product and marking.

19.2.4 Autoclaved aerated concrete masonry units

EN 771–4 Specification for masonry units – Part 4: Autoclaved aerated concrete masonry units

When published, this standard will specify the characteristics and performance requirements of autoclaved aerated concrete masonry units. It defines performance related to the following:

* strength
* density
* dimensional accuracy
* thermal conductivity
* moisture movement

In addition, guidance will be given on freeze/thaw resistance. The standard will not specify standard sizes for autoclaved aerated concrete units and will not cover products intended for use as a dpc.

The harmonised standard will also include provision for the evaluation of conformity of the product and marking.

19.2.5 Manufactured stone masonry units

EN 771–5 Specification for masonry units – Part 5: Manufactured stone masonry units

When published, this standard will specify characteristics and performance requirements of manufactured stone masonry units. It will cover units produced by using casting or pressing techniques with or without textured surfaces. Examples of production techniques include casting, splitting, washing, blasting or tooling. It will cover both homogenous masonry units and those consisting of different facing and backing concrete mixes, but not units with an adhesive bonded decorative face.

It will define performance related to the following:

* strength
* density
* dimensional accuracy

- thermal conductivity
- water absorption
- moisture movement

The standard will not cover standard sizes or units intended to be used as a dpc.

The standard will include provision for the evaluation of conformity of the product and marking.

19.2.6 Natural stone masonry units

EN 771–6 Specification for masonry units – Part 6: Natural stone masonry units

When published, this standard will specify the characteristics and performance requirements of natural stone masonry units whose width is equal to or greater than 80mm.

The following aspects of performance are defined:

- strength
- petrographic description
- density
- porosity
- dimensional accuracy
- thermal conductivity
- water absorption
- frost resistance

It will also contain reference to fire and acoustic properties. The standard will also include provision for the evaluation of conformity of the product and marking.

19.3 Methods of test for masonry units

EN 772–1	Methods of test for masonry units – Part 1: Determination of compressive strength
EN 772–2:1998	Methods of test for masonry units – Part 2: Determination of percentage area of voids in aggregate concrete masonry units (by paper indentation)
EN 772–3:1998	Methods of test for masonry units – Part 3: Determination of net volume and percentage of voids of clay masonry units by hydrostatic weighing
EN 772–4:1998	Methods of test for masonry units – Part 4: Determination of real and bulk density and of total and open porosity for natural stone masonry units
EN 772–5	Methods of test for masonry units – Part 5: Determination of the active soluble salts content of clay masonry units
EN 772–6	Methods of test for masonry units – Part 6: Determination of bending tensile strength of aggregate concrete masonry units
EN 772–7:1998	Methods of test for masonry units – Part 7: Determination of water absorption of clay masonry damp proof course units by boiling in water

EN 772–8	Nothing currently allocated to this standard number.
EN 772–9:1998	Methods of test for masonry units – Part 9: Determination of volume and percentage of voids and net volume of calcium silicate masonry units by sand filling
EN 772–10:1998	Methods of test for masonry units – Part 10: Determination of moisture content of calcium silicate and autoclaved aerated concrete units
EN 772–11	Methods of test for masonry units – Part 11: Determination of water absorption of aggregate concrete, manufactured stone and natural stone units due to capillary action and the initial rate of water absorption of clay masonry units.
EN 772–12	Nothing currently allocated to this standard number
EN 772–13	Methods of test for masonry units – Part 13: Determination of net and gross dry density of masonry units (except for natural stone)
EN 772–14	Methods of test for masonry units – Part 14: Determination of moisture movement of aggregate concrete and manufactured stone masonry units
EN 772–15	Methods of test for masonry units – Part 15: Determination of water vapour permeability of autoclaved aerated concrete masonry units
EN 772–16	Methods of test for masonry units – Part 16: Determination of dimensions
EN 772–17	Nothing currently allocated to this standard number
EN 772–18	Methods of test for masonry units – Part 18: Determination of freeze-thaw resistance of calcium silicate masonry units
EN 772–19	Methods of test for masonry units – Part 19: Determination of moisture expansion of large horizontally perforated clay masonry units
EN 772–20	Methods of test for masonry units – Part 20: Determination of flatness of faces of aggregate concrete, manufactured stone and natural stone masonry units
EN 772–21	Nothing currently allocated to this standard number
EN 772–22	Methods of test for masonry units – Part 22: Determination of freeze-thaw resistance of clay masonry units

19.3.1 Determination of compressive strength

EN 772–1 Methods of test for masonry units – Part 1: Determination of compressive strength

The method of determining the compressive strength of masonry units varies with the material being tested.

Initially it was hoped that the test procedure would be the same for each material covered by the six Masonry Unit Standards, but this proved to be impractical.

19.3.2 Determination of percentage area of voids in aggregate concrete masonry units (by paper indentation)

EN 772–2:1998 Methods of test for masonry units – Part 2: Determination of percentage area of voids in aggregate concrete masonry units (by paper indentation)

This procedure for determining the percentage area of voids in aggregate concrete masonry units uses thick sheets of paper impressed on the test surface. Before the measurement is made any roughness is removed from the test face of each unit. A minimum size unit is usually measured. A bottom sheet is placed on the lower platen of a compression testing machine, the unit is placed on top, and a top sheet can then be placed on the unit. A load of approximately 3KN is then applied. The load is then removed and the indentations marked in outline on both sheets of paper. The total cross sectional area of voids is either determined by area using a planimeter or by weighing.

19.3.3 Determination of net volume and percentage of voids of clay masonry units by hydrostatic weighing

EN 772–3:1998 Methods of test for masonry units – Part 3: Determination of net volume and percentage of voids of clay masonry units by hydrostatic weighing

This is a relatively simple procedure for obtaining the net volume of a clay masonry unit (essentially the masonry unit is weighed in air and then weighed in water) and subtracting this from the gross volume (obtained by measurement of its dimensions) to obtain the volume of voids.

19.3.4 Determination of real and bulk density and of total and open porosity for natural stone masonry units

EN 772–4:1998 Methods of test for masonry units – Part 4: Determination of real and bulk density and of total and open porosity for natural stone masonry units

In this procedure the stone samples (at least five) are dried to a constant mass. The bulk and impermeable volume and the bulk and real density of the units can then be calculated. A Le Chatelier flask is used for the measurement of real density.

19.3.5 Determination of the active soluble salts content of clay masonry units

EN 772–5 Methods of test for masonry units – Part 5: Determination of the active soluble salts content of clay masonry units

When published, the standard will specify a method for determining the active soluble salts content of clay

masonry units. The method adopted is based on water extraction from a crushed representative sample of the clay masonry units, and determines the amounts of soluble magnesium, sodium and potassium ions, released under the test conditions, which may be correlated with the potentially damaging effect of salts of those ions on cementitious mortars in certain circumstances, or even on the units themselves. In prEN771–1 these salts are described as active soluble salts.

19.3.6 Determination of bending tensile strength of aggregate concrete masonry units

EN 772–6 Methods of test for masonry units – Part 6: Determination of bending tensile strength of aggregate concrete masonry units

When published, this standard will specify a method for determining the bending tensile strength of aggregate concrete masonry units having a width less than 100mm and a ratio of length to width greater than 10, in accordance with prEN771–3. This test is similar to that described in BS 6073.

19.3.7 Determination of water absorption of clay masonry damp proof course units by boiling in water

EN 772–7:1998 Methods of test for masonry units – Part 7: Determination of water absorption of clay masonry damp proof course units by boiling in water

This is a well established test method for determining the water absorption of damp proof course clay masonry units by boiling the specimens in water for a fixed period. At lest six specimens are taken and dried to constant mass. They are then weighed and immersed in water which is boiled for 5 hours. They are then wiped and weighed. The ratio of increase in mass of the saturated specimens to the dry mass is calculated.

19.3.8 Determination of volume and percentage of voids and net volume of calcium silicate masonry units by sand filling

EN 772–9:1998 Methods of test for masonry units – Part 9: Determination of volume and percentage of voids and net volume of calcium silicate masonry units by sand filling

This is a very simple test for determining the volume and percentage of voids and net volume of calcium silicate masonry units. The voids in the unit are determined by measuring the volume of sand required to fill them.

19.3.9 Determination of moisture content of calcium silicate and autoclaved aerated concrete units

EN 772–10:1998 Methods of test for masonry units – Part 10: Determination of moisture content of calcium silicate and autoclaved aerated concrete units

This test method is applicable to both calcium silicate and autoclaved aerated concrete units. A minimum of six representative samples of the units are taken and weighed. They are then dried at a temperature of 105 ± 5 °C to constant mass. After drying to constant mass the specimens are weighed. The mean value of the moisture content is then calculated to the nearest 1%.

19.3.10 Determination of water absorption of aggregate concrete, manufactured stone and natural stone units due to capillary action and the initial rate of water absorption of clay masonry units

EN 772–11 Methods of test for masonry units – Part 11: Determination of water absorption of aggregate concrete, manufactured stone and natural stone units due to capillary action and the initial rate of water absorption of clay masonry units

When published this standard will specify a method of determining the water absorption coefficient due to capillary action for aggregate concrete, natural stone and manufactured stone masonry units and the initial rate of water absorption for clay masonry units.

The principle of the test is that, after drying the unit to constant mass, a face of the unit is immersed in water for a specific period of time and the increase in mass is determined.

For clay masonry units the initial rate of absorption of the bed face is measured. In the case of aggregate concrete, natural stone and manufactured stone masonry units the water absorption of the face of the unit to be exposed is measured, as described in the relevant product standard.

19.3.11 Determination of net and gross dry density of masonry units (except for natural stone)

EN 772–13 Methods of test for masonry units – Part 13: Determination of net and gross dry density of masonry units (except for natural stone)

When published this standard will specify a method of determining the net and gross dry density of masonry units. This procedure is applicable to all types of unit with the exception of natural stone,

The procedure involves drying the unit to a constant mass and calculating the net and gross volume. The net and gross dry density of the masonry units are then calculated.

19.3.12 Determination of moisture movement of aggregate concrete and manufactured stone masonry units

EN 772–14 Methods of test for masonry units – Part 14: Determination of moisture movement of aggregate concrete and manufactured stone masonry units

When published, this standard will specify a method of measuring the moisture movement of aggregate concrete

and manufactured stone masonry units between two specified extreme moisture conditions.

This test will measure, on masonry units of the same concrete compositions and the same sampling:

(a) the expansion between the initial condition and after soaking in water for 4 days;

(b) the shrinkage between the initial condition and after drying for 21 days in a ventilated oven at 33 °C.

19.3.13 Determination of water vapour permeability of autoclaved aerated concrete masonry units

EN 772–15 Methods of test for masonry units – Part 15: Determination of water vapour permeability of autoclaved aerated concrete masonry units

When published this standard will specify a method of determining the steady state water vapour permeability of autoclaved aerated concrete masonry units at the upper and lower part of the hygroscopic range. The test can only be carried out on products from which disc shaped specimens of uniform thickness can be made. Although there are no requirements given for conditioning the specimens it will take longer for the test arrangement to reach equilibrium if the specimens are not in an approximately air dry condition at the start of the test.

Specimens to be tested are sealed on the open mouth of circular cups in which the water vapour pressure is maintained constant at appropriate levels by means of saturated salt solutions. The cups are placed in a temperature controlled environment with a constant water vapour pressure different from inside the cups. The rate of moisture transfer is determined from the weight change of the cups under steady conditions.

19.3.14 Determination of dimensions

EN 772–16 Methods of test for masonry units – Part 16: Determination of dimensions

When published this standard will specify a method for determining the overall dimensions, thickness of shells and webs and depth of voids of masonry units.

A minimum of six units will be sampled for testing and any superfluous material adhering to the unit as a result of the manufacturing process is removed. A set procedure is then specified for measuring the dimensions of the unit and calculating and expressing the results.

9.3.15 Determination of freeze-thaw resistance of calcium silicate masonry units

EN 772–18 Methods of test for masonry units – Part 18: Determination of freeze–thaw resistance of calcium silicate masonry units

The test for determining the freeze–thaw resistance of calcium silicate masonry units involves conditioning them in water before subjecting them to repeated cycles of freezing and thawing. Damage caused by subjecting the units to freezing and thawing is assessed by visual inspection. Where visual damage is noted the compressive strength of the units can be compared with similar samples not subjected to a freeze-thaw regime.

19.3.16 Determination of moisture expansion of large horizontally perforated clay masonry units

EN 772–19 Methods of test for masonry units – Part 19: Determination of moisture expansion of large horizontally perforated clay masonry units

The test for determining the moisture expansion of large horizontally perforated clay masonry units involves measuring the length change of test specimens caused by subjecting them to boiling water for a 24 hour period. The minimum number of specimens to be tested is 6 and they are cut out of a shell from each masonry unit parallel to the perforations. The specimens need to be 150–250mm long and at least 40mm wide. After preparation the specimens are dried in a standard way and measured using vernier callipers and an invar reference bar. The specimens are then immersed in boiling water for 24 hours following which they are allowed to cool at room temperature and further readings taken. The moisture expansion can then be calculated to the nearest 0.1mm/m.

19.3.17 Determination of flatness of faces of aggregate concrete, manufactured stone and natural stone masonry units

EN 772–20 Methods of test for masonry units – Part 20: Determination of flatness of faces of aggregate concrete, manufactured stone and natural stone masonry units

The flatness of the faces of aggregate concrete, manufactured stone and natural stone masonry units is determined by using a graduated straight edge and a set of feeler gauges. A sample of at least 6 specimens needs to be taken and the mean maximum deviation from flatness is expressed to the nearest 0.1mm.

19.3.18 Determination of freeze–thaw resistance of clay masonry units

EN 772–22 Methods of test for masonry units – Part 22: Determination of freeze-thaw resistance of clay masonry units

When published this standard will specify a method for determining the freeze/thaw resistance of clay masonry units.

To carry out the test, a panel of clay masonry is assembled either from units which have been immersed in water for a prescribed period of time and which are separated from one another by a specified rubber jointing material, or from units and rapid hardening mortar which, when sufficiently hardened, is immersed in water for a prescribed period of time. The panel is subsequently

cooled until all of the water which has been absorbed is frozen and the water near to one face is repeatedly thawed and re-frozen. Damage caused by the freezing and thawing action is assessed and used to determine the freeze/thaw resistance of the bricks.

Units which are undamaged after 100 cycles are given a durability designation F.2; units which are undamaged after 5 cycles are given the durability designation F.1. All other units are given the designation F.0.

19.4 Standards for masonry mortars

EN 998–1 Specification for mortar for masonry – Part 1: Rendering and plastering mortar with inorganic binding agents
EN 998–2 Specification for mortar for masonry – Part 2: Masonry mortar

19.4.1 Rendering and plastering mortar with inorganic binding agents

EN 998–1 Specification for mortar for masonry – Part 1: Rendering and plastering mortar with inorganic binding agents

When published this standard will provide information on rendering and plastering mortar and is applicable to external rendering and internal plastering.

The properties of rendering and plastering mortars depend mainly upon the type or types of binder used and the proportions in which they are used. The standard is not intended to cover the use of calcium–sulphate as the principal binding agent, but does recognise its use as an additional binder together with air lime.

The standard does not provide guidance on the following:

- Fire resistant and acoustical mortars
- Mortars for structural repair and surface treatment of building elements

It does not cover the use of site–made rendering or plastering mortar.

19.4.2 Specification for mortar for masonry – Part 2: Masonry mortar

EN 998–2 Specification for mortar for masonry – Part 2: Masonry mortar

When published this standard will deal with the specification of masonry mortar. It provides information on properties relating to fresh, unhardened mortar and to hardened mortar.

For fresh mortars it defines the performance related to workable life, chloride content, consistence, air content, water retentivity and density. For hardened mortars it defines the performance related to compressive strength, bond strength, durability and thermal properties, correction time (for thin layer mortars) and density.

This standard does not cover site–mixed mortars which are outside the scope of the TC125 series of standards for masonry.

19.5 Methods of test for masonry mortars

The following list represents all the methods of test for mortar currently in preparation and which will in due course replace the various sections of BS 4551:

EN 1015–1:1999 Methods of test for mortar for masonry – Part 1: Determination of particle size distribution (by sieve analysis)

EN 1015–2:1999 Methods of test for mortar for masonry – Part 2: Bulk sampling of mortars and preparation of test mortars

EN 1015–3:1999 Methods of test for mortar for masonry – Part 3: Determination of consistence of fresh mortar (by flow table)

EN 1015–4:1999 Methods of test for mortar for masonry – Part 4: Determination of consistence of fresh mortar (by plunger penetration)

EN 1015–5 Nothing currently allocated to this standard number

EN 1015–6:1999 Methods of test for mortar for masonry – Part 6: Determination of bulk density of fresh mortar

EN 1015–7:1999 Methods of test for mortar for masonry – Part 7: Determination of air content of fresh mortar

EN 1015–8 Methods of test for mortar for masonry – Part 8: Determination of water retentivity of fresh mortar

EN 1015–9:1999 Methods of test for mortar for masonry – Part 9: Determination of workable life and correction time of fresh mortar

EN 1015–10:1999 Methods of test for mortar for masonry – Part 10: Determination of dry bulk density of hardened mortar

EN 1015–11:1999 Methods of test for mortar for masonry – Part 11: Determination of flexural and compressive strength of moulded mortar specimens

EN 1015–12 Methods of test for mortar for masonry – Part 12: Determination of adhesive strength of hardened rendering and plastering mortars on substrates

EN 1015–13 Nothing currently allocated to this standard number

EN 1015–14 Methods of test for mortar for masonry – Part 14: Determination of durability of hardened masonry mortars (with cement comprising greater than 50% of the total binder mass)

EN 1015–15 Nothing currently allocated to this standard number

EN 1015–16 Nothing currently allocated to this standard number

EN 1015–17 Methods of test of mortar for masonry – Part 17: Determination of water-soluble chloride content of fresh mortars

EN 1015–18 Methods of test of mortar for masonry – Part 18: Determination of water absorption coefficient due to capillary action of hardened rendering mortar

EN 1015–19:1999 Methods of test of mortar for masonry – Part 19: Determination of water vapour permeability of hardened rendering and plastering mortars.

EN 1015–20 Methods of test for mortar for masonry – Part 20: Determination of durability of hardened masonry mortars (with cement comprising less than or equal to 50% of the total binder mass)

EN 1015–21 Methods of test for mortar for masonry – Part 21: Determination of the compatibility of one-coat rendering mortars with backgrounds through the assessment of adhesive strength and water permeability after conditioning

19.5.1 Determination of particle size distribution (by sieve analysis)

EN 1015–1:1999 Methods of test for mortar for masonry – Part 1: Determination of particle size distribution (by sieve analysis)

Two methods of determining the particle size distribution of dry mixed or non–hardened wet mixed mortars are given in this standard. A method of wet sieving is provided for mortars containing normal weight aggregates and a method of dry sieving is provided for mortars containing lightweight aggregates. The test sieve aperture sizes are in the range 8mm to 0.063 mm. The method described are generally similar to those given in BS 4551–1, part of which this EN standard is intended to replace.

19.5.2 Bulk sampling of mortars and preparation of test mortars

EN 1015–2:1999 Methods of test for mortar for masonry – Part 2: Bulk sampling of mortars and preparation of test mortars

This standard specifies methods for taking a bulk sample of fresh mortar, and the preparation of a bulk test sample from this. It also describes the procedure for producing test mortars from dry constituents and water.

19.5.3 Determination of consistence of fresh mortar (by flow table)

EN 1015–3:1999 Methods of test for mortar for masonry – Part 3: Determination of consistence of fresh mortar (by flow table)

Consistence may be regarded as a measure of the fluidity of fresh mortar. It gives an indication of the deformability of the fresh mortar when it is subjected to a certain type of stress. The flow value is measured by the mean diameter of a test sample of the fresh mortar after it has been subjected to a set procedure. Essentially, this involves placing the sample on a defined flow table disc using a truncated conical mould and raising the table and allowing it to fall in order to subject the specimen to 15 vertical impacts. The flow value determined by this procedure should, for the same type of mortar, correlate linearly with the plunger penetration value (EN 1015–4) with increasing water content, but the slope will differ with different types of mortars.

19.5.4 Determination of consistence of fresh mortar (by plunger penetration)

EN 1015–4:1999 Methods of test for mortar for masonry – Part 4: Determination of consistence of fresh mortar (by plunger penetration)

This method of determining the consistence of fresh mortar requires the use of a plunger test apparatus. The apparatus allows a specified plunger rod to drop freely through a set height into the fresh mortar sample. The mean value of plunger penetration is determined from the results on each mortar sample and expressed to the nearest mm. As indicated in 19.5.3, it should be possible to correlate the result with flow table method of measuring consistence.

19.5.5 Determination of bulk density of fresh mortar

EN 1015–6:1999 Methods of test for mortar for masonry – Part 6: Determination of bulk density of fresh mortar

The bulk density of fresh mortar is determined by the quotient of its mass and the volume which it occupies when it is introduced, or introduced and compacted in a prescribed manner into a measuring vessel of a given capacity. Before the test is carried out the flow value is determined. Where the flow value is less than 140mm the mortar is placed in the measuring vessel and compacted by placing the vessel on a vibrating table. Where the flow value is 140–200mm the mortar in the vessel may be compacted either by using a compacting table or by using the serock method. Where the flow value is greater than 200mm the vessel can be filled without the need for compaction.

19.5.6 Determination of air content of fresh mortar

EN 1015–7:1999 Methods of test for mortar for masonry – Part 7: Determination of air content of fresh mortar

This European Standard gives two methods for determining the air content of fresh mortars including those containing mineral binders and both dense and lightweight aggregates. Method A is known as 'The pressure method' and is used for air contents less than 20%. Method B is known as 'The alcohol method' and is used for air contents of 20% or more.

To determine the air content a volume of mortar is placed in a measuring vessel. Water is then introduced on top of the mortar surface and, either by means of air pressure or the use of an alcohol-water mix, water is forced into the mortar displacing air from within any pores. This causes the water level to fall which gives a measure of the volume of air displaced from the mortar.

It should be noted that Method A requires the use of a test apparatus consisting of a metal bowl sample container and cover. This apparatus needs to be calibrated as described in the standard and a calibration curve established.

19.5.7 Determination of water retentivity of fresh mortar

EN 1015–8 Methods of test for mortar for masonry – Part 8: Determination of water retentivity of fresh mortar

When published this standard will specify a method for determining the water retentivity of freshly mixed mortars containing mineral binders and normal, as well as lightweight, aggregates when subjected to suction. The method is applicable to masonry mortars, rendering mortars and plastering mortars.

To carry out the test, fresh mortar is brought to a defined level of consistence and is then subjected to a standard suction treatment using standardised filter paper as substrate. The water retentivity of the defined mortar standard is determined by the mass of water retained in the mortar after the standard suction treatment and is expressed as a percentage of its original water content.

19.5.8 Determination of workable life and correction time of fresh mortar

EN 1015–9:1999 Methods of test for mortar for masonry – Part 9: Determination of workable life and correction time of fresh mortar

This Standard specifies three methods for determining the workable life and correction time of freshly mixed mortars. Method A is for general purpose masonry or rendering mortars including those containing mineral binders and both dense and lightweight aggregates. Methods B and C are for determining the workable life and correction time for thin–layer mortars.

The workable life is measured on a sample of the mortar which has been brought to a defined flow value. Essentially the method measures the time in minutes at which the sample reaches a set limit of stiffness during the test.

19.5.9 Determination of dry bulk density of hardened mortar

EN 1015–10:1999 Methods of test for mortar for masonry – Part 10: Determination of dry bulk density of hardened mortar

The procedure specified is applicable to lightweight,

general purpose and thin layer mortars with regular shaped specimens. The specimens need to be prepared in a specified way and dried to a constant mass. Each specimen is then immersed in water until its apparent mass is constant and the water displacement method used to determine its volume. The dry bulk density can then be calculated as the ratio of the recorded dry mass to the volume.

19.5.10 Determination of flexural and compressive strength of moulded mortar specimens

EN 1015–11:1999 Methods of test for mortar for masonry – Part 11: Determination of flexural and compressive strength of moulded mortar specimens

This test procedure is very similar to that which has been widely used in Europe for many years. Essentially a mortar prism 160 × 40 × 40mm is tested by three point loading to failure in order to establish the flexural strength. The two parts left over from the flexural test are then used to measure the compressive strength.

The flexural strength is calculated in N/mm² using the equation:

$$f = 1.5 \frac{F\,l}{bd^2}$$

where

f = the flexural strength
F = the maximum load in Newtons
l = the distance between the access of the support rollers (mm)
b = width of specimen (mm)
d = depth of specimen (mm)

The compressive strength is calculated as the maximum load carried by the specimen divided by its cross-sectional area.

19.5.11 Determination of adhesive strength of hardened rendering and plastering mortars on substrates

EN 1015–12 Methods of test for mortar for masonry – Part 12: Determination of adhesive strength of hardened rendering and plastering mortars on substrates

When published, this standard will specify a method for the determination of the adhesive strength between rendering and plastering mortars and a substrate. The adhesive strength is determined as the maximum tensile stress applied by a direct load perpendicular to the surface of the rendering or plastering mortar on a substrate. The tensile load is applied by means of a pull-head plate glued to the test area of the mortar surface. The adhesive strength obtained is the quotient between the failure load and the test area.

19.5.12 Determination of durability of hardened masonry mortars (with cement comprising greater than 50% of the total binder mass)

EN 1015–14 Methods of test for mortar for masonry – Part 14: Determination of durability of hardened masonry mortars (with cement comprising greater than 50% of the total binder mass)

When published, this standard will specify a laboratory method of test (under onerous conditions) for determining the durability of masonry mortars containing mineral binders, with cement comprising greater than 50% of the total binder mass, and normal, as well as lightweight, aggregates. The method may also be applicable to mortar beds taken from existing structures.

The method determines the resistance of mortars prepared in the laboratory or sampled from fresh batches placed between standard substrates with negligible suction, to freeze/thaw action, sulfate attack or a combination of both.

The mortar sample needs to be prepared in accordance with EN 1015–2 before it is laid between pairs of water saturated bricks of negligible suction properties. After hardening and separation of the mortar bed, it is cut into specimens. These are subjected to cyclic freeze/thaw action whilst saturated with water or a sulfate solution. Further specimens are subjected to sulfate saturation only. Visual assessment is used at the end of the test to determine the degree of damage suffered by the specimen.

19.5.13 Determination of water-soluble chloride content of fresh mortars

EN 1015–17 Methods of test of mortar for masonry – Part 17: Determination of water-soluble chloride content of fresh mortars

When published this standard will specify a method for determining the water-soluble chloride content of fresh mortars. The test is carried out by preparing an aqueous extract containing water-soluble chlorides from the mortar sample. A known volume of standard silver nitrate solution is than used to precipitate the dissolved chloride. Any sulfide present is oxidised to sulfate or decomposed and does not interfere. After boiling, the precipitate is washed with dilute nitric acid and discarded. The filtrate and washings are cooled to less than 25 °C. The excess silver nitrate is then filtrated with a standard ammonium thiocyanate solution using an iron (III) salt as indicator. This method gives the total halogen content except for fluoride and expresses the result as percentage of C1 of sample.

19.5.14 Determination of water absorption coefficient due to capillary action of hardened rendering mortar

EN 1015–18 Methods of test of mortar for masonry – Part 18: Determination of water absorption coefficient due to capillary action of hardened rendering mortar

When published, this standard will specify a method for determining the water absorption coefficient due to capillary action of rendering mortars containing mineral binders and normal as well as lightweight aggregates. The water absorption coefficient due to capillary action is measured using mortar prisms or cylindrical specimens under prescribed conditions at atmospheric pressure. After drying to constant mass, the face of the specimen is immersed in 5 to 10 mm of water for a specific period of time and the increase in mass determined.

19.5.15 Determination of water vapour permeability of hardened rendering and plastering mortars

EN 1015–19:1999 Methods of test of mortar for masonry – Part 19: Determination of water vapour permeability of hardened rendering and plastering mortars.

This method enables the steady state water vapour permeability to be established. It is applicable to both rendering and plastering mortars. The test is suitable for mortars from which disc shaped specimens can be made of a uniform thickness between 10 and 30 mm.

The specimens are sealed in the open end of circular cups and the water vapour pressure is maintained constant by the use of saturated salt solutions. The cups are placed in a temperature controlled environment with a constant vapour pressure different from that in the cups. It is then possible to determine the rate of moisture transfer by noting the change in the weight of the cups under steady state conditions.

9.5.16 Determination of durability of hardened masonry mortars (with cement comprising less than or equal to 50% of the total binder mass)

EN 1015–20 Methods of test for mortar for masonry – Part 20: Determination of durability of hardened masonry mortars (with cement comprising less than or equal to 50% of the total binder mass)

No drafting work has been completed on this standard.

9.5.17 Determination of the compatibility of one-coat rendering mortars with backgrounds through the assessment of adhesive strength and water permeability after conditioning

EN 1015–21 Methods of test for mortar for masonry – Part 21: Determination of the compatibility of one-coat rendering mortars with backgrounds through the assessment of adhesive strength and water permeability after conditioning

When published these test methods enable an evaluation of the compatibility of one-coat rendering mortars with a particular background by assessing the adhesive strength and water permeability after conditioning.

The satisfactory behaviour of one-coat render is very dependent upon the compatibility between the render and the type of substrate to which it is applied. The determination of adhesive strength and water permeability is carried out on test panels prepared according to the manufacturer's recommendations.

The water permeability apparatus consists of a conical section, which is sealed to the surface of the render, and

a device for maintaining a constant level and for measuring water flow. Following conditioning of the specimen the test is carried out for 48 hours. The water quantity required to maintain the level after 48 hours is determined for each test specimen and expressed in $ml/cm^2.48h$.

The adhesive strength is determined in accordance with EN 1015–2 and the mean value determined from the individual values of 5 measures done for each test specimen. The mode of failure should also be recorded as follows:

A: adhesion failure render/substrate
B: cohesion failure in the render
C: cohesion failure in the substrate

19.6 Standards for ancillary components for masonry

EN 845–1 Specification for ancillary components for masonry – Part 1: Ties, tension straps, hangers and brackets
EN 845–2 Specification for ancillary components for masonry – Part 2: Lintels
EN 845–3 Specification for ancillary components for masonry – Part 3: Bed joint reinforcement

19.6.1 Ties, tension straps, hangers and brackets

EN 845–1 Specification for ancillary components for masonry – Part 1: Ties, tension straps, hangers and brackets

When published this standard will specify requirements for ties, tension straps, hangers and brackets for interconnecting masonry and for connecting masonry to other parts of works and buildings including walls, floors, beams and columns. The standard is not intended to cover the following:

- fire performance
- anchors and fasteners other than as part of an ancillary component
- shelf angles
- wall starter plates for tying–in to existing walls

The standard will specify in detail a range of wall ties including the embedment length. Information will also be provided on fixings and materials and protective coatings.

19.6.2 Lintels

EN 845–2 Specification for ancillary components for masonry – Part 2: Lintels

When published this European standard will specify requirements for prefabricated lintels designed to span over clear openings up to a maximum of 4.5 m. It will cover lintels made from steel, autoclaved aerated concrete, manufactured stone, concrete, prestressed concrete, fired clay units, calcium silicate units or a combination of these materials.

It is not intended that this standard should cover the following:

- lintels made completely on site
- lintels in which the tensile parts are made on site
- timber lintels
- natural stone lintels which are not reinforced.

The standard will give guidance on single lintels, composite lintels and combined lintels.

Information is to be provided on the assessment of loads on lintels and their installation (including bearing length).

19.6.3 Bed joint reinforcement

EN 845–3 Specification for ancillary components for masonry – Part 3: Bed joint reinforcement

When published this standard will specify the requirements for welded wire meshwork, woven wire meshwork and expanded metal meshwork for use as a reinforcement in the bed joints of masonry. Where products are intended for use in cavity wall construction, the standard covers only the performance of the meshwork as reinforcement in bed joints and not its performance as wall ties across the cavity.

19.7 Methods of test for ancillary components

EN 846–1   Nothing currently allocated to this standard number

EN 846–2   Methods of test for ancillary components for masonry – Part 2: Determination of bond strength of bed joint reinforcement in mortar joints

EN 846–3   Methods of test for ancillary components for masonry – Part 3: Determination of shear load capacity of welds in prefabricated bed joint reinforcement

EN 846–4   Methods of test for ancillary components for masonry – Part 4: Determination of load capacity and load–deflection characteristics of straps

EN 846–5   Methods of test for ancillary components for masonry – Part 5: Determination of tensile and compressive load capacity and load displacement characteristics of wall ties (complete test)

EN 846–6   Methods of test for ancillary components for masonry – Part 6: Determination of tensile and compressive load capacity and load displacement characteristics of wall ties (single end test)

EN 846–7   Methods of test for ancillary components for masonry – Part 7: Determination of shear load capacity and load displacement characteristics of shear ties and slip ties (complete test for mortar joint connections)

EN 846–8   Methods of test for ancillary components for masonry – Part 8: Determination of load capacity and load deflection characteristics of joint hangers

EN 846–9   Methods of test for ancillary components for masonry – Part 9: Determination of flexural resistance and shear resistance of lintels

EN 846–10   Methods of test for ancillary components for masonry – Part 10: Determination of load capacity and load deflection characteristics of brackets

EN 846–11   Methods of test for ancillary components for masonry – Part 11: Determination of dimensions and bow of lintels

EN 846–12   Nothing currently allocated to this standard number

EN 846–13   Methods of test for ancillary components for masonry – Part 13: Determination of resistance to impact, abrasion and corrosion of organic coatings

19.7.1 Determination of bond strength of bed joint reinforcement in mortar joints

EN 846–2 Methods of test for ancillary components for masonry – Part 2: Determination of bond strength of bed joint reinforcement in mortar joints

When published this standard will specify a method for determining the bond strength of prefabricated bed joint reinforcement in a masonry specimen made from specified units and mortar. The test method is applicable to designed prefabricated bed joint reinforcement complying with prEN845-3, but is not applicable to other forms of reinforcement such as bars.

To carry out the test the prefabricated bed joint reinforcement is embedded in mortar in a small wall of bonded masonry units. The small wall is subjected to a compressive load but restrained from horizontal movement. The reinforcement is then subjected to tension in order to determine its bond strength. At least 5 specimens need to be tested.

19.7.2 Determination of shear load capacity of welds in prefabricated bed joint reinforcement

EN 846–3 Methods of test for ancillary components for masonry – Part 3: Determination of shear load capacity of welds in prefabricated bed joint reinforcement

When published this standard will specify a method for determining the shear strength of the welds in prefabricated bed joint reinforcement. To carry out the test samples of the welds in truss type prefabricated bed joint reinforcement are straightened and tested in a normal tensile test machine. Samples of the welds in ladder type reinforcement are held in a special clamp at one end and then tested in a normal tensile test machine.

19.7.3 Determination of load capacity and load-deflection characteristics of straps

EN 846–4 Methods of test for ancillary components for masonry – Part 4: Determination of load capacity and load–deflection characteristics of straps

When published, this standard will specify methods for determining the load capacity and load deflection characteristics of restraint straps fixed to timber joists, rafters and timber wall plates and masonry walls. For the test, straps are fixed to masonry walls, timber joists or rafters or to other floor/roof materials and loaded in a manner representative of their intended use.

19.7.4 Determination of tensile and compressive load capacity and load displacement characteristics of wall ties (complete test)

EN 846–5 Methods of test for ancillary components for masonry – Part 5: Determination of tensile and compressive load capacity and load displacement characteristics of wall ties (complete test)

When published this standard will specify the couplet method for determining the tensile and compressive load capacity and load displacement characteristics of wall ties embedded in mortar joints. The test is intended for ties used for connecting together two leaves of masonry and for the mortar–bedded end of ties for connecting masonry leaves to other structures.

To carry out the test the tie is embedded in a mortar typical of the type for which the tie is specified between a couplet of masonry units. The tie is then subjected to tension or compression until failure occurs.

19.7.5 Determination of tensile and compressive load capacity and load displacement characteristics of wall ties (single end test)

EN 846–6 Methods of test for ancillary components for masonry – Part 6: Determination of tensile and compressive load capacity and load displacement characteristics of wall ties (single end test)

When published this standard will specify a method for determining the tensile and compressive load capacity and load displacement characteristics of wall ties screwed, nailed, grouted or otherwise attached to frame elements or to inner leaf materials. The test is intended for ties for connecting masonry leaves to frame structures and to the inner leaves of cavity walls other than by embedding the inner connection in a mortar joint.

To carry out the test the tie is screwed, nailed, grouted or attached using other devices, such as keys in slots, to a representative section of the frame element or inner leaf material using normal site techniques. The tie is then subjected to tension or compression until failure occurs.

19.7.6 Determination of shear load capacity and load displacement characteristics of shear ties and slip ties (complete test for mortar joint connections)

EN 846–7 Methods of test for ancillary components for masonry – Part 7: Determination of shear load capacity and load displacement characteristics of shear ties and slip ties (complete test for mortar joint connections)

When published this standard will specify the couplet method for determining the horizontal and vertical shear load resistance and load deflection behaviour of shear ties and slip ties embedded in mortar joints. The test is intended for ties for connecting together two leaves of masonry forming a collar jointed wall or two walls at right angles. It is also applicable to ties used for connecting the edges of infill panel walls to frames which encircle them.

To carry out the test, one end of the tie is embedded in a mortar joint, typical of the type for which the tie is specified, between a couplet of typical masonry units. The tie is then clamped at its free end and subjected to shear against a reactive support for the couplet. Slip ties may be tested by the same method.

It should be noted that this method measures the capacity of the tie alone and does not measure the contribution to the total shear resistance given by two masonry faces separated by a vertical mortar joint. This value should be obtained by wallette tests if required.

19.7.7 Determination of load capacity and load deflection characteristics of joist hangers

EN 846–8 Methods of test for ancillary components for masonry – Part 8: Determination of load capacity and load deflection characteristics of joist hangers

When published, this standard will specify a method for determining the strength and load deflection characteristics of joist hangers fixed to a masonry wall and supporting a timber joist. Essentially, the test involves fixing joist hangers to a wall and loading through the joists in a manner representative of their intended use. The loading system must be capable of applying a vertical load to the specimen and means must be provided to measure the displacement of the hanger in relation to the floor. A minimum of five specimens need to be tested.

19.7.8 Determination of flexural resistance and shear resistance of lintels

EN 846–9 Methods of test for ancillary components for masonry – Part 9: Determination of flexural resistance and shear resistance of lintels

When published this standard will specify methods for determining the flexural and shear resistances and load deflection characteristics of single span, single or composite lintels used for supporting uniformly distributed loads over openings in masonry construction.

To carry out the tests, specimen lintels are simply supported and subjected to vertically applied loads in order to determine flexural strength, shear resistance and deflection.

19.7.9 Determination of load capacity and load deflection characteristics of brackets

EN 846–10 Methods of test for ancillary components for masonry – Part 10: Determination of load capacity and load deflection characteristics of brackets

When published this standard will specify a method for determining the load capacity and load deflection characteristics of brackets, used for the support of masonry, fixed to a backing wall or frame structure.

To carry out the test, specimen brackets are attached to a suitable backing material and subjected to vertically applied loads. Provision is made to determine strength, deflection and recovery.

19.7.10 Determination of dimensions and bow of lintels

EN 846–11 Methods of test for ancillary components for masonry – Part 11: Determination of dimensions and bow of lintels

When published, this standard will specify methods for determining the dimensions and straightness or bow of single span, single, combined or the prefabricated component of composite lintels conforming with prEN 845–2:1992. To carry out the test, specimen lintels or the constituent prefabricated components of lintels (e.g. components forming the tension zone) are measured in order to determine their overall length, width and height, as well as their bow. For lintels of non-rectangular cross-sections, a diagram is made showing their configuration and dimensions.

19.7.11 Determination of resistance to impact, abrasion and corrosion of organic coatings

EN 846–13 Methods of test for ancillary components for masonry – Part 13: Determination of resistance to impact, abrasion and corrosion of organic coatings

When published this standard will specify a method for determining the level of performance of those organic coatings classified in EN 845–1 and EN 845–2 as type 2 applied as a protective system to zinc coated steel plate used in the fabrication of ancillary components for masonry.

The performance of organic coatings is determined from tests for impact resistance, abrasion resistance and an accelerated corrosion test. The impact and abrasion resistance are determined using a heavy rigid pendulum apparatus whereby the spring loaded tip of the pendulum strikes the test specimen at a specified distance from bottom dead centre. The accelerated corrosion performance is determined from the electrical resistance of the organic coating during a period of exposure to an alkaline solution.

19.8 Test methods for masonry

The test methods for masonry provided by TC 125 were requested by the drafters of EC6 and cover the compressive strength of masonry, the flexural strength of masonry, and the shear strength of masonry.

In the UK the traditional method of establishing the characteristic compressive strength of masonry has been by testing storey height wall panels. The new CEN method uses small wallette specimens which are much cheaper and easier to test. The ratio of wall strength to wallette tests is about 0.9.

The approach to flexural strength in the CEN Standard is very similar to that described in BS 5628. Essentially the flexural strength of the masonry is determined normal and parallel to the bed joint. The values are then used in EC6 for design based upon the yield line method.

The shear strength of masonry along a bed joint is dependent on the stress normal to that joint. Where this stress is zero then the bond between the unit and mortar is paramount. The two tests methods proposed by CEN cover specimens containing a damp proof course and those that do not.

EN 1052–1:1999 Methods of test for masonry – Part 1: Determination of compressive strength

EN 1052–2 Methods of test for masonry – Part 2: Determination of flexural strength

EN 1052–3 Methods of test for masonry – Part 3: Determination of initial shear strength

EN 1052–4 Methods of test for masonry – Part 4: Determination of shear strength including damp proof course

EN 1052–5 Methods of test for masonry – Part 5: Determination of bond strength

19.8.1 Determination of compressive strength

EN 1052–1:1999 Methods of test for masonry – Part 1: Determination of compressive strength

The compressive strength of masonry is determined from axial tests on wallette specimens. The materials used should reflect those to be used in practice, as should the construction and bonding pattern. At least 3 specimens are required for test, each specimen being approximately 800 mm wide and 1,000 mm high. For the first 3 days after construction the specimens should be cured, e.g. by covering with a polyethylene sheet. Tests are made to check the mortar strengths in order to set the age at which the specimens should be tested. Each specimen is tested to destruction between the platens of a testing machine such that failure occurs between 15 to 30 minutes after commencement of load application. Both the mean and characteristic compressive strengths of the masonry may be calculated and expressed to the nearest 0.1 N/mm².

19.8.2 Determination of flexural strength

EN 1052–2 Methods of test for masonry – Part 2: Determination of flexural strength

The flexural strength of masonry is derived from the strength of small specimens tested to destruction under four point loading. The characteristic value is calculated from the maximum stresses achieved by the samples and is considered to be the flexural strength of the masonry.

For each of the 2 principal axes of loading, at least 5 specimens need to be tested. The specimens need to be built within 30 minutes of conditioning the units using mortar not more than 1 hour old. Immediately after building, each specimen is precompressed using a uniformly distributed mass to give a vertical stress between 2.5×10^{-3} N/mm^2 and 5×10^{-3} N/mm^2. The specimens are then cured under polythene until they are tested at 28 days.

For the test, each specimen is placed in a test frame under four point loading and the flexural stress is increased at a rate between 0.03 N/mm^2/min and 0.3 N/mm^2/min.

The general procedure is very similar to that described in BS 5628: Part 1.

19.8.3 Determination of initial shear strength

EN 1052–3 Methods of test for masonry – Part 3: Determination of initial shear strength

When published this standard will specify a method for determining the in plane initial shear strength of horizontal bed joints in masonry using a specimen tested in shear.

The initial shear strength of masonry is derived from the strength of small masonry specimens tested to destruction. The specimens are tested in shear under four-point load, with pre-compression perpendicular to the bed joints. Four different failure modes are considered to give valid results. The initial shear strength is obtained by extrapolation of the linear regression curve to zero normal stress.

19.8.4 Determination of shear strength including damp proof course

EN 1052–4 Methods of test for masonry – Part 4: Determination of shear strength including damp proof course

When published this standard will specify a method for determining the in plane shear strength of horizontal bed joints in masonry incorporating sheet damp proof course material.

The shear strength is derived from the strength of small masonry specimens tested to destruction. The specimens are tested in double shear under three–point load with precompression perpendicular to the bed joints. The shear strength is defined by the initial shear strength and the coefficient of friction.

19.8.5 Determination of bond strength

EN 1052–5 Methods of test for masonry – Part 5: Determination of bond strength

When published this test method will enable the bond strengths of masonry to be determined. The test procedure requires the use of a bond wrench which enables the determination of the bond strength of horizontal bed joints.

This is a destructive method of test carried out on small masonry specimens. Each specimen is rigidly held and a clamp is applied to the top unit. A bending moment is applied to the clamp by a lever until the top unit is pulled from the remainder.

At least 10 bed joints need to be available for testing in the form of stack jointed prisms. A detailed procedure is provided for noting the measurements and observations, including the need for replicate tests – depending upon the mode of failure. The procedure for calculating the characteristic bond strength is also specified.

19.9 Other design guidance

EN 1745   Masonry and masonry products – Methods for determining design thermal values

EN CBQD–1 Design, preparation and application of external renderings and internal plastering – Part 1: External renderings

EN CBQD–2 Design, preparation and application of external renderings and internal plastering – Part 2: Internal plastering

19.9.1 Methods for determining design thermal values

EN 1745 Masonry and masonry products – Methods for determining design thermal values

In this draft standard, rules are provided for the determination of design values for thermal conductivity and thermal resistance of both masonry and masonry products. Information is also provided on how the basic values for the calculation of design thermal values are determined. The calculation methods used to derive design values from basic values are also described.

The standard covers solid masonry units, masonry units with formed voids, and composite masonry units. Three procedures may be used for the determination of thermal resistance and/or thermal conductivity, namely:

1. Use tabulated lambda and/or R-values.
2. Measure the lambda and/or R-values.
3. Calculate the equivalent lambda and/or R-value.

19.9.2 External renderings

EN CBQD–1 Design, preparation and application of external renderings and internal plastering – Part 1: External renderings

When published this standard will cover the design, preparation and application of external renderings. It will cover the use of cement, lime, masonry cement and polymer modified binder based external renderings on all common backgrounds. It includes rendering on both new and old backgrounds and the maintenance and repair of existing work.

This standard is not intended to cover the following:

- liquid retaining structures
- backgrounds to cladding systems
- structural repair of concrete
- thermal insulation systems
- gypsum based renders used externally

19.9.3 Internal plastering

EN CBQD–2 Design, preparation and application of external renderings and internal plastering – Part 2: Internal plastering

When published this standard will deal with the design, preparation and application of plaster based on gypsum, anhydrite, cement, lime, masonry cement, silicate, polymer, etc., binders (and various combinations of each) for internal plastering on all types of background under normal conditions. It will include plastering onto both new and old backgrounds and the maintenance and repair of existing work. It will give guidance on materials, backgrounds, surface preparation, choice of plastering systems, method of application and inspection and testing of plastering. Plastering mixes with special properties to enhance thermal insulation, fire resistance, acoustic insulation and to increase the absorption of radiation are included.

The standard will not cover external finishes, painting, impregnations, wipe-on or slurry finishes, structural repair of concrete or fibrous plasterwork.

Chapter 20
Design in Accordance with Eurocode 6

20.1 Introduction

20.1.1 General

The Structural Eurocodes comprise a group of standards for the structural and geotechnical design of buildings and civil engineering works.

The Commission of the European Communities (CEC) initiated the work of establishing a set of harmonized technical rules for the design of building and civil engineering works which would initially serve as an alternative to the different rules in force in the various Member States and would ultimately replace them. These technical rules became known as the Structural Eurocodes.

Work is in hand on the following Structural Eurocodes, each generally consisting of a number of parts:

EN 1991   Eurocode 1: Basis of design and actions on structures.
EN 1992   Eurocode 2: Design of concrete structures.
EN 1993   Eurocode 3: Design of steel structures.
EN 1994   Eurocode 4: Design of composite steel and concrete structures.
EN 1995   Eurocode 5: Design of timber structures.
EN 1996   Eurocode 6: Design of masonry structures.
EN 1997   Eurocode 7: Geotechnical design.
EN 1998   Eurocode 8: Design of structures for earthquake resistance.
EN 1999   Eurocode 9: Design of aluminium alloy structures.

In view of the responsibilities of authorities in member countries for safety, health and other matters covered by the essential requirements of the Construction Products Directive (CPD), certain safety elements in this ENV 1996–1–1 have been assigned indicative values which are identified by ☐

('boxed values'). The authorities in each member country are expected to review the 'boxed values' and *may* substitute alternative definitive values for these safety elements for use in national application.

Supporting European, or International Standards, were not available by the time this prestandard was issued. A National Application Document (NAD), giving any substitute definitive values for safety elements, referencing compatible supporting standards and providing guidance on the national application of this prestandard, has been issued by BSI as part of DD ENV 1996–1–1: 1996. Throughout this chapter reference to DD ENV 1996–1–1 is to the code as prepared by CEN; the National Application Document is referred to as the UK NAD.

20.1.2 Scope of Eurocode 6

Eurocode 6 applies to the design of buildings and civil engineering works in unreinforced, reinforced, prestressed and confined masonry.

Eurocode 6 is only concerned with the requirements for resistance, serviceability and durability of structures. Other requirements, for example, concerning thermal or sound insulation, are not considered.

Execution is covered to the extent that is necessary to indicate the quality of the construction materials and products that should be used and the standard of workmanship on site needed to comply with the assumptions made in the design rules. Generally, the rules related to execution and workmanship are to be considered as minimum requirements which may have to be further developed for particular types of buildings or civil engineering works and methods of construction.

20.1.3 Scope of Part 1–1 of Eurocode 6

Part 1–1 of Eurocode 6, DD ENV 1996–1–1, gives a general basis for the design of buildings and civil engineering works in unreinforced, reinforced, prestressed and confined masonry made with the following masonry units laid in mortar made with natural sand, or crushed sand, or lightweight aggregate:

- fired clay units, including lightweight clay units;
- calcium silicate units;

- concrete units, made with dense or lightweight aggregates;
- autoclaved aerated concrete units;
- manufactured stone units;
- dimensioned natural stone units.

DD ENV 1996–1–1 deals with reinforced masonry where the reinforcement is added to provide ductility, strength or serviceability. The principles of the design of prestressed masonry and confined masonry are given, but application rules are not provided.

The following subjects are dealt with in DD ENV 1996–1–1:

- Section 1: General.
- Section 2: Basis of design.
- Section 3: Materials.
- Section 4: Design of masonry.
- Section 5: Structural detailing.
- Section 6: Construction.

20.1.4 Further parts of Eurocode 6

DD ENV 1996–1–1 of Eurocode 6 will be supplemented by further Parts which will complement or adapt it for particular aspects of special types of building or civil engineering works, special methods of construction and certain other aspects of design which are of general practical importance.

Further Parts of Eurocode 6 which include the following:

- Part 1–2: Structural fire design.
- Part 1–3: Detailed rules on lateral loading.
- Part 1–X: Complex shape sections in masonry structures*.
- Part 2: Design, selection of materials and execution of masonry.
- Part 3: Simplified and simple rules for masonry structures.
- Part 4: Constructions with lesser requirements for reliability and durability*.

Depending on the character of the individual clauses, distinction is made in DD ENV 1996–1–1 between principles and application rules.

The principles comprise:

- general statements and definitions for which there is no alternative;
- requirements and analytical models for which no alternative is permitted unless specifically stated.

The principles are defined in the text of DD ENV 1996–1–1 by the letter P, following the clause number, for example, (1)P.

The application rules, all clauses not indicated as being principles, are generally recognised rules which follow the principles and satisfy their requirements.

It is permissible to use alternative rules to the application rules given in DD ENV 1996–1–1, provided that it is shown that the alternative rules accord with the relevant principles and have at least the same reliability as the rules given. No guidance is given in the Code as to how this compatibility is to be achieved, although it is understood that the inference should not be drawn that National Code Clauses can be used in lieu of application rules per se.

The aim of this Guidance Document is to facilitate the use of DD ENV 1996–1–1 by practising engineers prior to its conversion into a formal EN Eurocode. It is intended to explain and elaborate the Clauses in DD ENV1996–1–1 and includes a number of worked examples.

The purpose of an ENV is to enable a standard to be used before it has the full force of a European document. After a period of time, CEN invites its Member Standards Bodies, for example BSI, to vote on the next step. After publication of an EN, a period of coexistence with National equivalent Codes will be permitted. How long that period will be, and whether at the end of it National Codes will have to be withdrawn, is yet to be decided by TC250, the CEN Committee responsible for all of the Eurocodes.

Throughout this Guidance Document, clauses from DD ENV1996–1–1 are printed in bold type, for example **Clause 4.4.2, Eqn (2.3)** and **Figure 5.9**.

20.2 Basis of design, including loads and material properties

20.2.1 General

DD ENV 1996–1–1, 6, like the other structural Eurocodes, is a partial factor design code, in which factors are applied both to loads and to material properties in order to reach appropriate design values. This chapter explains how to calculate the design values of loads (known as 'actions' in the Eurocodes), and the design values of material properties, for both the ultimate and serviceability limit states, when using the rather confusing algebraic formulation that is contained in Section 2 of DD ENV 1996–1–1.

20.2.2 Loading codes

For the time being, British loading codes will continue to be used in the UK as sources for the weights of general building materials, for imposed loads and for wind loads. The UK NAD lists the codes to be used as:

BS 648	Schedule of weights of building materials
BS 6399	Loading for buildings:
Part 1	Code of practice for dead and imposed loads
Part 2	Code of practice for wind loads
Part 3	Code of practice for imposed roof loads
CP3	Basic data for the design of the buildings: Chapter V Loading: Part 2. Wind loads

---

* It is unlikely that these Parts will be developed in the foreseeable future.

Table 20.1 Types of action

Type of action	Symbol*	Definition	Examples
Permanent	G_k	Invariable during the design life of the structure	Self-weight of the structure and its fittings Self-weight of fixed equipment
Variable	Q_k	Fluctuating, moveable or indefinite loads	Imposed floor and storage loads Snow Wind Concentrated load from a man Traffic loads
Accidental	A_k	Abnormal loads	Explosions Vehicle or ship impact Local drifted snow Earthquake**

* In EC6 the suffix 'k' indicates a characteristic value
** Rules for seismic design are given in EC8.

Table 20.2 Types of design situation

Design situation	Limit state		Associated types of action
Fundamental design situations	Ultimate	Strength Stability	Permanent + variable
Accidental design situations	Ultimate	Strength Stability	Accidental + permanent + variable Permanent + variable following an accident
Serviceability calculations	Serviceability		Permanent + variable

20.2.3 Limit states

DD ENV 1996–1–1 requires designers to consider two kinds of limit state, beyond which a structure can no longer perform satisfactorily.

Ultimate limit states are associated with fracture, collapse or buckling; and with loss of equilibrium or stability. Thus, they are associated with any form of structural failure which may endanger the safety of people. They involve calculations of strength and stability.

Serviceability limit states are associated with deflection and cracking. Thus, they are associated with any form of structural behaviour which may render the structure unsatisfactory in terms of its functioning or comfort. Appearance (for example damage to non-structural elements or finishes) may also be a criterion. Serviceability limit states involve calculations of deflection and cracking.

Limit states are checked by examining the effects on a structure and its individual components of loads and imposed deformations, and by examining the material properties of the components concerned and their capacity to resist these effects.

In DD ENV 1996–1–1, as in all the structural Eurocodes, the design values of loads acting on a structure depend on four factors. These are:

- the type of action
- the design situation under consideration
- the characteristic values of the actions
- the combination of actions under consideration.

20.2.4 Types of action

There are three types of action, which are defined in Table 20.1.

20.2.5 Design situations

DD ENV 1996–1–1 addresses three types of design situation, which are defined in Table 20.2.

The two design situations for ultimate limit states may each be sub-divided into verifications of strength and of stability or static equilibrium, because all these relate to safety rather than serviceability.

20.2.6 Characteristic values of actions

Most actions do not have a single known value, but come in a range of values. In general the range can be statistically described, and can be defined by a single characteristic value, which is set at a level which will be exceeded only with a certain calculated probability; this is explained in more detail below.

303

Table 20.3 Principal factors used to convert characteristic values to design values

Symbol	Type of factor	Purpose
γF	Partial safety factor for actions	Increases characteristic loads by a safety factor to allow for uncertainties
Ψ	Coefficient for representative values of actions	Reduces variable loads (for example snow), when they are in combination with other variable loads, to appropriate values

Table 20.4 Partial safety factors for loads in ultimate limit state calculations from DD ENV 1996–1–1

Verification	Fundamental Situations γ_G		γ_Q		Accidental Situations γ_{Ga}	γ_A
	Favourable	Unfavourable	Favourable	Unfavourable		
Strength	1.0	1.35	0.0	1.5	1.0	1.0
Overall Stability	0.9	1.1	0.0	1.5	1.0	1.0

Permanent actions

For masonry structures, permanent actions are confined to the weights of the building materials. For general materials it is usually considered adequate to take as characteristic values the weights quoted in BS 648 or the manufacturers' literature.

Variable actions

Variable actions normally have two possible values: an upper characteristic value, which is the maximum value that a designer needs to consider; and a lower value which is usually zero. For example, the snow loads specified in BS 6399: Part 3 are upper characteristic values corresponding to a snowfall which is likely to be exceeded only once in fifty years. A designer will use these upper values to verify the strength of a roof, but in checking wind uplift on a roof he will use the lower value of zero.

Accidental actions

Accidental actions may be specified as upper characteristic values, as in earthquake loading, or else as design values which may be used without further modification, as for local drifted snow or vehicle impact loads.

20.2.7 Partial factors for actions

The characteristic values of actions are converted to design values by means of partial factors. Two principal kinds of factor are used, and these are shown in Table 20.3.

20.2.8 Partial safety factors for actions

Partial safety factors, commonly called 'gamma factors' because they are identified by the Greek letter γ, are used as safety factors in ultimate limit state design situations. The subscript 'F' indicates a general action or load. More

specific subscripts such as 'G' for permanent actions or 'Q' for variable actions may be used instead.

The value of γ_F depends on the type of action, the design situation under consideration and, in some cases, on the type of structure. For example, in normal structures permanent loads (G_k) are increased by a factor of $\gamma_G = 1.35$ for strength verifications, whereas variable loads (Q_k) are increased by a factor of $\gamma_Q = 1.5$ because they are known with less certainty.

Table 20.4 gives the values of the various gamma factors for use in masonry structures. The factors γ_G and γ_{GA} are used to modify permanent loads in fundamental and accidental design situations respectively, and the factor γ_Q to modify variable loads in fundamental ones. In accidental design situations the factor γ_A is in theory used to modify accidental loads, A_k, but since these are normally specified by their design value, A_d, γ_A is set at 1.0.

The factors for favourable effects are applied only to loads which relieve stress or tend to restore equilibrium; they may be distinguished from the factors for unfavourable effects by writing them as $\gamma_{G,inf}$ and $\gamma_{Q,inf}$ ('inf' = inferior or lower). The factors for unfavourable effects may be distinguished as $\gamma_{G,sup}$ and $\gamma_{Q,sup}$ ('sup' = superior or higher). Most commonly the factors for unfavourable effects are used, and such distinctions are unnecessary.

20.2.9 Coefficients for representative values of actions

In many design situations several variable loadings occur simultaneously. When considering the shear resistance of a ground floor masonry wall panel, for example, the designer will have three simultaneous variable loads to consider: wind, snow and imposed loads. As variable loads, these are all specified as upper characteristic values, but DD ENV 1996–1–1 recognises that it is unlikely, if not impossible, for the maximum value of all three to occur simultaneously. The Code, therefore, provides a set of coefficients, psi identified by the Greek

Table 20.5 Coefficients for representative actions

Variable Action	Building type	Ψ_0	Ψ_1	Ψ_2
Imposed floor loads	Dwellings	0.5	0.4	0.2
	Other occupancy classes*	0.7	0.6	0.3
	Parking	0.7	0.7	0.6
Imposed roof loads Wind loads	All**	0.7	0.2	0.0

* That is Institutional and educational, public assembly, offices, retail, industrial and storage as defined in BS 6399: Part 1: 1984, Table 1
** As listed and defined in BS 6399: Part 1: 1984, Table 1

letter Ψ, which are used to obtain *representative values* of variable actions when they occur in combination with other variable actions or in a rarely-occurring accidental design situation.

The psi factors for use in each country are specified in the relevant NAD. The values for the UK are shown in Table 20.5.

Ψ_0 Combination coefficient

In *fundamental design situations* (permanent and variable loads only) the designer considers the full characteristic values of one variable action (for example wind) and multiplies all the others by the factor Ψ_0. This reduces the values of these secondary variable actions to the maximum values which they are likely to have in combination with the full value of the principal or *dominant* variable action. This process is repeated, taking each of the other variable actions in turn as the dominant action and factoring the remaining secondary actions. From the resulting total actions, the critical *design value* may be selected.

Where all the variable actions in a combination have the same values of Ψ_0, the design value may be obtained simply by selecting as the dominant action the one which itself produces the worst effect. Where different values of Ψ_0 are involved, however, it will usually be necessary to try each variable action in turn as the dominant one in order to discover the maximum value.

Ψ_1 Frequent coefficient

This gives a greater reduction than Ψ_0, producing values for variable loads which may be regarded as average values for dominant loads in *accidental design situations*, and for secondary loads in *serviceability calculations*.

Ψ_2 Quasi-permanent coefficient

This gives the greatest reduction of all, producing values which may be regarded as the minimum values of secondary variable loads when they are in combination with accidental loads. It is used only in accidental design situations, and for wind and snow its value is zero.

20.2.10 Expressions for design values of actions

In the light of the information which has been given, it should be possible to make sense of the expressions give in DD ENV 1996-1-1 and Eurocode 1 for calculating the design values of actions. In these expressions, the letters j and i identify individual permanent and variable actions, and the Greek letter Σ has the conventional meaning of 'the sum of'. The reference numbers, for example (2.17), all refer to equations in DD ENV 1996-1-1.

The design value of a combination of actions is calculated from the appropriate expression below. When more than one variable action is involved, the design value is the most critical value of the expression, taking each variable action in turn as the dominant one and the others as secondary ones. This critical value is termed the *design load* for ultimate limit states, or the *service load* for serviceability limit states.

20.2.11 Ultimate limit states

Fundamental Combination

$$\Sigma \gamma_{G,j} \ G_{k,j} \ + \ \gamma_{Q,1} \ Q_{k,1} \ + \ \underset{i>1}{\Sigma} \ \gamma_{Q,i} \ \Psi_{0,i} \ Q_{k,i} \qquad (2.17)$$

Sum of factored permanent loads   Factored dominant variable load   Sum of other factored variable loads

Alternative expressions for Fundamental Combination*

The design value of a combination of loads may be taken as the larger of the two expressions which follow.

$$\Sigma \gamma_{G,j} \ G_{k,j} \ + \ 1.5 Q_{k,1} \qquad (2.19)$$

Sum of factored permanent loads   1.5 × the most unfavourable variable load

$$\Sigma \gamma_{G,j} \ G_{k,j} \ + \ 1.35 \ \underset{i \geq 1}{\Sigma} \ Q_{k,i} \qquad (2.20)$$

Sum of factored permanent loads   1.35 × all the variable loads

*This is a simplified but approximate method which assumes values for Ψ_0. The design value obtained may be from 8% too low to 22% too high.

Accidental design situations

$$A_d + \Sigma\gamma_{GA,j} \ G_{k,j} + \Psi_{1,1} \ Q_{k,1} + \sum_{i>1}\Psi_{2,1} \ Q_{ki} \quad (2.18)$$

Design value of accidental load | Sum of factored permanent loads | Factored dominant variable load | Sum of other factored variable loads

20.2.12 Serviceability limit states

$$\Sigma G_{k,j} + Q_{k1} + \sum_{i>1}\Psi_{0,i} \ Q_{k,i} \quad (2.22)$$

Sum of deformations produced by unfactored permanent loads | Deformation produced by unfactored dominant variable load | Sum of deformations produced by other factored variable load

Components of variable loading

Wind, snow and the imposed floor and ceiling loads, should each be treated as a separate variable action. The reduction factors for floor loads in multi-storey buildings that are permitted by BS 6399 Part 1 are used in designs to DD ENV 1996–1–1, in the absence of the loading Eurocodes.

The total of the imposed floor loads acting on a member 'under consideration is regarded as a single variable action.

20.2.13 Material properties

The characteristic values of material strengths are defined, preferably, as values which are likely to be equalled or exceeded in 95% of the cases. DD ENV 1996–1–1 defines characteristic strength as the value of the strength corresponding to a 5% fractile of all strength measurements of the masonry. It is noted that the value may be taken from the results of specific tests or from an evaluation of test data or other specified values. However, a characteristic strength may also be defined as a representative value when sufficient data is not available to arrive at the characteristic value on a rigorous statistical basis.

To allow for the expectation that, in a number of cases, the strength of a material may be below the characteristic value, a Partial Factor is used to convert the characteristic value into a design strength. Partial safety factors, γ_M, are given in DD ENV 1996–1–1 for a number of materials.

For masonry strengths, the partial safety factor, γ_M, depends upon the level of manufacturing control of the masonry units and the execution control of the work on site.

Manufacturing control is defined in **Clauses 3.1.1(3)** and **(4)**.

Execution control is described in **Clause 6.9** and **Annex G**.

The definitions of the two categories of manufacturing control are similar to those with which designers are familiar in the UK, as given in BS 5628: Part 1.

For execution control, three levels are foreseen in DD ENV 1996–1–1, whereas BS 5628 uses only two; since no detailed definitions are, in any event, given in DD ENV 1996–1–1 on how to decide on the levels of execution control, this matter has had to be dealt with in detail in the UK NAD. There, it is stated that only two categories of execution control will be used in the UK, reflecting the current practice of BS 5628: Part 1.

Table 2.6 gives the values of γ_M to be used from the UK NAD, for masonry strengths, that is compression, flexure, shear and for anchorage, tensile and compressive resistance of wall ties and straps, anchorage bond of reinforcing steel and tensile and compressive strength of steel.

20.2.14 Stability and robustness

DD ENV 1996–1–1 requires the structure to be stable and robust in regard to its intended 'normal' use and, in addition, to be designed in such a way that it will not be damaged by events such as explosions, impact or consequences of human error, to an extent disproportionate to the original cause. This is stated as Principles in Clauses 2.1(2)P, 4.1.1(3)P, and 4.1.2(1)P.

Clauses 4.1.1(6)P and **4.1. 1(7)** require the designer responsible for the overall stability of the structure to ensure that the design and details of all parts and components of the construction are compatible with the intended design concept and that this responsibility is not diluted as a consequence of the design and detailing being undertaken by more than one designer.

The means of providing stability and robustness is given in **Clause 4.1.1(4)** which requires a suitably braced structure to be developed from a consideration of the layout of the structure (on plan and section), and the

Table 20.6 Partial safety factors for material properties (γ_M)

				Category of execution	
			A	B	C
Masonry (see note)	Category of manufacturing control of masonry units	I	2,0	2,5	Not used
		II	2,3	2,8	Not used
Anchorage and tensile and compressive resistance of wall ties and straps			2,5	2,5	Not used
Anchorage bond of reinforcing steel			1,5	2,0	Not used
Steel (referred to as γ_s)			1,15	1,15	Not used

Note: The value of γ_M for concrete infill should be taken as that appropriate to the category of manufacturing control of the masonry units in the location where the infill is being used.

Table 20.7 Mortar designations, proportion of constituents and class

UK designation	cement:lime:sand	Mix proportions masonry	cement:sand with plasticizer	compressive strength (fm) N/mm²	DD ENV 1996–1–1 class
(i)	1 : ¼ : 3	–	–	12 N/mm²	M12
(ii)	1 : ½ : 4½	1 : 2½ to 3½	1 : 3 to 4	6 N/mm²	M6
(iii)	1 : 1 : 6	1 : 4 to 5	1 : 5 to 6	4 N/mm²	M4
(iv)	1 : 2 : 9	1 : 5½ to 6½	1 : 7 to 8	2 N/mm²	M2

interaction of the masonry elements with other elements of the structure. The possible effect of construction imperfections should also be taken into account by assuming a hypothetical inclination of the whole structure to the vertical of $1/(100\sqrt{h_{tot}})$ radians, where h_{tot} is the total height of the structure in metres. The effect of the latter hypothetical inclination may conveniently be simulated by a system of horizontal forces H_i applied at each floor and roof level, where

$$H_i = \frac{0.01 \, N_i}{\sqrt{h_{tot}}}$$

In this expression N_i is the combined vertical characteristic dead and imposed load at the particular floor or roof level to which H_i is applied (that is N_i is not intended to be the summation of all vertical loading acting above each floor level in turn).

Although DD ENV 1996-1-1 does not indicate whether the effect of the foregoing construction imperfections should be considered in unison with other horizontal actions, the designer is recommended to assume that they do *not* act simultaneously on the structure and that the construction imperfections may be ignored, if their effect is less onerous than that caused by other horizontal actions, for example wind. It should be noted, however, that these provisions apply only to the stability and robustness of the *whole* structure. Construction imperfections in individual walls are catered for elsewhere in DD ENV 1996-1-1 (See Clause 4.4.7.2) by the provisions for accidental eccentricity. Regarding the requirements relating to accidental damage given in Clauses 2.1(2)P, 2.1(3) and 4.1.2, these are consistent with National Building Regulations and associated guidance material (for example Codes of Practice and Approved Document A applicable in England and Wales) concerning disproportionate collapse. Building Regulations and associated guidance or 'deemed to satisfy' documents, as applicable to England and Wales, Scotland and Northern Ireland, should therefore be adhered to in this regard.

20.3 Materials

20.3.1 Masonry units

20.3.1.1 Types

All masonry units will eventually be manufactured in accordance with the relevant part of EN 771 as listed below with its equivalent British Standard. In the meantime, UK masonry units will continue to be manufactured and specified to British Standards. Masonry units should comply with a relevant product standard, either BS or EN.

EN 771–1 Clay units (BS3921)
EN 771–2 Calcium Silicate units (BS187)
EN 771–3 Aggregate Concrete units (BS6073)
EN 771–4 Autoclaved Aerated Concrete units (BS6073)
EN 771–5 Manufactured Stone units (BS6073 or BS6457)
EN 771–6 Natural Stone units –

Units meeting the requirements of the relevant product Standard will be classified, in terms of manufacturing control, as being Category II. Category I may be assumed where the additional requirements of **Clause 3.1.1(3)** are met, that is where the manufacturer operates a quality control scheme which shows that the probability of failing to reach the specified compressive strength is not more than 2½%. This figure has been modified in the UK NAD from the boxed value of 5% given in DD ENV 1996-1-1 to align it with the Special Category requirement of BS 5628: Part 1. It should be noted that the terms Category I and Category II have also been adopted by the European Commision to represent the levels of attestation given in the masonry unit product standards and should not be confused with the categories of manufacturing control as defined in DD ENV 1996-1-1.

In addition, masonry units are grouped according to the extent of any voids or perforations (see **Table 3.1**). In general, units with not more than 25% voids are classified as Group 1. Units with percentage of voids between 25 and 70 are classified as Group 2a, 2b or 3.

All UK bricks currently manufactured to British Standards, including frogged and perforated bricks, will fall within Group 1. Group 2 will cover what are commonly known as hollow or cellular units. Group 3 units are specific to certain European countries and are not generally available in the UK. Individual manufacturers will be able to confirm the Group to which their products conform.

20.3.1.2 Compressive strength

Due to the large number of different types and sizes of units available throughout Europe, and to differences in testing, the tested compressive strength must be

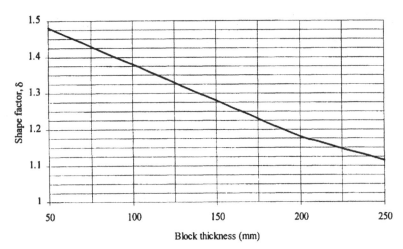

Figure 20.1 Shape factor, δ, for 215mm high blocks

Table 20.8 Value of constant K to be used in calculation of f_k with general purpose mortar

Type of bonding	Type of masonry units			
	Group 1	Group 2a	Group 2b	Group 3
No longitudinal joint through all or part of length of wall (see Figure 20.2(a))	0.70	0.55	0.50	0.40
Longitudinal joint through all or part of the length of masonry (see Figure 20.2(b))	0.50	0.45	0.40	0.40

normalised to make allowance for different shapes and conditioning regimes at test. The normalised compressive strength, f_b, used for the determination of the characteristic strength of masonry is equivalent to the compressive strength of an 'air dry' 100mm cube of material. In general, the normalised compressive strength for any masonry unit may be found using the relationship

$$f_b = m_c \times \text{(manufacturers declared compressive strength)} \times \delta$$

where

m_c is the adjustment for moisture content
δ is the shape factor

Units may be tested air dry or water saturated. The manufacturer should always quote the directly obtained results, stating the relevant conditions of test. The specified compressive strengths of units currently produced to British Standards are wet strengths and will therefore require to be converted to air dry values by multiplying by the factor $m_c = 1.2$. This conversion factor will also need to be applied where the compressive strength tests to EN 772–1 are carried out on saturated units. If the tests have been carried out on air dry units, as defined by the relevant part of EN 771, then the conversion factor is not required (that is $m_c = 1.0$).

Shape factor, δ
Most units will be tested whole but some types may have to be cut, or cubes may be made from the parent mixture.

Table 3.2 gives the shape factor, δ, which should be applied to the tested unit compressive strength. For 102mm × 65mm bricks, δ = 0.85. For concrete blocks, δ will depend on the unit thickness. Figure 20.1 is based on the values given in **Table 3.2** and may be used to find δ for 215mm high blocks.

Specification
In the job specification, if the designer does not know or need to specify the shape, type or Group of unit (as is some European practice) then he may specify the required normalised compressive strength only. If, on the other hand, the designer needs to specify the shape of the unit (for example for coursing) or the type (for example for aesthetics) then he may specify either the unit compressive strength directly or the required normalised compressive strength coupled with the shape, size and material of the units.

Manufactured and Natural Stone
The design method for vertical load contained in DD ENV 1996–1–1 applies only to manufactured and dimensional natural stone where the masonry units have a rectangular parallelepiped shape. For aesthetic reasons, manufactured stone and dimensional natural stone masonry is often constructed with units having different face dimensions. In principle, all of these units have the same normalised compressive strength, but in practice there will be minor variations due to the calculation

process. For vertical loading, it is conservative to design the resulting wall as if constructed entirely of the unit having the smallest normalised strength. The standard 150mm concrete cube crushing test, tested air dry, can be taken to yield the normalised strength directly. There is no similar standard test for the compressive strength of natural stone, but any cube of 100mm side or greater can be taken to yield the normalised strength directly, again when tested air dry.

20.3.2 Mortar

Cements used for mortar should be class 42.5 to BS12 (Portland), BS146 (Portland blastfurnace), BS4027 (Sulfate resisting) or BS5224 (Masonry cement). Limes should be in accordance with BS890, sand to BS1200 and plasticizers to BS4887.

Mortars may be specified by strength or mix proportions. When specifying by strength, the compressive strength in N/mm^2 should follow the letter M. For example M4 is a $4N/mm^2$ strength mortar. Table 20.8 shows a list of strength classes to be ascribed to mortar designations (that is specified mix proportions) as given in the UK NAD. Pre-mixed or ready to use mortars should comply with EN 998-2 or BS4721.

Note that the site and preliminary values as given in BS5628 have been combined into one figure, that is DD ENV 1996-1-1 does not differentiate between site and laboratory tests.

In general, the mortar class should not be less than M2 in unreinforced masonry. Where bed joint reinforcement is to be used, the minimum class should be M4. For prestressed masonry or masonry containing vertical reinforcement, the mortar class should be M6 or stronger.

Lightweight mortars and thin layer mortars should also be at least M5.

Selection of mortars for durability should continue to be based on clauses 22 and 23 of BS5628:Part 3:1985 until the publication of ENV 1996-2.

20.3.3 Concrete infill

Concrete used for infill should be normal weight concrete containing natural aggregates as defined in EN 206 or BS 5328 Parts 1 and 2 and should be classified according to the compressive cylinder or cube strength at 28 days. The mixes may be designed or prescribed. Tables 3.3 and 3.4 give the characteristic compressive strength, f_{ck}, and shear strength, f_{cvk}, of concrete for the classes normally used to infill masonry. Although DD ENV 1996-1-1 permits concrete infill strengths down to $12N/mm^2$, the UK NAD states that for UK applications concrete infill should have a characteristic compressive strength of at least $20N/mm^2$ at 28 days (that is class C20/25 to EN 206) for durability reasons.

Cements complying with the class 42.5 or 52.5 requirements of BS12 (Portland), BS146 (Portland blastfurnace) or BS4027 (Sulfate resisting) may be used for concrete infill. Masonry cement or high alumina cement should not be used. Admixtures for concrete infill

should only be used in accordance with Clause 2.11 of BS5628: Part 2.

The maximum aggregate size should be limited to 20 mm, except where the least dimension of the void to be filled is less than 100 mm or where less than 25 mm cover to the reinforcement is required, in which cases the aggregate size should not exceed 10 mm.

DD ENV 1996-1-1 requires that the workability of concrete should be such as to ensure that all voids will be completely filled. The UK NAD comments that this will normally mean using a slump class S2, S3 or S4 in accordance with EN 206, or slumps in the range of 75 mm to 175 mm in accordance with BS 5328 Part 2. Where a high workability concrete is to be used, consideration should be given to the possibility of cracking of the concrete due to shrinkage.

20.3.4 Reinforcing steel

Until the availability of EN10080, carbon steel used for the reinforcement of masonry should be in accordance with the requirements of BS4449 or BS4482. Stainless steel used for reinforcement of masonry should be in accordance with the requirements of EN10088. Reinforcing steel may be plain or high bond (that is deformed bars). Bed joint reinforcement should be in accordance with EN 845-3.

Although in general it may be assumed that reinforcing steel will have sufficient ductility for design purposes, Clause 3.4.1 defines limits for High and Normal ductility steel as these are referred to in Section 4. High ductility bars should be used when bending dimensions to BS4966 are specified; steel to BS4449 meets the requirements for high ductility.

The modulus of elasticity of reinforcing steel may be taken as $200kN/mm^2$.

Reinforcing steel shall have sufficient durability for the exposure class in which it is to be used (as defined in Clause 5.2.2.2). The UK NAD states that austenitic stainless steel types 304S31 or 316S33 to BS6744 or 304S15, 304S31 or 316S33 (excluding free machining specifications) to BS970 Part 1, meet the requirements of DD ENV 1996-1-1 in respect of durability. Carbon steel may require additional protection, in the form of galvanising, epoxy coatings or specified depths of concrete cover. In the absence of a CEN standard for galvanising, BS729 should be used.

20.3.5 Prestressing steel

Pending publication of EN 10138, prestressing steel should be in accordance with BS5896 or BS4486 and Clause 3.3 of ENV 1992-1-1.

20.3.6 Mechanical properties of unreinforced masonry

20.3.6.1 General

DD ENV 1996-1-1 identifies four mechanical properties of masonry which are used in design:

Table 20.9 Characteristic compressive strength of unreinforced masonry using Group I units and general purpose mortar

(a) 65 mm high × 102 mm thick standard format bricks: Wall without longitudinal joint
δ = 0.85 K = 0.7 (that is wall thickness = brick thickness)

Mortar designation		Compressive strength of unit (N/mm²) to BS 3921 or 187								
UK	ENV1996	5	10	15	20	27.5	35	50	70	100
(i)	M12	3.6	5.9	7.7	9.3	11.4	13.3	16.8	20.9	26.3
(ii)	M6	3.2	5.0	6.5	7.8	9.6	11.2	14.1	17.6	22.1
(iii)	M4	2.9	4.5	5.8	7.0	8.6	10.1	12.8	15.9	20.0
(iv)	M2	2.4	3.8	4.9	5.9	7.3	8.5	10.7	13.3	16.8

(b) 65 mm high × 102 mm thick standard format bricks: Wall with longitudinal joint
δ = 0.85 K = 0.5 (that is wall thickness > brick thickness)

Mortar designation		Compressive strength of unit (N/mm²) to BS 3921 or 187								
UK	ENV1996	5	10	15	20	27.5	35	50	70	100
(i)	M12	2.6	4.2	5.5	6.6	8.1	9.5	12.0	14.9	18.8
(ii)	M6	2.3	3.5	4.6	5.6	6.8	8.0	10.1	12.5	15.8
(iii)	M4	2.0	3.2	4.2	5.0	6.2	7.2	9.1	11.3	14.3
(iv)	M2	1.7	2.7	3.5	4.2	5.2	6.1	7.7	9.5	12.0

(c) 215 mm high × 100 mm thick solid blocks: Wall without longitudinal joint
δ = 1.38 K = 0.7 (that is wall thickness = block thickness)

Mortar designation		Compressive strength of unit (N/mm²) to BS 6073								
UK	ENV1996	2.8	3.5	4	5	7	10	15	20	35
(i)	M12	3.3	4.0	4.5	5.1	6.4	8.1	10.5	12.7	18.2
(ii)	M6	3.0	3.4	3.7	4.3	5.4	6.8	8.8	10.7	15.3
(iii)	M4	2.7	3.1	3.4	3.9	4.9	6.1	8.0	9.6	13.9
(iv)	M2	2.3	2.6	2.8	3.3	4.1	5.2	6.7	8.1	11.7

(d) 215 mm × 200 mm thick solid blocks: Wall without longitudinal joint
δ = 1.18 K = 0.7 (that is wall thickness = block thickness)

Mortar designation		Compressive strength of unit (N/mm²) to BS 6073								
UK	ENV1996	2.8	3.5	4	5	7	10	15	20	35
(i)	M12	2.9	3.5	4.0	4.6	5.8	7.3	9.5	11.4	16.5
(ii)	M6	2.7	3.1	3.4	3.9	4.9	6.1	8.0	9.6	13.9
(iii)	M4	2.4	2.8	3.1	3.5	4.4	5.5	7.2	8.7	12.5
(iv)	M2	2.0	2.4	2.6	3.0	3.7	4.7	6.1	7.3	10.5

(e) 215 mm × 250 mm thick solid blocks: Wall without longitudinal joint
δ = 1.115 K = 0.7 (that is wall thickness = block thickness)

Mortar designation		Compressive strength of unit (N/mm²) to BS 6073								
UK	ENV1996	2.8	3.5	4	5	7	10	15	20	35
(i)	M12	2.7	3.3	3.8	4.5	5.6	7.0	9.2	11.0	15.9
(ii)	M6	2.6	3.0	3.3	3.8	4.7	5.9	7.7	9.3	13.3
(iii)	M4	2.3	2.7	2.9	3.4	4.2	5.3	7.0	8.4	12.1
(iv)	M2	2.0	2.3	2.5	2.9	3.6	4.5	5.8	7.1	10.1

Linear interpolation is permitted within each table
Linear interpolation is permitted between tables (c) and (d) and (d) and (e) for other block thicknesses

Compressive strength	f
Shear strength	f_v
Flexural strength	f_x
Stress/strain relationship	σ/ε

Although direct tensile strength can be developed in masonry, it should generally be ignored for design purposes.

20.3.6.2 Characteristic compressive strength f_k

The characteristic compressive strength of masonry may be determined by

(i) tests in accordance with EN 1052–1, or
(ii) use of an established relationship between f_k and the strength of units and mortar based on existing test data (for example Table 2 of BS5628 Part 1), or
(iii) use of the formulae in DD ENV 1996-1-1.

In practice, the use of the formulae will be the most straightforward approach.

20.3.6.2.1 f_k with general purpose mortar
Eqn (3.1) gives a relationship between unit strength, mortar class and the characteristic compressive strength of masonry using general purpose mortar.

$$f_k = K \ f_b^{0.65} \cdot f_m^{0.25} \qquad (3.1)$$

where

f_b is the normalised compressive strength of the masonry units in N/mm²;

f_m is the specified compressive strength of the general purpose mortar in N/mm² (but not exceeding 20N/mm² or $2f_b$, whichever is the smaller); and

K is a constant which may be taken from Table 20.8.

Tables 20.9 and 20.10 are based on **Eqn (3.1)** and give the characteristic compressive strength of masonry built using Groups 1 and 2a units respectively with general purpose mortar.

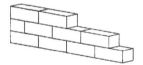

(a) Wall without longitudinal joints

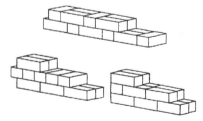

(b) Walls with longitudinal joints

Figure 20.2 Typical wall bonds

Table 20.10 Characteristic compressive strength of unreinforced masonry using Group 2a units and general purpose mortar

(a) 215 mm high × 100 mm thick cellular/hollow blocks: Wall without longitudinal joint
$\delta = 1.38$ K = 0.55 (that is wall thickness = block thickness)

Mortar designation		Compressive strength of unit (N/mm²) to BS 6073				
UK	ENV1996	3.5	7	10	15	21
(i)	M12	3.2	5.0	6.3	8.3	10.3
(ii)	M6	2.7	4.2	5.3	6.9	8.6
(iii)	M4	2.4	3.8	4.8	6.3	7.8
(iv)	M2	2.0	3.2	4.1	5.3	6.6

(b) 215 mm high × 200 mm thick cellular/hollow blocks: Wall without longitudinal joint
$\delta = 1.18$ K = 0.55 (that is wall thickness = block thickness)

Mortar designation		Compressive strength of unit (N/mm²) to BS 6073				
UK	ENV1996	3.5	7	10	15	21
(i)	M12	2.8	4.5	5.7	7.5	9.3
(ii)	M6	2.4	3.8	4.8	6.3	7.8
(iii)	M4	2.2	3.5	4.4	5.7	7.1
(iv)	M2	1.9	2.9	3.7	4.8	5.9

(c) 215 mm high × 250 mm or greater thick cellular/hollow blocks: Wall without longitudinal joint
$\delta = 1.115$ K = 0.55 (that is wall thickness = block thickness)

Mortar designation		Compressive strength of unit (N/mm²) to BS 6073				
UK	ENV1996	3.5	7	10	15	21
(i)	M12	2.6	4.4	5.5	7.2	8.9
(ii)	M6	2.3	3.7	4.6	6.0	7.5
(iii)	M4	2.1	3.3	4.2	5.5	6.8
(iv)	M2	1.8	2.8	3.5	4.6	5.7

Linear interpolation is permitted within each table.
Linear interpolation is permitted between Tables (a) and (b) and (b) and (c) for block thicknesses
Where
f_{vko} is the shear strength under zero compressive stress (obtained from Table 20.11)
σ_d is the design compressive stress perpendicular to the shear plane, and
f_b is the compressive strength of the units normalised for the direction of application of the load on the test

When Group 2 aggregate concrete units are used with the vertical voids filled with in-situ concrete, the value of f_b should be obtained by assuming the units to be solid (Group 1) with a compressive strength equal to the lesser of the strength of the infill concrete or the net compressive strength of the unit (that is the compressive strength based on the net area of the unfilled unit).

20.3.6.2.2 f_k with thin layer mortar
Units used for masonry constructed using thin layer mortar should be manufactured to a suitable dimensional accuracy. When clay or aggregate concrete units are used,

Table 20.11 Value of constant K to be used in calculation of f_k with thin layer mortar

Type of bonding	Type of masonry units			
	Group 1*	Group 2a	Group 2b	Group 3
No longitudinal joint through all or part of length of wall (see Figure 2(a))	0.70	0.60	0.50	–

Table 20.12 Value of constant K to be used in calculation of f_k with lightweight mortar

Type of masonry unit	Density of lightweight mortar	
	600–700 kg/m³	701–1500 kg/m³
lightweight aggregate and autoclaved aerated concrete	0.80	0.80
clay, calcium silicate and dense aggregate concrete	0.55	0.70

Eqn (3.1) may be used to calculate the characteristic compressive strength of masonry made using thin layer mortar with the value of K taken from Table 20.11.

When calcium silicate or autoclaved aerated concrete units are used the characteristic compressive strength of masonry constructed using thin layer mortar should be obtained using **Eqn (3.2)**.

$$f_k = 0.8\, f_b^{0.85} \qquad (3.2)$$

In all cases, the normalised compressive strength of the units, f_b, should not be taken to be greater than 50 N/mm² and the mortar should have a compressive strength of at least 5 N/mm² (M5). In addition, there should be no longitudinal vertical mortar joint through all or part of the length of the wall (see Figure 20.2).

20.3.6.2.3 f_k with lightweight mortar

The characteristic compressive strength of masonry constructed using lightweight mortar should be calculated using **Eqn (3.3)** provided that f_b is not taken to be greater than 15 N/mm² and that there is no longitudinal mortar joint through all or part of the length of the wall.

$$f_k = K \cdot f_b^{0.65} \qquad (3.3)$$

where K is a constant that should be taken from Table 20.12.

20.3.6.2.4 f_k with shell bedding

The term shell bedding applies to masonry laid on two equal strips of general purpose mortar at the outside edges of the unit. Where shell bedding is to be used, the thickness of the wall should be equal to the width or length of the masonry units so that there are no longitudinal mortar joints through all or part of the length of the wall (see Figure 20.2).

For Group 1 masonry units, Eqn (3.1) may be used to calculate f_k with the value of constant K as given in Figure 20.3.

Eqn 3.1 may also be used for Group 2a or Group 2b units provided that the normalised compressive strength,

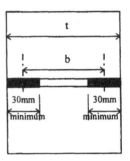

The ratio of b/t must not exceed 0.8.
For b/t $\leq$ 0.5,   K = 0.60
For b/t = 0.8,   K = 0.30
For intermediate values of b/t, K may be obtained by linear interpolation.

Figure 20.3 Value of constant K to be used in calculation of f_k with shell bedding

f_b, has been obtained from tests carried out on shell bedded units.

20.3.6.3 Characteristic shear strength f_{vk}

The characteristic shear strength of masonry, f_{vk}, may be determined by

(i)   tests in accordance with EN 1052–(3) , or
(ii) use of an established relationship between f_{vk} and the initial shear strength, f_{vko}, or
(iii) use of the formulae in DD ENV 1996–1–1.

Note: The relationship given in BS5628 Part 1 should not be used to determine the characteristic shear strength of masonry to be designed to DD ENV 1996–1–1.

Using general purpose mortar, with all joints filled, the characteristic shear strength of unreinforced masonry may be taken as the lesser of

$$f_{vk} = f_{vko} + 0.4\sigma_d \qquad (3.4)$$

or $f_{vk} = 0.065\, f_b$ (but not less than f_{vko})

or $f_{vk} =$ the limiting value in **Table 3.5**.

Where:

f_{vko} is the shear strength under zero compressive stress (obtained from **Table 3.5**),
σ_d is the design compressive stress perpendicular to the shear plane, and
f_b is the compressive strength of the units normalised for the direction of application of the load on the test specimens being perpendicular to the bed face.

For masonry with unfilled perpend joints, but with adjacent faces butted together, the characteristic shear

Table 20.13 Values of f_{vko} and limiting f_{vk} for thin layer mortars

Masonry Unit	Mortar Class	f_{vko} N/mm²	Limiting f_{vk} N/mm²
Group 1	M10 – M20	0.3	1.7
Group 2a	M10 – M20	0.3	1.4[1]
Group 2b	M10 – M20	0.2	1.4[1]
Group 3	M10 – M20	0.3	–

(1) Where the longitudinal compressive strength can be expected to be less than $0.15 f_b$, the limiting value of f_{vk} shall be taken as the measured longitudinal compressive strength (with $\delta = 1.0$).

Table 20.14 Values of f_{vko} and limiting f_{vk} for lightweight mortars

Masonry Unit	f_{vko} N/mm²	Limiting f_{vk} N/mm²
Group 1 clay units	0.2	1.5
Group 1 natural stone units	0.15	1.0
All other Group 1 units	0.15	1.5
Group 2a clay units	0.2	1.2[1]
All other Group 2 units	0.15	1.2[1]
Group 3 clay units	0.2	–

(1) Where the longitudinal compressive strength can be expected to be less than $0.15 f_b$, the limiting value of f_{vk} shall be taken as the measured longitudinal compressive strength (with $\sigma = 1.0$).

strength of unreinforced masonry may be taken as the lesser of

$$f_{vk} = 0.5 f_{vko} + 0.4\sigma_d \qquad (3.5)$$

or $f_{vk} = 0.045 f_b$ (but not less than f_{vko})

or $f_{vk} = 0.7 \times$ the limiting value in **Table 3.5**.

For shell bedded masonry, the characteristic shear strength may be taken as the lesser of

$$f_{vk} = (g/t)f_{vko} + 0.4\sigma_d \qquad (3.6)$$

or $f_{vk} = 0.05 f_b$ (but not less than f_{vko})

or $f_{vk} = 0.7 \times$ the limiting value in **Table 3.5**.

Where
g   is the total width of the two mortar strips, and
t   is the thickness of the wall

For masonry constructed using thin layer mortar, f_{vk} may be calculated using the relevant formula above but with f_{vko} and limiting f_{vk} values taken from Table 20.14.

For masonry constructed using lightweight mortar, f_{vk} may be calculated using the relevant formula above but with f_{vko} and limiting f_{vk} values obtained from Table 20.14.

20.3.6.4 Characteristic flexural strength f_{xk}

The characteristic flexural strength of masonry, f_{xk}, may be determined by tests in accordance with EN1052-2 or be established from an evaluation of test data based on the flexural strength of masonry obtained from appropriate combinations of units and mortar. The UK NAD recommends that in the absence of any test data

Table 20.15 Characteristic flexural strength of masonry, f_{xk}

Mortar class	Plane of failure parallel to the bed joint (f_{xk1})			Plane of failure perpendicular to the bed joint (f_{xk2})		
	M12	M6–M4	M2	M12	M6–M4	M2
Clay bricks having a water absorption						
less than 7%	0.7	0.5	0.4	2.0	1.5	1.2
between 7% and 12%	0.5	0.4	0.35	1.5	1.1	1.0
over 12%	0.4	0.3	0.25	1.1	0.9	0.8
Calcium silicate bricks		0.3	0.2		0.9	0.6
Concrete bricks		0.3	0.2		0.9	0.6
Concrete blocks of compressive strength‡ in N/mm²						
2.8 (100 mm wall thickness*)					0.40	0.4
3.5 (100 mm wall thickness*)					0.45	0.4
7.0 (100 mm wall thickness*)		0.25	0.2		0.60	0.6
2.8 (250 mm wall thickness*)					0.25	0.2
3.5 (250 mm wall thickness*)		0.15	0.1		0.25	0.2
7.0 (250 mm wall thickness*)					0.35	0.3
10.5 (Any thickness*)					0.75	0.6
14.0 and over		0.25	0.2		0.90†	0.7†

* The thickness should be taken to be the thickness of the wall, for a single leaf wall, or the thickness of the leaf, for a cavity wall.
† When used with flexural strength in parallel direction, assume the orthogonal ratio $\mu = 0.3$
‡The compressive strengths of concrete blocks should be taken as that specified to British Standards (that is not normalised) when using this table.

obtained by following EN 1052–2, values of the characteristic flexural strength of unreinforced masonry should be taken from BS5628: Part 1. However, Table 3 of BS 5628: Part 1 gives strengths according to mortar designation. Where the mortar class is known, the flexural strengths, both for failure parallel to the bed joint (f_{xk1}) and for failure perpendicular to the bed joint (f_{xk2}), may be obtained from Table 20.15. The characteristic flexural strength, f_{xk1}, should be taken as zero where failure of the wall in question would lead to a major collapse of the structure as a whole.

20.3.7 Mechanical properties of reinforced, prestressed and confined masonry

In addition to the mechanical properties required for the design of unreinforced masonry, the following will also have to be considered for reinforced, prestressed and confined masonry:

Compressive strength of concrete infill	f_c
Shear strength of concrete infill	f_{cv}
Tensile and compressive yield strength of the reinforcing steel	f_y
Tensile strength of prestressing steel	f_p
Anchorage bond strength of the reinforcing steel	f_{bo}

The characteristic anchorage bond strength, f_{bok}, for reinforcement embedded in concrete or mortar should be obtained from Tables 3.6 and 3.7. However, for UK applications the maximum values of f_{bok} are limited by the UK NAD to those given for M10–14 mortar or C20/25 concrete.

For prefabricated bed joint reinforcement, the bond strength should be determined by tests to EN 846–2. Alternatively, the bond strength of the longitudinal wires alone may be used.

20.3.8 Deformation properties of masonry

20.3.8.1 Stress/strain relationship

For the purpose of design, the stress/strain (σ/ε) relationship may be assumed to be rectangular parabolic (as shown in Figure 3.3).

20.3.8.2 Modulus of elasticity, E

The modulus of elasticity of masonry may either be determined from tests in accordance with EN 1052–1, with E calculated at one third of the maximum load; where tests are not available, E may be taken as 1000 f_k for use in structural analysis relating to the ultimate limit state. For calculations relating to the serviceability limit state, E should be taken as 600 f_k.

20.3.8.3 Shear modulus, G

In the absence of test data, G may be taken as 0.4E.

20.3.8.4 Creep

Table 3.8 gives a range of values from 0 to 2.0 for the final creep coefficient to be used in design depending on the type of masonry unit. The UK NAD gives a single design value of 1.5 for the final creep coefficient irrespective of unit type.

20.3.8.5 Moisture expansion or shrinkage

In the absence of test values, the values given in Table 3.8 should be used for calculation purposes.

20.3.8.6 Thermal expansion

The coefficient of thermal expansion should be taken as 6×10^{-6} for clay units; for all other types of units the UK NAD gives a value of 10×10^{-6}.

20.3.9 Ancillary components

20.3.9.1 Damp proof courses

Damp proof courses are required to resist the passage of water, to be formed of durable materials which are not easily punctured and to be able to resist the design compressive stresses without exuding. DPCs can be of bituminous materials, engineering bricks or slates in cement mortar or any other material that will prevent the passage of moisture.

20.3.9.2 Wall ties, straps, brackets, hangers and support angles

Wall ties, straps, hangers, brackets and support angles are required to be in accordance with the requirements of EN 845–1. Until this is published designers should specify wall ties to BS1243 and DD140 Part 2 and hangers to BS6178 for UK applications.

20.3.9.3 Prefabricated lintels

Pending the availability of EN 845–2, prefabricated lintels should meet the requirements of BS 5977.

20.3.9.4 Prestressing devices

Anchorages, couplers, ducts and sheaths should be in accordance with the requirements of Clause 3.4.

20.4 Design of Masonry

20.4.1 General

This section of DD ENV 1996–1–1 and of this chapter covers the design aspects of masonry. The various aspects of design are described and discussed in the following paragraphs. The paragraph numbers, here, do not correspond directly to those in DD ENV 1996–1–1, but full references are given.

20.4.2 Unreinforced masonry walls subjected to vertical loading

20.4.2.1 Principles of design

Clause 4.1 lays out the principles which need to be followed for the design of masonry. In Clause 4.1.1 design models and structural behaviour are considered. The models are set up to examine each relevant limit state and are to be based on the geometry of the structure, its materials, the environment in which it is built, the actions on it and how these are imposed. For each relevant limit state, the behaviour of the whole, or parts, of the structure may form the model.

Clause 4.1.3 specifically considers the design of structural members. It requires that their design be verified in the ultimate limit state and that the structure be designed so that cracks or deflections which might damage facing materials, partitions, finishes or technical equipment, or which might impair watertightness are avoided or minimised.

The design of members need not be verified separately at the serviceability limit state, which is deemed to be satisfied when the ultimate limit state is successfully verified. However, the serviceability of masonry members should not be unacceptably impaired by the behaviour of other structural elements, such as the deformation of floors. Where necessary, special precautions should be taken to ensure the overall stability of the structure or individual elements of it during construction.

20.4.2.2 Design load

Clause 4.2 discusses the effect of actions on structures. Table 20.16 summarises these and refers to other clauses in DD ENV 1996-1-1 where further detail on these actions exists, and to other standards with relevant information.

When considering the ultimate limit state, Clause 4.2.5 requires the design values which produce the most severe conditions to be taken into account. The most severe design value arises from a combination of actions as given in Clause 2.3.2.2, with the appropriate partial safety factors taken from Clause 2.3.3.1.

20.4.2.3 Design strength of masonry.

The design strength of masonry should be taken as the characteristic strength divided by the appropriate partial safety factor given in Clause 2.3.3.2.

Design strength in compression, $f_d = f_k/\gamma_m$
shear, $f_{vd} = f_{vk}/\gamma_m$
flexure, $f_{xd} = f_{xk}/\gamma_m$

Table 20.16 Actions to be considered

Action	Relevant clauses/ standards
Characteristic permanent action G_k	2.2.2.2
Characteristic variable action Q_k	2.2.2.2 and 2.2.2.3
Characteristic wind action W_k	ENV 1991
Characteristic lateral earth pressure	ENV 1991 and ENV 1997

20.4.2.4 Design

Clause 4.4.1 requires the resistance of unreinforced masonry walls to vertical loading be based on the geometry of the wall, the effect of applied eccentricities and the material properties of the masonry. It is to be assumed that plane sections remain plane, the tensile strength of the masonry perpendicular to the bed joints is zero and the stress/strain relationship is of the form shown in **Figure 3.2**.

At the ultimate limit state, the design vertical load on a masonry wall, N_{Rd}, should be less than or equal to the design vertical load resistance of the wall, N_{Sd}, such that,

$$N_{Sd} \leq N_{Rd}$$

The design should allow for the long term effects of loading, second order effects, and eccentricities calculated from a knowledge of the layout of the walls, the interaction of the floors and the stiffening of the walls, and from construction deviations and differences in the material properties of individual components.

Clause 4.4.2(1) gives the design vertical load resistance of a single leaf wall per unit length, N_{Rd}, as

$$N_{Rd} = \frac{\Phi_{i,m} t f_k}{\gamma_m} \qquad (4.5)$$

where:

$\Phi_{i,m}$ is the capacity reduction factor, Φ_i or Φ_m, as appropriate, allowing for the effects of slenderness and eccentricity of loading, obtained from Clause 4.4.3.

f_k is the characteristic compressive strength of the masonry obtained from Clause 3.6.2.

γ_m is the partial safety factor for the material obtained from Clause 2.3.3.2.

t is the thickness of the wall taking into account the depth of any recesses in the joints greater than 5 mm.

Clause 4.4.2 advises that the design strength of a wall may be at its lowest in the middle one fifth of the wall height when Φ_m should be used, or at the top or bottom of the wall, when Φ_i should be used.

Clause 4.4.2(8) states that walls that satisfy the ultimate limit state when verified in accordance with the above equation may be deemed to satisfy the serviceability limit state.

Clause 4.4.2(3) allows for a modification of the characteristic compressive strength of the masonry when the cross sectional area of the wall is less than $0.1\,m^2$. In these cases f_k should be multiplied by $(0.7 + 3A)$ where A is the loaded gross horizontal cross sectional area of the member expressed in square metres.

With cavity walls, the load carried by each leaf should be assessed and the design vertical load resistance of each leaf verified using the above equation. When only one leaf of a cavity wall is loaded the loadbearing capacity of the wall should be based on the horizontal cross sectional area of that leaf alone but using the effective thickness calculated using **Eqn (4.17)** for the purposes of determining the slenderness ratio. The UK NAD states that the

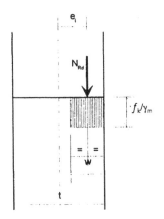

Figure 20.4 Stress block assumed in Clause 4.4.3 and Annex A.

and it can be seen that the width of the stress block, w, is a function of the wall thickness and the resultant eccentricity, such that

$$w = t (1 - 2e_i/t)$$

Hence, the design vertical load resistance of the wall, ignoring any slenderness effects, is $N_{rd} = t(1 - 2e_i/t)(f_k/\gamma_m)$. Comparing this with Eqn (4.5), and substituting for N_{rd} gives

$$\phi_i = 1 - 2e_i/t \qquad (4.7)$$

20.4.2.5.2 Reduction for slenderness in the middle height of the wall

A.1(1) gives three equations from which it is possible to reproduce **Table A.1**. A.1(2) confirms that **Figure 4.5** is a graphical representation of the same equations.

Eqn (A.1) can be split into two factors. The first factor, A1, is similar to the reduction factor for eccentricity at the top or bottom of the wall. This can be verified by comparing **Eqn (A.2)** with **Eqn (4.7)**. The second factor, the base of natural logarithms raised to the power $-u^2/2$, which is defined in **Eqn (A.3)**, is the reduction factor to allow for buckling effects due to slenderness.

A.1(3) offers a method for taking account of various values of the modulus of elasticity. When the modulus is taken as $1000f_k$ in **Eqn (A.5)**, it can be seen that substituting for the value of λ in (A.4) and multiplying throughout by the square root of one thousand produces Eqn (A.3).

From the equations in A.1(3), the effect of changes in the modulus of elasticity from 1000 f_k by ±20% is shown in Figure 20.5 for eccentricity ratios 0.05 and 0.33. Compared with changes in eccentricity or slenderness ratio, changes in modulus of elasticity are of relatively small effect. This justifies the simplification in A.1(1),

effective thickness of walls stiffened by piers should be taken from Figure 3 and Table 5 of BS 5628: Part 1: 1992.

A double leaf wall may be designed as a cavity wall or alternatively as a single leaf wall if the two leaves are tied together resulting in common action under load (see Clause 5.4.2.3).

20.4.2.5 The reduction factor for slenderness and eccentricity

20.4.2.5.1 The reduction factor for eccentricity at the top or bottom of the wall

The reduction factor for slenderness and eccentricity in Clause 4.4.3 is based on the assumption that the design load acting at a resultant eccentricity, e_1 or e_m, is resisted by a rectangular stress block, as shown in Figure 20.4. From static equilibrium the centreline of the stress block is coincident with the line of action of the design load,

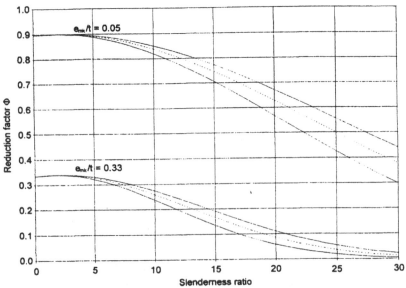

Note: For modulus of elasticity of masonry units, E, equal to 1000f_k the values shown by dotted lines match Figure 20.5. The lower and upper solid lines at each eccentricity ratio are for E equal to 800f_k and 1200f_k respectively.

Figure 20.5 Slenderness and eccentricity reduction factor within the middle height of the wall showing effect of varying modulus of elasticity, using A.1 (3).

316

although in extreme cases some correction may be desirable. For walls with high slenderness ratio and low eccentricity ratio the effect is approximately in direct proportion to the change in E from $1000f_k$.

DD ENV 1996–1–1 gives no specific guidance for cases where units are shell bedded. A stress block approach can be used based on the net bearing area and the compressive strength enhanced in the ratio of gross area to net area. Alternatively, for design purposes, a simpler and more conservative assumption will be to use the characteristic strength based on the gross unit area and the reduction factors from Eqn (4.7) or Figure 4.2 as appropriate.

20.4.2.6 Effective height of walls

Clause 4.4.4 of DD ENV1996–1–1 gives guidance on determining the effective height of walls to be used in calculating the slenderness ratio to enable the capacity reduction factor to be derived from Figure 4.2. The approach is similar to that in BS 5628: Part 1 and the designer is guided on the height of wall which may be considered as being unrestrained in the sense that it is free to buckle between notional pin ends. Factors which affect the extent to which the effective height differs from the actual height are the moment restraint at the top and bottom of the wall due to the connecting floors and also any stiffening of the vertical edges of the wall by intersecting walls or by connection to columns or other elements.

The effective height h_{ef} is derived from the clear storey height h using the formula

$$h_{ef} = \rho_n h \tag{4.12}$$

where ρ is a reduction factor on which guidance is given as to the values to be used for the number of edges, n, of the wall that are restrained or stiffened. For example in the case of a wall with two free vertical edges but restrained at both top and bottom n = 2.

Floors are assumed to provide restraint to the wall when they either span onto the wall from both sides at the same level or from one side with a minimum bearing of 2/3 of the thickness of the wall but not less than 85 mm. If the eccentricity of load at the head of the wall is greater than one quarter of the thickness, ρ_2 is taken as 1.0. If the eccentricity is lower than this, where the restraint is provided by reinforced concrete floors, ρ_2 may be taken as 0.75. In the case of timber floors DD ENV1996–1–1 recommends the use of $\rho_2 = 1.0$ but this is a boxed value and the UK NAD allows ρ_2 to be taken as 0.75, the same as in BS 5628: Part 1.

The provisions for walls to be regarded as having stiffened edges are quite different in DD ENV1996–1–1 from those in BS 5628: Part 1. In DD ENV1996–1–1 the concept of the stiffened edge falls somewhere between the BS 5628: Part 1 concept of simple and enhanced resistance. If the walls are similar and they are either bonded together or otherwise adequately connected, an edge can be considered as stiffened. The stiffening wall needs to be at least one-fifth of the storey height in length

and be at least 0.3 times as thick as the supported wall with an absolute minimum of 85 mm. The latter value is boxed but has been confirmed in the UK NAD.

Reduction factors ρ_3 and ρ_4 are given for the cases of walls with one or both sides stiffened and in each case there are different provisions for short and long walls. In both cases the reduction factor (ρ_3 or ρ_4) is related to the height/length ratio of the wall and to ρ_2. The formulae are depicted graphically in Annex B, which gives curves for both $\rho_2 = 0.75$ and $\rho_2 = 1.0$.

20.4.2.7 Effective thickness of walls

Clause 4.4.5 deals with effective thickness; in most cases the effective thickness of a wall is taken to be the actual thickness. However, if the wall is weakened by the presence of chases and recesses greater in size than those permitted in Table 5.3, then either the residual thickness should be used or alternatively the chase or recess should be considered as a free edge to the wall. Similarly, if there are openings greater in height or width than one quarter of the wall height or width respectively and having an area greater than one tenth of that of the wall, they should be considered as providing a free edge.

In the case of cavity walls, Clause 4.4.5(2) permits the effective thickness to be calculated using the equation

$$t_{ef} = \sqrt[3]{t_1^3 + t_2^3} \tag{4.17}$$

where t_1 and t_2 are the thicknesses of the two leaves. This formula determines a thickness such that the second moment of area about its axis is the sum of the second moments for the two individual leaves about their axes. This is conservative for those walls where the tying is such that the wall really behaves as a composite, as no component is included for the second moment of area of each leaf about the axis of the wall. The clause makes no qualification about the relative stiffness of the leaves, but the next clause, 4.4.5(3), states that, when the effective thickness would be overestimated if the loaded leaf has a higher E value than the other leaf, then the relative stiffness should be taken into account in calculating t_{ef}. This might be done by modifying Eqn (4.17) to be

$$t_{ef} = \sqrt[3]{t_1^3 + \frac{E_2}{E_1} t_2^3}$$

where E_2 is the modulus of elasticity of the unloaded leaf and which is lower than that for the loaded leaf, E_1.

The results obtained from using Eqn (4.17) are given for some common U.K. unit thicknesses in Table 20.17.

Table 20.17 Effective thickness of cavity walls made from leaves of different thickness using DD ENV 1996–1–1

Leaf thickness (1) mm	Leaf thickness (2) mm			
	90	100	140	200
90	113	120	151	205
100	120	126	155	208
140	151	155	176	220
200	205	208	220	252

In the case where only one leaf of a cavity wall is loaded, the effective thickness may be enhanced using **Eqn (4.17)** but only if the ties have sufficient flexibility that the loaded leaf is not adversely affected by the unloaded leaf (**Clause 4.4.5(4)**).

DD ENV1996–1–1 gives no guidance on the effective thickness of walls stiffened by piers. The UK NAD refers the user to BS 5628: Part 1.

20.4.2.8 *Eccentricity at right angles to the wall*

20.4.2.8.1 Calculation of structural eccentricity (Clause 4.4.7.1 and Annex C)

ENV 1996–1–1 requires that structural eccentricity at right angles to the wall, or out-of-plane eccentricity as it is called, shall be assessed. This is stated as a Principle in **Clause 4.4.7.1(1)P**.

Specific guidance on how to do the assessment is not given, but **Clause 4.4.7.1(2)** permits calculation of the eccentricity from the material properties given in **Section 3**, the joint behaviour and the principles of structural mechanics.

A full or partial frame analysis may be used provided that it takes into account that excessive wall moments will need to be redistributed. DD ENV 1996–1–1 implies that they should be limited to values which result in eccentricities no greater than 0.4 times the wall thickness. At joints where the moments are limited the design method given in C 1(4) may be applied, but it should be noted that there will be some rotation, which may cause a crack on the tension face of the wall. For the other joints, the analysis is likely to be conservative, unless the moments are reduced to take account of the fact that the joints are not fully rigid. There is little design guidance on reduction factors for joint fixity although one method is suggested by Hendry. The final set of reduced design moments should be in equilibrium with the applied external loads and reactions.

Informative **Annex C** gives a simplified method for calculating the out-of-plane eccentricity, and where the eccentricity becomes excessive, it permits the assumption that the resultant eccentricity is not greater than 0.4 times the wall thickness. Note that the simplified joint analysis only covers the effect of vertical loads and that the eccentricity resulting from horizontal loads on the wall is included subsequently when using **Eqns (4.8)** and **(4.10)**.

C.1(1) offers a simple cruciform model for the analysis of moments in a wall at its junction with a floor bearing onto the wall from one or both sides.

The method enables the moment at either the top or the bottom of a wall panel to be calculated. A second, separate, analysis is needed to find the moment at the other junction, unless the floor and wall members have loads, spans and stiffness that are repetitive. Note that the moments calculated from **Eqn (C.1)** are independent of the axial vertical load in the wall above the joint. The moments from one level to another of a repetitive multi-storey building will be similar, but the eccentricities will diminish as the design vertical loads increase.

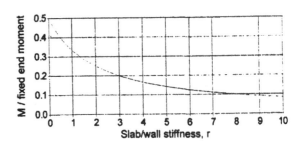

Figure 20.6 External wall moment, M, as a proportion of slab fully fixed end moment to C.1(1)

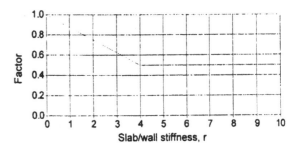

Figure 20.7 Joint fixity reduction factor 1–k/4 to C.1(2)

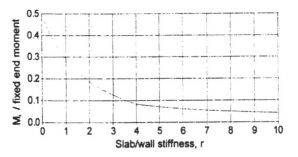

Figure 20.8 Reduced wall moment in external wall, M, as a proportion of slab fully fixed end moment

In repetitive multi-storey buildings the formulae (C.1) and (C.2) can be conveniently expressed in terms of the slab/wall stiffness ratio,

$$r = (E_{slab}I_{slab}h_{wall})/(E_{wall}I_{wall}l_{slab})$$

When the slab/wall stiffness ratio is high, the calculated values of wall moments are less sensitive to changes in the slab/wall stiffness ratio, and great accuracy is not essential in assessing the stiffness value. This can be seen from the values for an external wall as plotted in Figure 20.6.

C.1(2) takes account of the fact that full joint rigidity is not achieved in practice. It gives a formula for a reduction factor, (1–k/4), that may be applied to the wall moments. The value of joint fixity reduction factor versus slab/wall stiffness ratio is plotted in Figure 20.7

The value of k is limited to a maximum of 2, so the minimum value of the reduction factor is 0.5, which is considered by some researchers to be conservative, especially at high stiffness ratios. A value of k, limited to 3, gives a closer approximation to the research results summarised by Hendry and in the design method given in BDA Engineering File Note No.3. The reduced external wall moment values thus calculated from Clauses C.1(1) and (2), are shown in Figure 20.8 for the joints in a repetitive multi-storey structure for a range of slab/wall stiffness ratio, for comparison with Figure 20.6.

Although the calculated moments are independent of the load in the wall, the final values may need to be reduced further at some lightly loaded joints. At low values of design vertical load the calculated eccentricity can become excessive and the assumption of uncracked sections, rigid joints and elastic behaviour of the materials has to be called into question. C.1(2) and (3) imply that the method is only valid where the average design vertical stress across the thickness of the wall is greater than 0.25 N/mm². In many cases this limit results in a sudden and unjustified transition from a calculated eccentricity value to an assumed resultant eccentricity value of 0.4 times the wall thickness, as given in C.1(4). At low stress levels the values determined using Annex C are very conservative compared with practical design experience of the eccentricities calculated according to BS 5628: Part 1. Consequently, to achieve a smooth transition from the cruciform method to the simple bearing assumption of C.1(4), the UK NAD permits the use of the values calculated from Eqns (C.1) and (C.2) at average design vertical stress levels less than 0.25 N/mm² in circumstances where slight joint rotation, leading to slight cracking, can be accepted.

In Eqn (C.1) the short term modulus of elasticity should be used for floor members to correspond with the use of short term values as recommended for the masonry. When calculating the second moments of area for use in Eqn (C.1) it is a conservative assumption to ignore wall openings, but any factors which reduce the second moment of area of the floors should be taken into account.

The critical load case for the vertical load design of many walls is that under permanent and variable actions using the unfavourable effect partial safety factors given in Table 2.2. The method in C.1(1) uses unfavourable effect values in all cases. However, where loads on adjacent spans or levels differ, worse design combinations of wall moments and vertical loading can be found. To be consistent with Clause 2.3.1, the design should be sufficiently generous to allow for these cases. For example in slender walls, where the design is governed by the moments in the middle fifth of the height of the wall, a more onerous case occurs when the design actions on the floor at the base of the wall panel use favourable effect partial safety factors. The vertical load is un-changed but the wall moment at the base of the panel is reduced, resulting in a larger middle-height moment and eccentricity.

The method in C.1(1) is adequate for most cases, but caution in its use is advisable where the wall panels, loads and spans in a building do not follow a simple common repetitive pattern. In such cases a more rigorous analysis is recommended. BDA Engineering File Note No. 3 gives a more rigorous model based on a paper by Haller.

20.4.2.8.2 Creep eccentricity

Eccentricity due to creep is taken into account in Eqn (4.9) for calculating the eccentricity within the middle fifth of the wall height. Outside this zone no guidance is given but it can be assumed that it reduces linearly to zero at the top and bottom of the wall.

Clause 4.4.3(2) permits the creep eccentricity to be taken as zero for all walls with a slenderness ratio up to 15 and for more slender walls built with clay masonry units or natural stone masonry units.

The value of eccentricity due to creep is calculated from Eqn (4.11), which uses the final creep coefficient from Table 3.8. The table gives design values for masonry made with general purpose mortar, and Clause 3.8.4 indicates that these values may be used in the absence of test data for masonry made with either general purpose mortar or thin layer or lightweight mortar.

20.4.2.8.3 Accidental eccentricity (Clause 4.4.7.2)

DD ENV 1996–1–1 requires that, in order to allow for construction imperfections, an accidental eccentricity shall be assumed for the full height of the wall. This is stated as a Principle in Clause 4.4.7.2(1)P.

According to Clause 4.4.7.2(2) it is permissible, with an average level of Category of execution, to assume a value of $h_{ef}/450$, where h_{ef} is the effective height of the wall, calculated from Clause 4.4.4.

This accidental eccentricity is intended to take into account normal tolerances on materials and workmanship. It has no direct equivalent in BS 5628: Part 1, and it should not be confused with the provisions relating to accidental damage or with the additional eccentricity in Appendix B of BS 5628: Part 1.

The assumed value of accidental eccentricity is used in Eqns (4.8) and (4.10), and it may take a positive or negative sign to increase or reduce the absolute value of the resultant eccentricity, e_i or e_{mk}, at the particular level in the wall. For design purposes it is only meaningful to consider the case where the absolute value of the resultant eccentricity is increased.

20.4.2.8.4 Eccentricity due to horizontal loads

Eqns (4.8) and (4.10) include terms for eccentricity resulting from horizontal loads, but no guidance is given on the calculation of values. At any particular level in a wall, the structural eccentricity arising from horizontal loads can be found by dividing the design bending moment resulting from the horizontal loads by the design vertical load at that level.

However, it should be noted that Clause 4.4.3(3) states that the values of eccentricity from horizontal loads should not be applied to reduce e_i or e_m. Consequently, at each level in the wall it is sufficient to consider only the moments arising from horizontal loads where they increase the resultant eccentricity. From this it follows that some redistribution of the moments arising from horizontal loads can be beneficial in keeping the increase in resultant eccentricity as small as possible.

20.4.2.9 Chases, recesses and openings

Clause 4.4.2(7) notes that chases and recesses reduce the loadbearing capacity of walls. Their effect may be assumed to be insignificant if they fall within the limits given in **Clause 5.5** which in turn refers to **Table 5.3** and **Table 5.4**. **Table 5.3** provides considerable detail on the size of vertical chases and recesses in masonry units allowed without calculation whilst **Table 5.4** gives similar information for horizontal and inclined chases. Where there are chases and recesses outside these limits part of **Clause 4.4.2** and the whole of **Clause 4.4.4** provide design guidance. This can be summarised as follows.

(i) When walls have openings with a clear height of more than 1/4 of the storey height or a clear width of more than 1/4 of the wall length or a total area of more than 1/10 of that of the wall, the wall should be considered as having a free edge at the edge of the opening for the purposes of determining the effective height.

(ii) If a wall is weakened by vertical chases and/or recesses outside the limits given in **Table 5.3**, the reduced thickness of the wall should be used for t, or a free edge should be assumed at the position of the vertical chase or recess. A free edge should always be assumed when the thickness of the wall remaining after the vertical chase or recess has been formed is less than half the wall thickness.

(iii) Horizontal or inclined chases should be treated either as openings passing through the wall or, alternatively, the strength of the wall should be checked at the chase position taking account of the load eccentricity relative to the residual wall thickness. The UK NAD permits a chase within 5° of the vertical to be assumed to be vertical.

(iv) As a general guide, the reduction in vertical loadbearing capacity may be taken to be proportional to the reduction in cross sectional area due to any vertical chase or recess, provided the reduction in area does not exceed 25%.

20.4.3 Concentrated loading

20.4.3.1 General

When only a small central area of a pier is loaded, the capacity of the pier has been shown to increase several fold. The masonry beneath the loaded area develops a state of triaxial stress which enhances its ultimate strength compared with that of a uniformly loaded pier. Similar enhancement occurs when load is applied to only a short length of a wall, but other factors may in that case become significant.

20.4.3.2 Design according to DD ENV 1996–1–1

The design of masonry walls subjected to concentrated vertical loads in DD ENV 1996–1–1 requires a greater number of parameters to be considered than is required by BS 5628: Part 1. As a result of a significant amount of recent research work and studies on work carried out previously, DD ENV 1996–1–1 has attempted to optimise the key parameters affecting the bearing strength enhancement factor.

The main parameters that influence the magnitude of the enhancement factor are:

(a) The ratio of the area of concentrated load to the total cross-sectional area of the wall.
(b) The proximity of the concentrated load to the wall ends.
(c) The position of the concentrated load in relation to the wall-edge, given in terms of the eccentricity 'e' of the load.

Other factors that also require consideration in design are:

(d) The type of masonry unit.
(e) Any spreader beam beneath the load.
(f) Multiple loads.

Clause 4.4.8 has one principle to address, which is that, under the ultimate limit state, the design load resistance of an unreinforced wall subjected to concentrated loads shall be greater than the design concentrated load on the wall. Unlike BS 5628: Part 1, DD ENV 1996–1–1 uses a single equation to develop the enhancement factor, but different rules apply depending upon the type of masonry unit used.

For walls built with Groups 2a, 2b and 3 masonry units and when shell bedding is used the design compressive stress should not exceed f_k/γ_m locally under the bearing of a concentrated load, that is there is no strength enhancement.

For a wall built with Group 1 masonry units, the strength enhancement factor is given by:

$$(1 + 0.15x)\ (1.5 - 1.1\ A_b/A_{ef})$$

However, the factor should not be taken to be less than 1.0. nor greater than:

1.25   where $x = 0$ and
1.5   where $x = 1.0$

Linear interpolation is permitted for values of x between 0 and 1.0.

The parameters of the equation are shown on Figure 20.9 and are as follows:

$$x = \frac{2a_1}{H}$$

but not greater than 1.0
where:

a_1 is the distance from the end of the wall to the nearer edge of the bearing area

H is the height of the wall to the level of the load

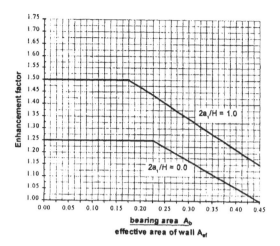

Figure 20.9 Enhancement factor for concentrated loads

A_b is the bearing area, taken to be not greater than 0.45 A_{ef}

A_{ef} is the effective area of the wall

L_{ef} is the effective length as determined at the mid height of the wall or pier

t is the thickness of the wall, taking into account the depth of recesses in joints greater than 5 mm

A number of other rules apply to this clause:

(i) The eccentricity of the load from the centre line of the wall should not be greater than $t/4$.

(ii) In all cases, the effect of the concentrated loads should be checked at the middle height of the wall below the bearings, including the effects of any other superimposed vertical loading, particularly for the case where concentrated loads are sufficiently close together for their effective lengths to overlap.

(iii) The concentrated load should bear on a Group 1 unit or other solid material of length equal to the required bearing length plus a length on each side of the bearing based on a 60° spread of load to the base of the solid material; for an end bearing the additional length is required on one side only. Note that the angle of load spread is different from that assumed in BS 5628.

(iv) Where the concentrated load is applied through a suitable spreader beam of width t, height greater than 200mm and length greater than three times the bearing length of the load, the enhancement factor beneath the loaded (not spreader) area should not exceed 1.5.

20.4.4 Unreinforced masonry shear walls

20.4.4.1 General

The horizontal forces to which buildings are subjected, for example due to wind, are resisted by the walls upon which they are applied; these walls are in turn supported by floors acting as diaphragms and walls acting as buttresses, all of which eventually transfer those forces to

the foundations. The verification of shear walls is very simple and is based on the use of the characteristic shear strength as derived from **Clause 3.6.3** over the length of the buttress that is deemed to be in compression. If there is tension at the base of the wall, as there might be close to the supported wall, the area subjected to tension is excluded from the calculation. The important consideration for the designer is the correct interpretation of the length of the wall which is effective as a buttressing wall. Openings in the shear wall and reductions in thickness will reduce the shear capacity. It may be appropriate to consider a shear wall as having a maximum length limited by the distance to any opening and, unless chases or recesses are smaller than described in **Clause 5.5**, it may be easiest to work using the reduced wall thickness or treat them as openings. It is also possible that where there is sufficient precompression, a length of the supported wall or any other intersecting wall may be considered to act together with the shear wall and restrictions on the length of what acts as a flange are provided. Caution needs to be exercised in certain situations where, for example, the shear force is being resisted by shear walls which are not symmetrically laid out in relation to the line of action of the force, and hence some rotation might be introduced, and also where floors are not able to act as rigid diaphragms, for which some limited guidance is given.

20.4.4.2 Analysis of shear walls

There are several combinations of design loads that need to be considered in order to determine which is the worst case. In all cases it is the design vertical and horizontal loads applied to the overall structure which are considered. This subject is covered in **Clause 4.5.2**.

The case where the maximum vertical load is applied to the shear wall needs to be considered, particularly if any bending is taking place, since this may introduce tension and hence reduce the length of wall available to resist shear stresses. The situation where the minimum vertical load is applied to the shear wall needs to be considered in relation to the maximum shear load on the wall, since this is the case when the frictional component of the shear resistance is at its least. If the area of any intersecting wall acting as a flange is to be considered, it is important to verify that the connection between the walls is adequate to resist the vertical shear forces being transferred across it.

Clause 4.5.2(5) states that, when designing a wall to resist the maximum horizontal load calculated using a linear elastic analysis, its stiffness may have been reduced by cracking and then, providing the other elements resisting that shear force and the connections to them are adequate to resist it, 15% of the load on the wall in question may be redistributed to those other elements.

20.4.4.3 Verification of shear walls

Verification of shear walls is covered in **Clause 4.5.3**. The length of the wall that may be considered in resisting the shear force is calculated, bearing in mind any openings or recesses. Similarly, the thickness of the wall that is to be taken as resisting the shear force is determined. The wall is analysed in vertical bending using a triangular stress distribution and from this the length of the wall which is in compression is calculated. The area of the wall resisting the shear force is taken as the product of the length in compression and the thickness of that part of the wall. The shear resistance is then taken as the product of the area and the design shear stress, f_{vk}/γ_m where f_{vk} is based on the design average compressive stress in the compressed area.

In verifying that the connections between elements are able to transfer the vertical shear stresses across them, the shear strength at zero precompression f_{vko}/γ_m may be used unless there is more relevant information, either specific to the project, for example from tests, or from a National database.

Clause 4.5.3(8) states that if the walls are satisfactorily verified at the ultimate limit state, they may be assumed to satisfy the serviceability limit state.

20.4.5 Unreinforced walls subjected to lateral loads

20.4.5.1 General

The design of masonry panels to sustain lateral wind loads is not carried out in all European countries, as many regions traditionally construct walls which are sufficiently thick to provide adequate transverse strength. Design is only deemed necessary in countries where walls subjected to predominantly lateral loading may be of insufficient thickness to provide adequate lateral resistance. In some of the localities subject to wind blown winter rainfall relatively slender cavity wall construction has been adopted as an alternative to the thicker walls utilised elsewhere. These walls need to be designed to sustain the high wind loads to which they are subjected. In addition, failures of freestanding walls subjected to similar lateral wind loads has highlighted the need for single skin walls under these conditions to be adequately designed. So, due to the wide variety of conditions and building practices which exist across the EU, only general design procedures for laterally loaded wall panels are given. The lateral loading design method is deemed to be appropriate only when loads are distributed, and of short duration – that is excluding seismic, impact and concentrated loadings. Local experience would then be utilised in specific regions.

Clause 4.6.1 deals with general aspects of the design of wall panels subjected to lateral loading. Precise methods are not available but two approximate techniques are recommended. The first method utilises coefficients dependent on the support conditions of the panels, the second approach is based on the arching action of a wall between supports.

20.4.5.2 The effect of chases and recesses

Clause 4.6.1 also considers the influence of chases and recesses in reducing the flexural strength of walls.

Provided the chases and recesses fall within specified limits their effect may be ignored. Outside these predetermined limits the flexural strength of the wall should be checked at the location of the chase or recess using the reduced thickness of the wall.

Further detail on this matter is given in **Clause 5.5** which requires that chases and recesses do not impair overall wall stability. Where walls are constructed using shell units, chases and recesses should not exceed half the shell thickness unless verified by calculation. If lintels or other structural elements are included in a wall then chases and recesses should not pass through them and, unless specifically allowed for by the designer, they should not be included in reinforced masonry. With cavity walls the requirement is that each leaf be examined separately if chases or recesses are included.

Clause 5.5 provides specific limitations for vertical chases and then for horizontal or inclined chases. Their depth should be taken as the chase or recess depth plus the depth of any void reached within the unit. DD ENV 1996–1–1 positively advises against horizontal or inclined chases or recesses, but where essential it recommends they be positioned within one eighth of the storey height, above or below floor level. The UK NAD notes that a chase within 5 degrees of the vertical may be assumed vertical. The size of chases and recesses allowed in masonry units without calculation are given in **Tables 5.3 and 5.4**. The notes on these tables provide additional information. The UK NAD however requires that where note 5 from **Table 5.3** applies to walls exceeding 2.0 m in length, the cumulative width limit should be applied to any 2.0 m length of wall. Also in walls less than 175 mm in thickness, local widening of chases to a maximum of 150 mm wide by 100 mm high is permitted so long as the length of the wall is at least 900 mm and not more than one such widening is made in any 900 mm length of wall. In **Table 5.4**, where the length of the chase is less than, or equal to 1250 mm, the limiting length should be taken as either the length of the chase or half the length of the wall, whichever is smaller.

20.4.5.3 The effect of damp proof courses

Clause 4.6.1 requires that where damp proof courses are used in walls allowance should be made for any effect on the flexural strength. **Clause 5.6** elaborates on this, requiring that damp proof courses be capable of transferring the horizontal and vertical design loads without suffering or causing damage and that they should have sufficient surface frictional resistance to prevent movement of the masonry resting on them.

20.4.5.4 Support conditions and continuity

Clause 4.6.2.1 requires support conditions and wall continuity to be taken into account when assessing the lateral resistance of masonry walls. The reaction along the edge of a wall due to a design load may normally be assumed to be uniformly distributed when designing the means of support. Restraint at a support may be provided by ties, bonded masonry returns, floors or roofs.

20.4.5.5 Cavity wall construction

With cavity construction, Clause **4.6.2.1** allows full continuity over a support to be assumed even if only one leaf is continuously bonded across the support, provided the following conditions from **Clause 5.4.2.2** are met:

(i) Both leaves of the cavity wall are effectively tied together.

(ii) The number of wall ties connecting the two leaves together is not less than either 2 ties per m², or those calculated from Clause 4.6.2.4. That clause requires the number of ties per m² to equal or exceed

$$\gamma_m \frac{W_{sd}}{F_t}$$

where

W_{sd} is the design horizontal action from wind per unit area to be transferred.

F_t is the characteristic compressive or tensile resistance of a wall tie, as appropriate to the design condition, determined by tests in accordance with EN846–4, EN846–5 or EN846–6.

γ_m is the partial safety factor for walls taken from **Clause 2.3.3.2**. For all categories of execution γ_m of 2.5 may be used in this application.

When bed joint reinforcement is used to connect two leaves together, each tying element should be treated as a wall tie.

(iii) Wall ties should be corrosion resistant for the relevant exposure class (see **Clause 5.5.5.5**) for the wall.

(iv) Wall ties should be provided at a free edge to connect both leaves together.

(v) Where an opening penetrates a wall and the frame for the opening is not capable of transferring the horizontal design action directly to the structure, those wall ties which would have been placed in the opening should be redistributed uniformly along the vertical edges of the opening.

(vi) Due allowance should be made in the selection of the wall ties for any differential movement between the leaves, or between a leaf and a frame.

(vii) In seismic regions special consideration is necessary as specified in ENV 1998.

Clause 4.6.2.1 goes on to recommend that, where the leaves are of different thicknesses, the thicker leaf should normally be the continuous leaf unless it is clear that full continuity can be assumed from the stiffness and strength of a continuous thinner leaf. The clause then explains that the lateral load to be transmitted from a panel to its support may be assumed to be transferred by the ties to the one leaf which is continuous across a support, provided there is adequate connection between the two leaves as noted above, particularly along the edges of the panels. In all other cases partial continuity may be

assumed. Clause 4.6.2.1(4) permits the design lateral strength of a cavity wall to be taken as the sum of the design lateral strengths of the two leaves, provided that the wall ties or other connectors between the leaves are capable of transmitting the actions to which the cavity wall is subjected.

20.4.5.6 Method of design for a wall supported along edges

Clause 4.6.2.2 gives two formulae which may be used to calculate M_d the design moment of a wall. It is implicit in these formulae that masonry is a non-isotopic material, and there is an orthogonal strength ratio depending on the unit and mortar used. When the plane of failure is perpendicular to the bed joints, that is in the f_{xk2} direction,

$$M_d = \alpha W_k \gamma_f L^2 \quad \text{per unit height of the wall.} \qquad (4.24)$$

When the plane of failure is parallel to the bed joints, that is in the f_{xk1} direction

$$M_d = \mu \alpha W_k \gamma_f L^2 \quad \text{per unit height of the wall.} \qquad (4.25)$$

Where

γ_f is the partial safety factor for actions on building structures obtained from **Clause 2.3.3.1**. Boxed values for the partial safety factors should be substituted by those given in the UK NAD.

μ is the orthogonal ratio of the characteristic flexural strengths of the masonry, f_{xk1}/f_{xk2}.

α is a bending moment coefficient which depends on the orthogonal ratio μ, the edge fixity and the height to length ratio of panels*.

L is the length of panel between supports.

W_k is the characteristic wind load per unit area.

20.4.5.7 Effect of vertical loads

Clause 4.6.2.2(3) allows the orthogonal strength ratio to be altered when a vertical load acts to increase the flexural strength f_{xk1} in that direction. When this occurs a modified flexural strength of $f_{xk1} + \gamma_m \sigma_{dp}$ is permitted.

Where

f_{xk1} is the characteristic flexural strength with the plane of failure parallel to the bed joints obtained as described above from **Clause 3.6.4**.

γ_m is the partial safety factor for the material, obtained from **Clause 2.3.3.2**.

σ_{dp} is the permanent vertical stress on the wall at the level under consideration.

20.4.5.8 Design moment of lateral resistance of a masonry wall

Clause 4.6.2.2 gives the design moment of lateral resistance, M_{Rd}, as

---

* According to **Clause 4.6.2.2** this coefficient must be determined from an appropriate theory; the UK NAD recommends that values from Table 9 of BS 5628: Part 1 be used.

$$M_{RD} = f_{xk} \frac{Z}{\gamma_m} \qquad (4.27)$$

Where

f_{xk} is the characteristic flexural strength of masonry determined in accordance with **Clause 3.6.4**, appropriate to the plane of bending.

Z   is the section modulus of the wall

Where piers are built into a wall, DD ENV 1996-1-1 recommends the outstanding length of flange from the face of the pier should be taken as:

h/10 for walls spanning vertically between restraints;
2h/10 for cantilever walls
never more than half the clear distance between piers

where h is the clear height of the wall.

Clause 4.6.2.2(7) states that walls that satisfy the ultimate limit state when verified in accordance with **Eqns (4.24)**, **(4.25)** and **(4.27)** may be deemed to satisfy the serviceability limit state.

Clause 4.6.2.2 notes that, in laterally loaded panels or freestanding walls built of masonry set in mortar designations M2 to M20, and designed in accordance with this Clause, the dimensions should be limited to avoid undue movements resulting from deflections, creep, shrinkage, temperature effects and cracking.

20.4.5.9 Movement and movement joints in laterally loaded walls

Clauses 4.1.3 and **5.7** refer to the avoidance of undue movement resulting from deflections, creep, shrinkage, temperature effects and cracking; this is referred to in **Clause 4.6.2.2(8)**.

Clause 4.1.3(2) recommends structures be designed so that cracks or deflections which might damage facing materials, partitions, finishes or technical equipment or which might impair watertightness are avoided or minimised.

Clause 5.7 requires that the effect of movement in masonry elements of a structure should not adversely affect the performance of the structure. This, the clause explains, can be achieved by providing vertical and horizontal movement joints to allow for the effects of thermal and moisture movement, creep and deflection and the possible effects of internal stresses caused by vertical or lateral loading, so that the masonry does not suffer damage. Designers are referred to **Table 3.8** where information on the creep, moisture expansion, shrinkage and thermal expansion of unreinforced masonry made with general purpose mortar is provided. **Table 3.8** examines the following types of masonry units: clay, calcium silicate, dense aggregate concrete, manufactured stone, lightweight aggregate concrete, autoclaved aerated concrete and natural stone. For each of these unit types, a range of possible as well as a recommended design value is given for the final creep coefficient, the final moisture expansion or shrinkage and the coefficient of thermal expansion. Some of the boxed values given in **Table 3.8** must be altered in accordance with the UK NAD.

In determining the maximum spacing of vertical movement joints, **Clause 5.7(3)** notes that special consideration should be given to the effects of the following:

- the drying shrinkage of calcium silicate units, aggregate concrete units, autoclaved aerated concrete units and manufactured stone units;
- the irreversible moisture expansion of clay units;
- variations in temperature and humidity;
- insulation provided to the masonry;
- the provision of prefabricated bed joint reinforcement.

Clause 5.7(4) states that precautions should be taken to allow for vertical movement of external walls. The uninterrupted height between horizontal movement joints in the outer leaf of external cavity walls should be limited to avoid the loosening of the wall ties.

The UK NAD provides that, pending publication of ENV 1996-2, **Clauses 5.7(3)** and **(4)** will be satisfied by following the recommendations of **Clause 20.3** in BS 5628: Part 1.

Finally, **Clause 5.7(5)** notes that the width of vertical and horizontal movement joints should allow for the maximum movement expected and, if expansion joints are to be filled, then they should be filled with an easily compressed material.

20.4.5.10 Method for design of arching between supports

Where walls are firmly restrained within their plane, arching action will occur if the panel is subject to bending stresses.

Clause 4.6.2.3(1) states that when a masonry wall is built solidly between supports capable of resisting an arch thrust, or when a number of walls are built continuously past supports, the wall may be designed assuming that a horizontal or vertical arch develops within the thickness of the wall. The paragraph notes that walls subjected to mainly lateral loads should be designed to arch horizontally but vertical arching may be included when accidental actions are considered.

Clause 4.6.2.3(2) notes that calculation should be based on a three pin arch and the bearing at the supports and central pin should be assumed as 0.1 times the thickness of the wall.

Clause 4.6.2.3(3) explains that the arch thrust should be assessed from knowledge of the applied lateral load, the strength of the masonry in compression and the effectiveness of the junction between the wall and the support resisting the thrust. It cautions designers to note that a small change in length of a wall in arching can considerably reduce the arching resistance so care should be taken if the masonry is built of units which may shrink in service.

Clause 4.6.2.3(4) gives the arch rise as

$$0.9t - d$$

where:

t   is the thickness of the wall, and

d   is the deflection of the arch under the design lateral load which may be taken as zero for walls having a length to thickness ratio of 25 or less.

Reference is made to **Clause 6.5.3(2)**, which does not exist. It is **Clause 4.5.3(2)** that is intended. The wall thickness should be taken as the residual thickness, discounting the part of the wall that is removed by the chase or recess. **Tables 5.3** and **5.4** are not applicable in this instance.

Clause 4.6.2.3(5) assumes the maximum design and thrust per unit length of wall is

$$1.5 \frac{f_k}{\gamma_m} \frac{t}{10}$$

and where the lateral deflection is small the design lateral strength is given by

$$q_{lat} = \frac{f_k}{\gamma_m} \left(\frac{t}{L} \right)^2$$

Where:

q_{lat} is the design lateral strength per unit area of wall
t   is the thickness of the wall, not reduced by vertical chases
f_k is the characteristic compressive strength of the masonry
L   is the length of the wall
γ_m is the partial safety factor for the material.

Clause 4.6.2.3(6) concludes the design procedure by allowing that walls which satisfy the ultimate limit state when verified using the equation for design lateral strength given in **Clause 4.6.2.3(5)** may be deemed to satisfy the serviceability limit state.

20.4.6 Reinforced masonry

20.4.6.1 Members subjected to bending, bending and axial load or axial load

20.4.6.1.1 General
The use of reinforcement in masonry is not new and was indeed used by Brunel on the Wapping to Rotherhithe tunnel under the River Thames. There has been an increase in interest in the U.K. in the use of reinforced masonry, in particular for retaining wall structures. The level of interest in continental Europe is generally lower than in the UK, other than in seismic areas, and the approach of DD ENV 1996–1–1 is based more on principles than on detail. The guidance in this section is not as extensive as in BS 5628: Part 2 and so, for example, elements subjected to axial load and bending are considered to be another application of the basic principles and their design is not explained in detail.

The basic assumptions which are made are similar to those for reinforced concrete design and indeed as adopted in the UK for reinforced masonry design. The assumptions are:

*   plane sections remain plane;
*   the reinforcement is subjected to the same variations in strain as the adjacent masonry;

*   the tensile strength of the masonry is zero;
*   the maximum compressive strain of the masonry is chosen according to the material;
*   the maximum tensile strain in the reinforcement is chosen according to the material;
*   the stress-strain relationship of masonry is taken to be parabolic, parabolic rectangular or rectangular (see **Clause 3.8.1(2)**);
*   the stress-strain relationship of the reinforcement is derived from **Figure 4.6**:
*   for cross-sections subject to pure longitudinal compression, the compressive strain in the masonry is limited to –0,002 (see **Figure 3.3**);
*   for cross-sections not fully in compression, the limiting compressive strain is taken as –0.0035 (see **Figure 3.3**). In intermediate situations, the strain diagram is defined by assuming that the strain is –0.002 at a level 3/7 of the height of the section from the most compressed face (see **Figure 4.9**).

20.4.6.1.2 Effective and limiting spans
The effective and limiting spans in DD ENV 1996–1–1 are the same as those in BS 5628: Part 2 and the limitations are introduced for the same reasons, which are to ensure that deflections and cracking will be within acceptable limits.

20.4.6.1.3 Slenderness ratio of vertically loaded elements
The limiting slenderness ratio is 27 which, although boxed, is the same as in BS 5628: Part 2: designers should consider whether the more restrictive limitation of 18, which is for cantilever walls or columns in BS 5628: Part 2 and also in DD ENV 1996–1–1 for laterally loaded walls, should be applied.

20.4.6.1.4 Flanged members
Guidance is given on the breadth of section to be considered in the design of elements where the reinforcement is concentrated locally. In the case of T shaped sections the guidance is similar to that in BS 5628: Part 2, except the additional width defined as the actual width of the flange is explicitly mentioned. In addition, guidance on widths for L shaped sections is included, the limitations being based on half of those for T shaped sections.

20.4.6.1.5 Verification of reinforced masonry members subjected to bending and/or axial load
The explanation of the design approach for reinforced masonry subjected to bending and/or axial load is somewhat theoretical in that a very general cross section is illustrated together with an indication of all of the strain distributions which are possible and consistent with the assumptions made about sections loaded axially with and without applied moment.

For most situations it will be possible to use a simplified approach to determine the moment of resistance of a section: this assumes a rectangular compression stress block, the force in which balances that in the reinforcement. The compressive strain in the masonry is limited to 0.0035 and, as plane sections are assumed to remain plane and the strain in the steel is what is needed to generate the necessary stress, the neutral axis is then fixed. This approach has been used very successfully in the UK where the depth of the neutral axis has been limited to half the effective depth. If the parabolic stress strain relationship is used then the equivalent uniform compressive stress in the rectangular stress block would be 0.75 f_k/γ_m. However, there is considerable experimental evidence to justify the use of f_k/γ_m. There is limited experience of the use of the simplified stress block approach in Europe as a whole and consequently, although f_k/γ_m has been adopted for the stress in the compressive stress block, the depth of the neutral axis has been limited to 0.4 times the effective depth and the depth of the compression block as 0.8 of this depth. This approach, which is aimed at eliminating the risk of generating over-reinforced sections, is quite conservative and restricts the moment which can be resisted to:

$$0.256 \; \frac{f_k}{\gamma_m} b \; d^2$$

In the case of cantilever retaining walls the designer is referred to Appendix F, where it is acknowledged that the UK experience with retaining wall design is the most extensive and relevant and although the principle of the simplified design approach is retained the maximum bending moment is increased to:

$$0.4 \;\; \frac{f_k}{\gamma_m} \; b \; d^2$$

The simplified approach is based upon the use of the formulae:

$$M_{Rd} = A_s \frac{f_{yk}}{\gamma_s} Z \tag{4.36}$$

and

$$z = d \left(1 - 0.5 \; \frac{A_s}{b} \; \frac{f_{yk}}{d} \; \frac{\gamma_m}{f_k} \; \frac{\gamma_m}{\gamma_{ms}} \right) \le 0.95d \tag{4.37}$$

where the design moment of resistance of the section is limited as above.

A simple iterative approach is to estimate the steel area needed using Eqn (4.36) taking z as the limiting value which is for the balanced section in either the general approach, which is 0.84d, or the cantilever wall approach, which is 0.72d. The value of A_s determined can then be used in Eqn (4.27) to obtain a second estimate of z to be further substituted in Clause 4.36 to obtain a better estimate of A_s. A limited number of such iterations should be sufficient to obtain sufficiently close successive estimates.

An alternative is to express M_{Rd} as a function of Q and b and d.

$$M_{Rd} = Q \; b \; d^2$$

Where Q is given by

$$Q = 2 \frac{z}{d} \left(1 - \frac{z}{d} \right) \frac{f_k}{\gamma_m}$$

The relationship between Q, f_k/γ_m and z/d, called c, can be expressed graphically, as shown in Figure 20.10.

The designer can establish the required value of Q and can then, within the permitted ranges of z/d, determine values of z/d and f_k/γ_m which are consistent. Once these are fixed the required steel area can be found from:

$$A_s = M_{Rd} \; \frac{\gamma_s}{f_{yk}} \; \frac{1}{z}$$

This approach allows some flexibility in the choice of f_k value in a tentative stage in the design.

20.4.6.2 Reinforced masonry members subjected to shear

20.4.6.2.1 General
The analysis of reinforced masonry shear walls is the same as for unreinforced walls. The basic principle is that the design shear resistance, which can include that from the basic section, ignoring the effect of shear reinforcement, together with that from the shear reinforcement, provided that the minimum recommended amount of steel is included, must exceed the design shear load.

20.4.6.2.2 Verification of members ignoring shear reinforcement
In this case it should be verified that:

$$V_{Sd} \le V_{Rd1}$$

Where

$$V_{Rd1} = f_{vk} \; \frac{b \; d}{\gamma_m}$$

There is a note to indicate that where required the characteristic shear strength can be enhanced to allow for the presence of longitudinal reinforcement. In the UK there have been two approaches to this provision, both supported by experimental data, which are relevant to the case where bars are in bed or vertical joints surrounded by mortar and where they are in pockets, cores or cavities, surrounded by concrete infill. In the former case, provided the mortar is of strength class M5 or greater, f_{vk} may be taken as 0.35 N/mm². The effect of increasing the percentage of reinforcement in this sort of element has been demonstrated as being low. However, where the reinforcement is in pockets, cores or cavities and is surrounded by concrete infill, increasing the percentage of reinforcement is effective in increasing the shear resistance which may be taken to be

$$f_{vk} = 0.35 + 17.5\rho$$

Where

$$\rho = \frac{A_s}{b \; d}$$

provided that f_{vk} is not taken to be greater than 0.7 N/mm².

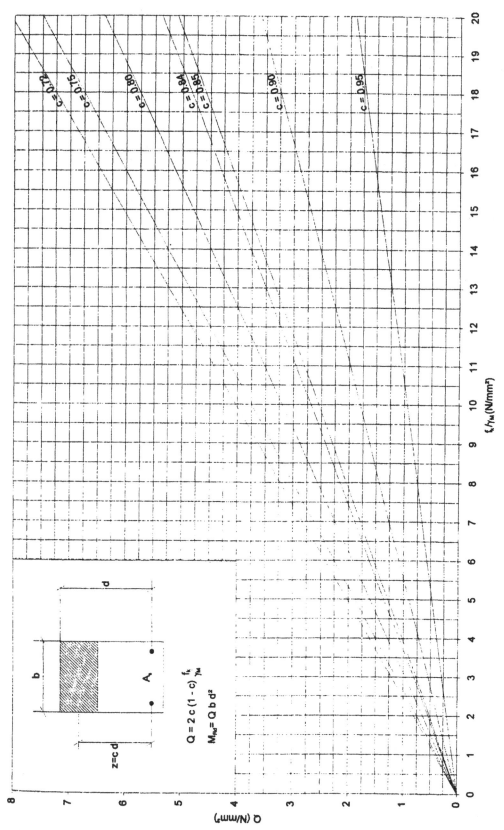

Figure 20.10 Moment of resistance factor, Q

Where simply supported beams or cantilevers have a ratio of shear span to effective depth of less than 2, then the f_{vk} value may be enhanced by a factor up to 4 to reflect the fact that such short shear spans act as struts reacted at a support. This is consistent with BS 5628: Part 2.

20.4.6.2.3 Verification of members taking into account steel reinforcement

In this case the basic shear resistance of the section is enhanced by the resistance that can be attributed to the reinforcement. The approach follows closely that for reinforced concrete in ENV 1992 and this is more appropriate to the use of bent up bars in reinforced concrete sections than the more commonly used arrangement of stirrups in reinforced masonry elements. It is of course possible to use bent up bars in grouted cavity beams but this is not common, consequently in most cases the angle $\propto$ in Eqn (4.45) given below is 90°.

$$V_{Rd2} = 0.9d \, \frac{A_{SW}}{s} \, \frac{f_{yk}}{\gamma_s} (1 + \cot \propto) \sin \propto \qquad (4.45)$$

and the equation reduces to

$$V_{Rd2} = 0.9d \, \frac{A_{SW}}{s} \, \frac{f_{yk}}{\gamma_s}$$

20.4.6.3 Reinforced masonry deep beams subjected to vertical loading

20.4.6.3.1 General

DD ENV 1996–1–1 contains some simple and helpful guidance on the design of deep beams which are defined as those where the height of masonry above the opening is at least half the effective span, which is 1.15 times the clear span.

A lever arm is established based on either the effective span alone or a combination of the masonry height and the effective span.

20.4.6.3.2 Verification of deep beams subjected to vertical loading

As for other verifications, the design resistance must exceed the design moment applied and from this moment the steel area is determined using the steel stress and assumed lever arm. This moment is subject to the overall limit of

$$\frac{0.4 \, f_k \, b \, d^2}{\gamma_m}$$

and there are requirements to check the compressive block against buckling and to provide crack control reinforcement in the lower half of the beam. For the purposes of checking the maximum bending moment and the shear resistance at the face of a support, d is taken to be 1.25 z.

20.4.6.3.3 Composite lintels

The guidance on the design of deep beams may be applied to situations where the tension element is

provided by a prefabricated element as long as the requirements for anchorage and bearing are satisfied.

20.4.7 Prestressed Masonry

20.4.7.1 General

The structural performance of masonry can be enhanced by the application of compressive prestress. This is because although masonry has a high compressive strength its tensile strength is low. If sufficient prestress is applied the tensile stresses in the masonry can be eliminated entirely under serviceability conditions so that the compressive strength of the whole masonry cross-section can be utilised. In prestressed members, unlike reinforced members, cracking need not occur under working loads.

The fundamental principles behind prestressed masonry have a long history. In the nineteenth century new masonry was sometimes prestressed with tightened iron or steel rods and, of course, the use of prestressed rods and straps is an age old method of repairing cracked and deformed masonry buildings and chimneys. The first recorded application of prestressed masonry using modern high tensile steel prestressing tendons was in the 1950s and since then there has been a steady increase in its usage.

There are two methods of prestressing masonry:

(a) Post-tensioning
The tendons are tensioned against the masonry when it has achieved sufficient strength using mechanical end anchorages. The tendons may be:
   (i) unrestrained against lateral movement in cavities or voids in the masonry,
   (ii) restrained against lateral movement at discrete points by projecting masonry units or continuously by ducts built into the masonry,
   (iii) bonded to the surrounding masonry by grout or concrete infill.

(b) Pre-tensioning
The tendons are tensioned against an independent anchorage and released only when the masonry and/or infill concrete has achieved sufficient strength. The transfer of the prestress force to the masonry is provided by bond alone. Pre-tensioning is usually only appropriate for prefabricated products.

The most widespread use of prestressing in masonry is for the post-tensioning of vertically spanning walls and columns. The principles set out in DD ENV 1996–1–1 are directly applicable to this (**Clause 4.8.1(2)**). The structural performance of such masonry can be enhanced by constructing the masonry member with a geometric cross-section, that is a cross-section with the structural units arranged to give greater structural efficiency than that obtained from the equivalent section in a solid rectangular wall or column or a cavity wall. The geometric cross-section is proportioned to have a high section modulus (for enhanced performance at the serviceability limit state), a high effective depth (for enhanced performance at the ultimate limit state) and a high radius of gyration (for enhanced performance under axial compression), as appropriate. The designer must ensure that the cross-section has satisfactory shear strength.

DD ENV 1996-1-1 covers the design of both post-tensioned and pre-tensioned masonry prestressed in one direction only. Design is based on a consideration of the serviceability and ultimate limit states using principles established in prestressed concrete construction (**Clause 4.8.1 (1) P**). DD ENV 1996-1-1 recommends that the serviceability limit state should be assessed first in bending and then the bending, axial and shear strength should be verified at the ultimate limit state.

20.4.7.2 Prestressed masonry members under the serviceability limit state.

At the serviceability limit state there are two cases which need to be examined: at transfer of prestress and under the design loads after losses (**Clause 4.8.2.2**).

The loads and forces involved must not cause cracking on the tension side (nominally face 1) or failure on the compression side (nominally face 2).

The partial safety factor for load, γ_f, is taken as 1.0 at transfer and under the design loads (**Clause 4.8.2.2 (2) P**). The material partial safety factors, γ_m, are, in effect, replaced by compression and tension stress limitations for the masonry, and for the tendons, by limiting the jacking force to a percentage (70% in BS 5628: Pt2) of their characteristic breaking load.

The analysis of a section at the serviceability limit state, following 4.8.2.2(3)P, results in a linear distribution of stress. Taking compression as positive:

$$\sigma_{1i} = + \frac{N + P}{A_m} + \frac{Pey_1}{I}$$

at transfer:

$$\sigma_{2i} = + \frac{N + P}{A_m} - \frac{Pey_2}{I}$$

where, under working loads:

σ_{1i} is the face 1 stress at transfer
σ_{2i} is the face 2 stress at transfer
P is the prestress force
e is the eccentricity of prestress force
I is the second moment of area of cross-section
N is the axial load
A_m is the cross-sectional area of masonry
y_1 is the distance from centroid of section to face 1
y_2 is the distance from centroid of section to face 2

After losses the applied prestressing force at transfer, P, becomes the effective prestressing force, P_e, so with the addition of the design serviceability moment, M_s, and the design serviceability axial load N_s, the equations for σ_{1i} and σ_{2i} become:

$$\sigma_1 = + \frac{N + P_e + N_s}{A_m} + \frac{(P_e e - M_s) y_1}{1}$$

$$\sigma_2 = + \frac{N + P_e + N_s}{A_m} + \frac{(P_e e - M_s) y_2}{1}$$

where

σ_1 is the face 1 stress at design load
σ_2 is the face 2 stress at design load.

There may be an immediate stage where M_s is applied from, say, backfill on a retaining wall and walls above are built to give N_s later (**Clause 4.8.2.2 (1)**).

Acceptable limiting values of σ_{1i}, σ_{2i}, σ_1 and σ_2 have to be chosen. For example BS 5628: Part2 requires $\sigma_{1i} < 0.4f_{ki}$, $\sigma_{2i} > 0$, $\sigma_1 > 0$ and $\sigma_2 < 0.33f_k$, where f_{ki} is the characteristic strength of the masonry at transfer.

In **Clause 4.8.2.2 (4)** DD ENV 1996-1-1 suggests that although design for flexural cracking and crushing will be satisfied by the above, a deflection check may need to be carried out. This can be done using the second moment of area I for the uncracked cross-section of the member together with the long term modulus of elasticity, E, of the masonry from **Clause 3.8.2**. The effect on deflection of prestressing is usually counter to the deflection caused by the applied loads and in some circumstances the deflection due to prestress may be larger than that due to load resulting in a net upwards deflection.

20.4.7.3 Prestressed members under the ultimate limit state

20.4.7.3.1 Partial safety factors

At the ultimate limit state the partial safety factors for actions, γ_f, are obtained from **Table 2.2**. For prestressed masonry actions are mainly of two types: those due to load and movement and those due to prestressing. For a permanent action such as earth, γ_f is taken as 1.35 if the effect is unfavourable and 1.0 if it is favourable. An unfavourable variable action, such as a traffic load, is assigned a γ_f factor of 1.5. The partial safety factors for the prestressing force, γ_f, are 0.9 if the effect is favourable and 1.2 if the effect is unfavourable.

The partial safety factors for material γ_m, (γ_s for steel), are obtained from **Table 2.3**, which covers the full range of masonry construction but only masonry unit manufacturing category I and execution category A are applicable to prestressed masonry. This means that the masonry γ_m is 1.7 and γ_s is 1.15.

40.4.7.3.2 Bending strength under the ultimate limit state

While flexural behaviour at the ultimate limit state is similar to reinforced masonry, in that, for example, the masonry in tension is cracked and the masonry in compression is in a non-linear state of stress, the analytical treatment, particularly of unbonded post-tensioned masonry, is more complicated than that for reinforced masonry. The lack of bond causes differential longitudinal movement between the tendons and the adjacent masonry and, when the voids in the masonry are large, there can be differential lateral movement also.

Using the assumptions given in **Clause 4.8.3.2(2)P**, the analytical treatment involves an interactive method which can only be satisfactorily carried out by computer. Many members, however, particularly walls, have rectangular compression zones and unbonded tendons and for these cases an empirical method is given in BS 5628: Part 2. In this method the design moment of resistance at the ultimate limit state, M_{Rd}, is based on the stress in the tendon at ultimate σ_{pu}:

$$M_{Rd} = \sigma_{pu}.A_p \left(\frac{d - x}{2} \right)$$

where:
A_p is the cross-sectional area of tendons
d is the effective depth of tendons
x is the depth of compression zone of member.

To comply with **Clause 4.8.3.2 (2) P** assumptions, in this equation σ_{pu} and x are found from:

$$\sigma_{pu} = \sigma_{pe} + 700 \frac{d}{l}\left(1 - 1.2 \frac{f_{kp}}{f_k} \cdot \frac{A_p}{Bd} \right)$$

$$x = 1.2 \frac{f_{kp}}{f_k} \cdot \frac{A_p}{B}$$

where:
σ_{pe} is the effective prestress after losses;
l is the distance between end anchorages;
f_{kp} is the characteristic strength of the tendons;
B is the width of compression zone.

The equations for σ_{pu} and x incorporate a γ_m for masonry of 1.7 and a tendon modulus of 200 kN/mm². The tendon stresses are limited to 70% of their characteristic strength, f_{kp}. With these values for σ_{pu} and x the value calculated for M_{Rd} above will satisfy **Clause 4.8.3.2 (3) P**.

20.4.7.3.3 Axial strength at the ultimate limit state
Clause 4.8.3.2 (4) requires axially loaded prestressed members to be designed as unreinforced members according to **Clause 4.4** and highlights the possible need to limit the prestress force to that which the member can carry without buckling due to slenderness. Post-tensioned members with unbonded tendons have this potential to buckle under the action of the prestress force alone when the tendons are placed in large voids in the masonry and their lateral movement is unrestricted.

20.4.7.3.4 Shear strength at the ultimate limit state
In prestressed masonry shear failure is due to the principal tensile stresses reaching the diagonal tensile strength of the masonry so forming a diagonal crack. To obviate this **Clause 4.8.3.2 (5) P** requires the design shear resistance to be greater than the design value of the applied shear load but the clause gives no guidance on how this should be done.

BS 5628: Part 2 Clause 5.2.3.1, however, sets out a design method in terms of applied shear stress and shear strength and this can be used to satisfy DD ENV 1996–1–1. This method has the advantage that the equation given for design shear strength already incorporates the partial safety factor for prestress of 0.9 required by the ENV. Furthermore, characteristic

diagonal strengths for both brick and block masonry are given in the British Standard.

20.4.7.4 Other design matters

In **Section 4.8.4** DD ENV 1996–1–1 goes on to consider the maximum initial prestressing force, anchorage bearing and bursting stresses and prestress losses. Little detail is given but, as for shear, the guidance given in BS 5628: Part 2 satisfies the ENV.

20.4.7.4.1 Maximum initial prestress and bearing stress
Regarding the level of the initial prestressing force (**Clause 4.8.4.2 (1) P**), neither the British Standard nor the ENV gives a figure. BS 5628: Part 2 does, however, specify that the jacking force should not exceed 70% of the characteristic load of the tendon (Clause 5.4.1). Since most prestressing of masonry in the United Kingdom is carried out using high tensile steel bars with nut anchorages the initial prestressing force and the jacking force can be regarded as identical so for design purposes the 70% figure can be used as the maximum permitted level of the initial prestressing force. For other anchorage systems which exhibit draw-in if the jacking force is limited to 70% of the characteristic tendon strength then the initial prestresssing force will be less than this.

BS 5628: Part 2 Clause 5.5 (detailing of prestressed masonry) gives helpful guidance on anchorage bearing and bursting stresses. In current United Kingdom practice tendon anchorage is usually made in a reinforced concrete end block which is often part of a foundation or capping beam. The reinforced concrete is designed to distribute the concentrated prestressing forces so as to keep the bearing stresses on the masonry to within the limits specified in Clause 5.5.1, that is 1.5 f_k /γ_{mm} where the prestressing loads are perpendicular to the bed joints or 0.65 f_k /γ_{mm} where they are parallel to the bed joints. The design of reinforced concrete end blocks to resist bursting forces should be carried out to BS 8110: Part 1: 1997 Clause 4.11. End blocks for prestressed masonry often have to span over voids in the masonry and in such cases they must also be designed to resist bending and shearing forces. An important aspect of this is that the tensile strain at the interface between the reinforced concrete end block and the masonry must be restricted so as not to cause tensile cracking in the masonry.

20.4.7.4.2 Loss of prestress
The clauses under 5.4.2 in BS 5628: Part 2 give detailed guidance on:

- relaxation of tendons;
- elastic deformation of the masonry;
- moisture movement of masonry;
- creep of masonry;
- tendon losses during anchoring;
- friction effects;
- thermal effects.

These sources of prestress loss are the same as those identified in DD ENV 1996–1–1 Clause 4.8.4.3. As far as post-tensioned masonry is concerned prestress loss will normally stem only from tendon relaxation, masonry moisture content changes and creep in the masonry. Change due to elastic deformation of the masonry as the prestress is being applied is automatically compensated for in post-tensioned masonry if all the tendons are stressed to the correct level at the same time because elastic deformation takes place concurrently with the stressing operation. Elastic deformation is effectively eliminated, if multiple tendons are used by the necessity of prestressing individual tendons or groups of tendons, to give the required level of stress at all points in the masonry. With the pretensioning system elastic masonry deformation at the time of the application of prestress can be significant and needs to be assessed.

Anchorage draw-in can be avoided in post-tensioned masonry if high tensile steel bar tendons with nut anchorages are used but some other tendon systems use wedges for anchorage and these are prone to draw-in at transfer. With wedges, draw-in can be up to 5 mm so a change in prestress from this source is particularly important for short walls. There will be no loss due to friction if the tendons are straight and are placed in straight voids in the masonry or in straight ducts. The results of temperature changes are reversible.

Prestress loss can always be reduced by adjusting the tendon stress at a later date. One advantage of using unbonded post-tensioned tendons is that they lend themselves very readily to this. Most advantage is gained by adjusting at an early stage while the contractor is still on site.

20.5 Structural detailing

20.5.1 Introduction

This section of DD ENV 1996–1–1, which is largely self-explanatory, gives the structural detailing rules which must be followed if the design methods in the earlier sections are to be applicable.

20.5.2 Masonry materials

Until ENV 1996–2 is published, the selection of masonry materials to provide the durability required of the wall should follow BS 5628: Part 3.

20.5.3 Minimum thickness of walls

The UK NAD sets the minimum thickness of a loadbearing wall at 75 mm, to align with the current UK practice, and that of a veneer wall at 70 mm.

20.5.4 Bonding of masonry

Clause 5.1.4 sets out bonding patterns which should suit UK practice. However, Clause 5.1.4(3) permits variations where these can be substantiated.

20.5.5 Mortar joints

Clause 5.1.5 makes it clear that the design rules do not apply to bed and perpend joints between 3 mm and 8 mm thick. This means that the design rules cannot be used with natural and manufactured stone masonry having joints of 6 mm.

The tolerance on the level of bed joints can be taken from BS 5606.

The concept that perpends can be regarded as filled, when in fact only 40% of the area is actually filled, is new to the UK. This relaxation should be used with caution. In areas of severe exposure, the designer should consider specifying *fully* filled perpends for exposed walls.

20.5.6 Protection of reinforcing steel

For the use of unprotected carbon steel, Table 5.1 contains provisions which do not follow current UK practice and which are considered to be unsatisfactory in the UK. Accordingly they have been deleted by the UK NAD.

Similarly, the UK NAD replaces the whole of Table 5.2 with the equivalent Table 14 in BS 5628: Part 2: 1995, giving the necessary equivalents between the exposure classes in the two documents to enable this to be done.

20.5.7 Minimum area of reinforcement

The UK NAD confirms the boxed values in Clause 5.2.3, with the exception that the minimum area of reinforcement to control cracking given in Clause 5.2.3 (3) is reduced to 0.01%.

20.5.8 Size of reinforcement

In the absence of EN 845–3 the UK NAD, in 5.4.2 (1), sets the minimum size of the longitudinal wires for prefabricated bed joint reinforcement, to be used in design applications, as 3 mm.

20.5.9 Anchorage and laps

There is no established UK method for determining the anchorage and lap lengths for prefabricated bed joint reinforcement. Until EN 846–2 is published the recommendations of the manufacturer should be followed (see Clauses 5.2.5.1 (8) and 5.2.5.2 (5)). The UK NAD deletes Clause 5.2.5.2 (3). The lap length between two reinforcing bars should not be less than the anchorage length required to develop full bond in the smaller of the two bars lapped, nor less than 25 × (bar diameter) plus 150 mm for tension reinforcement and 20 × (bar diameter) plus 100mm for compression reinforcement.

20.5.10 Shear reinforcement

The UK NAD acknowledges the generality of most of the guidance in Clause 5.2.6, but confirms the boxed values only for beams.

20.5.11 Confined masonry

There is no UK experience of confined masonry, and so the UK NAD offers no guidance on the boxed values in **Clause 5.2.9**. Designers contemplating the use of confined masonry should consult design guidance available in Southern Europe.

20.5.12 Connection of walls

The general guidance in **Clause 5.4** should be supplemented by the more detailed guidance and requirements in British Standards, as indicated in the UK NAD.

20.5.13 Vertical chases and recesses

There is an obvious omission in **Clause 5.5.2** that no tolerance is given on verticality. This is partially corrected in the UK NAD, which states that a chase within 5 degrees of vertical can be assumed to be vertical.

20.5.14 Thermal and long-term movement

Clause 5.7 is confined to general statements which will be amplified in EN 1996–2 when published. In the meantime, the UK NAD refers to Clause 20.3 of BS 5628: Part 3:1985.

20.5.15 Masonry below ground

Clause 5.8 is confined to general statements which will be amplified in EN 1996–2 when published. The UK NAD offers no further guidance. The manufacturer's advice should be sought, particularly when contemplating the use of concrete masonry products in sulphate bearing ground.

20.5.16 Prestressing details

Clause 5.3 of DD ENV 1996–1–1 covers the detailing of prestressed masonry and shows an example in **Figure 5.15**. The figure clearly shows a reinforced concrete ring beam which has a width equal to the overall wall thickness. This is good practice. The width of anchorage blocks should be equal to the overall thickness of the masonry section that is to be prestressed. The anchorage medium also needs to be stiff enough to spread the prestress into the masonry as evenly as possible to utilise fully the whole of the masonry section and to prevent over-stressing near the anchorage and local shear failures in the masonry.

20.6 Construction

20.6.1 General

Section 6 of DD ENV 1996–1–1 gives guidance on the *minimum* requirements for the standard of workmanship which should be included in contract specifications so as to ensure that the assumptions and design formulae given in the preceding sections of DD ENV1996–1–1 are valid.

When writing contract specifications, designers will need to consider whether

- more stringent numerical limits than given in DD ENV1996–1–1 (for example for tolerances) are necessary for the particular contract;
- more detailed specification of the topics covered in DD ENV1996–1–1 is required;
- topics not covered in DD ENV1996–1–1 need to be included.

More detailed guidance reflecting normal UK practice may be found in the following British Standards:

- BS 8000: Part 3: 1989 Workmanship on building sites, Code of practice for masonry (this document covers brick and block masonry only)
- BS 5628: Part 3: 1985 Use of masonry: Section 4 – Workmanship
- BS 5390: 1976 Code of practice for stone masonry

Note that of these the UK NAD lists BS 5628: Part 3 as a Normative Reference and BS5390 as an Informative Reference.

20.6.2 Masonry units

Clause 6.1 deals with matters to do with materials. The contractor must supply units which comply with the designer's specification, which itself must include

(a) the type of unit, that is the material, the group and the category of manufacturing control (refer to **Clause 3.1.1** and to the commentary thereon in this guidance document), and the size;
(b) the compressive strength of the unit (refer to **Clause 3.1.2.1** and to the commentary thereon in this guidance document).

The Contractor should also provide a production certificate stating the compressive strength and category of manufacturing control, otherwise samples should be taken and tested (refer to **Clause 6.1(3)**).

The significance of the category of manufacturing control is in the selection of the appropriate partial safety factor γ_m (refer to values for Table 2.3 given in the UK NAD). Note that Category II of manufacturing control is the norm; in the UK NAD Clause 3.1.1(3) states that Category I may be used only when the units are manufactured under a quality control scheme which demonstrates that the probability of failing to reach the specified compressive strength does not exceed 2.5%. Category I therefore corresponds with the 'Special' category and Category II with the 'Normal' category of manufacturing control in BS 5628: Part 1.

20.6.3 Handling and storage of masonry units and other materials

Clause 6.2 gives minimum requirements to ensure that the properties of materials/components are not adversely affected by the methods of handling and storage after delivery to the site.

The UK NAD states that the appropriate national standard for cutting and bending of reinforcement, referred to in Clause 6.2.4(3), is BS 4466.

20.6.4 Mortar and concrete infill

Clause 6.3 requires the contractor to use mortar and/or concrete infill which comply with the designer's specification (refer to Clauses 3.2 and 3.3 and to commentary thereon in this guidance document).

Minimum workmanship requirements are given in this section of DD ENV1996–1–1 to ensure that the required properties are realised; additional requirements as to workability are given in the UK NAD.

20.6.5 Construction of masonry; connection of walls; fixing reinforcement

The three Clauses, 6.4, 6.5 and 6.6, give minimum construction practice requirements for achieving the interaction of the units, mortar and other components that is necessary for the structural performance of the masonry. Other matters, such as appearance, are for the designer to consider in addition to these requirements.

Note that the bonding pattern, jointing requirements, connection details and reinforcement (if any) must be specified by the designer (refer to Section 5).

20.6.6 Protection of newly constructed masonry

Minimum requirements are given in Clause 6.7 for ensuring that the masonry is protected from adverse effects of the elements, or from overloading, between the time of construction and the time at which the masonry achieves sufficient inherent resistance and the structure becomes self-stable.

20.6.7 Permissible deviations in masonry

The numerical values shown in boxes in Clause 6.8 are confirmed in the UK NAD.

Note that the limits given are the maximum deviations which may be permitted unless specifically allowed for in the calculations. More stringent limits may be considered necessary for reasons, such as appearance, pertaining to a particular building, but designers should first consider whether such limits are in practice achievable and whether the probable increase in cost is justified.

20.6.8 Category of execution

Clause 6.9 is the one which will have most significance for designers at the time of preparing calculations and specifications. Clauses 6.9(1), 6.9(2), 6.9(3) and 6.9(4) must be satisfied in all cases. Clause 6.9(5) requires that the category of execution must be determined as being A, B or C for the purpose of selecting the partial safety factor γ_m (refer to the values for Table 2.3 given in the UK NAD). Category C is not used in the UK NAD; the requirements for Categories A and B are as follows:

Category A may be assumed when both of the following conditions are satisfied:

- regular inspection of the work is made by appropriately qualified persons independent* of the constructor's site staff to verify that the work is being executed in accordance with the drawings and specification;
- preliminary compressive strength tests carried out on the mortar to be used indicate compliance with the strength requirements given in Table 5 of the UK NAD and regular testing of the mortar used on site shows that compliance with the strength requirements given in Table 5 is being maintained.

Category B should be assumed when either or both of the conditions for Category A are not satisfied.

The UK NAD requires that Category A should be used where reinforced or prestressed masonry is constructed, other than in the particular cases of deep beams, composite lintels and panels containing bed joint reinforcement used to enhance lateral strength or to control cracking.

20.6.9 Other construction matters

Minimum requirements are given in Clause 6.10 for workmanship in the formation of movement joints, the limitation of daily height of construction and the construction of reinforced masonry walls.

The locations and method of formation of movement joints should be specified by the designer (refer to Section 5). However, in the UK it is common for determination of the appropriate daily height of construction to be delegated to the contractor, unless there are particular constraints inherent in the design.

20.6.10 Prestressing steel and accessories

Minimum requirements are given in Clause 6.11 for the specification of methods of work and handling of tendons and other components in the construction of prestressed masonry.

Clause 6.11.4 (1) P refers to the necessity of a pre-arranged programme for prestressing. This is essential, particularly when long walls are prestressed. The tendons have to be tensioned to give a uniform level of prestress along a wall. It is not normally possible to simultaneously prestress all the tendons, especially on a long wall, so individual tendons will need to be restressed because as each tendon is stressed it causes a small contraction in the wall which reduces the stress in adjacent tendons already anchored off.

It is efficient to adopt a pattern of prestressing with at least a pair of jacks, working from the end of the wall towards the middle, or *vice versa*, rather than simply

---

* In the case of Design-and-Build contracts, the designer may be considered as a person independent of the site organisation for the purposes of inspecting the work.

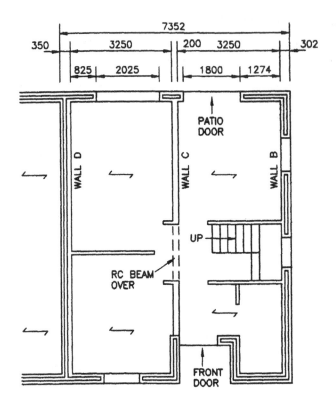

Figure 20.11 Ground floor plan of four storey domestic house

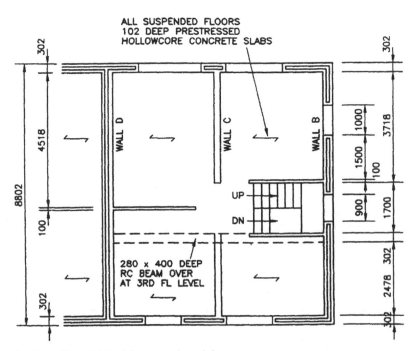

Figure 20.12 First floor plan (second floor similar) of four storey domestic house

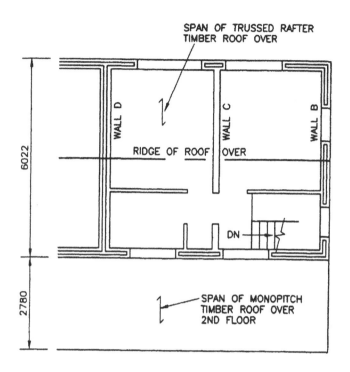

Figure 20.13 Third floor plan of four storey domestic house

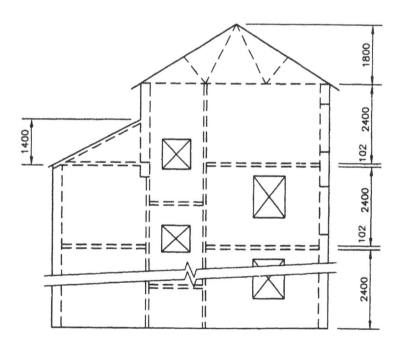

Figure 20.14 Gable wall elevation (Wall B) of four storey domestic house

adopting a sequential approach starting at one end. It is usually found that a uniform prestress can be applied with three or four cycles of stress.

20.7 Example calculations

20.7.1 Unreinforced masonry design example 1 – 4 storey domestic house (see Figures 20.11, 20.12, 20.13 and 20.14)

20.7.1.1 Actions

Permanent
Roof-finishes and trussed rafters at 600c/c = 0.83kN/m² (on plan)
Ceiling – Insulation on plasterboard = 0.25kN/m²
Floors – floating chipboard finish on = 2.30kN/m² 102 mm deep prestressed concrete slabs (hollow cored type) with plaster finish
Stairs – 100mm reinforced concrete waist = 5.23kN/m² and steps and finishes
External walls – 302.5mm thick; = 4.74kN/m² 102.5 mm outer brick skin, 150 mm inner blockwork skin, plaster finish
Internal walls (loadbearing) 200mm = 3.5kN/m² blockwork, plaster finish both sides
Internal partition (non-loadbearing) = 1.27kN/m² 100mm blockwork, plaster finish both sides
Internal party wall – 350mm thick = 5.00kN/m² 150mm blockwork skins, finish on both sides

Variable
Roof – 0.75kN/m² plus 0.25kN/m² ceiling = 1.00kN/m²
Floors = 1.50kN/m²

20.7.1.2 Design of wall D (350mm thick cavity wall)

Permanent load G_k on wall at ground storey:

from third floor
$= 2.3 \times \dfrac{3.25}{2}$ = 3.74kN/m on each leaf

from second floor
$= 2.3 \times \dfrac{3.25}{2}$ = 3.74kN/m on each leaf

from first floor
$= 2.3 \times \dfrac{3.25}{2}$ = 3.74kN/m on each leaf

from internal partitions
$= 1.27 \times \dfrac{3.25}{2} \times \dfrac{7.2}{4.8}$ = 3.10kN/m on each leaf

from own weight of wall
$= \dfrac{5}{2} \times 10.5$ = 26.25kN/m on each leaf

$\sum G_k$ = 40.57kN/m²

Variable load Q_k on wall at ground storey:

from third floor
$1.5 \times \dfrac{3.25}{2}$ = 2.44kN/m on each leaf

from second floor
$1.5 \times \dfrac{3.25}{2}$ = 2.44kN/m on each leaf

from first floor
$1.5 \times \dfrac{3.25}{2}$ = 2.44kN/m on each leaf

$\sum Q_k$ = 7.32kN/m on each leaf

As there is only one variable load in this example ψ_0 is not used, and the design value of the combination of loads is

$$\sum \gamma_{Gj} G_{kj} + 1.5 Q_{ki}$$

where $\gamma_G = 1.35$

$(1.35 \times 49.13) + (1.5 \times 7.32) = 77.31$ kN/m

Effective height of wall $h_{ef} = \rho_n h$ (see **Clause 4.4.4.3**) where ρ_n is a reduction factor to allow for edge restraint on the wall.
With reference to **Clause 4.4.4.3** and **Annex B**, for

$h/L = 2550/4700 = 0.54$

and $\rho_2 = 0.75$. From **Graph B2** $\rho_4 = 0.64$. Therefore $h_{ef} = 0.64 \times 2550 = 1630mm$.
Check suitability of stiffening wall (see **Clause 4.4.4.2**):

thickness = 100 (> 85mm)

which exceeds

$0.3 \times t_{ef} = 56$ mm

where

$t_{ef} = \sqrt[3]{150^3 + 150^3}$
$= 189$ mm

length of stiffening wall also exceeds $1/5 \times 2550$.
Check

$$\frac{L}{t} = \frac{4700}{189} = 25.74 \; (<30)$$

(assuming $t = t_{ef}$ in the case of a cavity wall.)
Design vertical load resistance (See **Clause 4.4.2(4)**)

$$N_{Rd} = \frac{\phi_{im} t_{fk}}{\gamma_m} \; \text{(for each leaf)}$$

Slab/wall stiffness ratio,

$$r = \frac{E_{slab} \, I_{slab} \, h_{wall}}{E_{wall} \, I_{wall} \, L_{slab}} = 1.17$$

where:
$E_{slab} = 30.5 \times 1000$
$E_{wall} = 1000 \, f_k = 5800$

$I_{slab} = \dfrac{1000 \times 100^3}{12} = 83.33 \times 10^6$

336

$$I_{wall} = \frac{1000 \times 150^3}{12} = 281 \times 10^6$$

$h_{wall} = 2550$

$L_{slab} = 3400$

f_k = K × $f_b^{0.65}$ × $f_m^{0.25}$ from Table 3 of Section 3 of this document, by interpolation

= 5.8N/m² for 10N/m² blocks and M4 (BS5628 (iii) mortar)

Floor fixed end moment

$$= (1.35 \times 2.3 + 1.5 \times 1.5) \times \frac{3.25 \times 3.4}{12}$$

= 4.8kN/m

Following **Annex C**

for slab,

$$\frac{EI}{L} = 0.75 \times 10^9$$

for one leaf of wall,

$$\frac{EI}{h} = 639 \times 10^6$$

$$M_1 = \frac{4.8 \times 10^6 \times 639 \times 10^6}{(0.75 \times 10^9) + 2(639 \times 10^6)}$$

= 1.512 × 10⁶Nmm/m

$$N_1 = \frac{77.31 \times 10^3}{150 \times 10^3}$$

= 0.515N/mm² (>0.25N/mm²)

$$k = \frac{0.75 \times 10^3}{2 \times 639 \times 10^6}$$

= 0.586

Therefore $M_{1(reduced)}$

$$M_1 \left(1 - \frac{k}{4} \right) = 1.512 \times 0.85$$

= 1.28 kNm/m

$$\frac{M_{1(reduced)}}{N_1} = \frac{1.28 \times 103}{77.31}$$

= 16.55 mm at top and bottom of wall

Accidental eccentricity (**Clause 4.4.7.2**)

$$\frac{h_{ef}}{450} = \frac{1630}{450}$$

= ±3.62 mm

over full height of wall.

At top and bottom of the wall therefore:

e_i = 16.55 + 3.62

= 20.17mm

> 0.05t (0.05t = 7.5mm)

e_{hi} = 0

$$\phi i = 1 - \left(2 \times \frac{20.79}{150} \right)$$

= 0.72

At the middle 1/5 of the wall height:

$M_1 = M_2$

$M_M = 1.280/5$

= 0.256kNm/m

$N_M \approx 77.31$kN/m

$$e_M = \left(\frac{0.256 \times 1000}{77.31} \right) + 3.62$$

= 6.93mm

Slenderness of wall

$$\frac{h_{ef}}{t_{ef}} = \frac{1630}{189}$$

= 8.62 (<15)

therefore e_k = 0.

Therefore

e_{mk} = 6.93

$$\frac{e_M}{t} = \frac{6.93}{150} = 0.05$$

From **Figure 4.2** ϕ_m = 0.83

Therefore top and bottom of wall governs (ϕ_1).

Wall capacity

$$N_{Rd} = \frac{0.73 \times 150 \times 5.8}{2.0}$$

= 317.55kN/m (for category I manufacturing control and category A execution control γ_m = 2.0)

or

$$= \frac{317.55 \times 2}{2.8}$$

= 226.82kN/m (for category II manufacturing control and category B execution control γ_m ≈ 2.8.)

Note both values of N_{Rd} exceed the value of N_{Sd} (77.31 kN/m) by a substantial margin, therefore consider using 3.5 N/mm² units in M4 mortar.

By reducing the f_k value of the masonry to 2.95 N/mm² the slab/wall stiffness ratio will increase resulting in reduced moment transfer into the wall and correspondingly less eccentricity.

$$N_{Rd} > \frac{0.72 \times 150 \times 2.95}{2.0}$$

= 159.3 kN/m (for category I manufacturing control and category A execution control.)

or

$$> \frac{159.3 \times 2}{2.8}$$

= 113.78k N/m (for category II manufacturing control and category B execution control.)

Both values of N_{Rd} exceed N_{Sd} of 77.31.

20.7.1.3 Design of wall C (200mm thick solid brickwork)

Permanent load G_k on wall at ground floor:

From third floor
$$= 2.3 \times 3.25 \times 2 \times 4.57 \qquad = 68.32 \text{kN}$$

From second floor
$$= 2.3 \times 3.25 \times 2 \times 4.57 \qquad = 68.32 \text{kN}$$

From first floor
$$= 2.3 \times 3.25 \times 2 \times 4.57 \qquad = 68.32 \text{kN}$$

Additional load from stairs
$$= (5.23 - 2.3) \times \frac{3.25}{2} \times 0.85 = 4.05 \text{kN}$$

from partitions
$$1.27 \times \frac{3.25}{2} \times 5.1 + 1.27 \times \frac{3.25}{4} \times 5.1 = 15.78 \text{Kn}$$

from self weight of wall
$$= 3.5 \times 3.72 \times 7.65 \qquad = 99.60 \text{kN}$$

$$\sum G_k \qquad\qquad\qquad = 324.39 \text{kN}$$

that is 87.20kN/m.

Variable load Q_k on wall at ground storey:
From third, second and first storeys
$$= 1.5 \times 3.25 \times 3 \qquad = 14.62 \text{kN/m}$$

$$\sum Q_k \qquad\qquad\qquad = 14.62 \text{kN/m}$$

Design value of combined loads
$$= (1.35 \times 87.20) + (1.5 \times 14.62)$$
$$= 117.72 + 21.93$$
$$= 139.65 \text{kN/m}$$

Wall stiffened on one edge.

$$\frac{L}{t} = \frac{3720}{200}$$

which exceeds 15, therefore ignore edge stiffening, Clause 4.4.4.3(3).

Effective height of wall h_{ef} $\qquad = 0.75 \times 2550$
$$\qquad\qquad = 1910 \text{mm}$$

Slab/wall stiffness ratio,

$$r = \frac{E_{slab}\, I_{slab}\, h_{wall}}{E_{wall}\, I_{wall}\, L_{slab}} = 0.65$$

where:

$E_{slab} = 30.5 \times 1000$
$E_{wall} = 1000\, f_k = 4400$
$I_{slab} = 83.33 \times 10^6$

$$I_{wall} = \frac{1000 \times 200^3}{12} = 666 \times 10^6$$

$h_{wall} = 2550$
$L_{slab} = 3400$
$f_k = 4.4 \text{N/m}^2$ for 7N/m≤ blocks and M4 mortar

Floor fixed end moment
$$= 1.5 \times 1.5 \times \frac{3.25 \times 3.4}{12}$$

$$= 2.07 \text{kN/m}$$

(assuming live load on one side only).
Following **Annex C**

$$\frac{EI}{L_{slab}} = 0.75 \times 10^9$$

$$\frac{EI}{h_{wall}} = 0.639 \times 10^9$$

$$M_1 = \frac{2.07 \times 10^6 \times 639 \times 10^6}{2 \times 0.75 \times 10^9 + 2 \times 639 \times 10^6}$$

$$= 0.6 \times 10^6 \text{ Nmm/m}$$
$$= 0.6 \text{ kNm/m}$$

$$N_1 = \frac{139.65 \times 10^3}{200 \times 10^3}$$

$$= 0.70 \text{N/mm}^2 \ (>0.25 \text{N/mm}^2)$$

$$k = \frac{0.75 \times 10^9 \times 2}{639 \times 10^6 \times 2}$$

$$= 1.17$$

Therefore $M_{1(reduced)}$

$$M_1 \left(1 - \frac{k}{4}\right) = 0.6 \times 0.71$$

$$= 0.43 \text{kNm/m}$$

$$\frac{M_{1(reduced)}}{N_1} = \frac{0.43 \times 10^3}{139.65}$$

$$= 3.08 \text{ mm (at top and bottom of wall)}$$

Accidental eccentricity
$e_a = \pm 4.24$ (over the full height of the wall)

At top and bottom of wall:

$$e_i = 3.08 + 4.24$$
$$= 7.32 \text{mm} \ (< 0.05t = 10 \text{mm})$$

therefore assume
$e_i = 10 \text{mm}$
$e_{hi} = 0$

$$\phi i = 1 - \frac{2 \times 10}{200}$$

$$= 0.9$$

At middle 1/5 of wall:

$$M_1 = M_2$$

$$M_M = \frac{0.43}{5}$$

$$= 0.086 \text{kNm/m}$$
$$N_M = 139.65 \text{kN/m}$$

$$e_M = \frac{0.086 \times 1000}{139.65} + 4.24$$

$$= 4.85 \text{mm}$$

Slenderness of wall

$$\frac{h_{ef}}{t} = \frac{1910}{200}$$

$$= 9.55 \ (<15)$$

therefore assume $e_k = 0$.
Therefore

$e_{mk} = 4.85$mm, which is less than 0.05t, therefore assume
$e_{mk} = 10$mm
From **Figure 4.2**
$\phi_M = 0.85$

Wall capacity
$$N_{Rd} = \frac{0.85 \times 200 \times 4.4}{2.0}$$

= 374kN/m (for category I manufacturing control and category A execution control) or

$$N_{Rd} = \frac{374 \times 2}{2.8}$$

= 267kN/m (for Category II manufacturing control and Category B execution control.) Both exceed N_{Sd} of 139.65kN/m. Note there is scope for reducing the 7N/m² blocks to 3.5N/m² blocks which would provide a minimum N_{Rd} of 170kN/m for Category II manufacturing control and Category B execution control.

20.7.1.4 Design check on bearing of 400 × 280mm reinforced concrete beam at third floor level on 200mm loadbearing internal wall

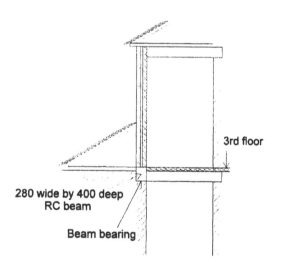

3rd floor

280 wide by 400 deep RC beam

Beam bearing

Figure 20.15 Arrangement of loads on beam and bearing

Permanent load Gk at beam bearing:
From roof and ceiling
$$= (0.83 + 0.25) \times \frac{8.6}{2} \times \frac{6.7}{2} \qquad = 15.55\text{kN}$$

From self weight walls
$$4.74 \times 2.15 \times \frac{6.7}{2} \qquad = 34.14\text{kN}$$

From self weight beam
$$23.8 \times 0.4 \times 0.28 \times \frac{6.7}{2} \qquad = 4.46\text{kN}$$

From floor and stairs
$$\left(\frac{23 \times 3.25}{2} + \frac{5.23 \times 3.25}{2} \right) \ 0.85 \ = 10.40\text{kN}$$

From internal partition
$$1.27 \times 2.4 \times \frac{3.25}{4} \qquad = 2.47\text{kN}$$
$$\sum G_k \qquad = 67.02\text{kN}$$

Variable load Q_k acting on bearing
From roof and ceiling
$$1.0 \times \frac{8.6}{2} \times \frac{6.7}{2} \qquad = 14.40\text{kN}$$

From third floor and stairs
$$= 1.5 \times 3.25 \times 0.85 \ (\text{approx.}) \qquad = 4.08\text{kN}$$

Hence the major dominant variable load is that due to the roof.

Design value of the combined loads
$$\sum \gamma_{Gj} \, G_{kj} + \gamma_{Q1} \, Q_{k1} + \sum \gamma_{Qi} \, \Psi_{0i} \, Q_{ki}$$

$$= (1.5 \times 67.02) + (1.5 \times 14.4) + (1.5 \times 2.5 \times 4.08)$$
$$= 115.14\text{kN}$$

Design compressive stress locally under reinforced concrete beam

$$\frac{115.14 \times 10^3}{200 \times 280} = 2.06 \text{ N/mm}^2$$

Strength enhancement factor:
$a_1 = 0$ therefore
$x = 0$ therefore
factor $= 1.25$

$$1.25 \ \frac{f_k}{\gamma_m} = 1.25 \times \frac{4.4}{2.0}$$

$= 275$N/mm² (assuming 7N/mm² blocks and M4 mortar).

Check compressive stress in wall at mid-height below concentrated load.

The only additional superimposed load acting at the mid-height of the wall is due to the self weight of the wall, that is

$$3.5 \times 1.225 = 4.28\text{kN/m}$$

therefore design load on length L_{ef} is

$$(1.35 \times 4.28 \times 0.71) + 115.14 = 119.24\text{kN}$$

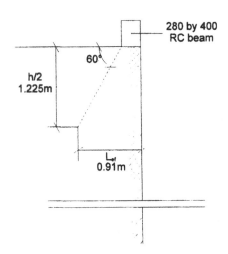

Figure 20.16 Spread of load form reinforced concrete beam

As this design load is considerably lower than the design load on the lowest storey of wall C, no further verification is required.

20.7.1.5 Design of inner loadbearing leaf of ground storey wall B (302.5mm cavity wall)

Permanent load G_k on wall at ground storey
From third floor

$$2.3 \times \frac{3.25}{2} \times (8.0 - 1.7) \qquad = 23.55\text{kN}$$

From second floor

$$2.3 \times \frac{3.25}{2} \times (8.0 - 1.7) \qquad = 23.55\text{kN}$$

From first floor

$$2.3 \times \frac{3.25}{2} \times (8.0 - 1.7) \qquad = 23.55\text{kN}$$

From stairs

$$5.23 \times \frac{3.25}{2} \times 1.7 \times 3 \qquad = 43.34\text{kN}$$

From internal partitions

$$1.27 \times \frac{3.25}{2} \times 7.65 \times 2 \qquad = 31.57\text{kN}$$

From self weight wall (150mm concrete inner leaf only)
$$2.5 \times 5.8 \times 9.8 + 2.5 \times 2.2 \times 7.07 = 180.98 \text{ kN}$$

$$\sum G_k \qquad = 326.54\text{Kn}$$

that is 40.82kN/m

Variable load Q_k on wall at ground storey:

From third, second and first floors

$$1.5 \times \frac{3.25}{2} \times 3 \qquad = 7.31\text{kN/m}$$

$$\sum Q_k \qquad = 7.31\text{kN/m}$$

Only one variable load, therefore Ψ_0 is not used.

$$\sum \gamma_{Gki} \, G_{kj} + 1.5 \, Q_{kj} = (1.35 \times 40.82) + (1.5 \times 7.31)$$
$$= 66.08\text{kN}$$

The critical length of wall is the 1500mm length between the stiffening wall to the stairs and the window. The window height is greater than one-quarter of the storey height, therefore the wall is assumed to have a free edge at the window. Therefore

$$\frac{h}{L} = \frac{2550}{1550}$$

$$= 1.64$$

where L is measured to the centre of the stiffening wall. Therefore
$\rho_3 = 0.64$ (from **Annex B**) and
$h_{ef} = 0.64 \times 2550$
$= 1630\text{mm}$

Slab/wall stiffness ratio r = 1.17 (similar to wall D above.)
Floor fixed end moment (as wall D) = 4.8kNm/m

$M_1 = 1.512\text{kNm/m}$

$$N_1 = \frac{66.08 \times 10^3}{150 \times 10^3}$$

$$= 0.44\text{N/mm}^2 \ (>0.25\text{N/mm}^2)$$
k $= 0.586$
and so $M_{1(reduced)}$

$$M_1 \left(1 - \frac{k}{4} \right) = 1.28\text{Knm/m}$$

$$\frac{M_{1(reduced)}}{N_1} = \frac{1.28 \times 103}{66.08}$$

$$= 19.37\text{mm (at top and bottom of wall)}$$

Accidental eccentricity

$$= \frac{h_{ef}}{450}$$

$$= \frac{1630}{450}$$

$$= \pm 3.62\text{mm}$$

At top and bottom of wall:
$\begin{aligned} e_i &= 19.37 + 3.62 \\ &= 22.99 \ (>0.05\text{t}) \end{aligned}$

$e_{hi} = 0$

therefore

$$\phi_i = 1 - \frac{2 \times 22.99}{150}$$

$$= 0.69$$

At middle 1/5 of wall height:
$M_1 = M_2$, and so

$$M_M = \frac{1.28}{5}$$

$$= 0.256 \text{kNm/m}$$

$$N_M \approx 66.08 \text{kN/m}$$

$$e_m = \frac{0.256 \times 1000}{66.08} + 3.62$$

$$= 7.49 \text{mm}$$

therefore

$$e_{mk} = 7.49 \text{mm} \ (e_k = 0, \text{ as } h_{ef}/t_{ef} < 15)$$

From **Figure 4.2**

$$\phi_i = 0.83$$

$$t_{ef} = \sqrt[3]{150^3 + 102^3}$$

$$= 164 \text{mm}$$

For

$$\frac{e_{mk}}{t} = \frac{7.49}{150}$$

$$= 0.05$$

and

$$\frac{h_{ef}}{t_{ef}} = \frac{1630}{164}$$

$$= 9.94 \ (<15)$$

therefore $e_k = 0$ and the top of wall governs (ϕ_i).
Wall capacity

$$N_{Rd} = \frac{0.69 \times 150 \times 2.95}{2.0}$$

$$= 152.66 \text{kN/m} \text{ (for Category I manufacturing control and Category A execution control.)}$$

$$\text{or} = \frac{152.66 \times 2}{2.8}$$

$$= 109.04 \text{kN/m} \text{ (for Category II manufacturing control and Category B execution control.)}$$

Both exceed N_{Sd} of 66.08kN/m

20.7.1.6 Design of inner leaf of top storey of Wall B (302.5 cavity wall)

Permanent load G_k on wall at mid-height of top storey:
from roof and ceiling

$$= (0.83 + 0.25) \times 1.0 \qquad = 1.08 \text{kN/m}.$$

from self weight of wall (150mm thick concrete inner leaf only)

$$= 2.5 \times \left(\frac{20.5 + 1.8}{2}\right) \qquad = 5.38 \text{kN/m}$$

$$\sum G_k \qquad = 6.46 \text{kN/m}$$

Variable load Q_k on wall from roof $= 1.00 \text{kN/m}$
Design load on wall $\sum \gamma_{Gki} G_{kj} + 1.5 Q_{kj}$ (only one variable load, therefore ψ_0 is not used)

$$= (1.35 \times 6.46) + (1.5 \times 1.0) = 10.22 \text{kN/m}$$

Effective height of wall (which is restrained at ceiling level)

$$h_{ef} = \rho_{nh} \text{ where } \rho_n = \rho_2 = 1.0 \ (h \leq L)$$
$$= 1.0 \times 2500$$
$$= 2500 \text{mm}$$

Slab/wall stiffness ratio $r = 1.17$ (similar to wall D)
Floor fixed end moment (as wall D) = 4.8 kNm/m

Slab, $\dfrac{EI}{L} = 0.75 \times 10^9$

Wall, $\dfrac{EI}{h} = 0.639 \times 10^9$

Annex C refers

$$M_1 = \frac{4.8 \times 10^6 \times 639 \times 10^6}{(0.75 \times 10^9) + 2.0 \ (639 \times 10^6)}$$

$$= 1.512 \text{kNm/m}$$

$$N_1 = \frac{10.22 \times 10^3}{150 \times 10^3} \approx 0.068 \ (< 0.25 \text{N/mm}^2)$$

$$k = 0.586$$
$$M_1 \times (1 - k/4) = 1.28 \text{kNm/m}$$

$$\frac{M_{1(reduced)}}{N_1} = \frac{1.28 \times 10^3}{10.22}$$

$$= 128 \text{mm (at bottom of wall)}$$

Accidental eccentricity $= \dfrac{h_{ef}}{450} = \dfrac{2500}{450} = \pm 5.55 \text{mm}$

At bottom of wall:

$$e_i = 128.2 + 5.55 = 130.8 \text{mm (exceeds 04t) and}$$
$$e_{hi} = 0$$

Reference **Annex C(4)** and UK NAD
Maximum $e_i = 0.4t$
therefore $\phi_i = 0.2$

Wall capacity (N_{rd}) $= \dfrac{0.2 \times 150 \times 2.95}{2.0}$

$= 44.25 \text{kN/m}$, which exceeds 10.22kN/m for Category 1 manufacturing control and Category A execution control and assuming a 3.5N/mm^2 block with M4 mortar.

At mid 1/5th of wall height:

Assume moment at A as $N_1 \times 0.4t$

$$= 10.22 \times 0.4 \times 0.150$$
$$= 0.613 \text{kNm/m}$$

$$\frac{2h}{5} = 1.0 \text{m}$$

$$0.67h = 1.67 \text{m}$$
$$M_M = 0.67 \times 1.28 = 0.86 \text{kNm/m}$$

$$N_M \approx 66.08 \text{kN/m}$$

$$e_M = \frac{0.86 \times 1000}{66.08} + 5.55 = 18.56 \text{ mm}$$

therefore $e_{Mk} = e_M + e_k \geq 0.05t$

$$\frac{h_{ef}}{t_{ef}} = \frac{2500}{164} = 15.24 \ (>15)$$

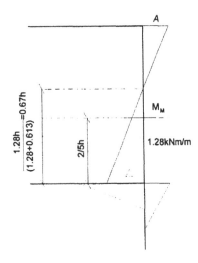

Figure 20.17 Bending moment diagram

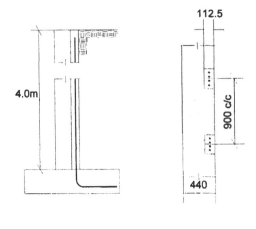

Figure 20.18 Design of brickwork wall stem

therefore $e_k = 0.002\phi_\infty \dfrac{h_{ef}}{t_{ef}} \sqrt{t_{em}}$,

where $\phi\infty' = 1.5$ (refer to UK NAD)

$e_k = 0.002 \times 1.5 \times 15.24 \times \sqrt{(150 \times 18.56)}$
$= 2.41\text{mm}$

therefore

$e_{Mk} = 18.56 + 2.41 = 20.97\text{mm}$

$\dfrac{e_{MK}}{t} = \dfrac{20.97}{150}$

From **Figure 4.2** $\phi_M = 0.53$

Wall capacity $N_{rd} = \dfrac{0.53 \times 150 \times 2.95}{2.0}$

$= 117.26\text{kN/m}$ which exceeds 10.22kN/m for Category 1 manufacturing control and Category A execution control and assuming a 3.5N/mm² block with M4 mortar.

The condition at the top of the wall, from inspection, is not as critical.

20.7.2 Reinforced masonry design example 1 – reinforced brickwork cantilever pocket-type retaining wall

A reinforced brickwork earth retaining wall 4.0m high has to support a well graded dense sand and gravel backfilling. The bricks to be used have a crushing strength of 35 N/mm² laid in 1:½:4½ mortar (designation (ii)). The reinforcement is grade 460 high yield deformed steel placed in pockets with 35 N/mm² concrete infill. The base will be a reinforced concrete spread foundation. The retained soil is horizontal behind the wall.

The design total lateral force at the ultimate limit state has been derived from the soil pressure, surcharge loading and partial safety factors as 73.63 kN/m and the design bending moment is 107.62 kNm/m.

Limiting span

From DD ENV 1996–1–1 **Table 4.1** the limiting ratio of the span to effective depth is 18.

Hence the effective depth must exceed 4000/18 which is 222mm.

Bending capacity

Using a 440mm thick brickwork reinforced pocket-type wall (2-bricks thick) d provided is (440 – 56) = 384mm, say 380mm with 112mm depth reinforcement pockets formed in rear brickwork wall face.

Design wall as flanged beam, assume pocket width along wall length of 235mm (1–brick)

Flange depth = d/2 = 380/2 = 190mm (t_f)

From **Clause 4.7.1.5** width of flange is lesser of:

(a) 235 + (12 × 190) = 2515mm
(b) 900mm (actual pocket spacing to be used)
(c) 4000/3 = 1333mm

therefore flange width = 900mm

The compressive strength of the bricks to BS3921 is 35N/mm². The restrictions on perforation size in BS3921 ensure that they are Group 1 units according to **Table 3.1.**

The normalised compressive strength of the bricks f_b is given by:

$f_b = 1.2 \times 0.85 \times 35$ N/mm²
$f_b = 35.7$ N/mm²

where 1.2 is the wet: air dried strength conversion factor from Clause 6.3 a) of the UK NAD and 0.85 is the shape factor δ taken from **Table 3.2.**

The designation (ii) mortar (to BS5628) is taken as being strength class M6 from Clause 6.3 d) of UK NAD.

The compressive strength of the brickwork is given by:

$f_k = K f_b^{0.65} f_m^{0.25}$

342

where K is taken as 0.7 for group 1 clay units in accordance with Table 3 of the UK NAD, hence

$$f_k = 0.7 \times 35.70^{0.65} \times 6^{0.25}$$
$$f_k = 11.19 \text{ N/mm}^2$$

The Category of manufacturing control for the bricks is I, the Category of execution control will be A and hence the partial safety factor for materials properties γ_m will be taken as 2.0 in accordance with Table 1 of the UK NAD. Hence f_k/γ_m is 5.59 N/mm²

The Moment of resistance factor (see Section 4.6) Q is given by

$$Q = \frac{M_{Rd}}{bd^2} = \frac{107.62 \times 0.9 \times 10^6}{900 \times 380^2} = 0.74$$

From **Figure 4.7**, interpolating c = 0.93 which is less than the 0.95 limiting value.

Hence the lever arm Z is 0.93 × 380 = 353mm
Hence from **Eqn (4.36)** in **Clause 4.7.1.6**

$$A_s \text{ required } = \frac{M_{Rd}\,\gamma_s}{f_y\, Z}$$

$$A_s \text{ required } = \frac{107.62 \times 0.9 \times 1.15 \times 10^6}{460 \times 343}$$

A_s required 706mm²

Using steel with a yield strength of 460N/mm² and taking γ_s as 1.15 which is the value in DD ENV 1996–1–1 and confirmed in the UK NAD.

Therefore use two T25 diameter bars which provides 982mm². If the bars are placed centrally in the pocket the cover is 112/2 – 25/2 = 44mm which is adequate for corrosion protection when a 35N/mm² concrete infill is used for exposure situation E3 in BS 5628: Part 2, Table 14 referred to by the UK NAD.

Check that the limiting compressive moment of the brickwork section is not exceeded.

$$M_{Rd \text{ Maximum}} = \frac{f_k}{\gamma_m}\, bt_f\, (d - 0.5t_f)$$

(See **Eqn (4.38)**)

$$M_{Rd \text{ Maximum}} = \frac{5.59}{10^6}\, 900 \times 190 \times (380 - 0.5 \times 190)$$
$$M_{Rd} = 199.81 \text{ kNm}$$

This exceeds the moment applied to the wall and hence the brickwork section is sufficient.

The design shear force is 73.63kN/m
Therefore shear stress due to design loads,

$$v = \frac{73.63 \times 10^3}{1000 \times 380} = 0.19 \text{N/mm}^2$$

And $\dfrac{A_s}{bd} = \dfrac{982}{900 \times 380} = 0.0029$

Characteristic shear strength of section, f_v = (0.35 + 17.5 × 0.0029) = 0.40N/mm²

f_v may be increased by factor (2.5 – 0.25 a/d) (See section 6.4 of the UK NAD)

where $a = \dfrac{107.62 \times 10^6}{73.63 \times 10^3} = 1462 \text{ mm}$

Therefore

$$f_{vk} = 0.40 \left(2.5 - 0.25\, \frac{1462}{380}\right) = 0.62 \text{ N/mm}^2$$

And
$$\frac{f_{vk}}{\gamma_m} = \frac{0.62}{2.0} = 0.31 \text{ N/mm}^2, \text{ where } \gamma_M = 2.0$$

in accordance with Table 1 of the UK NAD.

This exceeds 0.19 N/mm² and therefore no shear reinforcement is needed.

Consider now whether the main reinforcement can be curtailed.

Use T20 reinforcement to lap on. In order to curtail the T25's in the tension zone, **Clause 5.2.5.4** must be satisfied. The moment condition is appropriate and therefore the design moment capacity of the T20's must be at least twice the design moment due to applied lateral loads for curtailment.

Design moment of 2 no. T20 bars per pocket, A_s = 628 mm², assume c = 0.94

$$M = \frac{628 \times 460 \times 0.94 \times 380}{1.15 \times 10^6}$$

= 89.72 Knm per 900 mm pocket spacing

$$Q = \frac{M}{b\, d^2} = \frac{89.72 \times 10^6}{900 \times 380^2} = 0.69$$

and $\dfrac{f_k}{\gamma_M}$ = 5.59 N/mm²

c = 0.94 is the correct assumption to use in **Figure 20.5**.

Thus T25's may be stopped where the design moment due to applied lateral loads, M_d = (89.72 × 0.5/0.9) = 49.85 kNm/m.

From a consideration of the design lateral loads upon the wall stem the applied design moment of 49.85 kNm/m is achieved at 2.975 m from the top of the wall (from retained earth level).

Check that T25's will extend at least an effective depth or 12 bar diameters beyond the point where they are no longer needed. The T20's will resist a design moment of:

$$\left(\frac{89.72}{0.9}\right) = 99.69 \text{ kNm/m}$$

From a consideration of the design lateral loads this will occur at 3.845 m, therefore curtailment of T25's at 2.975 m below top of wall is satisfactory.

And 12 diameters above this level (3.845 – 12 × 25) = 3.545 m, therefore curtailment of T25's at 2.975 m below top of wall is satisfactory.

Check lap length for T20 bars.

Characteristic bond strength, f_{bo} = 3.4 N/mm² from Table 3 of the UK NAD.

Therefore lap length required to achieve full anchorage bond is,

$$= \frac{f_y \times \gamma_{Mb} \times A_s}{\gamma_s \times f_{b0} \times \pi \times \text{bar size}}$$

343

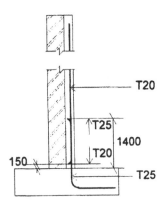

Figure 20.19 Steel arrangement in retaining wall

$$= \frac{460 \times 1.7 \times 314}{1.15 \times 3.4 \times \pi \times 20} = 999 \text{ mm}$$

where $\gamma_m = 1.7$ is taken from Table 1 of the UK NAD.

This is greater than $(25 \times \text{bar size} + 150) = 650$ mm and therefore governs the lap.

Therefore curtail T25's 1400 mm above foundation base and continue T20's down to within 150 mm of foundation base. This provides a lap length for T20's of 1250 mm which is satisfactory.

20.7.3 Reinforced masonry design example 2 – reinforced hollow concrete blockwork short and slender column subjected to single axis bending

Design a 2.7m high reinforced brickwork column to carry a vertical design load of 800kN and a design bending moment of 85kNm. If the height of the column is increased to 6.0m what effect does this have on the column section chosen?

Try column section 665mm by 440mm.

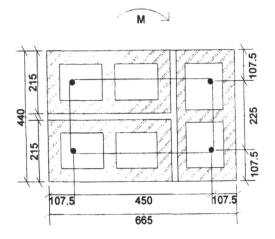

Figure 20.20 Steel positions in hollow blocks

The blocks to be used are dense aggregate concrete with a compressive strength of 10N/mm². The blocks are Group 2(a) according to Table 3.1. The infill concrete has a compressive strength of 25N/mm². The mortar is a 1:½:4½cement:lime:sand mix.

The guidance in DD ENV 1996–1–1 on the design of columns when the axial load exceeds 10% of the masonry strength is limited and the UK NAD refers to the user to Clause 4.3 of BS 5628 Part 2 although the partial safety factors, characteristic strengths, effective height, effective thickness and slenderness ratios should be taken from DD ENV 1996–1–1.

The normalised compressive strength of the unit is given by

$$f_b = 10 \times 1.2 \times 1.18 \text{ N/mm}^2$$

where 1.2 is the wet:dry strength conversion taken from section 6.3 of the UK NAD and 1.18 is the shape factor interpolated from **Table 3.2**.

$$f_b = 14.15 \text{ N/mm}^2$$

the strength of the unit based on the net area is 20.2 N/mm² .

As this strength is lower than that of the concrete infill the masonry is designed as having a compressive strength based on that of a Group 1 solid unit with a strength equal to that of the net area block strength. The UK NAD ascribes a mortar class of M6 to a 1:½:4½ mix, hence

$$f_k = 0.7 \times 20.2^{0.65} \times 6^{0.25}$$

$$f_k = 7.7 \text{ N/mm}^2$$

Assume top and bottom of columns have lateral supports restricting movement in both directions.
Therefore

h_{ef} = 2700mm
t_{ef} = minimum thickness = 440mm
Slenderness ratio = 2700/440 = 6.14

This is less than 12, therefore column is short and can be designed to Clause 4.2.3.1 of BS5628 Part 2.

Check whether only minimum reinforcement required from

$$N_d = \frac{f_k}{\gamma_M} b (t - 2ex)$$

Resultant eccentricity,

$$e_x = \frac{85}{800} = 0.106$$

$$N_d = \frac{7.7}{2} 440 (665 - 212) \frac{1}{10^3}$$

= 767.4 kN

This is less than design vertical load N, therefore more than minimum reinforcement is required.

Consider stress distribution across section
Assume $d_c = t - d_2$ and $f_{s2} = 0$
Try 2 no. T25 bars each face

From $N_d = \dfrac{f_k}{\gamma_M} b \, d_c + \dfrac{f_{s1}}{\gamma_s} A_{s1} - \dfrac{f_{s2}}{\gamma_s} A_{s2}$

$N_d = \dfrac{7.7}{2.0} \times 440 \times 510 + \dfrac{0.83 \times 460 \times 982}{1.15} - 0$

$= 863.9 \times 10^3 + 326.0 \times 10^3$

$= 1189.9 \times 10^3 \, N$

$= 1189.9 \, kN$

From

$M_d = \dfrac{0.5 f_k}{\gamma_M} b \, d_c \, (t - d_c) + \dfrac{0.83 f_y}{\gamma_s} A_{s1} \, (0.5t - d_1)$

$\qquad + \dfrac{f_{s2}}{\gamma_s} A_{s2} \, (0.5t - d_2)$

$M_d = \dfrac{0.5 \times 7.7}{2} \times 440 \times 510 \times (665 - 510)$

$\qquad + \dfrac{0.83 \times 460}{1.15} \times 982 \times (0.5 \times 665 - 155) + 0$

$= 67.0 \times 10^6 + 57.9 \times 10^6$

$= 124.9 \times 10^6 \, Nmm$

$= 124.9 \, kNm$

Thus both N_d and M_d exceed N and M respectively and the section is satisfactory.

Consider whether reinforcement area may be reduced. Try 2 no T20 bars in each face where $A_{s1} = A_{s2} = 628$ mm²

$N_d = 863.9 \times 10^3 + \dfrac{0.83 \times 460 \times 620}{1.15} - 0$

$= 1072.3 \times 10^3 \, N$

$= 1072.3 \, kN$

$M_d = 67.0 \times 10^6 \, Nmm + \dfrac{0.83 \times 460}{1.15}$

$\qquad \times 628 (0.5 \times 665 - 155) + 0$

$= 104.0 \times 10^6 \, Nm$

$= 104.0 \, kNm$

Both N_d and M_d exceed N and M respectively and therefore 4 no. T20 bars are satisfactory.

Link reinforcement

$$A_s = 4 \times 314 = 1256 \text{ mm}^2$$

0.25% of section $= \dfrac{0.25 \times 440 \times 665}{100} = 731.5 \text{ mm}^2$

Therefore links are required, see **Clause 5.2.7**, use R6 links at 300 crs.

If effective height of column is increased to 6.0m Slenderness ratio becomes 6000/440 = 13.6 and column becomes slender that is slenderness ratio greater than 12, but less than the limiting value of 27

An additional moment, M_a, must therefore be allowed for

$Ma = \dfrac{N(h_{ef})^2}{2000t} = \dfrac{800}{2000} \times 62 \times 0.665 = 21.7 \text{ kNm}$

Design moment, M, becomes 85 + 21.7 = 106.7kNm

From the above it will be seen that using 4 no. T25 bars, both N_d and M_d exceed the N and M (including M_a) respectively. Thus in the case of increasing the column height so that it becomes slender the effect is to require an increase in reinforcement from 4 no. T20 bars to 4 no. T25 bars.

20.7.4 Reinforced masonry design example 3 – reinforced brickwork simply supported beam

Design a simply supported reinforced brickwork beam required to span 3.5m carrying a characteristic dead load of 20.0kN/m (including self-weight) and a characteristic imposed load of 6.0kN/m.

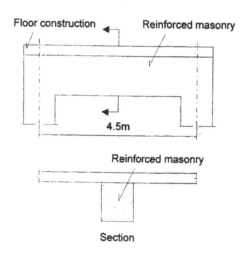

Figure 20.21 Elevation and section of reinforced masonry beam

The bricks to be used have a compressive strength of 35N/mm² and the mortar is a 1:½:4½. The characteristic compressive strength of the brickwork is derived in the same way as in Example 1 and is 11.19N/mm².

Consider the design load = $1.35 G_k + 1.5 Q_k$

Where 1.35 and 1.5 are the partial safety factors for the unfavourable effects of permanent and variable loads in accordance with DD ENV 1996–1–1 and confirmed by the UK NAD.

Design Load = 1.35 × 20 + 1.5 × 6
Design Load = 36.0kN/m

the simply supported design moment

$$M_d = \dfrac{36 \times 4.5^2}{8} = 91.1 \text{ kNm}$$

the design shear force is

$$V_{sd} = \frac{36 \times 4.5}{2} = 81 \text{ kN}$$

Assuming that there are non-structural beam depth requirements, the effective depth will be based upon the limiting span to depth ratios given in DD ENV 1996-1-1, the coordinating dimensions of brickwork coursing and possibly the shear strength of the beam.

In order to suit column width make beam width, b = 440mm. Use bricks as soldier courses for aesthetic reasons and so that brickwork in compression is loaded normal to its bed face.

(Note that if brickwork is laid horizontally such that it is loaded in compression on its perpend or stretcher face, f_k must be determined in relation to that direction of loading).

Try section as shown. Provision of vertical voids at 300 mm centres to accept links means shear strength is unlikely to control depth.

Beam depth = 553mm

Assume Exposure situation E1 with grade 25 concrete cover to reinforcement = 20mm

Effective depth, d = 553 − 103 − 20 − 10 − 12.5 = 407.5 mm, say 400 mm

Span/effective depth must not exceed 20, see **Table 4.1** thus, min. effective depth = 4500/20 = 225 mm

This is less than 400 and therefore satisfactory.

Design moment of resistance of beam in compression

$$M_d = 0.256 \frac{f_k \, b \, d^2}{\gamma_m}$$

$$= 0.256 \times \frac{11.19 \times 440 \times 400^2}{2.0}$$

= 100.8 kNm

This exceeds design moment, 95.2kNm

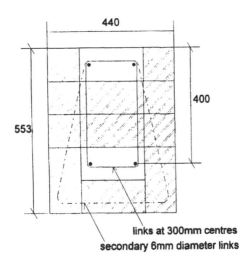

Figure 20.22 Steel arrangement in reinforced masonry beam

440

400

553

links at 300mm centres
secondary 6mm diameter links

Design moment of resistance of beam in tension

$$M_d = \frac{A_s \, f_y \, Z}{\gamma_M}$$

Assume lever arm, Z, = 0.75d. Equate M_d to M and solve for area of tensile steel A_s.

$$A_s = \frac{95.2 \times 10^6 \times 1.15}{460 \times 0.75 \times 400} = 749 \text{ mm}^2$$

Therefore use 2 no. T25 diameter bars (982mm²)

$$Z = d \left(1 - \frac{0.5 \times 982 \times 460 \times 2.0}{440 \times 400 \times 11.18 \times 1.15} \right) = 0.80d$$

Clearly this reinforcement is adequate and there is no need to iterate further. Alternatively the lever arm ratio approach used in Example 1 could be followed.

This lever arm is higher than the assumed value thus 2 T25 bars are satisfactory. This area of steel gives a reinforcement percentage of

$$\frac{982 \times 100}{440 \times 400} = 0.56\%$$

which exceeds the minimum value of 0.1% given in DD ENV 1996−1−1 and confirmed in the UK NAD.

Shear stress due to design loads,

$$V = \frac{V_{Sd}}{bd}$$

$$\frac{81 \times 10^3}{440 \times 400} = 0.46 \text{ N/mm}^2$$

Characteristic shear strength of masonry, $f_v = 0.35 + 17.5\rho$ in accordance with Section 6.4 of the UK NAD where

$$\rho = \frac{A_s}{b \, d} = 0.0056 \text{ (from above)}$$

Therefore $f_v = 0.35 + 0.10 = 0.45 \text{N/mm}^2$.

For a simply supported beam where shear span ratio, $\frac{a}{d}$ is less than 6, f_v may be increased by a factor

$$\left[2.5 - 0.25 \left(\frac{a}{d} \right) \right]$$

where

$$a = \frac{M}{V} = \frac{91.1}{81} = 11.2 \text{m}$$

a/d = 1.12/0.4 = 2.8; which is less than 6
Enhancement factor = 2.5 − 0.25 × 2.8 = 1.8
Therefore increased f_v = 1.8 × 0.45 = 0.81N/mm²
This is less than the maximum 1.75N/mm² and is therefore the value to use.

Design shear strength of beam =

$$\frac{f_v}{\gamma_M} = \frac{0.81}{2.0} = 0.4 \text{ N/mm}^2$$

This is less than the design shear stress and shear reinforcement is required.

To satisfy the criteria for shear

$$V_{Sd} \leq V_{Rd1} + V_{Rd2}$$

V_{Rd1} is the resistance due to the brickwork alone and is given by

$$V_{Rd1} = \frac{f_v}{\gamma_m} b \, d$$

$$V_{Rd1} = \frac{0.81}{2} \frac{440 \times 400}{10^3} \text{ kN}$$

= 71.3 kN

V_{Rd2} is given by

$$V_{Rd2} = 0.9d \frac{A_{sw}}{s} \frac{f_{yk}}{\gamma_s} (1 + \cot\alpha) \sin\alpha$$

where α is the angle of inclination of the shear reinforcement which in this case is 90°

Hence

$$V_{Rd2} = 0.9 \times 400 \times \frac{A_{sw}}{300} \frac{460}{1.15}$$

where A_{sw} is the cross sectional area of the steel reinforcement required at the 300mm pocket centres. As 300 mm is 0.75d this spacing is acceptable (see **Clause 5.2.6**).

$V_{Rd2} = 0.48 \, A_{sw}$ kN

Hence 81 = 71.3 + 0.48 A_{sw}
and A_{sw} = 20.2mm²
Check that the minimum area 0.1% is provided
Minimum area = 0.001 × 440 × 400
Minimum area = 176mm²
8 mm links in pairs at 300mm centres would be sufficient.

Index

Printed and bound by CPI Group (UK) Ltd, Croydon, CR0 4YY

01/11/2024

01782605-0018